Lecture Notes in Computer Science   14450

Founding Editors

Gerhard Goos
Juris Hartmanis

The series Lecture Notes in Computer Science (LNCS), including its subseries Lecture Notes in Artificial Intelligence (LNAI) and Lecture Notes in Bioinformatics (LNBI), has established itself as a medium for the publication of new developments in computer science and information technology research, teaching, and education.

LNCS enjoys close cooperation with the computer science R & D community, the series counts many renowned academics among its volume editors and paper authors, and collaborates with prestigious societies. Its mission is to serve this international community by providing an invaluable service, mainly focused on the publication of conference and workshop proceedings and postproceedings. LNCS commenced publication in 1973.

Biao Luo · Long Cheng · Zheng-Guang Wu ·
Hongyi Li · Chaojie Li

Editors

Neural
Information Processing

30th International Conference, ICONIP 2023
Changsha, China, November 20–23, 2023
Proceedings, Part IV

 Springer

Editors
Biao Luo 🄳
Central South University
Changsha, China

Long Cheng 🄳
Chinese Academy of Sciences
Beijing, China

Zheng-Guang Wu 🄳
Zhejiang University
Hangzhou, China

Hongyi Li 🄳
Guangdong University of Technology
Guangzhou, China

Chaojie Li 🄳
UNSW Sydney
Sydney, NSW, Australia

ISSN 0302-9743   ISSN 1611-3349 (electronic)
Lecture Notes in Computer Science
ISBN 978-981-99-8069-7   ISBN 978-981-99-8070-3 (eBook)
https://doi.org/10.1007/978-981-99-8070-3

This Springer imprint is published by the registered company Springer Nature Singapore Pte Ltd.
The registered company address is: 152 Beach Road, #21-01/04 Gateway East, Singapore 189721, Singapore

Paper in this product is recyclable.

Preface

Welcome to the 30th International Conference on Neural Information Processing (ICONIP2023) of the Asia-Pacific Neural Network Society (APNNS), held in Changsha, China, November 20–23, 2023.

The mission of the Asia-Pacific Neural Network Society is to promote active interactions among researchers, scientists, and industry professionals who are working in neural networks and related fields in the Asia-Pacific region. APNNS has Governing Board Members from 13 countries/regions – Australia, China, Hong Kong, India, Japan, Malaysia, New Zealand, Singapore, South Korea, Qatar, Taiwan, Thailand, and Turkey. The society's flagship annual conference is the International Conference of Neural Information Processing (ICONIP). The ICONIP conference aims to provide a leading international forum for researchers, scientists, and industry professionals who are working in neuroscience, neural networks, deep learning, and related fields to share their new ideas, progress, and achievements.

ICONIP2023 received 1274 papers, of which 256 papers were accepted for publication in Lecture Notes in Computer Science (LNCS), representing an acceptance rate of 20.09% and reflecting the increasingly high quality of research in neural networks and related areas. The conference focused on four main areas, i.e., "Theory and Algorithms", "Cognitive Neurosciences", "Human-Centered Computing", and "Applications". All the submissions were rigorously reviewed by the conference Program Committee (PC), comprising 258 PC members, and they ensured that every paper had at least two high-quality single-blind reviews. In fact, 5270 reviews were provided by 2145 reviewers. On average, each paper received 4.14 reviews.

We would like to take this opportunity to thank all the authors for submitting their papers to our conference, and our great appreciation goes to the Program Committee members and the reviewers who devoted their time and effort to our rigorous peer-review process; their insightful reviews and timely feedback ensured the high quality of the papers accepted for publication. We hope you enjoyed the research program at the conference.

October 2023

Biao Luo
Long Cheng
Zheng-Guang Wu
Hongyi Li
Chaojie Li

Organization

Honorary Chair

Weihua Gui   Central South University, China

Advisory Chairs

Jonathan Chan	King Mongkut's University of Technology Thonburi, Thailand
Zeng-Guang Hou	Chinese Academy of Sciences, China
Nikola Kasabov	Auckland University of Technology, New Zealand
Derong Liu	Southern University of Science and Technology, China
Seiichi Ozawa	Kobe University, Japan
Kevin Wong	Murdoch University, Australia

General Chairs

Tingwen Huang	Texas A&M University at Qatar, Qatar
Chunhua Yang	Central South University, China

Program Chairs

Biao Luo	Central South University, China
Long Cheng	Chinese Academy of Sciences, China
Zheng-Guang Wu	Zhejiang University, China
Hongyi Li	Guangdong University of Technology, China
Chaojie Li	University of New South Wales, Australia

Technical Chairs

Xing He	Southwest University, China
Keke Huang	Central South University, China
Huaqing Li	Southwest University, China
Qi Zhou	Guangdong University of Technology, China

Local Arrangement Chairs

Wenfeng Hu	Central South University, China
Bei Sun	Central South University, China

Finance Chairs

Fanbiao Li	Central South University, China
Hayaru Shouno	University of Electro-Communications, Japan
Xiaojun Zhou	Central South University, China

Special Session Chairs

Hongjing Liang	University of Electronic Science and Technology, China
Paul S. Pang	Federation University, Australia
Qiankun Song	Chongqing Jiaotong University, China
Lin Xiao	Hunan Normal University, China

Tutorial Chairs

Min Liu	Hunan University, China
M. Tanveer	Indian Institute of Technology Indore, India
Guanghui Wen	Southeast University, China

Publicity Chairs

Sabri Arik	Istanbul University-Cerrahpaşa, Turkey
Sung-Bae Cho	Yonsei University, South Korea
Maryam Doborjeh	Auckland University of Technology, New Zealand
El-Sayed M. El-Alfy	King Fahd University of Petroleum and Minerals, Saudi Arabia
Ashish Ghosh	Indian Statistical Institute, India
Chuandong Li	Southwest University, China
Weng Kin Lai	Tunku Abdul Rahman University of Management & Technology, Malaysia
Chu Kiong Loo	University of Malaya, Malaysia

| Qinmin Yang | Zhejiang University, China |
| Zhigang Zeng | Huazhong University of Science and Technology, China |

Publication Chairs

Zhiwen Chen	Central South University, China
Andrew Chi-Sing Leung	City University of Hong Kong, China
Xin Wang	Southwest University, China
Xiaofeng Yuan	Central South University, China

Secretaries

| Yun Feng | Hunan University, China |
| Bingchuan Wang | Central South University, China |

Webmasters

| Tianmeng Hu | Central South University, China |
| Xianzhe Liu | Xiangtan University, China |

Program Committee

Rohit Agarwal	UiT The Arctic University of Norway, Norway
Hasin Ahmed	Gauhati University, India
Harith Al-Sahaf	Victoria University of Wellington, New Zealand
Brad Alexander	University of Adelaide, Australia
Mashaan Alshammari	Independent Researcher, Saudi Arabia
Sabri Arik	Istanbul University, Turkey
Ravneet Singh Arora	Block Inc., USA
Zeyar Aung	Khalifa University of Science and Technology, UAE
Monowar Bhuyan	Umeå University, Sweden
Jingguo Bi	Beijing University of Posts and Telecommunications, China
Xu Bin	Northwestern Polytechnical University, China
Marcin Blachnik	Silesian University of Technology, Poland
Paul Black	Federation University, Australia

Anoop C. S.	Govt. Engineering College, India
Ning Cai	Beijing University of Posts and Telecommunications, China
Siripinyo Chantamunee	Walailak University, Thailand
Hangjun Che	City University of Hong Kong, China
Wei-Wei Che	Qingdao University, China
Huabin Chen	Nanchang University, China
Jinpeng Chen	Beijing University of Posts & Telecommunications, China
Ke-Jia Chen	Nanjing University of Posts and Telecommunications, China
Lv Chen	Shandong Normal University, China
Qiuyuan Chen	Tencent Technology, China
Wei-Neng Chen	South China University of Technology, China
Yufei Chen	Tongji University, China
Long Cheng	Institute of Automation, China
Yongli Cheng	Fuzhou University, China
Sung-Bae Cho	Yonsei University, South Korea
Ruikai Cui	Australian National University, Australia
Jianhua Dai	Hunan Normal University, China
Tao Dai	Tsinghua University, China
Yuxin Ding	Harbin Institute of Technology, China
Bo Dong	Xi'an Jiaotong University, China
Shanling Dong	Zhejiang University, China
Sidong Feng	Monash University, Australia
Yuming Feng	Chongqing Three Gorges University, China
Yun Feng	Hunan University, China
Junjie Fu	Southeast University, China
Yanggeng Fu	Fuzhou University, China
Ninnart Fuengfusin	Kyushu Institute of Technology, Japan
Thippa Reddy Gadekallu	VIT University, India
Ruobin Gao	Nanyang Technological University, Singapore
Tom Gedeon	Curtin University, Australia
Kam Meng Goh	Tunku Abdul Rahman University of Management and Technology, Malaysia
Zbigniew Gomolka	University of Rzeszow, Poland
Shengrong Gong	Changshu Institute of Technology, China
Xiaodong Gu	Fudan University, China
Zhihao Gu	Shanghai Jiao Tong University, China
Changlu Guo	Budapest University of Technology and Economics, Hungary
Weixin Han	Northwestern Polytechnical University, China

Xing He	Southwest University, China
Akira Hirose	University of Tokyo, Japan
Yin Hongwei	Huzhou Normal University, China
Md Zakir Hossain	Curtin University, Australia
Zengguang Hou	Chinese Academy of Sciences, China
Lu Hu	Jiangsu University, China
Zeke Zexi Hu	University of Sydney, Australia
He Huang	Soochow University, China
Junjian Huang	Chongqing University of Education, China
Kaizhu Huang	Duke Kunshan University, China
David Iclanzan	Sapientia University, Romania
Radu Tudor Ionescu	University of Bucharest, Romania
Asim Iqbal	Cornell University, USA
Syed Islam	Edith Cowan University, Australia
Kazunori Iwata	Hiroshima City University, Japan
Junkai Ji	Shenzhen University, China
Yi Ji	Soochow University, China
Canghong Jin	Zhejiang University, China
Xiaoyang Kang	Fudan University, China
Mutsumi Kimura	Ryukoku University, Japan
Masahiro Kohjima	NTT, Japan
Damian Kordos	Rzeszow University of Technology, Poland
Marek Kraft	Poznań University of Technology, Poland
Lov Kumar	NIT Kurukshetra, India
Weng Kin Lai	Tunku Abdul Rahman University of Management & Technology, Malaysia
Xinyi Le	Shanghai Jiao Tong University, China
Bin Li	University of Science and Technology of China, China
Hongfei Li	Xinjiang University, China
Houcheng Li	Chinese Academy of Sciences, China
Huaqing Li	Southwest University, China
Jianfeng Li	Southwest University, China
Jun Li	Nanjing Normal University, China
Kan Li	Beijing Institute of Technology, China
Peifeng Li	Soochow University, China
Wenye Li	Chinese University of Hong Kong, China
Xiangyu Li	Beijing Jiaotong University, China
Yantao Li	Chongqing University, China
Yaoman Li	Chinese University of Hong Kong, China
Yinlin Li	Chinese Academy of Sciences, China
Yuan Li	Academy of Military Science, China

Yun Li	Nanjing University of Posts and Telecommunications, China
Zhidong Li	University of Technology Sydney, Australia
Zhixin Li	Guangxi Normal University, China
Zhongyi Li	Beihang University, China
Ziqiang Li	University of Tokyo, Japan
Xianghong Lin	Northwest Normal University, China
Yang Lin	University of Sydney, Australia
Huawen Liu	Zhejiang Normal University, China
Jian-Wei Liu	China University of Petroleum, China
Jun Liu	Chengdu University of Information Technology, China
Junxiu Liu	Guangxi Normal University, China
Tommy Liu	Australian National University, Australia
Wen Liu	Chinese University of Hong Kong, China
Yan Liu	Taikang Insurance Group, China
Yang Liu	Guangdong University of Technology, China
Yaozhong Liu	Australian National University, Australia
Yong Liu	Heilongjiang University, China
Yubao Liu	Sun Yat-sen University, China
Yunlong Liu	Xiamen University, China
Zhe Liu	Jiangsu University, China
Zhen Liu	Chinese Academy of Sciences, China
Zhi-Yong Liu	Chinese Academy of Sciences, China
Ma Lizhuang	Shanghai Jiao Tong University, China
Chu-Kiong Loo	University of Malaya, Malaysia
Vasco Lopes	Universidade da Beira Interior, Portugal
Hongtao Lu	Shanghai Jiao Tong University, China
Wenpeng Lu	Qilu University of Technology, China
Biao Luo	Central South University, China
Ye Luo	Tongji University, China
Jiancheng Lv	Sichuan University, China
Yuezu Lv	Beijing Institute of Technology, China
Huifang Ma	Northwest Normal University, China
Jinwen Ma	Peking University, China
Jyoti Maggu	Thapar Institute of Engineering and Technology Patiala, India
Adnan Mahmood	Macquarie University, Australia
Mufti Mahmud	University of Padova, Italy
Krishanu Maity	Indian Institute of Technology Patna, India
Srimanta Mandal	DA-IICT, India
Wang Manning	Fudan University, China

Piotr Milczarski	Lodz University of Technology, Poland
Malek Mouhoub	University of Regina, Canada
Nankun Mu	Chongqing University, China
Wenlong Ni	Jiangxi Normal University, China
Anupiya Nugaliyadde	Murdoch University, Australia
Toshiaki Omori	Kobe University, Japan
Babatunde Onasanya	University of Ibadan, Nigeria
Manisha Padala	Indian Institute of Science, India
Sarbani Palit	Indian Statistical Institute, India
Paul Pang	Federation University, Australia
Rasmita Panigrahi	Giet University, India
Kitsuchart Pasupa	King Mongkut's Institute of Technology Ladkrabang, Thailand
Dipanjyoti Paul	Ohio State University, USA
Hu Peng	Jiujiang University, China
Kebin Peng	University of Texas at San Antonio, USA
Dawid Połap	Silesian University of Technology, Poland
Zhong Qian	Soochow University, China
Sitian Qin	Harbin Institute of Technology at Weihai, China
Toshimichi Saito	Hosei University, Japan
Fumiaki Saitoh	Chiba Institute of Technology, Japan
Naoyuki Sato	Future University Hakodate, Japan
Chandni Saxena	Chinese University of Hong Kong, China
Jiaxing Shang	Chongqing University, China
Lin Shang	Nanjing University, China
Jie Shao	University of Science and Technology of China, China
Yin Sheng	Huazhong University of Science and Technology, China
Liu Sheng-Lan	Dalian University of Technology, China
Hayaru Shouno	University of Electro-Communications, Japan
Gautam Srivastava	Brandon University, Canada
Jianbo Su	Shanghai Jiao Tong University, China
Jianhua Su	Institute of Automation, China
Xiangdong Su	Inner Mongolia University, China
Daiki Suehiro	Kyushu University, Japan
Basem Suleiman	University of New South Wales, Australia
Ning Sun	Shandong Normal University, China
Shiliang Sun	East China Normal University, China
Chunyu Tan	Anhui University, China
Gouhei Tanaka	University of Tokyo, Japan
Maolin Tang	Queensland University of Technology, Australia

Shu Tian	University of Science and Technology Beijing, China
Shikui Tu	Shanghai Jiao Tong University, China
Nancy Victor	Vellore Institute of Technology, India
Petra Vidnerová	Institute of Computer Science, Czech Republic
Shanchuan Wan	University of Tokyo, Japan
Tao Wan	Beihang University, China
Ying Wan	Southeast University, China
Bangjun Wang	Soochow University, China
Hao Wang	Shanghai University, China
Huamin Wang	Southwest University, China
Hui Wang	Nanchang Institute of Technology, China
Huiwei Wang	Southwest University, China
Jianzong Wang	Ping An Technology, China
Lei Wang	National University of Defense Technology, China
Lin Wang	University of Jinan, China
Shi Lin Wang	Shanghai Jiao Tong University, China
Wei Wang	Shenzhen MSU-BIT University, China
Weiqun Wang	Chinese Academy of Sciences, China
Xiaoyu Wang	Tokyo Institute of Technology, Japan
Xin Wang	Southwest University, China
Xin Wang	Southwest University, China
Yan Wang	Chinese Academy of Sciences, China
Yan Wang	Sichuan University, China
Yonghua Wang	Guangdong University of Technology, China
Yongyu Wang	JD Logistics, China
Zhenhua Wang	Northwest A&F University, China
Zi-Peng Wang	Beijing University of Technology, China
Hongxi Wei	Inner Mongolia University, China
Guanghui Wen	Southeast University, China
Guoguang Wen	Beijing Jiaotong University, China
Ka-Chun Wong	City University of Hong Kong, China
Anna Wróblewska	Warsaw University of Technology, Poland
Fengge Wu	Institute of Software, Chinese Academy of Sciences, China
Ji Wu	Tsinghua University, China
Wei Wu	Inner Mongolia University, China
Yue Wu	Shanghai Jiao Tong University, China
Likun Xia	Capital Normal University, China
Lin Xiao	Hunan Normal University, China

Qiang Xiao	Huazhong University of Science and Technology, China
Hao Xiong	Macquarie University, Australia
Dongpo Xu	Northeast Normal University, China
Hua Xu	Tsinghua University, China
Jianhua Xu	Nanjing Normal University, China
Xinyue Xu	Hong Kong University of Science and Technology, China
Yong Xu	Beijing Institute of Technology, China
Ngo Xuan Bach	Posts and Telecommunications Institute of Technology, Vietnam
Hao Xue	University of New South Wales, Australia
Yang Xujun	Chongqing Jiaotong University, China
Haitian Yang	Chinese Academy of Sciences, China
Jie Yang	Shanghai Jiao Tong University, China
Minghao Yang	Chinese Academy of Sciences, China
Peipei Yang	Chinese Academy of Science, China
Zhiyuan Yang	City University of Hong Kong, China
Wangshu Yao	Soochow University, China
Ming Yin	Guangdong University of Technology, China
Qiang Yu	Tianjin University, China
Wenxin Yu	Southwest University of Science and Technology, China
Yun-Hao Yuan	Yangzhou University, China
Xiaodong Yue	Shanghai University, China
Paweł Zawistowski	Warsaw University of Technology, Poland
Hui Zeng	Southwest University of Science and Technology, China
Wang Zengyunwang	Hunan First Normal University, China
Daren Zha	Institute of Information Engineering, China
Zhi-Hui Zhan	South China University of Technology, China
Baojie Zhang	Chongqing Three Gorges University, China
Canlong Zhang	Guangxi Normal University, China
Guixuan Zhang	Chinese Academy of Science, China
Jianming Zhang	Changsha University of Science and Technology, China
Li Zhang	Soochow University, China
Wei Zhang	Southwest University, China
Wenbing Zhang	Yangzhou University, China
Xiang Zhang	National University of Defense Technology, China
Xiaofang Zhang	Soochow University, China
Xiaowang Zhang	Tianjin University, China

Xinglong Zhang	National University of Defense Technology, China
Dongdong Zhao	Wuhan University of Technology, China
Xiang Zhao	National University of Defense Technology, China
Xu Zhao	Shanghai Jiao Tong University, China
Liping Zheng	Hefei University of Technology, China
Yan Zheng	Kyushu University, Japan
Baojiang Zhong	Soochow University, China
Guoqiang Zhong	Ocean University of China, China
Jialing Zhou	Nanjing University of Science and Technology, China
Wenan Zhou	PCN&CAD Center, China
Xiao-Hu Zhou	Institute of Automation, China
Xinyu Zhou	Jiangxi Normal University, China
Quanxin Zhu	Nanjing Normal University, China
Yuanheng Zhu	Chinese Academy of Sciences, China
Xiaotian Zhuang	JD Logistics, China
Dongsheng Zou	Chongqing University, China

Contents – Part IV

Human Centred Computing

Cross-Modal Method Based on Self-Attention Neural Networks
for Drug-Target Prediction 3
Litao Zhang, Chunming Yang, Chunlin He, and Hui Zhang

GRF-GMM: A Trajectory Optimization Framework for Obstacle
Avoidance in Learning from Demonstration 18
Bin Ye, Peng Yu, Cong Hu, Binbin Qiu, and Ning Tan

SLG-NET: Subgraph Neural Network with Local-Global Braingraph
Feature Extraction Modules and a Novel Subgraph Generation Algorithm
for Automated Identification of Major Depressive Disorder 31
Yan Zhang, Xin Liu, Panrui Tang, and Zuping Zhang

CrowdNav-HERO: Pedestrian Trajectory Prediction Based Crowded
Navigation with Human-Environment-Robot Ternary Fusion 43
Siyi Lu, Bolei Chen, Ping Zhong, Yu Sheng, Yongzheng Cui, and Run Liu

Modeling User's Neutral Feedback in Conversational Recommendation 56
Xizhe Li, Chenhao Hu, Weiyang Kong, Sen Zhang, and Yubao Liu

A Domain Knowledge-Based Semi-supervised Pancreas Segmentation
Approach 69
Siqi Ma, Zhe Liu, Yuqing Song, Yi Liu, Kai Han, and Yang Jiang

Soybean Genome Clustering Using Quantum-Based Fuzzy C-Means
Algorithm 83
*Sai Siddhartha Vivek Dhir Rangoju, Keshav Garg, Rohith Dandi,
Om Prakash Patel, and Neha Bharill*

DAMFormer: Enhancing Polyp Segmentation Through Dual Attention
Mechanism 95
*Huy Trinh Quang, Mai Nguyen, Quan Nguyen Van, Linh Doan Bao,
Thanh Dang Hong, Thanh Nguyen Tung, and Toan Pham Van*

BIN: A Bio-Signature Identification Network for Interpretable Liver
Cancer Microvascular Invasion Prediction Based on Multi-modal MRIs 107
Pengyu Zheng, Bo Li, Huilin Lai, and Ye Luo

Human-to-Human Interaction Detection 120
   Zhenhua Wang, Kaining Ying, Jiajun Meng, and Jifeng Ning

Reconstructing Challenging Hand Posture from Multi-modal Input 133
   Xi Luo, Yuwei Li, and Jingyi Yu

A Compliant Elbow Exoskeleton with an SEA at Interaction Port 146
   Xiuze Xia, Lijun Han, Houcheng Li, Yu Zhang, Zeyu Liu, and Long Cheng

Applications

Differential Fault Analysis Against AES Based on a Hybrid Fault Model 161
   Xusen Wan, Jinbao Zhang, Weixiang Wu, Shi Cheng, and Jiehua Wang

Towards Undetectable Adversarial Examples: A Steganographic
Perspective 172
   Hui Zeng, Biwei Chen, Rongsong Yang, Chenggang Li, and Anjie Peng

On Efficient Federated Learning for Aerial Remote Sensing Image
Classification: A Filter Pruning Approach 184
   *Qipeng Song, Jingbo Cao, Yue Li, Xueru Gao, Chengzhi Shangguan,
   and Linlin Liang*

ASGNet: Adaptive Semantic Gate Networks for Log-Based Anomaly
Diagnosis 200
   *Haitian Yang, Degang Sun, Wen Liu, Yanshu Li, Yan Wang,
   and Weiqing Huang*

Propheter: Prophetic Teacher Guided Long-Tailed Distribution Learning 213
   *Wenxiang Xu, Yongcheng Jing, Linyun Zhou, Wenqi Huang,
   Lechao Cheng, Zunlei Feng, and Mingli Song*

Sequential Transformer for End-to-End Person Search 226
   Long Chen and Jinhua Xu

Multi-scale Structural Asymmetric Convolution for Wireframe Parsing 239
   Jiahui Zhang, Jinfu Yang, Fuji Fu, and Jiaqi Ma

S3ACH: Semi-Supervised Semantic Adaptive Cross-Modal Hashing 252
   Liu Yang, Kaiting Zhang, Yinan Li, Yunfei Chen, Jun Long, and Zhan Yang

Intelligent UAV Swarm Planning Based on Undirected Graph Model 270
   Tianyi Lv, Qingyuan Xia, and Qiwen Zheng

Learning Item Attributes and User Interests for Knowledge Graph
Enhanced Recommendation 284
    Zepeng Huai, Guohua Yang, Jianhua Tao, and Dawei Zhang

Multi-view Stereo by Fusing Monocular and a Combination of Depth
Representation Methods 298
    Fanqi Yu and Xinyang Sun

A Fast and Scalable Frame-Recurrent Video Super-Resolution Framework 310
    Kaixuan Hou and Jianping Luo

Structural Properties of Associative Knowledge Graphs 326
    Janusz A. Starzyk, Przemysław Stokłosa, Adrian Horzyk, and Paweł Raif

Nonlinear NN-Based Perturbation Estimator Designs for Disturbed
Unmanned Systems 340
    Xingcheng Tong and Xiaozheng Jin

DOS Dataset: A Novel Indoor Deformable Object Segmentation Dataset
for Sweeping Robots 352
    Zehan Tan, Weidong Yang, and Zhiwei Zhang

Leveraging Sound Local and Global Features for Language-Queried
Target Sound Extraction 367
    Xinmeng Xu, Yiqun Zhang, Yuhong Yang, and Weiping Tu

PEVLR: A New Privacy-Preserving and Efficient Approach for Vertical
Logistic Regression 380
    Sihan Mao, Xiaolin Zheng, Jianguang Zhang, and Xiaodong Hu

Semantic-Pixel Associative Information Improving Loop Closure
Detection and Experience Map Building for Efficient Visual Representation ... 393
    Yufei Deng, Rong Xiao, Jiaxin Li, and Jiancheng Lv

Knowledge Distillation via Information Matching 405
    Honglin Zhu, Ning Jiang, Jialiang Tang, and Xinlei Huang

CenAD: Collaborative Embedding Network for Anomaly Detection
with Leveraging Partially Observed Anomalies 418
    Li Cheng, Bin Li, Renjie He, and Feng Yao

PAG: Protecting Artworks from Personalizing Image Generative Models 433
    Zhaorui Tan, Siyuan Wang, Xi Yang, and Kaizhu Huang

Attention Based Spatial-Temporal Dynamic Interact Network for Traffic
Flow Forecasting   445
    Junwei Xie, Liang Ge, Haifeng Li, and Yiping Lin

Staged Long Text Generation with Progressive Task-Oriented Prompts   458
    Xingjin Wang, Linjing Li, and Daniel Zeng

Learning Stable Nonlinear Dynamical System from One Demonstration   471
    *Yu Zhang, Lijun Han, Zirui Wang, Xiuze Xia, Houcheng Li,
    and Long Cheng*

Towards High-Performance Exploratory Data Analysis (EDA) via Stable
Equilibrium Point   483
    Yuxuan Song and Yongyu Wang

MVFAN: Multi-view Feature Assisted Network for 4D Radar Object
Detection   493
    Qiao Yan and Yihan Wang

Time Series Anomaly Detection with a Transformer Residual
Autoencoder-Decoder   512
    Shaojie Wang, Yinke Wang, and Wenzhong Li

Adversarial Example Detection with Latent Representation Dynamic
Prototype   525
    Taowen Wang, Zhuang Qian, and Xi Yang

A Multi-scale and Multi-attention Network for Skin Lesion Segmentation   537
    Cong Wu, Hang Zhang, Dingsheng Chen, and Haitao Gan

Temporal Attention for Robust Multiple Object Pose Tracking   551
    Zhongluo Li, Junichiro Yoshimoto, and Kazushi Ikeda

Correlation Guided Multi-teacher Knowledge Distillation   562
    Luyao Shi, Ning Jiang, Jialiang Tang, and Xinlei Huang

Author Index   575

Human Centred Computing

Cross-Modal Method Based on Self-Attention Neural Networks for Drug-Target Prediction

Litao Zhang[1], Chunming Yang[1,2(✉)], Chunlin He[1], and Hui Zhang[1]

[1] School of Computer Science and Technology, Southwest University of Science and Technology, Mianyang 621010, Sichuan, China
yangchunming@swust.edu.cn
[2] Sichuan Big Data and Intelligent Systems Engineering Technology Research Center, Mianyang 621010, Sichuan, China

Abstract. Prediction of drug-target interactions (DTIs) plays a crucial role in drug retargeting, which can save costs and shorten time for drug development. However, existing methods are still unable to integrate the multimodal features of existing DTI datasets. In this work, we propose a new multi-head-based self-attention neural network approach, called SANN-DTI, for dti prediction. Specifically, entity embeddings in the knowledge graph are learned using DistMult, then this information is interacted with traditional drug and protein representations via multi-head self-attention neural networks, and finally DTIs is computed using fully connected neural networks for interaction features. SANN-DTI was evaluated in three scenarios across two baseline datasets. After ten fold cross-validation, our model outperforms the most advanced methods. In addition, SANN-DT has been applied to drug retargeting of breast cancer via HRBB2 targets. It was found that four of the top ten recommended drugs have been supported by the literature. Ligand-target docking results showed that the second-ranked drug in the recommended list had a clear affinity with HRBB2, which provides a promising approach for better understanding drug mode of action and drug repositioning.

Keywords: Drug repositioning · Self-Attention · DTI prediction · Cross-Modal

1 Instructions

Drug repositioning involves identifying new therapeutic uses for existing drugs. This approach offers several advantages over traditional drug development approachs since the safety, efficacy, and administration routes of these drugs are already well-established. Therefore, it can reduce costs and time spent on research and development [1]. The primary challenge in drug repositioning is to identify potential drug-target interactions (DTIs) for novel indications or diseases [2]. To address this issue, various computational approaches have been developed to to predict new DTIs [3–5]. These approaches aim to narrow down

B. Luo et al. (Eds.): ICONIP 2023, LNCS 14450, pp. 3–17, 2024.
https://doi.org/10.1007/978-981-99-8070-3_1

the search space for potential drug and protein candidates, thereby reducing the time and cost associated with experimental validation [6].

Currently, the computational approaches for predicting DTIs can be divided into three categories: ligand-based approaches, docking approaches and chemogenomic approaches [7]. The ligand-based approach relies on the assumption that molecules with similar structures will have similar biological activities, and its accuracy largely depends on activity data for known ligands [8]. Docking approaches use the 3D structure of drugs and proteins to predict their binding mode and strength. However, few target proteins have clear 3D structure, so its application is limited [9]. Chemogenomics approaches use widely enriched information on drugs and targets to predict DTIs. With the growth of available data on drugs and targets, these approaches are gaining attention due to their high accuracy. For example, DeepDTI [10] proposed by Öztürk et al. applies a convolutional neural network to output the binding affinity score by taking the molecular structure of the drug and the sequence of the target protein as input. DTINet [11] proposed by Lou et al. uses a random walk algorithm to predict DTIs in a heterogeneous network containing information about drugs, targets, diseases, and side effects. DTiGEMS+ [12] proposed by Thafar et al. combines graph embedding, graph mining, and similarity-based techniques to capture structural and functional similarities between drugs and targets to predict DTIs.

In addition, the rapid development of knowledge graph embedding (KGE) has extended the application of deep learning to the field of knowledge graphs(KG), and related approaches have also been applied to drug repositioning [13–15]. These approaches extract triples from unstructured data or existing databases and formulate the problem as a link prediction in KG. For example, mohamed et al. combined protein-protein interaction data and text information in the biomedical literature to construct a knowledge graphs representing the relationship between proteins and drugs, and proposed a TriModel [16] to transform DTIs prediction into a representation learning and link prediction problem. KGE_NFM [17] proposed by Ye et al. integrated the representation of drug and protein structure information on the basis of KGE, combined with the neural factorization machine to predict DTIs, effectively reduced the noise in complex biological networks, and ensured the high stability of the approache.

Although these approaches show good results in DTIs prediction, they are limited by the order of interaction between different features and may not be able to capture higher-order interaction features, resulting in information loss and poor prediction performance. Meanwhile, the application in the realistic scenario of drug and protein cold start is also a challenge for deep learning approaches to predict DTIs [17]. This is because in this case there is little or no prior information about the drugs or proteins, which makes it difficult for deep learning models to accurately predict DTIs.

To address these challenges, this study proposes a new predictive model, SANN-DTI, which is based on multi-head self-attention neural networks. The model first learns the low-dimensional representation of various drug-related concepts in the KG by KGE approach, and then uses the multi-head self-attention neural network to integrate the learned multi-omics information and the

Table 1. Details and partitioning of each dataset.

Dataset	Drug	Target	DTIs	KG Entity	KG Relation
Yamanishi	791	989	5,126	25,487	95,672
BioKG	6,214	23,100	26,051	101,847	2,017,795

structural information of drugs and proteins to achieve the prediction of DTI. The main contributions of this study are as follows: (1) A self-attention network is designed for to integrate multi-modal information carried by source datasets. (2) Higher-order explicit interactions between features are achieved, leading to higher accuracy and wider DTIs prediction applications. (3) We explored the impact of different validation scenarios on DTIs predictions that could provide new insights into new drug target discovery.

In this study, two datasets were used to evaluate the performance of the proposed model in three validation scenarios. Experimental results show that our approache is superior to other baseline approaches, and SANN-DTI has better predictive performance. In addition, SANN-DTI was used to predict DTIs of approved drugs and ERBB2 targets, since ERBB2 has been confirmed to be closely related to the formation of Breast cancer. In summary, the model has good predictive ability of DTI, which provides a promising approache for better understanding of drug action and drug repositioning.

2   Materials and Approaches

2.1   Benchmark Datasets

Two benchmark datasets containing different types of heterogeneous data were used in the experiment: the Yamanishi [18] and BioKG [19] datasets. The details of each dataset are shown in Table 1.

The Yamanishi dataset contains four protein families (enzyme (E), ion channel (IC), G-protein-coupled receptor (GPCR), and nuclear receptor (NR)). The data was collected from several sources, such as DrugBank [20], KEGG [21], BRENDA [22], and SuperTarget [23]. Based on this dataset, Mohamed et al [16]. combined relevant heterogeneous data of drug ATC codes, BRITE identifiers, related diseases and pathways extracted in KEGG, DrugBank, InterPro and UniProt to construct a knowledge graph.

BioKG is a biological knowledge graph that integrates and connects data from a vast biomedical data repository, including genes, proteins, diseases, drugs, and pathways. The categories of directed relationships that connect two classes of entities include drug-drug interactions, protein-protein interactions, drug-protein interactions, disease-protein, disease-pathway, gene-protein, etc.

For each dataset, 90% of the DTI data was selected as the training set and 10% as the test set, and ten-fold cross-validation was performed to obtain experimental results. We followed the experimental scenario designed by ye et al [17].

to ensure fairness. After dividing the training set and test set, negative samples are constructed for the training set and test set respectively, so that the ratio of positive and negative samples in the train set and test set is about 1:10. For the cold start of drugs, ensure that the drugs covered by the test set are not present in the train set; For protein cold start, ensure that none of the target proteins covered in the test set are present in the train set.

2.2   Implementation Process of SANN-DTI

SANN-DTI consists of three main parts: (1) Extract heterogeneous information in the knowledge graph and structural information in drugs and proteins; (2) Dimensionality reduction and normalization of extracted feature representations; (3) Information integration and prediction of DTIs through multi-head self-attention neural networks. The overall flow of our SANN-DTI model is shown in Fig. 1.

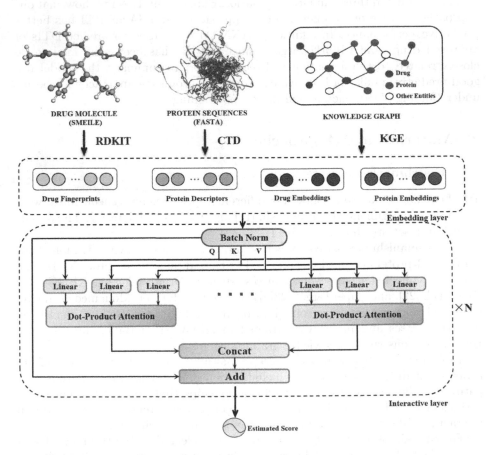

Fig. 1. The overall flow of the SANN-DTI model

Extract the Representation of Drus and Proteins. This step, we use the Morgan fingerprint computed by RDKit [24] as the structural representation of the drug and the descriptors computed by CTD [25] as the structural representation of the protein. Morgan fingerprint describes the topological features of a molecule based on the ring substructure of the molecule, which can effectively capture the chemical information of the molecule. We assign a radius of 2 to each atom. CTD computes synthesis descriptors, transition descriptors, and distribution descriptors based on different physical and chemical properties, with the same properties labeled as the same number. For a given protein sequence, 147 descriptors can be obtained. Then, the entities in KG are embedded into a continuous low-dimensional vector space by using the bilinear model DistMult [26]. DistMult mainly describes the semantic relevance of entities under relationships by bilinear transformation based on the relationships between entities. The model is not only simple and easy to calculate, but also can effectively describe the cooperation between entities. Its scoring function is defined as: $f_r(h,t) = h^T diag(M_r)t$. Where $h, t \in \mathbb{R}^d$ are entity vectors including the drug and the target, and $M_r \in \mathbb{R}^d$ is the relationship matrix.

Dimensionality Reduction and Normalization. Due to the high noise, sparse and non-uniform distribution of the extracted biological data features, it is not appropriate to directly apply it as input to the prediction network. In order to solve this problem, we use the principal component analysis [27] to reduce dimension and normalize, which aims to accelerate the convergence of the model and improve the accuracy of the model.

Multi-head Self-Attention Neural Network. This step is to integrate features from various data sources and make predictions through self-attention neural networks. Multi-head self-attention [28] is commonly used for natural language processing tasks such as machine translation, text classification, and question answering. The network can map different feature domains into the same low-dimensional feature space, and then send the mapped features into the attention layer to learn the cross-over features. The cross-over effect of different features is evaluated by the attention mechanism through mapping and projecting into different subspaces. Therefore, it has a good performance in modeling complex relations. The Key-Value attention network determines which feature combinations are meaningful. Taking features p and q as examples, we first define the relation between features vector $\mathbf{e_p}$ and $\mathbf{e_q}$ under an attention head:

$$\alpha_{\mathbf{p,q}} = softmax((\mathbf{W_{Query}e_p})(\mathbf{W_{Key}e_q})^T) \tag{1}$$

where $\mathbf{W_{Query}}, \mathbf{W_{Key}}$ is the transformation matrix that projects the feature vector into the new space. Then we update feature vector $\mathbf{e_p}$ with all the relevant features in the attention head.

$$\tilde{\mathbf{e}}_{\mathbf{p}} = \sum_{i=1}^{n} \alpha_{\mathbf{p,q}}(\mathbf{W_{Value}e_q}) \tag{2}$$

The above feature vector $\tilde{e}_p$ is the combination representation of feature e_p and related features under the interaction of an attention head, nd the value of n is the number of other features, so we can get the representation of new discrete features interaction in different subspaces. Then combine the features of all attention head learning:

$$\hat{e}_p = \tilde{e}_p^{(1)} \oplus \tilde{e}_p^{(2)} \oplus \cdots \oplus \tilde{e}_p^{(h)} \tag{3}$$

where $\oplus$ represents the concatenate operation and $\mathbf{h}$ represents the number of attention heads. In order to maintain the original learned combination features, we add a residual network.

$$\hat{e}_p^{res} = ReLU(\hat{e}_p + W_{res}e_p) \tag{4}$$

where $ReLU$ is the activation function, we added dropout after it to prevent overfitting. In this way, the features through the interaction layer are fully represented. In addition, the layers of this interaction can be superimposed to form higher-order combined features. Finally, we simply concatenate the combined features and then calculate the predicted value of DTIs y through a fully connected layer and the sigmoid activation function σ.

$$y = \sigma(\mathbf{w}^T(e_1^{Res} \oplus e_2^{Res} \oplus \cdots \oplus e_M^{Res}) + b) \tag{5}$$

2.3　Adjustment of Parameters

We use binary cross entropy as loss function, defined as:

$$Loss = -\frac{1}{N}\sum_{i=1}^{N}(y_i \log(\hat{y}_i) + (1 - y_i)\log(1 - \hat{y}_i)) \tag{6}$$

We choose adam as the optimizer, set the epoch to 1000 and set the early stop, the condition of the early stop is that the loss does not decrease within 20 epochs. The embedding dimension of the knowledge graph is 400. In addition, we use grid search to find suitable hyperparameters, the detailed parameters are shown in the Table 2.

2.4　Evaluation Metrics

The ratio of positive to negative samples in the negative sample of the data set is 1:10, which leads to the possibility that the negative sample may be much larger than the positive sample. Models may always predict negative samples and obtain high accuracy, which can be misleading. Therefore, the evaluation metrics we use is Area Under the Receiver Operating Characteristic curve (AUROC) and Area Under Precision-Recal Curve (AUPR). AUROC curve is a performance metrics for classification problems under various threshold Settings, plotted as horizontal and vertical coordinates using true false positive (TPR) rate and false positive rate (FPR), respectively. TPR and FPR are defined as follows:

Table 2. Experimental parameters setting.

Parameters	Yamanishi		BioKG	
	warm start	cold start	warm start	cold start
Learning rate	1e-4	1e-4	1e-4	1e-4
Batch_size	5000	5000	20000	20000
Dropout	0.3	0.5	0.5	0.3
L2_reg_embedding	1e-2	1e-1	1e-1	1e-1
Attention_head_num	2	2	2	2
Interactive_layer_num	2	3	2	3

$$TPR = \frac{TP}{TP + FN} \tag{7}$$

$$FPR = \frac{FP}{FP + TN} \tag{8}$$

Another evaluation metric, AUPR, is the area under the PR curve, which is the curve composed of recall rate and accuracy rate. The larger the AUC and AUPR, the better the model.

3 Experimental Results

3.1 Compared with Baseline Models

We compared SANN-DTI with baseline models in three sample scenarios of Yamanishi and BioKG datasets. All results were obtained using 10-fold cross validation and shown as box plots. Box plot show the median as the center line, the upper and lower quartiles as the box boundaries, whiskers as the maximum and minimum values, and dots as outliers. The great advantage of a box plot is that it is not affected by outliers.

Performance Evaluation of Yamanishi Dataset. We compared SANN-DTI with DeepDTI, Random Forest classifier(RF), DTiGEMS+, DistMult, TriModel, and KGE_NFM. The results of evaluation metrics are shown in Fig. 2. In the warm start scenario, we observe that our SANN-DTI has a higher center line and a smaller box than other approaches. This shows that the performance of SANN-DTI is high and stable compared to other approaches. Specifically, the AUPR of SANN-DTI (AUROC = 0.993, AUPR = 0.971) was about 4% higher than that of the suboptimal RF (AUROC = 0.982, AUPR = 0.935), and the AUROC of SANN-DTI is about 1% higher than that of the suboptimal TriModel (AUROC = 0.984, AUPR = 0.886). In the cold start scenario, we observed that our SANN-DTI performed best. Especially in AUPR, SANN-DTI (AUROC = 0.879, AUPR = 0.680) is 14% higher than the suboptimal RF (AUROC = 0.854, AUPR =

0.584) in the drug cold start scenario; In the protein cold start scenario, SANN-DTI (AUROC = 0.920, AUPR = 0.761) is about 14% higher than the suboptimal KGE_NFM (AUROC = 0.905, AUPR = 0.652). This shows that SANN-DTI can combine the semantic information of KG with traditional representations and interact into higher-order features to improve the prediction performance. DeepDTI does not perform well in cold start scenarios because of the absence of knowledge graph features.

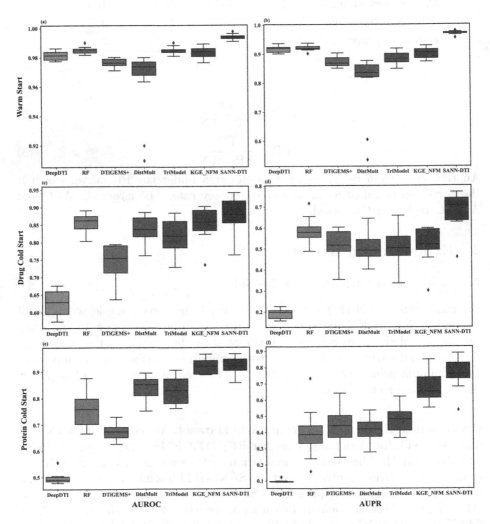

Fig. 2. Evaluation performance of Yamanishi dataset in three sample scenarios.

Performance Evaluation of BioKG Dataset. We compared SANN-DTI with DeepDTI, Random Forest Classifier(RF), MPNN_CNN, DistMult, Tri-Model and KGE_NFM. The results of evaluation metrics are shown in Fig. 3. The data scale of the BioKG dataset is larger than that of Yamanishi, and the assessment performance of the baselines under the three sample scenarios is slightly different. In the warm start scenario, SANN-DTI (AUROC = 0.993, AUPR = 0.932) performed best, DeepDTI (AUROC = 0.988, AUPR = 0.907) performed sub-best, and DIstMult (AUROC = 0.932, AUPR = 0.729) performed poorly. In the cold start scenario, SANN-DTI's performance is still ahead in most metrics; DeepDTI, which was previously poorly represented in ya datasets, has improved its performance in comparison. Interestingly, in the drug cold start

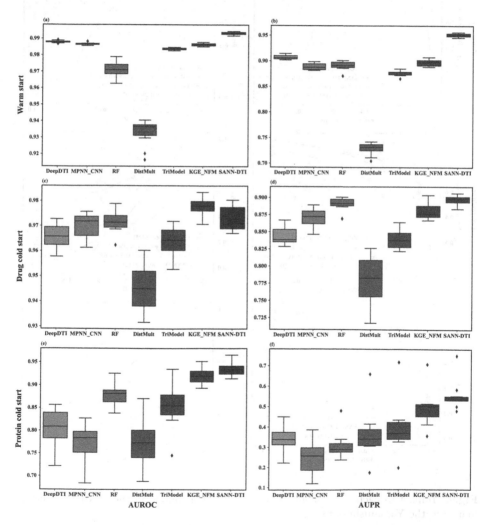

Fig. 3. Evaluation performance of BioKG dataset in three sample scenarios.

scenario, deep learning network models DeepDTI (AUROC = 0.966, AUPR = 0.844) and MPNN_CNN (AUROC = 0.970, AUPR = 0.871) relied only on traditional biomolecular information also performed as well as TriModel (AUROC = 0.964, AUPR = 0.839) using KG information. In the protein cold start scenario, because the KG contains the omics information of proteins and genes, the model using KG will be significantly ahead, such as SANN-DTI (AUROC = 0.554, AUPR = 0.871), SANN-DTI (AUROC = 0.554, AUPR = 0.871).

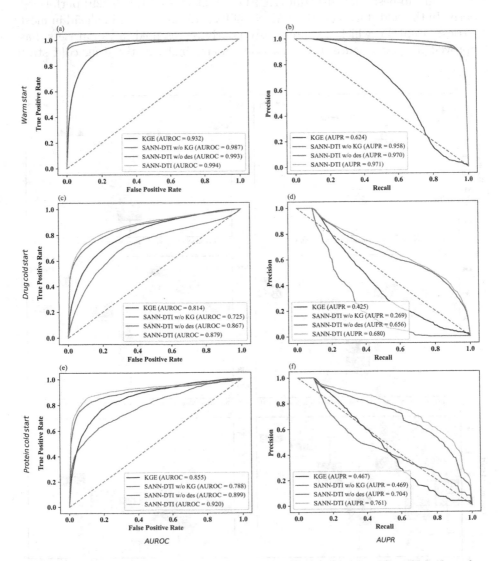

Fig. 4. Impact of each component in the SANN-DTI model on the predicted performance in the Yamanishi dataset.

Table 3. The top 10 drugs related to ERBB2 recommended by SANN-DTI

Rank	Drug ID	Drug Name	Score	Evidence
1	DB00128	Aspartic acid	0.999602	[29]
2	DB01136	Carvedilol	0.998383	Docking Simulation
3	DB11652	Tucatinib	0.998300	[30]
4	DB04325	Phenethylamine	0.997855	-
5	DB03915	2-Amino-3-Ketobutyric Acid	0.997346	-
6	DB00435	Nitric Oxide	0.996379	[31]
7	DB01168	Procarbazinee	0.995608	-
8	DB09073	Palbociclib	0.994060	[32]
9	DB01136	Pseudoephedrine	0.992460	-
10	DB05786	Irofulven	0.987707	-

KGE-NFM (AUROC = 0.492, AUPR = 0.918), TriModel (AUROC = 0.392, AUPR = 0.852).

3.2 Impact of Each Component on Predicted Performance

We removed the knowledge graph module (w/o KG) and the traditional structural information (w/o des) module respectively for validation on the Yamanishi dataset, and the results were shown in Fig. 4. KGE represents knowledge graph embedding method (DistMult) for prediction in knowledge graph. In the warm start scenario, the traditional biomolecular features and knowledge graph embedding features do not show advantages, and their evaluation metrics are similar. Because KGE method does not remove noise, its evaluation metrics is low. In the cold start scenario, the omics information embedded in knowledge graph is more important than the traditional biomolecular information. Specifically, if the knowledge graph embedding module was removed, AUROC on drug cold start and protein cold start decreased by 17% and 44%, and AUPR decreased by 67% and 38%, respectively; If the traditional molecular information module is removed, AUROC on drug cold start and protein cold start are decreased by 1% and 2% respectively, and AUPR is decreased by 3% and 7% respectively. These results show that our model can effectively integrate the features of biomolecules and knowledge graph and interact with higher-order features.

4 Case Study

Next, we used SAN-DTI to predict DTIs for approved drugs and ERBB2 targets. ERBB2 targets has been proved to be associated with breast cancer. The experiment is implemented on the warm start scenario of BioKG dataset. The top 10 recommended drugs with DTIs value are shown in Table 3. Among the predicted DTI, four DTIs are supported by the literature, and DB11652 has been used to treat breast cancer.

Table 4. Ten results of docking between Tucatinib and ERBB2

POSE	Affinity (kcal/mol)	RMSD lower bound	RMSD upper bound
1	-7.5	0	0
2	-7.4	3.98	9.307
3	-7.4	4.175	9.667
4	-7.2	7.356	10.578
5	-7.1	4.593	8.669
6	-7.1	3.388	6.202
7	-7.1	4.63	7.919
8	-7.1	4.406	9.945
9	-7	4.167	7.47
10	-7	3.201	9.124

In addition, we used the docking simulator AutoDock [33] to simulate the second ranked drug Tucatinib with the target ERBB2. Ten results of molecular docking are shown in Table 4. Affinity lower than -7 kcal/mol is a strong binding force. RMSD represents the degree of difference in the space position of heavy atoms of each pose relative to the optimal pose. RMSD upper bound is the

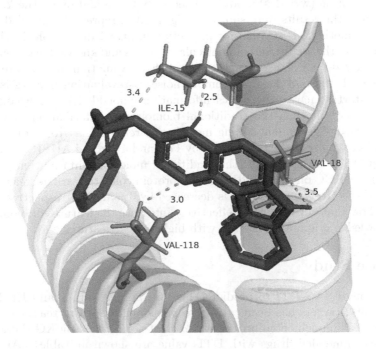

Fig. 5. The docked pose 1 between Tucatinib and ERBB2.

RMSD of strict correspondence of atoms between poses; RMSD lower bound is the optimal RMSD between pose atoms. The docking results indicate that Tucatinib can dock with the structure of ERBB2. Specifically, Tucatinib binds to ERBB2 by forming hydrogen bonds with residues VAL-118, VAL18, and ILE-15 on pose 1, as shown in Fig. 5. The above case show that our model has excellent predictive power for DTIs, providing a way to better understand drug action and drug repositioning.

5   Conclusion

In this study, we propose to integrate information from different sources based on multi-head self-attention networks to predict new DTI, namely SANN-DTI. The model extracts omics information from the knowledge graph via KGE and then combines this information with traditional drug and protein characterization via multi-head self-attention neural networks to obtain accurate and robust predictions of DTI. The powerful predictive power of SANN-DTI has been extensively validated on two benchmark datasets. The experimental results show that SANN-DTI is ahead of the most advanced methods in AUROC and AUPR in three assessment scenario settings. In addition, we analyze the impact of knowledge graphs and molecular information and explain their advantages. Therefore, four of the top ten recommended drugs were found to be supported by the literature. Moreover, the docking results of the second recommended drug with the HRBB2 target showed significant affinity, which provides help for potential drugs to target breast cancer and is conducive to the success of clinical trials. Overall, by narrowing the search space for DTIs, SANN-DTI is a powerful model for discovering novel DTIs, providing a promising approach for better understanding drug mode of action and drug repositioning.

Acknowledgements. This study is supported by the Sichuan Science and Technology Program (NO.2021YFG0031, 22YSZH0021) and Advanced Jet Propulsion Creativity Center (Projects HKCX2022-01-022).

References

1. Parvathaneni, V., Kulkarni, N.S., Muth, A., Gupta, V.: Drug repurposing: a promising tool to accelerate the drug discovery process. Drug Discov. Today **24**(10):2076–2085 (2019)
2. Pushpakom, S., et al.: Drug repurposing: progress, challenges and recommendations. Nature Rev. Drug Discov. **18**(1):41–58 (2019)
3. Pan, X., et al.: Deep learning for drug repurposing: methods, databases, and applications. Wiley Interdiscip. Rev.: Comput. Molecular Sci. **12**(4), e1597 (2022)
4. Luo, H., Li, M., Yang, M., Fang-Xiang, W., Li, Y., Wang, J.: Biomedical data and computational models for drug repositioning: a comprehensive review. Brief. Bioinform. **22**(2), 1604–1619 (2021)
5. Carracedo-Reboredo, P., et al.: A review on machine learning approaches and trends in drug discovery. Comput. Struct. Biotechnol. J. **19** 4538–4558 (2021)

6. Bagherian, M., Sabeti, E., Wang, K., Sartor, M.A., Nikolovska-Coleska, Z., Najarian, K.: Machine learning approaches and databases for prediction of drug-target interaction: a survey paper. Brief. Bioinform. **22**(1), 247–269 (2021)
7. Ezzat, A., Min, W., Li, X.-L., Kwoh, C.-K.: Computational prediction of drug-target interactions using chemogenomic approaches: an empirical survey. Brief. Bioinform. **20**(4), 1337–1357 (2019)
8. Jacob, L., Vert, J.P.: Protein-ligand interaction prediction: an improved chemogenomics approach. Bioinformatics **24**(19), 2149–2156 (2008)
9. Zheng, L., Fan, J., Yuguang, M.: Onionnet: a multiple-layer intermolecular-contact-based convolutional neural network for protein-ligand binding affinity prediction. ACS Omega **4**(14), 15956–15965 (2019)
10. Öztürk, H., Özgür, A., Ozkirimli, E.: Deepdta: deep drug-target binding affinity prediction. Bioinformatics **34**(17), i821–i829 (2018)
11. Luo, Y., et al.: A network integration approach for drug-target interaction prediction and computational drug repositioning from heterogeneous information. Nat. Commun. **8**(1), 573 (2017)
12. Thafar, M.A., et al.: DTiGEMs+: drug-target interaction prediction using graph embedding, graph mining, and similarity-based techniques. J. Cheminform. **12**(1), 1–17 (2020)
13. Gao, Z., Ding, P., Rong, X.: Kg-predict: a knowledge graph computational framework for drug repurposing. J. Biomed. Inform. **132**, 104133 (2022)
14. Mohamed, S.K., Nounu, A., Nováček, V.: Biological applications of knowledge graph embedding models. Brief. Bioinform. **22**(2), 1679–1693 (2021)
15. Zhang, R., Hristovski, D., Schutte, D., Kastrin, A., Fiszman, M., Kilicoglu, H.: Drug repurposing for Covid-19 via knowledge graph completion. J. Biomed. Inform. **115**, 103696 (2021)
16. Mohamed, S.K., Nounu, A., Nováček, V.: Discovering protein drug targets using knowledge graph embeddings. Bioinformatics **36**(2), 603–610 (2020)
17. Ye, Q., et al.: A unified drug-target interaction prediction framework based on knowledge graph and recommendation system. Nat. Commun. **12**(1), 6775 (2021)
18. Yamanishi, Y., Araki, M., Gutteridge, A., Honda, W., Kanehisa, M.: Prediction of drug-target interaction networks from the integration of chemical and genomic spaces. Bioinformatics **24**(13), i232–i240 (2008)
19. Walsh, B., Mohamed, S.K., Nováček, V.: Biokg: a knowledge graph for relational learning on biological data. In: Proceedings of the 29th ACM International Conference on Information and Knowledge Management, pp. 3173–3180 (2020)
20. Wishart, D.S., et al.: Drugbank: a knowledgebase for drugs, drug actions and drug targets. Nucleic Acids Res. **36**(suppl_1), D901–D906 (2008)
21. Kanehisa, M., et al.: From genomics to chemical genomics: new developments in kegg. Nucleic Acids Res. **34**(suppl_1), D354–D357 (2006)
22. Schomburg, I., et al.: Brenda, the enzyme database: updates and major new developments. Nucleic Acids Res. **32**(suppl_1), D431–D433 (2004)
23. Günther, S., et al.: Supertarget and matador: resources for exploring drug-target relationships. Nucleic Acids Res. **36**(suppl_1), D919–D922 (2007)
24. Landrum, G., et al.: Rdkit: open-source cheminformatics software. https://www.rdkit.org/, https://github.com/rdkit/rdkit**149**(150), 650 (2016)
25. Dubchak, I., Muchnik, I., Holbrook, S.R., Kim, S.H.: Prediction of protein folding class using global description of amino acid sequence. Proc. National Acad. Sci. **92**(19), 8700–8704 (1995)
26. Yang, B., Yih, W.T., He, X., Gao, J., Deng, L.: Embedding entities and relations for learning and inference in knowledge bases. arXiv preprint arXiv:1412.6575 (2014)

27. Abdi, H., Williams, L.J.: Principal component analysis. Wiley Interdisc. Rev.: Comput. Stat. **2**(4), 433–459 (2010)
28. Ashish V., et al.: Attention is all you need. In: Advances in Neural Information Processing Systems, 30 (2017)
29. Kawasaki, Y., et al.: Feedback control of ErbB2 via ERK-mediated phosphorylation of a conserved threonine in the juxtamembrane domain. Sci. Reports **6**(1), 1–9 (2016)
30. Borges, V.F., et al.: Tucatinib combined with ado-trastuzumab emtansine in advanced erbb2/her2-positive metastatic breast cancer: a phase 1b clinical trial. JAMA Oncol. **4**(9), 1214–1220 (2018)
31. Kundumani-Sridharan, V., Subramani, J., Owens, C., Das, K.C.: Nrg1β released in remote ischemic preconditioning improves myocardial perfusion and decreases ischemia/reperfusion injury via erbb2-mediated rescue of endothelial nitric oxide synthase and abrogation of trx2 autophagy. Arteriosclerosis, Thromb., Vasc. Biol. **41**(8), 2293–2314 (2021)
32. Llombart-Cussac, A., et al.: Fulvestrant-palbociclib vs letrozole-palbociclib as initial therapy for endocrine-sensitive, hormone receptor-positive, ErbB2-negative advanced breast cancer: a randomized clinical trial. JAMA Oncol. **7**(12), 1791–1799 (2021)
33. Trott, O., Olson, A.J.: Autodock vina: improving the speed and accuracy of docking with a new scoring function, efficient optimization, and multithreading. J. Comput. Chem. **31**(2), 455–461, (2010)

GRF-GMM: A Trajectory Optimization Framework for Obstacle Avoidance in Learning from Demonstration

Bin Ye[1], Peng Yu[1], Cong Hu[2], Binbin Qiu[3], and Ning Tan[1(✉)]

[1] School of Computer Science and Engineering, Sun Yat-sen University,
Guangzhou, China
tann5@mail.sysu.edu.cn
[2] Guangxi Key Laboratory of Automatic Detecting Technology and Instruments,
Guilin University of Electronic Technology, Guilin, China
[3] School of Intelligent Systems Engineering, Sun Yat-sen University, Shenzhen, China

Abstract. Learning from demonstrations (LfD) provides a convenient pattern to teach robot to gain skills without mechanically programming. As an LfD approach, Gaussian mixture model/Gaussian mixture regression (GMM/GMR) has been widely used for its robustness and effectiveness. However, there still exist many problems of GMM when an obstacle, which is not presented in original demonstrations, appears in the workspace of robots. To address these problems, this paper presents a novel method based on Gaussian repulsive field-Gaussian mixture model (GRF-GMM) for obstacle avoidance by optimizing the model parameters. A Gaussian repulsive force is calculated through Gaussian functions and employed to work on Gaussian components to optimize the mixture distribution which is learnt from original demonstrations. Our approach allows the reproduced trajectory to keep a safe distance away from the obstacle. Finally, the feasibility and effectiveness of the proposed method are revealed through simulations and experiments.

Keywords: GMM/GMR · Potential field · Obstacle avoidance

1 Introduction

Robotic manipulators have demonstrated their impressive application potential in manufacturing industry. However, professional knowledge and programming skills are usually required for their trajectory planning, which is not friendly to non-professional employees. Therefore, robots are expected to work independently through robotic learning algorithms. As a potential approach, LFD helps

This work is partially supported by the National Natural Science Foundation of China (62173352, 62006254), the Guangxi Key Laboratory of Automatic Detecting Technology and Instruments (YQ23207), and the Guangdong Basic and Applied Basic Research Foundation (2021A1515012314).

B. Luo et al. (Eds.): ICONIP 2023, LNCS 14450, pp. 18–30, 2024.
https://doi.org/10.1007/978-981-99-8070-3_2

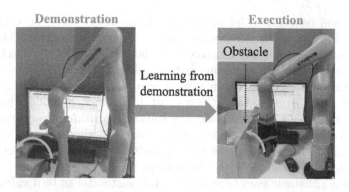

Fig. 1. Collision problem of learning from demonstration in the changing environment. A human operator demonstrates a motion trajectory for the robot in the free space. After learning from the demonstration, the robot reproduces the motion trajectory in its workspace. Due to the changes in the environment, an obstacle appears on the trajectory, which leads to the collision between the robot and the obstacle.

Table 1. Comparisons Among Different LfD Frameworks.

Framework	Method	Obstacle avoidance	Training time	Feature lost
GMM [3]	Expectation-maximization algorithm	✗	Short	Low
Model in [9]	Simulated annealing-RL	✓	Long	High
Model in [10]	Data resampling	✓	Long	Low
Model in [11]	CMA-ES	✓	Long	High
Ours	Gaussian repulsive field	✓	Short	Low

users teach robots complex motions through simple human demonstrations [1]. The aspect of LfD that attracts researchers is its capacity to implicitly learn motion constraints and features from demonstrations, and adapt to different environments.

As a multi-shot learning method, GMM learns from a few demonstrations of the same task so that experts are allowed to demonstrate the task more than once. It is worth noting that the experts here refer to skilled employees in factories, rather than those with professional knowledge and programming skills related to robots. The demonstration data is modelled with probability density functions, and the learning process is accomplished through the expectation-maximization (EM) algorithm [2]. In this way, the learnt model gets robust to noisy demonstrations and generates a high-quality motion trajectory for robots by GMR [3]. As a weighted summation regression method, GMR extracts model features to reconstruct trajectory which can be regard as a weighted superposition of the demonstrations. Nevertheless, there are still some drawbacks of GMM/GMR. For example, when the task specifies passing points from demonstrations, GMM/GMR hardly learns the motion feature and reproduces a low accuracy path. The lack of generalization capability leads to the failure of task. For the improvement of generalization, researchers have taken a large effort and proposed a lot of excellent frameworks based on GMM. Calinon *et al.* [4] proposed a parametric Gaussian mixture model (PGMM) that combines task-parameterized movement models by encoding the relationship between task parameters and motion features. Guenter *et al.* [5] combined GMM with reinforcement learning which is used to adapt the centers of GMM components for a slippery solution. Chen *et al.* [6] trained a model through the reinforcement-learning expectation-maximization algorithm for accurate inverse kinematics control. However, existing GMM-based methods always work terribly with complex task in changing environments, as illustrated in Fig. 1, where obstacles may appear on the reproduced trajectory.

Considering obstacle avoidance, existing LfD methods can be divided into two classes. The first class teaches the robot to avoid obstacles by human demonstrations, which means obstacles have been presented in original demonstrations. For example, Zhang *et al.* [7] taught robots the obstacle avoidance mechanism of human beings from demonstrations. Li *et al.* [8] used data preprocessing and retrieved the successful demonstration for GMM to learn. The second class focuses on dealing with the case where obstacles are not presented in original demonstrations. For instance, Wang *et al.* [9] used an optimization algorithm based on simulated annealing-reinforcement learning (SA-RL) to solve the under-/over-fitting problem and obstacle avoidance problem. Li [10] used a resampling technique to train the GMM model by continuously sampling points from the teaching trajectory. Osa *et al.* [11] proposed an idea of covariance matrix adaptation evolutionary strategy (CMA-ES) which updates the model parameters with demonstration-guided cost function. However, the above-mentioned methods need plenty of time to train the model, which leads to the problem that robots are unable to cope with sudden obstacles. And some obstacle avoidance trajectories are quite different from the original teaching trajectories that the model loses demonstrations feature [9,12]. Table 1 gives a review on different GMM frameworks for obstacle avoidance.

Trajectory optimization, a widely-employed approach in path planning, has been topics of research for many years [13]. While existing methods limitedly focus on robot-specific issues such as smoothness and energy minimization [14], human demonstrations are not considered during the optimization process. Inspired by [15,16], We propose a novel potential field, namely Gaussian repulsive field (GRF), to modify GMM, which takes the advantages of GMM and optimization-based motion planning. First, GMM is employed to learn the movement features from multiple motion demonstrations. As an obstacle appears on the reproduced trajectory, GRF generates Gaussian repulsive forces around GMM and optimizes the parameters of Gaussian components. The optimization process will repeat until the goal is achieved. Therefore, we achieve an optimal distribution that incorporates the motion features and the obstacle information. The proposed method ensures that the reproduced trajectory keeps a safe margin from the obstacle which is not mentioned in other literature. In addition, the proposed method has superior scalability, which means it can be directly and easily combined with other probabilistic motion models.

The main contributions of this paper are listed below:

- A distribution optimization algorithm using Gaussian repulsive force is proposed to modify the parameters of Gaussian components, thereby achieving the avoidance of obstacles.
- A novel GRF-GMM method is proposed for learning from demonstrations and avoiding the obstacle which is not presented in original demonstrations.
- A LfD framework is proposed by combining the proposed GRF-GMM method and a control system designed based on zeroing dynamics, thereby achieving the LfD of robotic manipulators.

2   Problem Statement

This paper investigates the GMM methods problem of robotic manipulators in the changing environment. More specifically, robotic manipulators are required to reproduce a collision-free motion trajectory to avoid the obstacle which is not presented in original demonstrations. Firstly, the expert provides demonstrations for robot. Then, the robot executes the reproduced trajectory $\{\xi_i\}_{i=1}^{N}$ learnt by GMM/GMR, where ξ_i denotes the i-th point and N is the length of the reproduced trajectory. However, the obstacle O_{bs} may appear on the reproduced trajectory in the changing environment. To avoid such a collision, the reproduced trajectory should be adjusted to keep a safe distance S_{safe} from the obstacle. We define an assessment variable S_{nearest} which calculates the nearest distance between the reproduced trajectory and the obstacle

$$S_{\text{nearest}} = \min\|P_{\text{obs}} - \xi_i\|_2, \quad i = 1, \dots, N, \tag{1}$$

where P_{obs} is the position of O_{bs} and $\| \cdot \|_2$ denotes the Euclidean distance between two points. For safe operation, the nearest distance S_{nearest} is supposed to surpass S_{safe} which is defined by prior knowledge. Inspired by the improved potential field method, the paper focuses on modifying GMM to generate a fine-tuned S_{nearest}, which is adjustable in our method, between the reproduced trajectory and the obstacle.

3   Method

This section elaborates that the proposed learning framework includes two main parts: a GMM model and a model distribution optimization algorithm. The GMM model is used to extract movement features from demonstrations, and the algorithm optimizes the distributions of the learning model with Gaussian repulsive field.

3.1   Gaussian Mixture Model/Gaussian Mixture Regression

Given M expert demonstrations, each demonstration has N points $\{\tau_i\}_{i=1}^N$. In the m-th demonstration ($m = 1, \cdots, M$), τ_i is divided into d dimensions that consist of a 1-dimensional temporal value $\tau_{t,i}$ and $(d-1)$-dimensional spatial values $\tau_{s,i}$. GMM can be regarded as the superposition of K Gaussian distributions

$$p(\tau_i) = \sum_{k=1}^K p(k)p(\tau_i|\Theta_k), \tag{2}$$

with

$$p(k) = \pi_k, \tag{3}$$

$$p(\tau_i|\Theta_k) = \mathcal{N}(\tau_i|\mu_k, \Sigma_k)$$

$$= \frac{1}{\sqrt{(2\pi)^D|\Sigma_k|}} e^{-\frac{1}{2}\left((\tau_i-\mu_k)^T \Sigma_k^{-1}(\tau_i-\mu_k)\right)}, \tag{4}$$

where τ_i is the i-th data point, $p(k)$ is the prior probability, $\Theta_k = \{\pi_k, \mu_k, \Sigma_k\}$ denotes the set of the prior probability, mean and covariance of the k-th Gaussian component and $p(\tau_i|\Theta_k)$ is the conditional probability density function of the k-th Gaussian component [3]. To achieve well trained parameters, a traditional k-means clustering method is utilized to initialize the model [17]. Then, maximum-likelihood estimation of the parameters is guaranteed to be enhanced iteratively using the standard EM algorithm [2]. Finally, a regression algorithm, e.g., Gaussian mixture regression (GMR), can be used to reproduce a trajectory.

3.2   Optimization Algorithm: GRF-GMM

Trajectory information is stored in the model parameters of GMM in the form of priors, means and variances. As mentioned above, GMM/GMR performs poorly when an obstacle appears on the reproduced trajectory T_{rep}. For the safety of robots, the executed trajectory should be sufficiently far away from the obstacle. This can be considered as an optimal problem and the optimization goal is to adjust the parameters of GMM so that the robot manipulator stays far enough away from the obstacle.

A possible approach for achieving the above optimization objective is the artificial potential field [16,18,19]. The idea of classical artificial potential field is to design the movement of robots into an abstract artificial gravitational field. The target point produces "gravity" on robots and the obstacle produces "repulsive force" on robots.

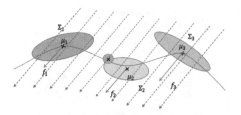

Fig. 2. Schematic of Gaussian repulsive force. The colorful ellipses represent the covariance of Gaussian components learnt from demonstrations. The blue curve is the reproduced trajectory. The gray circle is an obstacle which produces a force field. The red dotted lines represent the force field and the red solid lines are the forces acting on the components. All the forces acting at the centers of the components have the same direction but different magnitudes. (Color figure online)

In this part, based on the field which generates "repulsive force", we propose a Gaussian repulsive field (GRF) working between the obstacle and K Gaussian components rather than working straight on the trajectory. Figure 2 visualizes the forces acting on the Gaussian components. In the field, the direction of the forces are the same, but the magnitude of the forces are different. As the forces work, variations are generated on the parameters of the components. These forces remain until the model converges. Figure 3 exhibits the overall framework of our method.

Next we discuss the details of our optimization method (termed GRF-GMM). We calculate the forces using Gaussian functions rather than traditional artificial potential field:

$$S_k = \|P_{\text{obs}} - \mu_k\|_2, \quad k = 1, \dots, K, \tag{5}$$

$$f_k = \begin{cases} f_{\text{init}}, & \text{if } S_k = 0, \\ (P_{\text{obs}} - \mu_k)^T \Sigma_k (P_{\text{obs}} - \mu_k) \mathcal{N}(P_{\text{obs}}|\Theta_k), & \text{if } S_k \neq 0, \end{cases} \tag{6}$$

where S_k denotes the Euclidean distance between obstacle position P_{obs} and the mean μ_k of the k-th component, f_k represents a Gaussian repulsive force acting on the k-th Gaussian component, f_{init} means a constant force when the obstacle overlaps the k-th Gaussian component, $\mathcal{N}(P_{\text{obs}}|\Theta_k)$ is the probability that O_{bs} is in the distribution of k-th component.

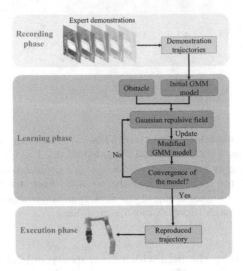

Fig. 3. Flowchart of the proposed LfD framework. In the first phase we collect multiple trajectories from experts. Subsequently, we build a learning model using GMM to process the demonstrations. As obstacle information is added, our method GRF-GMM repeats optimization until the convergence of the model. Finally, the robot executes the reproduced collision-free trajectory in the task space.

Unlike traditional potential field, the way we calculate produces a stable force when the obstacle is too close to the robot. The forces are directly proportional to Gaussian basic functions because they reflect the possibility of the obstacle actuality nearby. The Euclidean distance S_k in (5) is utilized to assess the correlation between O_{bs} and the k-th component Θ_k. However, the field losses dynamic force if S_k is zero and we correspondingly define an initial force for this situation. Meanwhile, we calculate the probability when O_{bs} appears on k-th distribution. In this way, the formulation of the Gaussian repulsive field generates a stable force to modify the distances between the components and the obstacle.

Algorithm 1. Optimization algorithm of GMM for obstacle avoidance

---

Input: Demonstration data set T_{demo} and obstacle O_{bs}
1: $\mu_k, \Sigma_k, \pi_k \leftarrow$ initialize the model using (3)
2: **repeat**
3:      $T_{rep} \leftarrow$ using GMR
4:      **for** k=1:K **do**
5:          $f_k \leftarrow$ update forces using (6)
6:          $D \leftarrow$ update direction using (9)
7:          $\mu_k \leftarrow$ update centers using (10)
8:          $\pi_k \leftarrow$ update weights using (11)
9:      **end for**
10:     $S_j \leftarrow$ using (8)
11: **until** $S_j - S_{safe} \geq \sigma$
Output: Θ_i and T_{rep}

Then, we are supposed to find the direction of the Gaussian repulsive force acting on k-th component as follows

$$j = \arg\min_{1 \le i \le n} ||P_{\text{obs}} - \xi_i||_2, \tag{7}$$

$$S_j = ||\xi_j - P_{\text{obs}}||_2, \tag{8}$$

$$D = \begin{cases} P(\xi_j), & \text{if } S_j = 0, \\ \xi_j - P_{\text{obs}}, & \text{if } S_j \ne 0, \end{cases} \tag{9}$$

where ξ_i denotes the i-th point of the reproduced trajectory T_{rep}, ξ_j means the nearest point from T_{rep} to the obstacle, $P(\cdot)$ is a function that calculates the normal vector in the tangential direction of the point ξ_j. $P(\xi_j)$ is selected as the direction if O_{bs} is on the reproduced trajectory. Otherwise we choose the direction from O_{bs} to T_{rep}. Our optimization process is to gradually repel Θ_i away as O_{bs} is on the reproduced trajectory or near it. Such a direction D will lead to a growing tendency for the distance between ξ_j and O_{bs}.

When the repulsive force and the direction are determined, model parameters of the k-th component iterate as follows at the i-th iteration:

$$\mu_k^{i+1} = \mu_k^i + \alpha f_k^i D, \tag{10}$$

$$\hat{\pi}_k^{i+1} = \pi_k^i \left(\frac{f_k^i}{\sum_{k=1}^{K} f_k^i} + \beta \right), \tag{11}$$

$$\pi_k^{i+1} = \frac{\hat{\pi}_k^{i+1}}{\sum_{k=1}^{K} \hat{\pi}_k^{i+1}}, \tag{12}$$

where μ_k^{i+1} and π_k^{i+1} are parameters of the k-th component at the i-th iteration after optimization, α and β are constant, respectively. Especially, α is used to modulate the magnitude of Gaussian repulsive force stopping an excessive force from destroying the integrity of the distribution. Equations (11) and (12) update the weight of each Gaussian component. This ensures that the Gaussian repulsive force have more significant impact on the distribution for a rapid convergence.

As shown in Algorithm 1, the iteration repeats until the convergence of the model. This framework judges the convergence of the model based on the fact that S_j is supposed to be larger than S_{safe}. Therefore, we expect to maintain the most motion features extracted from demonstrations as the obstacle avoidance is achieved.

4   Simulations and Experiments

To demonstrate the effectiveness of the proposed LfD method, simulations and experiments are conducted in this part. For simulation, the host configuration is V-REP PRO EDU 3.6, Intel(R) Core(TM) i5-9400 CPU @ 2.90 GHz, RAM 16.00 GB. In the simulation, the robot we use is KUKA LBR iiwa 7 R800. The algorithm is implemented with MATLAB R2021b and task is to reproduce a 2D handwriting letter. For experiment, the robot we use is the Kinova Gen3 manipulator and the task is pick and place in the dynamical environment.

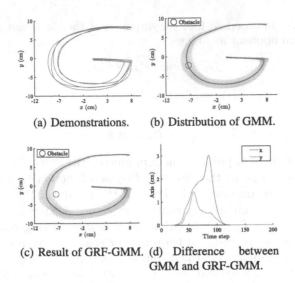

(a) Demonstrations.   (b) Distribution of GMM.

(c) Result of GRF-GMM. (d) Difference   between
                       GMM and GRF-GMM.

Fig. 4. Demonstrative trajectories and reproduced trajectories.

4.1   2D Handwriting Letter Task

The demonstration data is the 2D handwriting letters dataset from [22] and
the letter 'G' is selected. We sample 5 G-shape demonstrations, each with 200
data points, and preprocess them into a temporal sequence with the time interval
being 0.1 s. The number of GMM components are set to 14. GMM parameters are
initialized using a k-means cluster algorithm. The hyper-parameters of equation
(10) and (11) are respectively 1 and 10. The safe distance S_{safe} is defined as
2 cm by the requirement of task and prior knowledge. GMR is chosen as the
regression method to generate a trajectory from the movement distribution. As
Fig. 4(a) shows, the blue curves are 5 demonstrations. Figure 4(b) visualizes the
distribution of the reproduced trajectory by GMM. The red curve is the result
of GMM and red area denotes the variance of the reproduced trajectory. An
obstacle, with the position being $(-7.7, -2.2)$ cm, is on the way of the reproduced
trajectory.

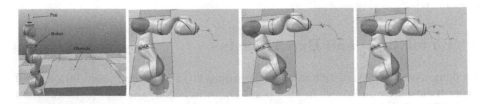

Fig. 5. Simulation in VREP. The first figure shows the scene in the VREP. The other
figures display the process that the robot catches the pen and writes a curve on the
table. The pen successfully avoids the obstacle and writes G-shape path during the
simulation.

Figure 4(c) represents the result of the proposed GRF-GMM: a red optimal distribution with a new blue reproduced trajectory. The result illustrates
that our distribution optimization method can deal with the obstacle avoidance
problem, which reveals the effectiveness of our method. Compared with GMM,
our method significantly changes the distributions around the obstacle. Besides,
our reproduced trajectory preserves the most original features, i.e., G-shape,
and great smoothness. Figure 4(d) displays the difference between the trajectory
reproduced by GMM and that reproduced by our GRF-GMM on different axes.

Finally, we build the scene shown in the Fig. 5 based on the virtual robot
experimentation platform (VREP). A pen is attached to the robot end-effector
and the pen will follow the robot to leave a mark on the path. A red cuboid
locates on the table as an obstacle. The simulation result shows that the robot
successfully completes the handwriting task and verifies the effectiveness of the
proposed LfD framework for robotic manipulators.

4.2   Experiment

To further reveal the effectiveness of the proposed method, we conduct experiments in real-world environment. In an open space, a simple task is demonstrated
by experts that we control the robot manipulator to pick the red cuboid up and
put it down somewhere else. Then we put an arched box as an obstacle on the
table. When we input the trajectory signal reproduced by GMM as the desired
path, the cuboid collides with the box as depicted in the Fig. 6(a). Then, we set
the collision position as the obstacle position for GRF-GMM.

Figure 6(b) shows a collision-free robot operation by means of the proposed
GRF-GMM method. As the robot catches the cuboid according to the trajectory
generated by GRF-GMM, the whole process is carried out safely. The result
reveals that the robot captures the information of the obstacle and adjust the
learning model using GRF-GMM.

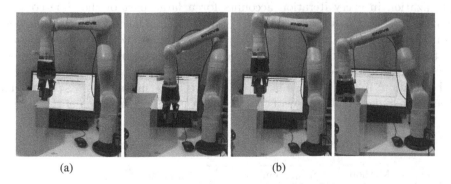

     (a)   (b)

Fig. 6. Physical experiment in real environment. The first picture shows the collision.
The others show that robot catches the cuboid and avoids the obstacle successfully.

4.3   Comparisons

To show the advantages of the proposed method, we compare the proposed GRF-GMM with modified LfD algorithm: GMM, kernelized movement primitives (KMP) [20]. For comparison, we set the same input demonstrations and initial conditions for each case. As GMM lacks obstacle avoidance, the way we apply GMM in obstacle avoidance is initializing the model and relearning the collision-free trajectory of GRF-GMM using EM algorithm. The comparison with GMM will show the improvement of our method. For KMP, we find the trajectory point of the collision and replace it as a desired point which is $S_{nearest}$ away from the obstacle to accomplish obstacle avoidance.

To keep consistency, we compare the algorithm results on the following aspects: time cost T_{cost} of the algorithms, nearest distance $S_{nearest}$ from the obstacle to the new trajectory, smoothness C_{smooth} of the new trajectory and the degree of deviations C_{dev} between the old trajectory and the new trajectory. T_{cost} is the running time of the different algorithms. C_{smooth} denotes the average jerk of the trajectory

$$J_i = \frac{\partial^3 \xi_i}{\partial t^3}, \tag{13}$$

$$C_{smooth} = \frac{\sum_{i=1}^{N} |J_i|}{N}. \tag{14}$$

C_{dev} is the average difference of points:

$$C_{dev} = \frac{\sum_{i=1}^{N} |\xi_i^{new} - \xi_i^{old}|}{N}. \tag{15}$$

Among them, $S_{nearest}$ is the most essential index in obstacle avoidance. We repeated the experiment several times with different obstacle position. The quantitative Comparison in shown in Table 2.

In terms of time cost, most methods can finish the task in 1 s. This result illustrates the potential of our method for online optimization. In GRF-GMM, optimization in every iteration accounts for a large part of the time cost. In terms of smoothness, although our GRF-GMM method has not achieved the best result, the trajectory produced by the GRF-GMM method is smooth enough for the robot according to the simulation and experiment results. Finally, we mainly focus on the performance of these methods in terms of nearest distance and deviation. In particular, a larger value of $S_{nearest}$ means a lower probability of collision and a smaller value of C_{dev} means a better preservation of the motion

Table 2. Comparison Among Different Methods.

Method	$S_{nearest}$ [cm]	T_{cost} [s]	C_{smooth} [cm/s^3]	C_{dev} [cm]
GMM	1.9833	0.1804	7.2438	0.5556
KMP	1.1721	0.7383	1.2780	0.5734
GRF-GMM	**2.0285**	0.6519	3.2906	**0.4911**

features of original demonstrations. As shown in the table, the proposed GRF-GMM is the only method that generates a nearest distance S_{nearest} more than S_{safe}, which is set to 2 cm. At the same time, the proposed GRF-GMM achieves the lowest value of C_{dev}, which implies the GRF-GMM method preserves the most motion features. In summary, these results have revealed the advantages of the proposed GRF-GMM method.

5 Conclusions

In this paper, to improve the generalization ability of GMM in the environment, a trajectory distribution optimization scheme based on GRF has been proposed. With the optimization mechanism and the idea of repulsive field, GRF-GMM can efficiently solve the obstacle avoidance problem in changing environments. The model obtained by our optimization method does not require further adjustment and is directly applied to the probability regression model GMR. Finally, simulative and experimental results have shown that the trajectory reproduced by the proposed GRF-GMM method for robot learning can avoid the obstacle that has not been presented in original demonstrations, and comparative studies have revealed the merits of the proposed method.

For further researches, there are some potential directions of the study. To reduce the run time for online learning, a gradient-decent method could be added to Gaussian force calculation for less optimization iterations [21]. Besides, dealing with dynamic obstacles is also one of the future research.

References

1. Zhu, Z., Hu, H.: Robot learning from demonstration in robotic assembly: a survey. Robotics **7**(2), 17 (2018)
2. Chen, D., Li, G., Zhou, D., Ju, Z.: A novel curved gaussian mixture model and its application in motion skill encoding. In: IEEE/RSJ International Conference on Intelligent Robots and Systems, pp. 7813–7818 (2021). https://doi.org/10.1109/IROS51168.2021.9636121
3. Calinon, S., Guenter, F., Billard, A.: On learning, representing, and generalizing a task in a humanoid robot. IEEE Trans. Syst., Man, Cybern. Part B (Cybernetics) **37**(2), 286–298 (2007)
4. Calinon, S., Alizadeh, T., Caldwell, D.G.: On improving the extrapolation capability of task-parameterized movement models. In: IEEE/RSJ International Conference on Intelligent Robots and Systems, pp. 610–616 (2013)
5. Guenter, F., Hersch, M., Calinon, S., Billard, A.: Reinforcement learning for imitating constrained reaching movements. Adv. Robot. **21**(13), 1521–1544 (2007)
6. Chen, J., Lau, H.Y., Xu, W., Ren, H.: Towards transferring skills to flexible surgical robots with programming by demonstration and reinforcement learning. In: Eighth International Conference on Advanced Computational Intelligence, pp. 378–384 (2016)
7. Zhang, H., Han, X., Fu, M., Zhou, W.: Robot obstacle avoidance learning based on mixture models. J. Robot. **2016**, 1–14 (2016)

8. Li, W., Cheng, H., Liang, Z., Xiao, J., Zhang, X.: Adaptive obstacle avoidance optimization algorithm based on learning from demonstration. In: IEEE 11th Annual International Conference on CYBER Technology in Automation, Control, and Intelligent Systems, pp. 447–452 (2021). https://doi.org/10.1109/CYBER53097.2021.9588300

9. Wang, Y., Hu, Y., El Zaatari, S., Li, W., Zhou, Y.: Optimised learning from demonstrations for collaborative robots. Robot. Comput.-Integr. Manufact. **71**, 102169 (2021)

10. Li, X., Cheng, H., Liang, X.: Adaptive motion planning framework by learning from demonstration. Ind. Robot. **46**(4), 541–552 (2019)

11. Osa, T., Esfahani, A.M.G., Stolkin, R., Lioutikov, R., Peters, J., Neumann, G.: Guiding trajectory optimization by demonstrated distributions. IEEE Robot. Autom. Lett. **2**(2), 819–826 (2017)

12. Chi, M., Yao, Y., Liu, Y., Zhong, M.: Learning, generalization, and obstacle avoidance with dynamic movement primitives and dynamic potential fields. Appl. Sci. **9**(8), 1535 (2019)

13. Koert, D., Maeda, G., Lioutikov, R., Neumann, G., Peters, J.: Demonstration based trajectory optimization for generalizable robot motions. In: IEEE-RAS 16th International Conference on Humanoid Robots, pp. 515–522 (2016)

14. Kalakrishnan, M., Chitta, S., Theodorou, E., Pastor, P., Schaal, S.: Stomp: stochastic trajectory optimization for motion planning. In: IEEE International Conference on Robotics and Automation, pp. 4569–4574 (2011)

15. Duan, A., et al.: Learning to avoid obstacles with minimal intervention control. Front. Robot. AI **7**, 60 (2020)

16. Tan, H., Erdemir, E., Kawamura, K., Du, Q.: A potential field method-based extension of the dynamic movement primitive algorithm for imitation learning with obstacle avoidance. In: IEEE International Conference on Mechatronics and Automation, pp. 525–530 (2011)

17. MacQueen, J.: Classification and analysis of multivariate observations. In: 5th Berkeley Symp. Math. Statist. Probability, pp. 281–297. University of California Los Angeles LA USA (1967)

18. Tian, Y., Zhu, X., Meng, D., Wang, X., Liang, B.: An overall configuration planning method of continuum hyper-redundant manipulators based on improved artificial potential field method. IEEE Robot. Autom. Lett. **6**(3), 4867–4874 (2021)

19. Mok, J., Lee, Y., Ko, S., Choi, I., Choi, H.S.: Gaussian-mixture based potential field approach for uav collision avoidance. In: 56th Annual Conference of the Society of Instrument and Control Engineers of Japan, pp. 1316–1319 (2017)

20. Huang, Y., Rozo, L., Silvério, J., Caldwell, D.G.: Kernelized movement primitives. Int. J. Robot. Res. **38**(7), 833–852 (2019)

21. Zucker, M., et al.: Chomp: covariant Hamiltonian optimization for motion planning. Int. J. Robot. Res. **32**(9–10), 1164–1193 (2013)

22. Sylvain, C.: Robot programming by demonstration: a probabilistic approach (2009)

SLG-NET: Subgraph Neural Network with Local-Global Braingraph Feature Extraction Modules and a Novel Subgraph Generation Algorithm for Automated Identification of Major Depressive Disorder

Yan Zhang, Xin Liu, Panrui Tang, and Zuping Zhang[✉]

School of Computer Science and Engineering, Central South University,
Changsha 410083, China
zpzhang@csu.edu.cn

Abstract. Major depressive disorder (MDD) is a severe mental illness that poses significant challenges to both society and families. Recently, several graph neural network (GNN)-based methods have been proposed for MDD diagnosis and achieved promising results. However, these methods encode entire braingraph directly, have overlooked the subgraph structure of braingraph, which leads to poor specificity to braingraphs. Additionally, the GNN framework they used is rudimentary, resulting in insufficient feature extraction capabilities. In light of the two shortcomings mentioned above, this paper designed a novel depression diagnosis framework named SLG-NET based on subgraph neural network. To the best of our knowledge, this study is the first attempt to apply subgraph neural network to the field of depression diagnosis. In order to enhance the specificity of our model to braingraphs, we propose a novel subgraph generation algorithm based on sub-structure information of brain. To improve feature extraction capabilities, a local and global braingraph feature extraction modules are proposed to extract braingraph properties at both local and global levels. Comprehensive experiments performed on rest-metamdd dataset show that the performance of proposed SLG-NET significantly surpasses many state-of-the-art methods, which show that the SLG-NET has the potential for auxiliary diagnosis of depression in clinical scenarios.

Keywords: Major depressive disorder · Subgraph neural network · Brain functional connectivity · Resting-state fMRI

1 Introduction

Major depressive disorder is a serious mental illness that poses an extraordinary economic burden on public systems [1]. Therefore, developing an efficient

B. Luo et al. (Eds.): ICONIP 2023, LNCS 14450, pp. 31–42, 2024.
https://doi.org/10.1007/978-981-99-8070-3_3

and reliable method to detect MDD is important. Usually, clinical diagnosis typically involves expert interviews and self-report measures [2]. However, these assessments can be influenced by subjective factors, leading to overdiagnosis or misdiagnosis [3].

Research has revealed that altered or aberrant resting-state functional connectivity (FC), measured through fMRI, can be observed in the FC networks of MDD [4], individuals diagnosed with MDD display heightened connectivity within the default mode network and reduced connectivity between the default mode network and central executive network [5]. All of these indicate the potential of brain FC network in distinguishing MDD from Health Control (HC).

Recently, various methods have been proposed to diagnose depression using graph convolutional and graph pooling techniques based on brain FC network. In 2023, Gallo et al. [3] and Venkatapathy et al. [6] utilized graph convolution network (GCN) to classify the FC network of MDD and HC for the diagnosis of depression. Although they achieved good results, the GCN framework they used is rudimentary, leading to a suboptimal feature extraction capability. In a study by [7]. A new method called SGP-SL was developed to detect MDD, SGP-SL using self-attention modules to refine the created graph based on local and global connections. The excellent results of SGP-SL indicate that self-attention mechanisms are crucial for diagnosing depression. However, it should be noted that SGP-SL does not take into account the subgraph structure of braingraph, which may result in the loss of specificity to braingraph for detecting MDD.

Although there has been significant progress in deep learning-based automated identification of MDD over the past few years, there are two remaining issues that need to be addressed:

1) Existing depression diagnosis model has insufficient specificity for braingraphs [5–7]. Brain regions can be classified based on their function and anatomical location, each sub-structure has distinct function. However, existing GNN-based research encoding the entire braingraph directly has neglected the critical sub-structure information of the brain;
2) Existing depression diagnosis model has insufficient feature extraction ability. The current depression diagnosis models have a rudimentary network framework [3,5,6,8], resulting in the loss of many important information in the braingraph.

In light of the two deficiencies mentioned above, this paper proposes a novel depression diagnosis framework named SLG-NET. The primary contributions of our research are briefly outlined as follows:

1) To improve the specificity of the present model to braingraphs for MDD diagnosis, we apply subgraph neural network to encode the sub-braingraph extracted by the proposed subgraph generation algorithm (S-BFS), S-BFS can preserve the crucial sub-structure information of braingraph.
2) To improve the feature extraction capabilities, we propose a local braingraph feature extraction (LFE) module and a global braingraph feature extraction

(GFE) module. The LFE module uses reinforcement pooling and self-attention pooling mechanisms to focus on identifying crucial local brain-graph properties. The GFE module applies self-supervised mutual information (SSMI) mechanism to enhance the awareness of the sub-braingraph embedding to the global braingraph structural properties;

3) Extensive experiments conducted on the rest-metamdd dataset demonstrated the proposed SLG-NET significantly surpasses many state-of-the-art methods with an accuracy of 74.15%. The high accuracy shows that the proposed SLG-NET has the potential for auxiliary diagnosis of depression in clinical settings;

2 Related Work

2.1 Construction of Braingraph

The process of constructing a braingraph is shown in Fig. 1. Our approach utilizes the Ledoit-Wolf estimator to assess the FC between ROIs and employs proportional thresholding to derive the connectivity pattern of the braingraph. The Automated Anatomical Labeling (AAL) atlas with N = 116 regions-of-interest (ROIs) is used in our method. The partition method of functional brain region is based on standard (7+1) system template defined in literature [8]. In addition, Cerebellum and Vermis were categorized as the ninth functional brain regions named the CV, these 116 ROIs are divided into nine functional brain regions as shown in Fig. 1 (Visualization of braingraph); the nodes with different colors correspond to the nine different functional brain regions.

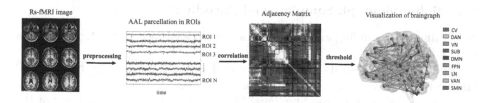

Fig. 1. An illustration of constructing a braingraph.

3 Method

As shown in Fig. 2: The overall process of SLG-NET consists of three steps: (1) Sub-braingraph sampling and encoding ; (2) Sub-braingraph selection and sub-braingraph's node selection by LFE module; (3) Sub-braingraph sketching by GFE module and classification. We provide complete detail of these three steps in Sects. 3.1, 3.2, and 3.3.

The graph of the FC network is denoted as $G = (V, X, A)$. $V = \{v_1, v_2, \cdots, v_N\}$ represents a set of N nodes (ROIs), the adjacency matrix is denoted as $\mathbf{A} = [a_{ij}] \in \{0, 1\}^{N \times N}$, the node feature matrix for G is denoted as $\mathbf{X} = [\mathbf{x}_1, \ldots, \mathbf{x}_N]^T \in \mathbb{R}^{N \times d}$.

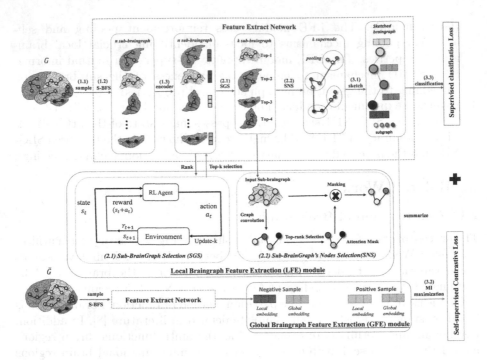

Fig. 2. An illustration of SLG-NET architecture.

3.1   Sub-braingraph Sampling and Encoding

Subbraingraph sampling and encoding can be divided into three steps as shown in Fig. 2 (1.1, 1.2, 1.3):

First (1.1): To identify the centers of each sub-braingraph, we first sort all the nodes in descending order based on their degree. Then select the top N nodes from each of the nine functional brain regions to serve as centers of the sub-braingraphs.

Second (1.2): For each of the selected central nodes, we generate a sub-braingraph using the Selective-Breadth First Search (S-BFS) algorithm, which prioritizes expanding nodes within the same functional region as the central node. Each sub-braingraph is limited to a maximum of s nodes. We obtain the set of sub-braingraphs as $\{g_1, g_2, \cdots, g_n\}$.

Third (1.3): We use a GNN-based encoder to acquire the node representations of the sub-braingraphs denoted by $\mathcal{E} : \mathbb{R}^{s \times d} \times \mathbb{R}^{s \times s} \rightarrow \mathbb{R}^{s \times d_1}$, where d_1 represents the node representation dimension. The generalized equation for computing the node representations $\mathbf{H}(g_i) \in \mathbb{R}^{s \times d_1}$ for nodes within sub-braingraph g_i is given by:

$$\mathbf{H}(g_i) = \mathcal{E}(g_i) = \{\boldsymbol{h}_j \mid v_j \in V(g_i)\} \tag{1}$$

To incorporate $\mathcal{E}$ as a message-passing framework, we introduce a unified formulation represented by Eq. 2, where $U(\cdot)$ is a state updating function, $AGG(\cdot)$ is the aggregation function, and $M(\cdot)$ is the message generation function.

$$\mathbf{h}_i^{(l+1)} = \mathrm{U}^{(l+1)}\left(\mathbf{h}_i^{(l)}, \mathrm{AGG}\left(\mathrm{M}^{(l+1)}\left(\mathbf{h}_i^{(l)}, \mathbf{h}_j^{(l)}\right) \mid v_j \in N\left(v_i\right)\right)\right) \quad (2)$$

We compute the attention coefficient $c_j^{(i)}$ to indicate how relevant v_j is to subgraph g_i, which is given by the following equation:

$$c_j^{(i)} = \sigma\left(\mathbf{a}_{\mathrm{intra}}^{\mathrm{T}}\, \mathbf{W}_{\mathrm{intra}}\, \mathbf{h}_j^{(i)}\right) \quad (3)$$

The weight vector $\mathbf{a}_{\mathrm{intra}} \in \mathbb{R}^{d_1}$ is used in the calculation, and $\mathbf{W}_{\mathrm{intra}} \in \mathbb{R}^{d_1 \times d_1}$ represents a weight matrix. To ensure that the attention coefficients are normalized across all nodes, we employ a softmax function. The representation $\mathbf{z}_i$ of subgraph g_i is obtained by:

$$\mathbf{z}_i = \sum_{v_j \in \mathrm{V}(g_i)} c_j^{(i)} \mathbf{h}_j^{(i)}. \quad (4)$$

3.2 Sub-braingraph Selection and Sub-braingraph's Node Selection by LFE Module

SLG-NET utilizes a Local Braingraph Feature Extraction (LFE) module to extract the important local braingraph properties. The LFE module can be divided into two sections as shown in Fig. 2 (2.1, 2.2).

Sub-Braingraph Selection (SGS) section (2.1): To select significant sub-braingraphs, we use top-k sampling involves adjusting the adaptive pooling ratio k. We project all sub-braingraph features to 1D footprints using a trainable vector p and then select the top $n' = \lceil k \cdot n \rceil$ sub-braingraphs. The val_i of sub-braingraph g_i is determined using the following equation:

$$val_i = \frac{\mathbf{z}_i \mathbf{p}}{\|\mathbf{p}\|}, idx = \mathrm{rank}\left(\{val_i\}, n'\right) \quad (5)$$

We use the $rank\left(\{val_i\}, n'\right)$ function to rank the importance of subgraphs and obtain the indices of the n'-largest values in $\{val_i\}$. To dynamically update the pooling ratio $k \in (0,1]$, a reinforcement learning (RL) algorithm is used. Updating k is modeled as a finite horizon Markov decision process (MDP). The process of MDP is divided into the following steps as shown in Fig. 2 (2.1):

1) State. At epoch t, with a pooling ratio of k, the state s_t can be expressed using the selected sub-braingraph indices idx defined in Eq. 5:

$$s_t = idx_t \quad (6)$$

2) Action. The value of k is updated by the RL agent, wherein it selects an action a_t relying on the received reward. We define action a as an addition or subtraction of a constant value $\Delta k \in [0,1]$ from the present value of k.

3) Transition. Upon adjusting k, we utilize $top - k$ sampling described in Eq. 5 to choose a new collection of sub-braingraphs for the following epoch.

4) Reward. For each a_t corresponding to s_t, we define a categorical reward function, denoted as $reward\,(s_t, a_t)$, which is directly dependent on the accuracy of classification acc_t at epoch t:

$$reward\,(s_t, a_t) = \begin{cases} +1, & \text{if } acc_t > acc_{t-1} \\ 0, & \text{if } acc_t = acc_{t-1} \\ -1, & \text{if } acc_t < acc_{t-1} \end{cases} \tag{7}$$

Eq. 7 states that if the acc_t achieved with a_t is greater than the preceding epoch, the reward will be positive. Conversely, if the accuracy decreases, the reward will be negative.

5) Termination. In the event that the variation in k over 20 consecutive epochs does not surpass Δk, the RL algorithm concludes. The termination criterion is expressed as follows:

$$\text{Range } (\{k_{t-20}, \cdots, k_t\}) \leq \Delta k \tag{8}$$

To train the MDP, the Q-learning [9] is implemented. Q learning resolves the Bellman optimality equation, as shown below:

$$Q^*\,(s_t, a_t) = reward\,(s_t, a_t) + \gamma \arg\max_{a'} Q^*\,(s_{t+1}, a') \tag{9}$$

The discount factor for future reward γ (where $\gamma \in [0, 1]$) is employed in our approach. Additionally, we implement an $\varepsilon - greedy$ policy, where the probability of exploration is determined by the value of ε.

$$\pi\,(a_t \mid s_t; Q^*) = \begin{cases} \text{random action,} & \text{w.p.}\varepsilon \\ \arg\max_{a_t} Q^*\,(s_t, a)\,, & \text{otherwise} \end{cases} \tag{10}$$

Thus, rather than choosing actions depending on maximum future reward, the RL agent chooses an action randomly with a probability of ε to seek out new states.

After identifying the significant sub-braingraphs, we proceed to the Sub-braingraph's Node Selection (SNS) section (2.2) using self-attention pooling [10] as shown in Fig. 2 (2.2). The self-attention score $Z \in \mathbb{R}^{N \times 1}$ is calculated using the following equation:

$$Z = \sigma\left(\dot{D}^{-\frac{1}{2}}\dot{A}\dot{D}^{-\frac{1}{2}}X\Theta_{att}\right) \tag{11}$$

where $\dot{A}$ represents the adjacency matrix with self-connections, $\dot{D}$ is the degree matrix, σ is the activation function, and X denotes the input features of the braingraph. We use Gao et al.'s node selection method [11]. The number of nodes to preserve by choosing the top $\lceil kN \rceil$ nodes based on the Z score.

$$\text{idx} = top - rank(Z, \lceil kN \rceil), Z_{\text{mask}} = Z_{\text{idx}} \tag{12}$$

We obtain the indices of the highest $\lceil kN \rceil$ values via the top-rank function. The input sub-braingraph is processed using the 'masking' operation, with Z_{mask} representing the feature attention mask and idx representing an indexing operation.

$$X' = X_{\text{idx},;}, X_{\text{out}} = X' \odot Z_{\text{mask}}, A_{\text{out}} = A_{\text{idx,idx}} \qquad (13)$$

In the above equation, $X_{\text{idx},;}$ denotes the feature matrix with row-wise indexing, while $\odot$ represents the elementwise product that has been broadcasted. $A_{\text{idx, idx}}$ denotes the adjacency matrix with both row-wise and column-wise indexing. The resulting feature matrix and adjacency matrix are represented by X_{out} and A_{out}, respectively.

3.3   Sub-braingraph Sketching by GFE Module and Classification

Sub-braingraph sketching by GFE module and classification can be divided into three steps as shown in Fig. 2 (3.1, 3.2, 3.3).

First (3.1): The conversion of the original graph to a sketched graph is shown in Fig. 2 (3.1), where selected sub-braingraphs are treated as supernodes. The sketched graph is donated as $G^{\text{sk}} = \left(V^{\text{sk}}, E^{\text{sk}} \right)$. The number of edges between distinct sub-braingraphs is represented by edge (g_i, g_j). The connectivity between supernodes is established by edge (g_i, g_j), and an edge $e(i, j)$ is added to the sketched graph when the number of edges between g_i and g_j meets a pre-defined threshold b_{com}.

$$\begin{aligned} V^{\text{sk}} &= \{g_i\}, \forall i \in idx; \\ E^{\text{sk}} &= \{e_{i,j}\}, \forall \text{ Edge } e(g_i, g_j) > b_{\text{com}} \end{aligned} \qquad (14)$$

To calculate the attention coefficient α_{ij} of sub-braingraph g_i on g_j, we use a multi-head attention mechanism defined in [12]. Subsequently, we derive the sub-braingraph embeddings z'_i using the following equation:

$$z'_i = \frac{1}{M} \sum_{m=1}^{M} \sum_{e_{ij} \in E^{\text{ske}}} \alpha_{ij}^m \mathbf{W}_{\text{inter}}^m \, \mathbf{z}_i \qquad (15)$$

where M denotes the number of independent attention heads, while α_{ij} and $\mathbf{W}_{\text{inter}} \in \mathbb{R}^{d_2 \times d_1}$ denote the attention coefficient and weight matrix, respectively. The sub-braingraph embeddings z'_i will be refined and enhanced by the GFE module.

Second (3.2): The GFE module applies a self-supervised mutual information (SSMI) mechanism to maximize the mutual information (MI) between local sub-braingraph representations and the global braingraph representation as shown in Fig. 2 (3.2). By using the GFE module, all the obtained sub-braingraph representations are enforced to be aware of the global structural characteristics.

A READOUT function is utilized to condense the sub-braingraph level embeddings into a fixed-length vector that denotes the global graph representation $\mathbf{r}$:

$$\mathbf{r} = \text{READOUT}\left(\{\mathbf{z}'_i\}_{i=1}^{n'} \right) \qquad (16)$$

In this case, we utilize an averaging READOUT function, we employ the Jensen-Shannon (JS) MI estimator [13] to maximize the estimated MI on the local/global pairs. Specifically, a discriminator function $\mathcal{D} : \mathbb{R}^{d_2} \times \mathbb{R}^{d_2} \to \mathbb{R}$ is used to analyze a subgraph/graph embedding pair, where $\mathbf{W}_{MI}$ denotes a scoring matrix and $\sigma(\cdot)$ represents the sigmoid function. A bilinear score function is applied as the discriminator:

$$\mathcal{D}\left(\mathbf{z}_i', \mathbf{r}\right) = \sigma\left(\mathbf{z}_i'^T \mathbf{W}_{MI} \mathbf{r}\right) \tag{17}$$

Our self-supervised mutual information (MI) approach is involving contrastive learning and uses the same methodology described in literature [9] for generating negative samples. To optimize the self-supervised MI objective and enhance the mutual information between z_i' and $\mathbf{r}$, we employ a standard binary cross-entropy (BCE) loss $\mathcal{L}_{MI}^G$, where n_{neg} represents the number of negative samples.

$$\mathcal{L}_{MI}^G = \frac{1}{n' + n_{\text{neg}}} \left(\sum_{g_i \in G}^{n'} \mathbb{E}_{\text{pos}} \left[\log\left(\mathcal{D}\left(\mathbf{z}_i', \mathbf{r}\right)\right)\right] + \sum_{g_j \in \hat{G}}^{n_{\text{neg}}} \mathbb{E}_{\text{neg}} \left[\log\left(1 - \mathcal{D}\left(\dot{\mathbf{z}}_j', \mathbf{r}\right)\right)\right] \right) \tag{18}$$

Third (3.3): By applying a softmax function, we convert the sub-braingraph embeddings to label predictions. The graph classification results are determined through voting amongst the sub-braingraphs. The loss $\mathcal{L}$ for SLG-NET is defined by merging the supervised classification loss $\mathcal{L}_{\text{Classify}}$ and the self-supervised MI loss $\mathcal{L}_{MI}^G$ in Eq. 18:

$$\mathcal{L} = \mathcal{L}_{\text{Classify}} + \beta \sum_{G \in \mathcal{G}} \mathcal{L}_{MI}^G + \lambda \|\Theta\|^2 \tag{19}$$

The contribution of the self-supervised MI loss is determined by the value of β; the parameter set Θ in SLG-NET, which consists of trainable parameters, is subject to L2 regularization with a coefficient of λ;

SLG-NET enhances the specificity of braingraphs by using S-BFS to generate sub-braingraphs from nine functional brain regions. While LFE module and GFE module are used to improve the feature extraction capabilities in both local and global levels. By combining S-BFS, LFE module, and GFE module, the SLG-NET performs better than existing models in terms of specificity to braingraphs and feature extraction capabilities.

4   Experiments

4.1   Dataset and Parameters Setting

The rest-metamdd dataset is the most comprehensive resting-state fMRI database of individuals with depression currently available [14]. This dataset was selected to evaluate the efficacy of the proposed SLG-NET. Further details on fMRI data preprocessing can be found in literature [5]. To implement the proposed method, common parameters for model training were set as follows: $Momentum = 0.8$, $Dropout = 0.4$, and $L2$ Norm Regularization weight $decay = 0.01$, we set $\gamma = 1$ in Eq. 9 and $\varepsilon = 0.8$ in Eq. 10. To assess the performance of competing models and the proposed SLG-NET, a 10-fold cross-validation strategy was employed. The parameter settings of the competing models can be obtained in literature [5, 15].

4.2 Overall Evaluation

The classification performance of the SLG-NET is evaluated by benchmarking them against the traditional SVM classifier, BrainNetCNN [16], Wck-CNN [17], XGBoost [18] and five GCN-based methods [5], which are cutting-edge connectome-based models. As shown in Table 1. The proposed SLG-NET significantly outperformed the other models show that the proposed SLG-NET has the potential for auxiliary diagnosis of depression in clinical scenarios.

Table 1. Comparing SLG-NET's Performance (average accuracy±standard deviation) with Other State-of-The-Art Methods in distinguishing MDD patients from HC.

Classifier	Acc	Sen	Spe	Pre	F1
SVM-RBF	54.63 ± 4.26	51.23 ± 7.23	48.27 ± 4.27	45.16 ± 5.27	48.95 ± 4.26
Wck-CNN (Medical image analysis, 2020)	59.19 ± 4.29	52.16 ± 5.38	63.65 ± 5.46	53.08 ± 5.38	53.46 ± 4.37
BrainNetCNN (NeuroImage, 2017)	57.29 ± 5.19	52.46 ± 6.57	51.16 ± 3.46	51.36 ± 4.28	54.46 ± 4.24
GroupINN (ACM SIGKDD, 2019)	59.28 ± 5.37	57.46 ± 2.17	57.86 ± 4.764	57.18 ± 5.58	56.86 ± 8.17
Hi-GCN (CIBM, 2020)	56.27 ± 3.89	53.26 ± 5.17	49.28 ± 6.18	51.37 ± 4.39	50.43 ± 4.37
E-Hi-GCN (Neuroinformatics, 2021)	58.18 ± 3.89	60.18 ± 4.23	56.57 ± 4.98	56.17 ± 5.18	58.16 ± 4.29
XGBoost (ACM SIGKDD, 2016)	61.28 ± 7.16	58.13 ± 6.15	57.18 ± 6.57	50.16 ± 3.57	51.23 ± 5.27
Population-based GCN (BSPC, 2023)	60.96 ± 3.37	59.28 ± 5.64	59.26 ± 3.46	61.67 ± 6.27	62.46 ± 5.23
GAE-FCNN (ArXiv:2107.12838, 2022)	60.18 ± 6.54	57.56 ± 5.48	58.17 ± 8.54	57.32 ± 5.46	54.16 ± 4.23
SLG-NET	**74.15 ± 3.56**	**72.18 ± 3.79**	**67.26 ± 4.57**	**67.46 ± 8.42**	**70.16 ± 4.27**

The rest-metamdd dataset's 74.15% accuracy falls below the average accuracy of 86.49%, 84.91%, and 84% reported in small-scale studies [7, 19, 20]. This decline is primarily attributed to increased clinical heterogeneity associated with larger samples. In our study, we utilized data consisting of small samples derived from multiple hospitals, resulting in varying performances across sites. The large sample size, accompanied by significant heterogeneity, is responsible for the poor accuracy of our model, but this does not mean that our model is inferior. On the contrary, increased clinical heterogeneity with larger samples has increased the reliability of our model.

In 2023, Gallo et al. [3] and Fang et al. [15] used the rest-metamdd dataset for the model's training and validation, achieving an accuracy of 61.47% and 59.73%. This result demonstrates SLG-NET significantly outperforms other large-scale research in terms of accuracy.

4.3   S-BFS, LFE, and GFE Modules Analysis

To verify the effectiveness of the S-BFS, LFE, and GFE modules, we compare SLG-NET with the variants of SLG-NET in Fig. 3 (a). The proposed SLG-NET outperformed other variations of the SLG-NET in different proportion ratios, supporting the effectiveness of the S-BFS, LFE, and GFE modules. We find that compared to SLG-NETnoSGS and SLG-NETnoSNS, the removal of GFE module significantly reduced accuracy. This result suggests that the GFE module, which promotes sub-braingraph embedding to be conscious of the global braingraph structural properties, is effective and complementary to SGS and SNS mechanisms, which primarily focus on local braingraph properties.

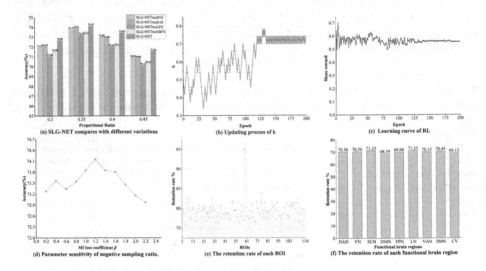

Fig. 3. S-BFS LFE and GFE modules analysis.

In order to determine whether the RL algorithm has ultimately converged during the GNN training process. Figure 3 (b) visualizes the overall process of updating k, revealing that the RL environment is initially unstable, after reaching the terminal condition described in Eq. 8, the updating of k tends to be stable, which indicates that the framework is gradually converging. Further analysis in Fig. 3 (c) shows that with a stable learning curve, the RL mechanism converges to a mean reward of 0.556, which means that our framework can adaptively locate the most noteworthy sub-braingraphs. In Fig. 3 (d), we analyze the impact of hyper-parameters β with different values to determine how self-supervised MI loss affects the SLG-NET. When MI loss coefficient $\beta = 1.2$, SLG-NET achieves better performance compared to other values; this proves that self-supervised training is quite beneficial for our framework. Furthermore, when the self-supervised MI loss coefficient β exceeds 1.2, the result shows that the self-supervised MI mechanism has a counterproductive effect on the proposed model's accuracy; this suggests that the supervised learning component (i.e., the subgraph neural network part) is also essential to the overall performance.

To determine which ROIs and functional brain regions contribute the most to depression diagnosis. Figure 3 (e) and Fig. 3 (f) show the retention rates of each node (ROI) and the functional brain region after SNS and SGS mechanisms; we found that the thalamus had the highest retention rate among ROIs, indicating its significant value for depression diagnosis. In addition, we found no significant difference in the retention rates of sub-braingraph in nine functional brain regions as shown in Fig. 3 (f), indicating the sub-braingraph in nine functional brain regions are equally important for diagnosing depression.

5   Conclusion

This paper designed a novel depression diagnosis framework named SLG-NET based on subgraph neural network. the proposed SLG-NET significantly surpasses many state-of-the-art methods, which shows that the proposed SLG-NET has the potential for auxiliary diagnosis of depression in clinical settings. Compared with other GNN-based depression diagnosis models [3,5–7,21], SLG-NET has advantages in the following two aspects:

1) The specificity of our model to braingraphs for MDD diagnosis is improved; SLG-NET can preserve the crucial sub-structure information of braingraph by encoding the sub-braingraph extracted by the proposed subgraph generation algorithm.
2) Our model has stronger feature extraction capabilities, by identifying important local braingraph properties with LFE module and enhancing the awareness of subgraph embedding to global braingraph structural properties with GFE module; our model takes into account both local braingraph properties and global braingraph structural properties.

References

1. Kishi, T., et al.: Antidepressants for the treatment of adults with major depressive disorder in the maintenance phase: a systematic review and network meta-analysis. Mol. Psychiatry 28(1), 402–409 (2023)
2. Lu, S., Shi, X., Li, M., Jiao, J., Feng, L., Wang, G.: Semi-supervised random forest regression model based on co-training and grouping with information entropy for evaluation of depression symptoms severity. Math. Biosci. Eng. 18(4), 4586–4602 (2021), https://www.aimspress.com/article/doi/10.3934/mbe.2021233
3. Gallo, S., et al.: Functional connectivity signatures of major depressive disorder: machine learning analysis of two multicenter neuroimaging studies. Mol. Psychiatry 1–10 (2023)
4. Cao, L.: Aberrant functional connectivity for diagnosis of major depressive disorder: a discriminant analysis. Psychiatry Clin. Neurosci. 68(2), 110–119 (2014)
5. Noman, F., et al.: Graph autoencoders for embedding learning in brain networks and major depressive disorder identification. arXiv preprint arXiv:2107.12838 (2021)

6. Venkatapathy, S., et al.: Ensemble graph neural network model for classification of major depressive disorder using whole-brain functional connectivity. Front. Psychiatry **14**, 1125339 (2023)
7. Chen, T., Guo, Y., Hao, S., Hong, R.: Exploring self-attention graph pooling with EEG-based topological structure and soft label for depression detection. IEEE Trans. Affect. Comput. **13**(4), 2106–2118 (2022)
8. Shi, Y., et al.: Multivariate machine learning analyses in identification of major depressive disorder using resting-state functional connectivity: A multicentral study. ACS Chem. Neurosci. **12**(15), 2878–2886 (2021)
9. Sun, Q., Li, J., Peng, H., Wu, J., Ning, Y., Yu, P.S., He, L.: Sugar: Subgraph neural network with reinforcement pooling and self-supervised mutual information mechanism. In: Proceedings of the Web Conference 2021, pp. 2081–2091 (2021)
10. Lee, J., Lee, I., Kang, J.: Self-attention graph pooling. In: International Conference on Machine Learning, pp. 3734–3743. PMLR (2019)
11. Cangea, C., Veličković, P., Jovanović, N., Kipf, T., Liò, P.: Towards sparse hierarchical graph classifiers. arXiv preprint arXiv:1811.01287 (2018)
12. Veličković, P., Cucurull, G., Casanova, A., Romero, A., Lio, P., Bengio, Y.: Graph attention networks. arXiv preprint arXiv:1710.10903 (2017)
13. Nowozin, S., Cseke, B., Tomioka, R.: f-gan: Training generative neural samplers using variational divergence minimization. In: Advances in Neural Information Processing Systems 29 (2016)
14. Yan, C.G., et al.: Reduced default mode network functional connectivity in patients with recurrent major depressive disorder. Proc. Natl. Acad. Sci. **116**(18), 9078–9083 (2019)
15. Fang, Y., Wang, M., Potter, G.G., Liu, M.: Unsupervised cross-domain functional MRI adaptation for automated major depressive disorder identification. Med. Image Anal. **84**, 102707 (2023)
16. Kawahara, J., et al.: BrainNetCNN: convolutional neural networks for brain networks; towards predicting neurodevelopment. Neuroimage **146**, 1038–1049 (2017)
17. Jie, B., Liu, M., Lian, C., Shi, F., Shen, D.: Designing weighted correlation kernels in convolutional neural networks for functional connectivity based brain disease diagnosis. Med. Image Anal. **63**, 101709 (2020)
18. Chen, T., Guestrin, C.: XGBoost: a scalable tree boosting system. In: Proceedings of the 22nd ACM SIGKDD International Conference on Knowledge Discovery and Data Mining. pp. 785–794 (2016)
19. Chen, T., Hong, R., Guo, Y., Hao, S., Hu, B.: Ms ^{2}-GNN: exploring GNN-based multimodal fusion network for depression detection. IEEE Trans. Cybern. 1–11 (2022). https://doi.org/10.1109/TCYB.2022.3197127
20. Kambeitz, J., et al.: Detecting neuroimaging biomarkers for depression: a meta-analysis of multivariate pattern recognition studies. Biol. Psychiat. **82**(5), 330–338 (2017)
21. Zhu, et al.: Cross-network interaction for diagnosis of major depressive disorder based on resting state functional connectivity. Brain Imaging Behav. **15**, 1279–1289 (2021)

CrowdNav-HERO: Pedestrian Trajectory Prediction Based Crowded Navigation with Human-Environment-Robot Ternary Fusion

Siyi Lu[1], Bolei Chen[1], Ping Zhong[1(✉)], Yu Sheng[1], Yongzheng Cui[1], and Run Liu[2]

[1] School of Computer Science and Engineering, Central South University, Changsha 410083, China
{siyilu,boleichen,ping.zhong,shengyu,214712191}@csu.edu.cn
[2] Research Center of Ubiquitous Sensor Networks, University of Chinese Academy of Sciences, Beijing 100049, China
liurun22@mails.ucas.ac.cn

Abstract. Navigating safely and efficiently in complex and crowded scenarios is a challenging problem of practical significance. A realistic and cluttered environmental layout usually significantly impacts crowd distribution and robotic motion decision-making during crowded navigation. However, previous methods almost either learn and evaluate navigation strategies in unrealistic barrier-free settings or assume that expensive features like pedestrian speed are available. Although accurately measuring pedestrian speed in large-scale scenarios is itself a difficult problem. To fully investigate the impact of static environment layouts on crowded navigation and alleviate the reliance of robots on costly features, we propose a novel crowded navigation framework with **Human-Environment-Robot** (HERO) ternary fusion named CrowdNav-HERO. Specifically, **(i)** a simulator that integrates an agent, a variable number of pedestrians, and a series of realistic environments is customized to train and evaluate crowded navigation strategies. **(ii)** Then, a pedestrian trajectory prediction module is introduced to eliminate the dependence of navigation strategies on pedestrian speed features. **(iii)** Finally, a novel crowded navigation strategy is designed by combining the pedestrian trajectory predictor and a layout feature extractor. Convincing comparative analysis and sufficient benchmark tests demonstrate the superiority of our approach in terms of success rate, collision rate, and cumulative rewards. The code is published at https://github.com/SiyiLoo/CrowdNav-HERO.

Keywords: Crowded Navigation · Human-Environment-Robot Ternary Fusion · Pedestrian Trajectory Prediction · Deep Reinforcement Learning

---

S. Lu and B. Chen—Contribute equally.

B. Luo et al. (Eds.): ICONIP 2023, LNCS 14450, pp. 43–55, 2024.
https://doi.org/10.1007/978-981-99-8070-3_4

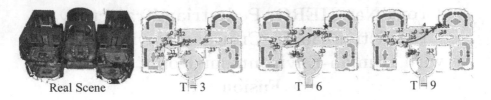

Real Scene   T = 3   T = 6   T = 9

Fig. 1. Illustrates the process of our crowded navigation. The red curve indicates the motion trajectory of the agent. The black circle and star indicate the starting position and the target, respectively. The numbers indicate the pedestrian ID. (Color figure online)

1   Introduction

Intelligent agents navigating safely and efficiently in complex pedestrian environments is an important part of the socialization of embodied artificial intelligence. Despite the increasing attention in recent years, significant challenges remain. We consider a more realistic pedestrian navigation environment that is different from previous works, which either only contain fully static environments with static obstacles or purely dynamic environments with pedestrians. Our environment contains both static and dynamic obstruction. In previous work, navigating in crowd and static environments are always treated as two separate problems. Therefore, fixed obstacles are not sufficiently considered in the crowded navigation task. However, the importance of scene layout and semantic information in static environments cannot be ignored for pedestrian navigation. For example, pedestrian motion is constrained by environmental layout, and customers tend to linger and gather under shelves. Although it is possible to treat pedestrians as dynamic obstacles from a robotic perspective, if only obstacle avoidance strategies are used to avoid collisions with pedestrians without considering human-robot interaction, the pedestrian experience will be greatly compromised. With the recent rapid development of interactive learning, reinforcement learning has shown outstanding potential in solving social interaction problems. However, much of the existing work [2,4,5,24] on socially aware crowd navigation is based on stringent assumptions, where it is unrealistic to use pedestrian velocity as the basic feature input for navigation strategies since crowd velocity detection is inherently a thorny problem.

In order to solve the two problems mentioned above, we propose a novel crowded navigation framework, CrowdNav-HERO. At first, a more realistic simulator, HERO-Sim, was built using MP3D datasets and Gibson environment to provide the basis for crowded navigation research in complex and realistic environments. Then, we build a pedestrian state prediction module based on Spatio-Temporal Graph Convolutional Network (STGCN) [15] which uses the set of simple multi-step human historical features instead of expensive single-step human velocity features. Finally, considering Human-Environment-Robot (HERO) ternary fusion in crowded navigation, the pedestrian trajectory predictor and complex obstacle map modeling are integrated into a dual-channel value estimation network to learn crowded navigation strategies. The superiority and generalization of our method are demonstrated through extensive benchmarks and comparisons with state-of-the-art work in point navigation tasks.

2   Related Work

2.1   Socially Aware Crowded Navigation

In crowded navigation, agents need to interact with humans safely and politely while humans' intentions are unknowable. It is important to learn the relationships among humans because humans usually move in groups [14]. SARL [5] proposes to represent the human state by utilizing a local map, which can't efficiently extract the relationship features among humans. RGL [4] models the agent and the crowd as a relational graph so as to know how much each agent pays attention to the other. SG-DQN [24] introduces a social attention mechanism to reduce the tremendous computational burden of complex interaction representations. Further, Diego et. al [2] design two intrinsic rewards based on SG-DQN that promote strategy learning and accelerate convergence. Recently, method [13] provides a representation of the personal zones of pedestrians based on the prediction of future humans' states. Although the trajectory prediction-based methods are prone to freezing phenomena [22], trajectory prediction is still a promising module in crowd navigation.

2.2   Simulator for Crowded Navigation

Simulation stands out as one of the foremost methodologies employed for assessing social navigation strategies. The linchpin of these simulations resides in the collision avoidance algorithm, which exerts precise control over pedestrian movement. Pioneering the field, the Social Force Model (SFM) [8] was introduced as a means to address the intricate dynamics inherent to crowded scenarios. Within SFM, the behaviors of pedestrians are intricately woven into the fabric of forces that emanate from their interactions. Complementing this approach are velocity-centric methodologies, exemplified by the Velocity Obstacle (VO) paradigm [7]. A more contemporary advancement, the Optimal Reciprocal Collision Avoidance (ORCA) algorithm [19], seeks to empower pedestrians in avoiding collisions within the intricate tapestry of multi-agent environments. These methodological pursuits are underpinned by the presupposition of discernible pedestrian intentions and the inherent homogeneity in pedestrian conduct. Consequently, crowd simulation software fashioned upon these algorithms emerges as a pivotal research platform for the study of navigation in densely populated settings.

Notable examples include CrowdSim[1], an ORCA-centric crowd simulation software, and pedsim_ros[2], a popular crowd simulation system rooted in the SFM framework. Within our research endeavor, we embark upon the creation

---

[1] https://github.com/vita-epfl/CrowdNav.
[2] https://github.com/srl-freiburg/pedsim_ros.

of a simulator founded on the CrowdNav paradigm, distinguished by intricate spatial layouts, and characterized by the incorporation of HRO ternary fusion. Concurrently, we introduce a non-homogeneous obstacle avoidance algorithm based on ORCA.

3   Problem Formulation

In a crowded domestic scenario with N pedestrians, crowded navigation is modeled as a Partially Observed Markov Decision Process (POMDP) using a tuple $(\mathcal{S}, \mathcal{O}, \mathcal{A}, P, R, \gamma, L)$. Here, $\mathcal{S}$ denotes the state space, $\mathcal{A}$ denotes the action space of the agent, $\mathcal{O}$ denotes the observation space, $P : \mathcal{S} \times \mathcal{A} \to \mathcal{S}$ is the transition function. $R : \mathcal{S} \times \mathcal{A} \to \mathbb{R}$ denotes the reward function, γ denotes the discount factor, and L denotes the maximum episode length. At time step t, the agent's state is denoted as $w^t = \{p_x, p_y, v_r, g_x, g_y, \rho, v_p, \theta\}$ and the i-th human's state is denoted as $u_i^t = \{\mathbf{p}^{t+1}, \mathbf{p}^{t+2} ..., \mathbf{p}^{t+q}, \rho\}$. The agent state consists of position $\mathbf{p} = (p_x, p_y)$, linear velocity v_r, target position $\mathbf{g} = (g_x, g_y)$, radius ρ, maximum speed v_p, and heading angle θ. The i-th human's state no longer contains the velocity but consists of the predicted future states $\{\mathbf{p}^{t+1}, \mathbf{p}^{t+2} ..., \mathbf{p}^{t+q}\} = STGCN(\mathbf{p}^{t-\lambda}, ..., \mathbf{p}^{t-1}, \mathbf{p}^t)$ and radius ρ.

In our settings, a static environment is represented by a global grid map $\mathcal{M}_t$ of size $C \times H \times W$, where each element corresponds to a physical world cell of size $25cm^2$. The map is labeled with a one-hot vector of C channels indicating the semantic category for each pixel. The agent's observation is restricted to a local map m_t of size $C \times M \times M$, and its steering δ is divided into 25 uniformly spaced orientations. The linear velocity v_r is randomly sampled from $[0, v_p]$. The optimal navigation policy, $\pi^* : S^t \mapsto a^t$, maps the current observation state $S^t = \{w^t, u_i^t, m_t\}(i = 1, 2, ..., N)$ to action $a^t = \{v_r, \delta\}$ at time step t, maximizing the expected return, which defined as:

$$\pi^* \left(S^t \right) = \gamma \int_{S^{t+\Delta t}} P \left(S^t, a^t, S^{t+\Delta t} \right) V^* \left(S^{t+\Delta t} \right) dS^{t+\Delta t}$$
$$+ \arg \max_{a^t} R \left(S^t, a^t \right) \tag{1}$$

$$V^* \left(S^t \right) = \sum_{k=t}^{T} \gamma^k R^k \left(S^k, \pi^* \left(S^k \right) \right).$$

where $R(S^t, a^t)$ is the reward received at time step t, $\gamma \in (0, 1)$ is the discount factor, V^* is the optimal value function, $P(S^t, a^t, S^{t+\Delta t})$ is the state transition from time t to time $t + \Delta t$. Similar to the formulation in [2], the following reward function is considered to encourage the RL of crowded navigation policy:

$$R = \begin{cases} -0.005\mathcal{T}, & if\ the\ agent\ is\ alive \\ -0.25, & if\ collide\ with\ an\ obstacle \\ -0.5, & if\ collide\ with\ a\ human \\ \alpha(d_{min} - d_{dis})\Delta t, & if\ too\ close\ to\ humans \\ +5, & if\ reach\ target. \end{cases} \tag{2}$$

Equation 2 has multiple terms, including penalties and rewards. $\mathcal{T}$ represents cumulative time steps, and α denotes the weight of sub-reward. d_{min} and d_{dis} are the closest human distance and comfortable distance, respectively. Δt is the duration of discomfort distance, set to a single time step. The first term encourages fewer navigation steps and punishes excessive exploration. The second and third terms penalize collisions with obstacles and pedestrians, respectively. The fourth term penalizes impolite navigation that violates human comfort space. The last term rewards successful navigation substantially.

4 HRO Ternary Fusion Simulator

Most existing simulators [2, 4, 24] ignore the effect of the environment on humans and agents [2, 4, 24] or assume that the obstacles in the workspace are naive circles or lines [23]. We construct an HRO ternary fusion simulator, HERO-Sim, to deal with it.

4.1 Simulator Setting

HERO-Sim consists of a crowd, an agent, and maps of the domestic environments. In specific navigation tasks, the robot is invisible to humans, thus humans can only interact with humans and obstacles in the simulator. In other words, humans will never notice and avoid the robot, and the robot has to take full responsibility for collision avoidance. In addition, to fully evaluate the impact of pedestrian density on the performance of the crowded navigation strategies, the number of pedestrians is set to be adjustable. Because pedestrians are widely distributed in all corners of the scene, we only consider pedestrians close to the agent while the agent making a decision. Therefore, if not specifically stated, only pedestrians located within the observation field of view of the agent m_t will be considered.

4.2 Static Environment Construction and Collision Avoidance

To make the simulation environments complex and realistic enough, sufficient semantic maps are extracted from realistic domestic scenarios [3, 20]. Specifically, we detect the contours of object instances in the map graphically and represent each object as a series of line segments that are connected at the beginning and end. It is worth noting that each object instance is a closed polygon, ensuring that agents do not enter or pass through them. These line segments are considered as constraints for ORCA to solve the linear programming for collision-free motion.

4.3 Crowd Interaction Optimization

Traditional simulators use ORCA [19] to guide pedestrian movement, assuming that pedestrians equally avoid collisions. However, researchers [18] show that human interaction-aware decision-making can be mathematically formulated as

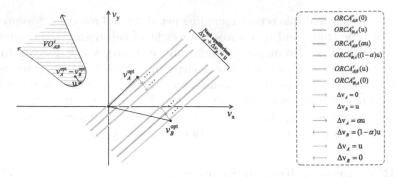

Fig. 2. The set $ORCA_{A|B}^{\tau}(\alpha u)$ of permitted velocity for human A for optimal reciprocal collision avoidance with human B is a half-plane delimited by the line perpendicular to u through the point $v_A^{opt} + \Delta v_A, (\Delta v_A = \alpha u)$. Different α can produce different velocity half-plane, Nash equilibrium can be formed as long as $\Delta v_A + \Delta v_B = u$.

searching for Nash equilibrium [9]. So we assume that the human in HERO-Sim acts rationally and tries not to change the original speed while avoiding collisions. As shown in Fig. 2, u is the smallest change required to the relative velocity of human A and human B to avert collision within τ time as defined in ORCA. We give the cost function of humans' decisions:

$$
\begin{aligned}
J_A\left(\Delta v_A, \Delta v_B\right) &= \begin{cases} \Delta v_A & \Delta v_A + \Delta v_B \geqslant u, \\ \infty & \text{else.} \end{cases} \\
J_B\left(\Delta v_B, \Delta v_A\right) &= \begin{cases} \Delta v_B & \Delta v_A + \Delta v_B \geqslant u, \\ \infty & \text{else.} \end{cases}
\end{aligned}
\tag{3}
$$

$J_A\left(\Delta v_A, \Delta v_B\right)$ is the cost function of human A, and $\Delta v_A{}^3$ is the velocity change of human A. In order to minimize the cost function of human A(i.e., solving problem $\min_{\Delta v_A} J_A\left(\Delta v_A, \Delta v_B\right)$). Since the cost function only has finite values in a specific interval, this problem can be transformed into a Linear Programming (LP) problem presented as Eq. 4. Clearly, its closed-form solution is given by $\Delta v_A = u - \Delta v_B$, i.e., $\Delta v_A + \Delta v_B = u$.

$$
\min_{\Delta v_A} \Delta v_A \; (\text{ s.t. } \Delta v_A + \Delta v_B \geqslant u)
\tag{4}
$$

In order to avoid collisions, velocity pairs $(v_A^{opt} + \Delta v_A, v_B^{opt} - \Delta v_B)$ are randomly selected from the set of Nash equilibrium to give human A and human B. Different Nash equilibria reflect different pedestrian obstacle avoidance strategy preferences. Random sampling pedestrians' obstacle avoidance strategy from the set of Nash equilibrium improves the diversity of pedestrians. Also consider each pedestrian's own personality, which makes them behave consistently among the episodes.

---

^{3} It is worth noting that we don't consider the direction of velocity.

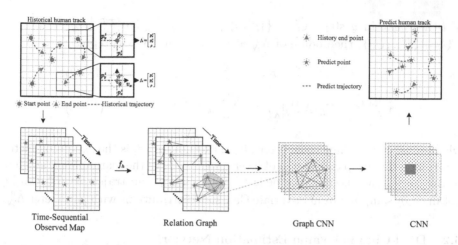

Fig. 3. Spatial-Temporal Pedestrian Trajectory Prediction

5 A Crowded Navigation Framework with HERO Ternary Feature Fusion

In this section, an STGCN-based [15] pedestrian trajectory predictor is first introduced. Then, a two-channel value estimation network is built in order to account for the coupled effects of pedestrians and the environment.

5.1 Spatial-Temporal Pedestrian Trajectory Prediction

Previous work [2,4,5,24] often use the current pedestrian state to predict the next pedestrian state before the agent's decision-making. But many features of the pedestrian state are difficult to obtain precise, such as velocity. So we replaced the use of expensive pedestrian features using simple pedestrian features from multiple moments in history. As shown in Fig. 3, at each time step, the features and relationships of pedestrians are denoted as nodes and edges respectively. We employ a two-layer GCN [11] with residual connections to construct the spatial feature encoder. The l-th layer node features are denoted as $G^{(l)} = [g_1^{(l)}, ..., g_N^{(l)}]^T$, and the node aggregation at layer l is then calculated as:

$$G^{(l+1)} = ReLU(D^{-\frac{1}{2}}\hat{A}D^{-\frac{1}{2}}G^{(l)}W_S^{(l)}). \tag{5}$$

where $\hat{A} = A + I$, I is the unit matrix, A is the adjacency matrix, D is the degree matrix, $W_S^{(l)}$ is the learnable weight matrix of layer l. Then, a multi-layer Temporal Convolutional Network (TCN) [1] is utilized to comprehensively manage time-series features. The input of the TCN is the GCN-encoded past λ-step features of N humans $T_N^\lambda = \{\tau_1^\lambda, ..., \tau_N^\lambda\}$, where $\tau_i^\lambda = \{\tilde{\mathbf{p}}_i^{t-\lambda+1}, ..., \tilde{\mathbf{p}}_i^t\}$, $i = 1, 2, ..., N$. We assume that pedestrian location (p_x, p_y) follows bi-variate Gaussian distribution [15] such that $\mathbf{p}_i^t \sim \mathcal{N}(\mu_i^t, \sigma_i^t, \rho_i^t)$. Denote $\hat{\mathbf{p}}_i^q$ as the predicted future distribution of i-th human in q steps, $\hat{\zeta}_i^q = \{\{\mu_i^t, \sigma_i^t, \rho_i^t\}, ..., \{\mu_i^{t+q-1}, \sigma_i^{t+q-1}, \rho_i^{t+q-1}\}\}$. Let $\hat{\Lambda}_N^q = \{\hat{\zeta}_1^q, ..., \hat{\zeta}_N^q\}$, the update of $\hat{\Lambda}_N^q$ at l-th layer in TCN is represented as:

$$\hat{\Lambda}_N^q = T_N^\lambda \qquad (l = 0)$$

$$\hat{\Lambda}_N^{q(l+1)} = \mathcal{F}_{d_l}(\hat{\Lambda}_N^{(l)})|_{\hat{\Lambda}_N^{q(l)}} = \sum_{k=1}^{K} \eta_k \hat{\Lambda}_N^{(q-(K-k)d_l)(l)} \quad (l > 0) \tag{6}$$

where K is the one-dimensional convolution kernel size, d_l is the expansion factor controlling the jump connection of layer l, $\eta_1 \sim \eta_K$ are the learnable parameters of the convolution kernel. The pedestrian's future q-step trajectory T_N^q can be obtained by sampling from bi-variate Gaussian distributions with parameter $\hat{\Lambda}_N^q$.

5.2   Dual-Channel Value Estimation Network

The inputs of our dual-channel value estimation network include the observation states of the agent and pedestrians, and an agent-centric local map, $S^t = \{w^t, u_i^t, m_t\}, i = 1, 2, ..., N$. In the socially aware channel, predicted pedestrian states $\hat{f}_h$ and agent's features f_r are input to the GCNs, followed by a ReLU activation σ. A two-layer GCN is employed for human-human and human-agent messaging to model the social awareness, followed by a Multi-Layer Perception (MLP) to summarize the crowded features:

$$F_{sa} = MLP(ReLU^{(l)}(GCN^{(l)}([f_r, \hat{f}_h]))), l = 1, 2 \tag{7}$$

In the map channel, a map predictor of UNet [17] architecture is employed to anticipate the local map m_{t+1} conditional on m_t and the current action a_t. The output of the map predictor is normalized and binarized to obtain the anticipated local map. This process is denoted as:

$$m_{t+1} \leftarrow Bi(Norm(UNet(Trans(m_t, a_t)))). \tag{8}$$

Another UNet network is utilized to extract spatial features in the local maps. The navigation target is embedded as a matrix of dimension $C \times M \times M$, which is concatenated with the local map matrix as the input. Previous network interpretability works [10] show that latent features from different layers of the UNet model contain different types of fine-grained details of objects. Therefore, features from different layers of the decoder are concatenated and fed into an MLP to summarize multi-granularity features:

$$F_{mg} = MLP(cat(UNet([m_t, \phi(O)]))). \tag{9}$$

Where $\phi(\cdot)$ denotes the target embedding. Finally, the outputs of the two channels are weighted by the learnable weights followed by a Gate Recurrent Unit (GRU) [6] to obtain the final value estimation:

$$V^* = GRU(ReLU(W([F_{sa}, F_{mg}]) + b)). \tag{10}$$

6 Experiments

6.1 Experimental Settings

HERO-Sim is used for training and testing, agent-centric local maps (-l) and historical cumulative global maps (-g) [21] are utilized in conjunction with navigation policies to implement multiple crowded Navigation policy variants. Navigating to the end without collision means success, any collision with pedestrians or obstacles and timeout greater than 50-time steps implies failure. The maximum velocity of both the humans and the agent is 1 m/s. The radius of the humans and the agent is set to 0.3 m. γ in formula (1) is set to 0.9. d_{dis} and α in formula (2) are set to 0.5 m and 1.0, respectively. The size of the agent-centric local map is set to 6.4 m × 6.4 m.

Classical method ORCA and three existing state-of-the-art methods, RGL [4], SG-D3QN [24] and IR-RE3 [2], are implemented as baseline methods to be compared with our method in terms of Success Rate (SR), Collision Rate with Obstacles (CO), Collision Rate with Pedestrians (CP), Average Navigation Step (AS), and Average Reward (AR). The start and end positions of the point navigation task are randomly sampled in the maps. By sampling evenly in the testing environments, 100 navigation episodes are employed for navigation evaluation. An example of point navigation is shown at the top of Fig. 1.

Table 1. Statistical experimental results of point navigation. * indicates the improved ORCA, which is able to avoid obstacles to a certain extent.

Method	SR (%) ↑	CO (%) ↓	CP (%) ↓	AS ↓	AR ↑
ORCA* [19]	0.18	0.77	0.04	9.16	0.083
RGL-l [4]	0.51	0.38	0.09	14.8	0.325
SG-D3QN-l [24]	0.58	0.34	0.06	**11.4**	**0.470**
IR-RE3-l [2]	0.58	0.37	0.05	11.8	0.426
Ours-l	**0.62**	**0.33**	**0.03**	12.1	0.467
RGL-g [4]	0.52	0.36	0.06	13.4	0.386
SG-D3QN-g [24]	0.61	0.34	0.05	12.1	0.477
IR-RE3-g [2]	0.59	0.38	0.03	**11.4**	0.476
Ours-g	**0.64**	**0.30**	**0.02**	12.8	**0.481**

6.2 Quantitative Evaluations for Crowded Navigation

To evaluate the performance of crowded navigation quantitatively, the various metrics of the 5 methods in the MP3D setting are presented in Table 1. As expected, the SR of ORCA* is low due to the violation of the reciprocal assumption. Since ORCA* has a short navigation distance, there are few opportunities to interact with pedestrians, so the CP metric is relatively low.By combining RGL with agent-centric local maps, RGL-l achieves an absolute 33% improvement in terms of SR relative to ORCA*, which is attributed to the learning of interaction mechanisms and multi-granularity maps. Benefiting from the usage of pedestrian trajectory prediction to remedy the lack of pedestrian speed features, our approach achieves the lowest CP metric. Although a longer AS means more robust and longer-lasting navigation, the AS decreases somewhat with increasing collision rates (CO and CP). In the point navigation task, our approach maintains a moderate level of AS a competitive AR. When replacing local maps with cumulative global maps, there is a small performance improvement for each method. The reason may be that the navigation goals provided by the point navigation task already provide global guidance and the small-scale maps are sufficient to support the local motion decision-making of the agent.

6.3   Quantitative Evaluation of Impact of Environment on Navigation

We quantitatively evaluate the impact of the static environment and the crowd on navigation through experimental performance in different environments, including a crowd-only simulator(-crowd), crowd-only dataset(-dataset), and simulator with the crowd and static environment(-crowd+env). It's difficult to unify experimental parameters in a different environment, so we only select the success rate and collision rate as evaluation indicators. Experience results in Table 2 show that the success rate is the highest in the simulation environment with only pedestrians. When the pedestrian trajectory from the real trajectory or add static environment, the success rate decreases, and the latter decreases greatly. The experimental results indirectly show the influence of the pedestrian path generation model and static obstacles on navigation.

Table 2. Statistical experimental results of point navigation on simulator only include crowd, real crowd trajectory dataset, and HERO-Sim.

Method	SR (%) ↑	CO (%) ↓	CP (%) ↓
ORCA-crowd [19]	0.43	-	0.57
ORCA-crowd+env [19]	0.18	0.77	0.04
RGL-crowd [4]	0.96	-	0.02
RGL-dataset	0.82	-	0.16
RGL-crowd+env	0.52	0.36	0.06
Ours-crowd	0.98	-	0.02
Ours-dataset	0.98	-	0.00
Ours-crowd+env	0.64	0.30	0.02

6.4 Quantitative Evaluations on Real Pedestrian Dataset

A dataset containing real crowd trajectory can help us verify the performance of our approach in the real world. There are two datasets widely used in both social robot navigation and human motion prediction literature: ETH [16] and UCY [12]. Real crowd trajectory datasets do not include static environments. It is mainly used to test the navigation performance after replacing the velocity feature with the trajectory prediction module. In order for the pedestrian and the robot to fully interact, we randomly generate the starting point and target point of the robot according to the crowd movement range and mainly compared it with RGL. The experience result shows in Table 3, our method significantly outperforms RGL in terms of SR and CP But navigation time is slightly lower than RGL. It is proved that the trajectory prediction modules can assist navigation methods to improve navigation safety indication.

Table 3. Statistical experimental results of crowded navigation on the real pedestrian dataset.

Method	SR (%) ↑	CP (%) ↓	AS ↓	AR ↑
RGL [4]	0.82	0.16	12.75	0.6050
Ours	1.00	0.00	15.20	0.8236

7 Conclusion

This paper explores the interplay between pedestrians and static environments in crowded navigation and aim to simplify the process of obtaining environmental features for robots by eliminating the reliance on costly features. To this end, we introduce a new simulator, HERO-Sim, which utilizes a Human-Environment-Robot ternary fusion approach for training and evaluation. Our simulator incorporates realistic static environments constructed using real environment datasets. Additionally, we improve the group interaction model by employing Nash equilibrium, which enhances the group's obstacle avoidance preferences and reduces the homogenization of group behavior. To account for the combined effects of pedestrians and complex static environments, we propose a novel crowded navigation framework, CrowdNav-HERO. This framework incorporates a STGCN-based trajectory prediction module, which replaces the need for velocity with a set of historical trajectory.

In the current research on social navigation, pedestrian behavior is assumed to be homogeneous. In future work, we will study the navigation problem in more complex real pedestrian scenarios.

Acknowledgment. This work received partial support from the National Natural Science Foundation of China (62172443), the Natural Science Foundation of Hunan

Province (2022JJ30760), and the Natural Science Foundation of Changsha (kq2202107, kq2202108). We are grateful for resources from the High Performance Computing Center of Central South University.

References

1. Bai, S., Kolter, J.Z., Koltun, V.: An empirical evaluation of generic convolutional and recurrent networks for sequence modeling. arXiv preprint arXiv:1803.01271 (2018)
2. Baselga, D.M., Riazuelo, L., Montano, L.: Improving robot navigation in crowded environments using intrinsic rewards. In: IEEE International Conference on Robotics and Automation, ICRA 2023, London, UK, May 29–June 2, 2023. pp. 9428–9434. IEEE (2023). https://doi.org/10.1109/ICRA48891.2023.10160876
3. Chang, A.X., et al.: Matterport3D: learning from RGB-D data in indoor environments. In: 2017 International Conference on 3D Vision, 3DV 2017, Qingdao, China, October 10–12, 2017, pp. 667–676 (2017)
4. Chen, C., Hu, S., Nikdel, P., Mori, G., Savva, M.: Relational graph learning for crowd navigation. In: 2020 IEEE/RSJ International Conference on Intelligent Robots and Systems (IROS), pp. 10007–10013 (2020)
5. Chen, C., Liu, Y., Kreiss, S., Alahi, A.: Crowd-robot interaction: Crowd-aware robot navigation with attention-based deep reinforcement learning. In: 2019 international conference on robotics and automation (ICRA), pp. 6015–6022 (2019)
6. Cho, K., et al.: Learning phrase representations using RNN encoder-decoder for statistical machine translation. In: Proceedings of the 2014 Conference on Empirical Methods in Natural Language Processing, EMNLP 2014, October 25–29, 2014, Doha, Qatar, A meeting of SIGDAT, a Special Interest Group of the ACL, pp. 1724–1734. ACL (2014)
7. Fiorini, P., Shiller, Z.: Motion planning in dynamic environments using velocity obstacles. Int. J. Robot. Res. **17**(7), 760–772 (1998)
8. Helbing, D., Molnar, P.: Social force model for pedestrian dynamics. Phys. Rev. E **51**(5), 4282 (1995)
9. Holt, C.A., Roth, A.E.: The nash equilibrium: a perspective. Proc. Natl. Acad. Sci. **101**(12), 3999–4002 (2004)
10. Islam, M.A., et al.: Shape or texture: Understanding discriminative features in CNNs. In: 9th International Conference on Learning Representations, ICLR 2021, Virtual Event, Austria, May 3–7, 2021 (2021)
11. Kipf, T.N., Welling, M.: Semi-supervised classification with graph convolutional networks. In: 5th International Conference on Learning Representations, ICLR 2017, Toulon, France, April 24–26, 2017, Conference Track Proceedings. OpenReview.net (2017), https://openreview.net/forum?id=SJU4ayYgl
12. Lerner, A., Chrysanthou, Y., Lischinski, D.: Crowds by example. In: Computer Graphics Forum, vol. 26, pp. 655–664 (2007)
13. Liu, S., et al.: Socially aware robot crowd navigation with interaction graphs and human trajectory prediction. arXiv preprint arXiv:2203.01821 (2022)
14. McPhail, C., Wohlstein, R.T.: Using film to analyze pedestrian behavior. Sociol. Methods Res. **10**(3), 347–375 (1982)
15. Mohamed, A., Qian, K., Elhoseiny, M., Claudel, C.: Social-STGCNN: a social spatio-temporal graph convolutional neural network for human trajectory prediction. In: Proceedings of the IEEE/CVF Conference on Computer Vision and Pattern Recognition, pp. 14424–14432 (2020)

16. Pellegrini, S., Ess, A., Schindler, K., Van Gool, L.: You'll never walk alone: modeling social behavior for multi-target tracking. In: 2009 IEEE 12th International Conference on Computer Vision, pp. 261–268 (2009)
17. Ronneberger, O., Fischer, P., Brox, T.: U-net: Convolutional networks for biomedical image segmentation. In: Medical Image Computing and Computer-Assisted Intervention-MICCAI 2015: 18th International Conference, Munich, Germany, October 5–9, 2015, Proceedings, Part III 18, pp. 234–241 (2015)
18. Turnwald, A., Olszowy, W., Wollherr, D., Buss, M.: Interactive navigation of humans from a game theoretic perspective. In: 2014 IEEE/RSJ International Conference on Intelligent Robots and Systems, pp. 703–708 (2014)
19. Van Den Berg, J., Guy, S.J., Lin, M., Manocha, D.: Reciprocal n-body collision avoidance. In: Robotics Research: The 14th International Symposium ISRR, pp. 3–19 (2011)
20. Xia, F., Zamir, A.R., He, Z., Sax, A., Malik, J., Savarese, S.: Gibson Env: real-world perception for embodied agents. In: Proceedings of the IEEE Conference on Computer Vision and Pattern Recognition, pp. 9068–9079 (2018)
21. Zhai, A.J., Wang, S.: PEANUT: predicting and navigating to unseen targets. arXiv preprint arXiv:2212.02497 (2022)
22. Zhang, X., et al.: Relational navigation learning in continuous action space among crowds. In: 2021 IEEE International Conference on Robotics and Automation (ICRA), pp. 3175–3181 (2021)
23. Zhou, Z., et al.: Navigating robots in dynamic environment with deep reinforcement learning. IEEE Trans. Intell. Transp. Syst. 23(12), 25201–25211 (2022)
24. Zhou, Z., Zhu, P., Zeng, Z., Xiao, J., Lu, H., Zhou, Z.: Robot navigation in a crowd by integrating deep reinforcement learning and online planning. Appl. Intell. 52(13), 15600–15616 (2022)

Modeling User's Neutral Feedback in Conversational Recommendation

Xizhe Li[1], Chenhao Hu[1], Weiyang Kong[1], Sen Zhang[1], and Yubao Liu[1,2($\boxtimes$)]

[1] Sun Yat–Sen University, Guangzhou, China
{lixzh33,huchh8,kongwy3,zhangs}@mail2.sysu.edu.cn
[2] Guangdong Key Laboratory of Big Data Analysis and Processing, Guangzhou,
China
liuyubao@mail.sysu.edu.cn

Abstract. Conversational recommendation systems (CRS) enable the traditional recommender systems to obtain dynamic user preferences with interactive conversations. Although CRS has shown success in generating recommendation lists based on user's preferences, existing methods restrict users to make binary responses, i.e., accept and reject, after recommending, which limits users from expressing their needs. In fact, the user's rejection feedback may contain other valuable information. To address this limitation, we try to refine user's negative item-level feedback into attribute-level and extend CRS to a more realistic scenario that not only incorporates positive and negative feedback, but also neutral feedback. Neutral feedback denotes incomplete satisfaction with recommended items, which can help CRS infer user's preferences. To better cope with the new setting, we propose a CRS model called Neutral Feedback in Conversational Recommendation (NFCR). We adopt a joint learning task framework for feature extraction and use inverse reinforcement learning to train the decision network, helping CRS make appropriate decisions at each turn. Finally, we utilize the fine-grained neutral feedback from users to acquire their dynamic preferences in the update and deduction module. We conducted comprehensive evaluations on four benchmark datasets to demonstrate the effectiveness of our model.

Keywords: Multi-round Conversational Recommendation · User's Neutral Feedback · Graph Representation Learning

1 Introduction

Conversational Recommender System (CRS), which has witnessed remarkable progress in eliciting the users' current preferences through multi-turn interactions, has become one of the trending research topics in recommender systems.

Supported by the National Nature Science Foundation of China (NSFC 61572537), and the CCF-Huawei Populus Grove Challenge Fund (CCF-HuaweiDBC202305).
X. Li and C. Hu—These authors have contributed equally to this work.

B. Luo et al. (Eds.): ICONIP 2023, LNCS 14450, pp. 56–68, 2024.
https://doi.org/10.1007/978-981-99-8070-3_5

Different from traditional recommender systems, CRS not only considers the historical user-item interactions, but also directly asks user questions about attributes on items to capture their preferences. In recent years, various methods [6,10,13,14,17] based on different settings have been proposed to improve the performance of CRS. In this work, we focus on the Multi-round Conversational Recommendation (MCR) setting, where the system aims to make a successful recommendation with the minimum conversation turns by continuously asking questions about attributes or recommending items.

In MCR scenario, most CRSs follow a "System Ask User Response" (SAUR) [22] paradigm. The system takes an action at each step, either asking questions of attributes or recommending items, and then the users response according to their preferences. However, most of them [4,6,10,11,21] expect a binary yes/no response from users after recommending. Under this setting, users often play a passive role and cannot clearly express their preferences, as they can only reply with an absolute tone of acceptance or rejection. Although users' feedback can help CRS to estimate their preferences, it is difficult to directly utilize the item-level feedback since the reasons for rejection can be varied [1].

Therefore, to better obtain user's preference, we try to refine negative item-level feedback into attribute-level and extend the MCR setting into a more realistic scenario. We propose a novel framework, named **N**eutral **F**eedback in **C**onversational **R**ecommendation (NFCR), where we split the negative feedback into rejection and partial satisfaction, and neural feedback corresponds to the case of partial satisfaction. Then the system can better optimizes the recommendation with use's response.

In summary, the contributions of this paper are as follows:

- We extend the existing CRS setting to a more realistic scenario and propose a novel framework named NFCR, which incorporates neutral feedback from users and enables them to give more specific answers to the recommendations.
- To better utilize neutral feedback, we first use graph representation learning to extract node features from historical data and then adopt adaptive decision and selection strategies to infer their preference with neutral feedback.
- We conducted a series of experiments on four real datasets widely used in CRS. The results show that our method outperforms the state-of-the-art methods.

2   Related Work

In recent years, various approaches based on different CRS settings [5] have been proposed. Multi-Armed Bandits based [3,20] and Meta-Learning based [8,19] algorithms are introduced to solve the cold-start problem. Besides, Question-Driven methods [2,22,24] focus on asking questions about items or their attributes/topics/categories to obtain user preferences. Another direction of CRS is User Understanding and Response Generation [12,15,23], which is aimed at providing high-quality recommendations and generating fluent dialogues.

This paper follows the most realistic conversation recommendation setting proposed so far, named Multi-round Conversational Recommendation (MCR). It focuses on finding a flexible strategy to determine whether to ask questions or make recommendations, with the goal of recommending successfully with fewer turns. SCPR [11] models conversation recommendation task as an interactive path reasoning problem on a knowledge graph. UNICORN [4] improves CRS by formulating three separate decision-making processes as a unified framework. MCMIPL [21] proposes a more realistic scenario setting that users may have more than one target item and use a union set strategy to extract user's multiple interests. CRIF [6] designs an inference module to infer user implicit preferences and adopt inverse reinforcement learning to select suitable actions. However, these works cannot effectively use negative feedback from users. Therefore, we propose a new scenario with user's neutral feedback to fill this gap.

3   Problem Definition

In NFCR, CRS maintains a set of users U and an item set $\mathcal{I}$. Users can have multiple interests as the settings in MCMIPL [21], where each attribute instance a in the attribute set $\mathcal{A}$ has its attribute types $\mathcal{C}$. Each conversation starts by the user u specifying an attribute instance a_0 that in all target items. Then CRS selects actions between asking questions about attributes and recommending a certain number of items (e.g., top-K) in the candidate item list I_{cand}. The user will response based on his/her preference. After that, the system updates the conversational state and decides the following actions based on the feedback. The conversation will repeat until CRS successfully recommending at least one acceptable item to the user or reaching the maximum number of turn T.

4   Proposed Methods

Fig. 1 illustrates the architecture of our method. It consists of four components: representation learning, action decision, selection strategies and update and deduction. The representation learning module is used to train the node embeddings. After user giving an attribute, the system the system starts working at the action decision module, which decides to ask or recommend. The selected action is optimized by a predefined rule to obtain a flexible policy. Then CRS chooses the attributes or items based on the selection strategies. At last, the update and deduction module uses user's feedback to update the conversation state. The system will enter the action decision module again until the user accepts the recommendation or the conversation reaches the maximum turn. We will present the detailed design of each module in the following.

4.1   Representation Learning

To obtain historical interest of users and capture complex relationships between items and attributes, we employ a two-layer Graph Convolutional Network

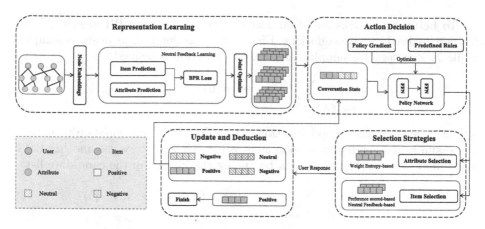

Fig. 1. Overview of the proposed framework, NFCF.

(GCN) [7] to encode the nodes into a hidden vector e_v for all $v \in \mathcal{V}$ with a graph $\mathcal{G} = (\mathcal{V}, \mathcal{E})$. The node set $\mathcal{V}$ is made up of three sets of nodes: users, items and attributes, and the edge set $\mathcal{E}$ consists of two kinds of edges: user-item edges (u, i) and item-attribute edges (i, a), which means that the user u has interacted with the item i and the item i contains the attribute a, respectively. We denote the embeddings of user u, item i and attribute a with $\mathbf{e}_u$, $\mathbf{e}_i$, $\mathbf{e}_a$, respectively.

Preference Prediction. To achieve personalized recommendations, we need to score the items and attributes. We use $\mathcal{A}_u^+$ and $\mathcal{A}_u^-$ to estimate user's preference on an item or an attribute, where $\mathcal{A}_u^+$ and $\mathcal{A}_u^-$ denote the set of attributes that the user u has accepted and rejected, respectively:

$$w(u, i) = tanh(\mathbf{e}_u^T \mathbf{e}_i) + \sum_{a \in \mathcal{A}_u^+} tanh(\mathbf{e}_u^T \mathbf{e}_a) - \sum_{a \in \mathcal{A}_u^-} tanh(\mathbf{e}_u^T \mathbf{e}_a) \tag{1}$$

$$w(u, a) = tanh(\mathbf{e}_u^T \mathbf{e}_a) + \sum_{a \in \mathcal{A}_u^+} tanh(\mathbf{e}_u^T \mathbf{e}_a) - \sum_{a \in \mathcal{A}_u^-} tanh(\mathbf{e}_u^T \mathbf{e}_a) \tag{2}$$

Neutral Feedback Learning In order to refine item-level feedback, we need to identify the items that can render neutral feedback from all candidate items. Given a session $S = \{u, \mathcal{A}_u^+ = \{a_1\}, \mathcal{A}_u^- = \emptyset\}$, we assume that the user's target item i_{tar} contains four attributes $\mathcal{A}_{i_{tar}} = \{a_1, a_2, a_3, a_4\}$. In our work, the item whose attribute set contains $\mathcal{A}_u^+$, as well as is the subset of $\mathcal{A}_{i_{tar}}$ can bring neutral feedback. We denote these items (e.g., $\mathcal{A}_{i_1} = \{a_1, a_2\}$) as $\mathcal{I}_{neu}$. Moreover, there is another type of items in the candidate list $\mathcal{I}_{cand}$, which may contain attributes in $\mathcal{A}_u^-$ or not in $\mathcal{A}_{i_{tar}}$. We consider them as potential rejected items and denote them with $\mathcal{I}_{pot}$. If they are selected into the recommendation list, not only will the user reject, but also the system fails to determine the reason for rejection.

To better rank items based on user preference, we use BPR algorithm [16] for personalized recommendations. Firstly, we consider $\mathcal{I}_{neu}$ as positive samples while $\mathcal{I}_{pot}$ as negative samples, and the loss function is defined as:

$$\mathcal{L}_{item1} = \sum_{i_{neu} \in \mathcal{I}_{neu} \wedge i_{pot} \in \mathcal{I}_{pot}} -ln\sigma(w(u, i_{neu}) - w(u, i_{pot})) \tag{3}$$

where σ denotes the sigmod function. In addition, the system should make a successful recommendation as soon as possible. Therefore, we use another BPR loss function to let the target item be among the highest in the candidate list:

$$\mathcal{L}_{item2} = \sum_{i_{tar} \in \mathcal{I}_{tar} \wedge i_{neu} \in \mathcal{I}_{neu}} -ln\sigma(w(u, i_{tar}) - w(u, i_{neu})) \tag{4}$$

where $\mathcal{I}_{tar}$ denotes the user's target items. Inspired by [10], we also build another pairs of nodes to identify the attributes highly correlated with the current conversation state, respectively. The loss function is defined as:

$$\mathcal{L}_{attr} = \sum_{a^+ \in \mathcal{A}_{i_{tar}} nA_u^+ \wedge a^- \in \mathcal{A}n\mathcal{A}_{i_{tar}}} -ln\sigma(w(u, a^+) - w(u, a^-)) \tag{5}$$

where a^+ and a^- is sampled from all the attributes related to the target item excluding $\mathcal{A}_u^+$ and the attributes unrelated to the target item, respectively.

Finally, the training objective of the recommendation task is set as follows:

$$\mathcal{L}_{rec} = \mathcal{L}_{item1} + \mathcal{L}_{item2} + \mathcal{L}_{attr} + \lambda \|\Theta\|^2 \tag{6}$$

where λ is the regularization term and $\|\Theta\|^2$ is to avoid overfitting.

4.2   Action Decision

In the action decision module, the system needs to decide actions at each turn. Since it is difficult to learn a policy function with a large action space [11], only ask and recommend are included in the output action space.

Inspired by [6,9], we use policy gradient [18] and inverse reinforcement learning (IRL) to tackle multi-round decision making problem. IRL in our work is based on a predefined rule that can show which action can get more reward. Since our goal is to find the target item, the rules are simply guided by the length of the candidate item list and the rank of the target item. We also incorporate the reward of neutral feedback into the rules by updating the user preference with neutral response to judge whether recommending will gain more. The policy π_ϕ and reward function $\hat{r}_\psi$ are updated cyclically according to the following steps:

- Agent learning: Policy π_ϕ interacts with environment to collect states and actions. We update it via maximizing the rewards $\hat{r}_\psi$ with the RL algorithms.
- Reward learning: We optimize the reward function $\hat{r}_\psi$ via comparing the actions selected by the system with those preferred by the predefined rules.

We model a preference predictor assuming that action ac_t (i.e., ask or recommend) preferred by the rules at turn t with the reward function $\hat{r}_\psi$ as follows:

$$\hat{P}_\psi[ac_t^1 \succ ac_t^2] = \frac{\exp \hat{r}_\psi(s_t^1, ac_t^1)}{\exp \hat{r}_\psi(s_t^1, ac_t^1) + \exp \hat{r}_\psi(s_t^2, ac_t^2)} \tag{7}$$

where $ac_t^1 \succ ac_t^2$ denotes that action ac_t^1 is more preferred than ac_t^2. Since the action space only contains two actions, aligning the predictor can be considered as a binary classification problem. We minimize the binary cross-entropy loss with the action pairs (ac_t^1, ac_t^2) in a dataset $\mathcal{D}$ to update the reward function:

$$\mathcal{L} = \sum_{(ac_t^1, ac_t^2) \in \mathcal{D}} -ln\hat{P}_\psi[ac_t^1 \succ ac_t^2] - ln(1 - \hat{P}_\psi[ac_t^2 \succ ac_t^1]) \tag{8}$$

4.3 Selection Strategies

Once action decision is finished, the system needs to select the attributes to ask or generate a recommendation list according to the selection strategies.

Attribute Selection If the selected action is asking, we try to choose attributes that can both eliminate the uncertainty of the candidate items and fit the current preferences of the user. Following [11], we calculate the score of each attribute instance with a weighted uncertainty estimation function:

$$score(u, a, \mathcal{I}_{cand}) = -prob(a) \cdot log(prob(a)) \tag{9}$$

$$prob(a) = \sum_{i \in \mathcal{I}_{cand} \cap \mathcal{I}_a} w(u, i) / \sum_{i \in \mathcal{I}_{cand}} w(u, i) \tag{10}$$

where $\mathcal{I}_{cand}$ is the candidate item set and $\mathcal{I}_a$ denotes the item that contains attribute instance a. Then we select the attribute type corresponding to the attribute instance a with the highest score to generate multiple choice questions and the user can choose the attribute instances he/she likes.

Item Selection In fact, we would rather recommend items in $\mathcal{I}_{neu}$ than in $\mathcal{I}_{pot}$ since we can update the user's preferences with the neutral feedback. Therefore, if the system decides to recommend items, we will construct the recommendation list based on both preference score and neutral feedback:

Firstly, we rank all the items in $\mathcal{I}_{cand}$ by the Eq.(1) to find the items that best match user's current preferences. Then we select top-$\frac{K}{2}$ items in $\mathcal{I}_{cand}$ as the first part of the list, denoted by $\mathcal{I}_{rank}$, where K is the size of the list.

Another $\frac{K}{2}$ items are constructed based on the current state. Specifically, given a session $S = \{u, \mathcal{A}_u^+, \mathcal{A}_u^-\}$, we select $\frac{K}{2}$ items that may trigger user's neutral feedback, whose attribute set contains $\mathcal{A}_u^+$ while excludes $\mathcal{A}_u^-$ from the candidate list. We denote this set of items as $\mathcal{I}_{neu^*}$ and use them to refine item-level negative feedback into attribute-level.

4.4   Update and Deduction

After the system makes an action and the user responds, the system needs to update the conversation state according to the user's feedback. If the system asks a question about attributes, it will update the attribute set $\mathcal{A}_u^+$ and $\mathcal{A}_u^-$ based on the user's answer. If the system recommends an item list and the user is not satisfied with it, the system can refine item-level feedback to attribute-level if the user response with neutral feedback. Once the attribute set has been updated, the system is also able to make corresponding adjustments to the candidate item list to fit the user's dynamic preferences. The new conversation state will also be used for training a flexible policy in the action decision module. Moreover, we use the items not accepted by the user (i.e., $\mathcal{I}_{rej}$) to adjust the corresponding embedding representation to better adapt the user's current preferences as follows:

$$e_u = e_u - \frac{1}{|\mathcal{I}_{rej}|} \sum_{i \in \mathcal{I}_{rej}} e_i \qquad (11)$$

5   Experiments

We conducted experiments on four real datasets widely used in MCR to verify the following research questions (RQ) to demonstrate the superiority of our method.

- **RQ1**. How does the multi-round CRS with NFCR perform as compared with state-of-the-art conversational recommendation methods?
- **RQ2**. Are the key components in NFCR really effective?
- **RQ3**. How can neutral feedback help CRS obtain user preferences?

5.1   DataSet

The details of the four datasets are presented in Table 1. For the Yelp, Lei et al. [10] build a 2-layer taxonomy for the 590 s-layer categories with 29 first-layer categories, which we define as the attribute instances and types, respectively. For another three datasets, Zhang et al. [21] selected entities and relationships in knowledge graph as attribute instances and types, respectively.

5.2   Experimental Settings

User Simulator. As CRS is an interactive system, we design a user simulator to train and evaluate it. Following the previous study [21], we simulate a conversation session for each observed user-item set interaction pair $(u, \mathcal{I}_{tar})$. Each item $i \in \mathcal{I}_{tar}$ is treated as the ground-truth. The session is initialized by the simulator specifying an attribute instance that all items in $\mathcal{I}_{tar}$ contain. User gives feedback following the rules: (1) When CRS recommends, the user will accept it only if the list contains at least one item in $\mathcal{I}_{tar}$. Otherwise, the user will give neutral or negative feedback based on his/her preference; (2) When CRS asks questions, the user will confirm he/she likes it only when the attribute instances are associated with any item in $\mathcal{I}_{tar}$; (3) We consider that the user will become impatient when the turn of conversation reaches the maximum number T.

Table 1. Summary statistics of datasets.

DataSet	Yelp	LastFM	Amazon-Book	MovieLens
#Users	27,675	1,801	30,921	20,892
#Items	70,311	7,432	17,739	16,482
#Interactions	1,368,609	76,693	478,099	454,011
#Attribute instances	590	8,438	988	1,498
#Attribute types	29	34	40	24
#Entities	98,576	17,671	49,018	38,872
#Relations	3	4	2	2
#Triplets	2,533,827	228,217	565,068	380,016

Implementation Details. Following [6], we split each dataset randomly by 7:1.5:1.5 for training, validation and testing. We set the size K of the recommendation list as 10, the maximum turn T of conversation as 15. The embedding dimension in representation learning is set as 64. We employ the Adam optimizer for multi-task training, with 0.0005 learning rate. The regulation parameter is 10^{-5}. To train the conversation module, we adopt the user simulator described in Sect. 5.2 to interact with the CRS.

Baselines. We use several state-of-the-art methods for MCR to estimate our proposed framework: **CRM** [17] employs policy gradient to learn the decision network and uses a belief tracker to estimate user's preferences. **EAR** [10] proposes a three-stage solution to enhance the interaction between the conversation and recommendation component. **SCPR** [11] models conversational recommendation as an interactive path reasoning problem on graph and adopt DQN to determine whether to ask or recommend. **UNICORN** [4] proposes a unified learning framework based on a dynamic weighted graph for three decision-making processes. **MCMIPL** [21] defines a more realistic CRS scenario where the user can have more than one target items. **CRIF** [6] is the state-of-art method on MCR. It infers users' implicit preferences with an inference module and employ inverse reinforcement learning to obtain a flexible decision network.

For a fair performance comparison, we adopt the new scenario for the last four baselines that user can give *neutral feedback* described in Sect. 3 for recommendation. Besides, we also compare with the original version of CRIF and MCMIPL to illustrate the effectiveness of neutral feedback. We name the two original methods with CRIF* and MCMIPL*, respectively.

Evaluation Metrics. Following previous studies [10], we adopt success rate at turn t (SR@t) [17] and average turn (AT) to evaluate the performance. SR@t is the cumulative ratio of successful recommendation by turn t. AT stands for the average number of turns for all sessions. Furthermore, Deng et al. [4] argued that SR@t and AT are insensitive for ranking. They proposed hDCG@(T,K), a two-

Table 2. Experimental results. (hDCG stands for hDCG@(15,10).)

Dataset	Yelp			LastFM			Amazon-Book			MovieLens		
Metric	SR@15	AT	hDCG	SR@15	AT	hDCG	SR@15	AT	hDCG	SR@15	AT	hDCG
CRM	0.223	13.83	0.073	0.597	10.60	0.269	0.309	12.47	0.117	0.654	7.86	0.413
EAR	0.263	13.79	0.098	0.612	9.66	0.276	0.354	12.07	0.132	0.714	6.53	0.457
SCPR	0.476	12.71	0.142	0.820	8.45	0.303	0.507	10.85	0.182	0.843	4.51	0.501
UNICORN	0.502	11.44	0.172	0.852	7.26	0.331	0.574	10.04	0.234	0.866	4.20	0.539
MCMIPL	0.604	10.53	0.205	0.866	7.12	0.353	0.588	10.6	0.237	0.901	3.39	0.615
CRIF	0.706	10.12	0.218	0.923	6.06	0.438	0.715	8.69	0.337	0.973	2.42	0.879
MCMIPL*	0.482	11.87	0.160	0.838	7.33	0.337	0.545	10.83	0.223	0.882	3.61	0.599
CRIF*	0.653	11.18	0.201	0.885	6.67	0.403	0.623	9.02	0.298	0.952	2.63	0.847
NFCR	0.753	8.98	0.278	0.924	6.02	0.412	0.795	6.55	0.382	0.988	2.13	0.908

level hierarchical version extended from the normalized discounted cumulative gain (NDCG@K), for a comprehensive evaluation.

5.3 Performance Comparison of NFCR with Existing Models (RQ1)

Overall Performance. Table 2 reports the performance between NFCR and all baselines among four datasets. On the whole, NFCR outperforms all the baselines by achieving a higher success rate and less average turn, which can also be demonstrated by the improvements on hDCG.

Compared to the original version of MCMIPL and CRIF (i.e., MCMIPL* and CRIF*), adapting neutral feedback in MCMIPL and CRIF both achieve a better performance, which indicates the effectiveness of incorporating user's neutral feedback in the CRS. However, our method still outperforms them. We infer that our model can refine more user preference because of the feature extraction of neutral feedback and the action decision policy.

Comparison at Different Conversation Turns. We also intuitively shows the Success Rate at each turn (SR@t) in Fig. 2. To better observe the differences among these methods, we report the relative success rate compared with the state-of-the-art baseline CRIF, since it has relatively stable performance. For a clear observation, we only report the results of our model and other five competitive baselines. We summarize several notable observations as follows:

(i) Our approach may not outperform other models at the first few turns. Because the system not only recommends items based on score, but also tries to trigger neutral feedback to capture user preferences, causing that the recommendation list may not exactly fit the user's preferences most.

(ii) As the number of turns increases, NFCR gradually shows its advantages. Our model can better capture user's dynamic preferences and optimize the subsequent conversation. Similarly, the versions with neutral feedback of CRIF and MCMIPL also outperform the original version in the middle stage of

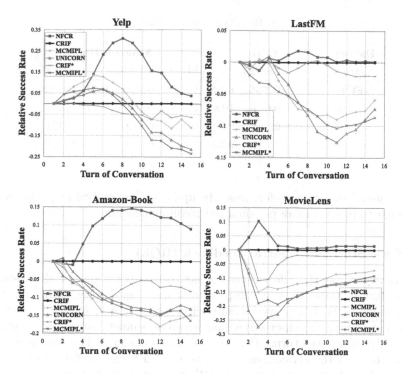

Fig. 2. Comparisons at different conversation turns.

the conversation. These results verify neutral feedback can help CRS obtain useful information and ensure a more stable performance.

5.4   Ablation Studies (RQ2)

The key design of NFCR is extracting the features of neutral feedback in graph representation learning and leveraging them during the conversation. Therefore, we conduct ablation experiments on the Yelp and Amazon-Book datasets to verify the effectiveness of these designs. The results are shown in Table 3.

Firstly, we evaluate the effectiveness of representation learning. We remove the first two loss functions to verify they are helpful for refining neutral feedback feature. As shown in the Table 3, without this module (i.e., -w/o NFL), the performance decreases significantly, which suggests that our design can better capture potential relationships between items and attributes.

Then we compare with the policy network that detaches reward of neutral feedback from the predefined rules (i.e., -w/o NF reward), and we do not update the user's embedding at the end of each turn (i.e., -w/o update), separately. The first experiment performs worse, which demonstrates the efficiency of adjusting predefined rules based on neutral feedback. Besides, we have observed that the items rejected by users during the conversation provide valuable information

Table 3. Results of the ablation study.

Dataset	Yelp			Amazon-Book		
Metric	SR@15	AT	hDCG	SR@15	AT	hDCG
NFCR	**0.753**	**8.98**	**0.278**	**0.795**	**6.55**	**0.382**
-w/o NFL	0.694	10.67	0.215	0.758	7.64	0.342
-w/o NF reward	0.714	9.92	0.221	0.719	8.51	0.324
-w/o update	0.734	9.16	0.249	0.770	7.12	0.367

into the users' dynamic preferences. By updating the embedding, the system can adjust to provide more personalized recommendations.

5.5 Case Study on Neutral Feedback (RQ3)

To intuitively show the effectiveness of neutral feedback during the conversation, we further present a case sampled from the Yelp dataset with NFCR and other two methods. The conversation is displayed in Fig. 3. As can be seen, our method can estimate the user's dynamic preferences much faster and recommend successfully with fewer turns. Besides, our method can make a comprehensive decision to select the actions, enabling the system to get useful information at each turn. Despite the success of MCMIPL in exacting multi-interests, its action decision network cannot adapt to user's neutral feedback, which tends to recommend more but fails to capture user's preferences. Although CRIF leverages the implicit feedback to infer user's dynamic preferences, without the feature extraction of the neutral feedback and a suitable rule, its policy network is unable to use current preferences to trigger user's neutral feedback when recommending.

6 Conclusions

In this work, we extend CRS to a more realistic scenario in which users can give neutral responses besides yes/no to the recommendations. Based on this scenario, we propose a novel framework NFCR to refine item-level negative feedback. We

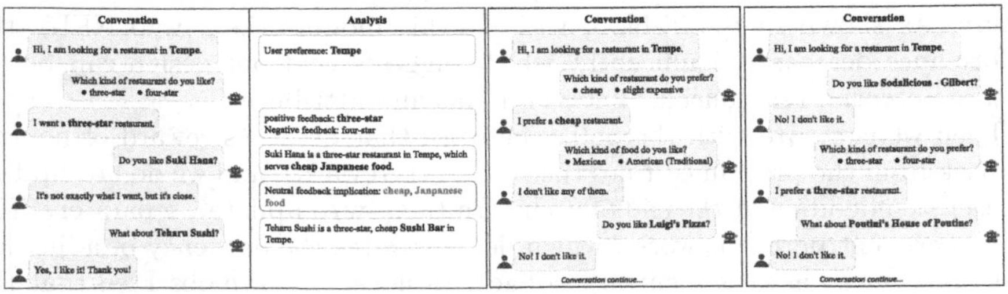

Fig. 3. Sample conversations generated by NFCR, CRIF and MCMIPL.

first extract feature of neutral feedback from the historical data. In addition, we improve the policy network by incorporating reward from neutral feedback into the predefined rules to leverage user's feedback to adjust the conversation. Compared with the state-of-the-art methods on four real datasets, the experimental results demonstrate the superiority of our proposed method.

References

1. Bi, K., Ai, Q., Zhang, Y., Croft, W.B.: Conversational product search based on negative feedback. In: Proceedings of the 28th ACM International Conference on Information and Knowledge Management, pp. 359–368 (2019)
2. Christakopoulou, K., Beutel, A., Li, R., Jain, S., Chi, E.H.: Q&R: a two-stage approach toward interactive recommendation. In: Proceedings of the 24th ACM SIGKDD International Conference on Knowledge Discovery & Data Mining, pp. 139–148 (2018)
3. Christakopoulou, K., Radlinski, F., Hofmann, K.: Towards conversational recommender systems. In: Proceedings of the 22nd ACM SIGKDD International Conference on Knowledge Discovery and Data Mining, pp. 815–824 (2016)
4. Deng, Y., Li, Y., Sun, F., Ding, B., Lam, W.: Unified conversational recommendation policy learning via graph-based reinforcement learning. In: Proceedings of the 44th International ACM SIGIR Conference on Research and Development in Information Retrieval, pp. 1431–1441 (2021)
5. Gao, C., Lei, W., He, X., de Rijke, M., Chua, T.S.: Advances and challenges in conversational recommender systems: a survey. AI Open 2, 100–126 (2021)
6. Hu, C., Huang, S., Zhang, Y., Liu, Y.: Learning to infer user implicit preference in conversational recommendation. In: Proceedings of the 45th International ACM SIGIR Conference on Research and Development in Information Retrieval, pp. 256–266 (2022)
7. Kipf, T.N., Welling, M.: Semi-supervised classification with graph convolutional networks. arXiv preprint arXiv:1609.02907 (2016)
8. Lee, H., Im, J., Jang, S., Cho, H., Chung, S.: MeLU: meta-learned user preference estimator for cold-start recommendation. In: Proceedings of the 25th ACM SIGKDD International Conference on Knowledge Discovery & Data Mining. pp. 1073–1082 (2019)
9. Lee, K., Smith, L., Dragan, A., Abbeel, P.: B-Pref: Benchmarking preference-based reinforcement learning. arXiv preprint arXiv:2111.03026 (2021)
10. Lei, W., et al.: Estimation-action-reflection: towards deep interaction between conversational and recommender systems. In: Proceedings of the 13th International Conference on Web Search and Data Mining, pp. 304–312 (2020)
11. Lei, W., et al.: Interactive path reasoning on graph for conversational recommendation. In: Proceedings of the 26th ACM SIGKDD International Conference on Knowledge Discovery & Data Mining, pp. 2073–2083 (2020)
12. Li, R., Ebrahimi Kahou, S., Schulz, H., Michalski, V., Charlin, L., Pal, C.: Towards deep conversational recommendations. In: Advances in Neural Information Processing Systems 31 (2018)
13. Li, S., Lei, W., Wu, Q., He, X., Jiang, P., Chua, T.S.: Seamlessly unifying attributes and items: conversational recommendation for cold-start users. ACM Trans. Inf. Syst. (TOIS) 39(4), 1–29 (2021)

14. Li, S., Xie, R., Zhu, Y., Ao, X., Zhuang, F., He, Q.: User-centric conversational recommendation with multi-aspect user modeling. arXiv preprint arXiv:2204.09263 (2022)
15. Liu, Z., Wang, H., Niu, Z.Y., Wu, H., Che, W., Liu, T.: Towards conversational recommendation over multi-type dialogs. arXiv preprint arXiv:2005.03954 (2020)
16. Rendle, S., Freudenthaler, C., Gantner, Z., Schmidt-Thieme, L.: BPR: bayesian personalized ranking from implicit feedback. arXiv preprint arXiv:1205.2618 (2012)
17. Sun, Y., Zhang, Y.: Conversational recommender system. In: The 41st International ACM SIGIR Conference on Research & Development in Information Retrieval, pp. 235–244 (2018)
18. Sutton, R.S., McAllester, D., Singh, S., Mansour, Y.: Policy gradient methods for reinforcement learning with function approximation. In: Advances in Neural Information Processing Systems 12 (1999)
19. Wei, T., et al.: Fast adaptation for cold-start collaborative filtering with meta-learning. In: 2020 IEEE International Conference on Data Mining (ICDM), pp. 661–670. IEEE (2020)
20. Zhang, X., Xie, H., Li, H., CS Lui, J.: Conversational contextual bandit: algorithm and application. In: Proceedings of the web conference 2020, pp. 662–672 (2020)
21. Zhang, Y., et al.: Multiple choice questions based multi-interest policy learning for conversational recommendation. In: Proceedings of the ACM Web Conference 2022, pp. 2153–2162 (2022)
22. Zhang, Y., Chen, X., Ai, Q., Yang, L., Croft, W.B.: Towards conversational search and recommendation: system ask, user respond. In: Proceedings of the 27th ACM International Conference on Information and Knowledge Management, pp. 177–186 (2018)
23. Zhou, K., Zhao, W.X., Bian, S., Zhou, Y., Wen, J.R., Yu, J.: Improving conversational recommender systems via knowledge graph based semantic fusion. In: Proceedings of the 26th ACM SIGKDD International Conference on Knowledge Discovery & Data Mining, pp. 1006–1014 (2020)
24. Zou, J., Chen, Y., Kanoulas, E.: Towards question-based recommender systems. In: Proceedings of the 43rd International ACM SIGIR Conference on Research and Development in Information Retrieval, pp. 881–890 (2020)

A Domain Knowledge-Based Semi-supervised Pancreas Segmentation Approach

Siqi Ma, Zhe Liu$^{(\boxtimes)}$, Yuqing Song, Yi Liu, Kai Han, and Yang Jiang

School of Computer Science and Communication Engineering, Jiangsu University,
Zhenjiang 212013, China
1000004088@ujs.edu.cn

Abstract. The five-year survival rate of pancreatic cancer is extremely low, and the survival time of patients can be extended by timely detection and treatment. Deep learning-based methods have been used to assist radiologists in diagnosis, with remarkable achievements. However, obtaining sufficient labeled data is time-consuming and labor-intensive. Semi-supervised learning is an effective way to alleviate dependence on annotated data by combining unlabeled data. Since the existing semi-supervised pancreas segmentation works are easier to ignore the domain knowledge, leading to location and shape bias. In this paper, we propose a semi-supervised pancreas segmentation method based on domain knowledge. Specifically, the prior constraints for different organ sub-regions are used to guide the pseudo-label generation for unlabeled data. Then the bidirectional information flow regularization is designed by further utilizing pseudo-labels, encouraging the model to align the labeled and unlabeled data distributions. Extensive experiments on NIH pancreas datasets show: the proposed method achieved Dice of 76.23% and 80.76% under 10% and 20% labeled data, respectively, which is superior to other semi-supervised pancreas segmentation methods.

Keywords: Semi-supervised learning · Medical image segmentation · Domain knowledge · Regularization

1 Introduction

Accurately segmenting organs and tumor regions is important and challenging in computer-aided diagnosis systems, which are often used to quantify the size and shape of organs of interest, population studies, disease quantification, and treatment planning, playing a crucial role in medical image analysis and surgical planning [7,13]. Data from the SEER [4] database in the United States showed that the average 5-year survival rate of pancreatic cancer patients enrolled from

Supported by National Natural Science Foundation of China (61976106, 62276116); Jiangsu Six Talent Peak Program (DZXX−122); Jiangsu Graduate Research Innovation Program (KYCX23_3677).

2011 to 2017 was only 11%. If detected and treated in time, the average survival time can reach more than one year. The appearance of pancreatic cancer is often accompanied by changes in pancreatic morphology [15]. Therefore, automatic pancreas segmentation can assist doctors to determine whether there are lesions, and contribute to the timely diagnosis of pancreatic diseases. However, obtaining high-quality labels requires the domain knowledge of radiologists. In addition, the lengthy and labor-intensive labeling process makes it easier for doctors' judgment to come out as biased, which lowers the quality of manual labels [6]. This supervised setting largely hinders the deployment of segmentation models in clinical scenarios. To alleviate the need for labeled data, semi-supervised medical image segmentation methods [6,23] have attracted the interest of researchers. These models are trained on data from limited labeled samples and a large number of unlabeled samples, alleviating the dependency of deep models on label data [17].

For semi-supervised learning, Wu et al. [20] employed collaborative training to exchange internal information among models, aiming to increase output consistency and alleviate missegmentation prediction of adhesive edges or thin branches of pancreas and other organs. In addition to consistency methods, uncertainty-aware methods are also favored. Xia et al. [22] proposed a uncertainty-aware multi-view co-training (UMCT) framework, which combined the information of multiple views and utilized the uncertainty estimation of each view to achieve accurate pancreas segmentation. Uncertainty-aware mean teacher (UA-MT) [24] encouraged consistent prediction of the same input by different perturbations, and introduced uncertainty information as a regularization term. However, the above models contain more parameters and occupy high computing resources. Inspired by the pyramid feature network, uncertainty rectified pyramid consistency (URPC) [11] minimized the discrepancy between each scale prediction for unlabeled data at a lower cost. Besides, Han et al. [5] proposed a pseudo-label generation method that generates pseudo labels with guidance from labeled class representations. However, this method is not appropriate for 3D images and ignores the fact: *sub-regions are different within the same organ*. In other words, different sub-regions of the same organ have different features.

To this end, combined with semi-supervised learning and medical domain knowledge, this paper proposes a domain knowledge-based semi-supervised pancreas segmentation method. Firstly, within the constraint of medical domain knowledge, the class representation method is used to guide the generation of pseudo-labels. Then, bidirectional information flow regularization is introduced to encourage models to align the labeled and unlabeled data distributions. The proposed method is evaluated on a public pancreas dataset, and the results show that our method is effective and superior to other methods.

The main contributions are as follows:

1) We propose a domain knowledge-based semi-supervised pancreas segmentation method that imposes region-level constraints on organs with uneven morphological distributions.
2) A bidirectional information flow regularization is designed to encourage model reverse learning, aligning the labeled and unlabeled data distributions.

3) Extensive experimental results verified the effectiveness of each component and the superiority of our method to other semi-supervised learning methods in terms of segmentation details.

2   Related Work

2.1   Semi-supervised Medical Image Segmentation

In recent years, various semi-supervised methods [8,23] have been designed to reduce the requirement for labeled data, which are mainly divided into three parts: pseudo label-based [1,2,5,22,26], consistency regularization [9,20,21] and other methods [19]. The core of the pseudo-labelling algorithm is to optimize the quality of the generated pseudo labels. Optimization methods usually include postprocessing [1,5] and uncertainty estimation [2,22,26]. Consistency regularization-based methods usually apply data and model perturbations to obtain consistency. Li et al. [9] introduced a transformation-consistent strategy in the self-ensembling model to enhance the regularization effect for pixel-level predictions. Wu et al. [20,21] designed a cycled pseudo-label scheme to encourage mutual consistency through the prediction discrepancies of multiple decoders. For other methods, based on the segmentation task, Wang et al. [19] introduced the image reconstruction task and the symbolic distance field (SDF) prediction task, which can help the network capture more semantic information and constrain the global geometry of the segmentation result. However, the above methods generally focus on universal image processing, ignoring the importance of integrating medical knowledge.

2.2   Domain Knowledge

Different from natural images, medical images can provide rich human anatomy information. Prior knowledge can teach models which features deserve to be focused on, as well as the location and size range. To make full use of the complementary information of multi-sequence cardiac magnetic resonance (MS CMR) images, Liu et al. [10] utilized the equilibration-steady state free precession (bSSFP) image sequences as prior knowledge for left ventricular localization and meanwhile used the late gadolinium enhanced (LGE) image sequences for accurate segmentation. Zhuang et al. [27] designed a unified framework combining three sequences (bSSFP, T2 and LGE) to align images to a public space for segmentation. Duan et al. [3] combined the multi-task deep learning method with graph propagation and applied shape prior knowledge to effectively overcome image artifacts and improve segmentation quality. However, these methods ignore making use of prior information to impose region-level constraints.

3   Methodology

Due to the small number of pixels occupied by the pancreas region, the model tends to pay attention to the background region and ignores the pancreas region, resulting in a performance reduction. In addition, the pancreas has the problem of uneven morphological distribution. If the model pays too much attention to the overall morphology in training, it is easy to ignore the differences of different parts, resulting in reduced generalization performance. To solve the above problems, we propose a semi-supervised pancreas segmentation method based on domain-knowledge and the overall framework is shown in Fig. 1. It includes a prior position acquisition module (Fig. 1(a)) and a semi-supervised pancreas segmentation strategy module (Fig. 1(b)), while the latter includes a pseudo-label generation method based on prior position and bidirectional information flow regularization.

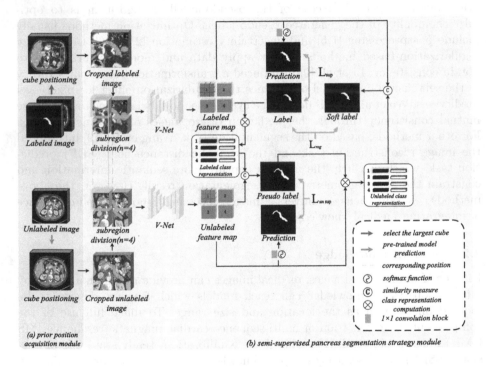

Fig. 1. The overview of the proposed method.

Task Setup. Before introducing the proposed method, a set of symbols is introduced to define the semi-supervised segmentation task. The dataset contains two subsets $D = D_L \cup D_U$, and $D_L \cap D_U = \emptyset$. Where, $D_L = \{(X_i, Y_i)\}_{i=1}^{N_L}$ is labeled dataset while $D_U = \{X_i\}_{i=1}^{N_U}$ is unlabeled dataset. X and Y denote image and label, respectively. The goal of semi-supervised learning is to enhance the performance of model trained on D_L using unlabeled dataset D_U.

The Prior Position Acquisition Method. Considering that there is a serious imbalance between foreground and background classes in pancreas segmentation, it is necessary to extract the region of interest to remove the interference of irrelevant information, as shown in Fig. 2. The detailed steps are as follows:

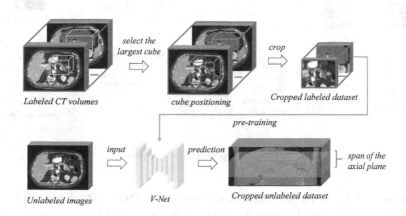

Fig. 2. Prior position acquisition strategy.

1) We pre-process the labeled images by first calculating the size of the cube containing the pancreatic region in each computed tomography (CT) volume, and then selecting the largest one so that it could contain all the pancreatic regions in the dataset. Following that, by cropping the labeled image set according to the largest cube, the following dataset $X, Y \in \mathbb{R}^{V' \times H' \times W' \times 1}$ will be obtained, V', H' and W' denote the slice number, length and width of truncated labeled data, respectively. Subsequently, the model is pre-trained using this dataset.

2) Due to the individual differences, the span of the axial plane fluctuates greatly. By observing that the position of the pancreas in 3D CT images has a certain range, the distribution information of labeled image data can be used to impose prior constraints on unlabeled images. The pre-trained model in 1) is used to predict unlabeled dataset D_U, roughly locate the span of the pancreas axial plane, then sample according to cube size H' and W' which mentioned in 1). Next, the processed unlabeled data is expanded bidirectionally to V', then the X^U will be obtained.

3) Considering that the pancreas has certain morphological characteristics in the human body, a related hypothesis is introduced: *the same location of organs shares similar characteristics in different patients.* Therefore, the labeled and unlabeled images are divided into n block subregions of the same size, and constraints are imposed between the subregions at their corresponding positions. The size of the divided subregions is V', $\frac{H'}{n}$, and $\frac{W'}{n}$, as shown in Fig. 1 ($n = 4$).

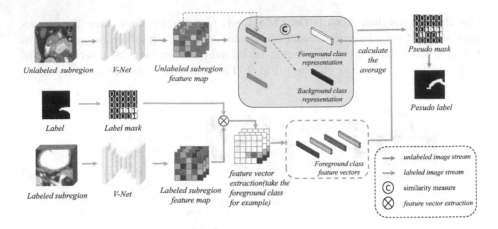

Fig. 3. Pseudo-label generation process.

Pseudo-label Generation Method Based on Prior Position. The flow chart of the pseudo-label generation method is shown in Fig. 3. A group of sub-region pairs (X_j, Y_j) (X_j is the green luminous subregion in Fig. 1(b)) and X_j^U (blue luminous subregion in Fig. 1(b)) are selected. For the former, the supervision loss can be calculated, and the formula is expressed as Eq. 1 and Eq. 2:

$$F_j = f_\theta(X_j) \tag{1}$$

$$L_{sup} = \frac{1}{n} \sum_{j=1}^{n} Loss(softmax(Conv_{1\times1}(F_j)), Y_j) \tag{2}$$

where, f denotes the segmentation network with the parameters θ, $F_j \in \mathbb{R}^{\frac{W'}{n} \times \frac{H'}{n} \times C}$ is the feature map before softmax function, C, $Conv_{1\times1}(\cdot)$ and $softmax(\cdot)$ represents the channel number, 1×1 convolution and $softmax$ function, respectively. j and n represent the serial number of subregions and number of subregions, respectively.

Based on previous work [5], we use class representations to learn foreground and background class features. Specifically, the features of labeled images F_j and corresponding masks Y_j are combined to generate class representations for guiding unlabeled images, which are shown in Eq. 3:

$$P_j^k = \frac{1}{B_L} \sum_{i=1}^{B_L} \frac{\sum_{x,y} F_{ij}^{(x,y)} \mathbb{I}(Y_{ij}^{(x,y)} = k)}{\sum_{x,y} \mathbb{I}(Y_{ij}^{(x,y)} = k)}, k \in \{f_g, b_g\} \tag{3}$$

where, B_L represents the batch size. $P_j^{f_g}, P_j^{b_g} \in \mathbb{R}^{1\times1\times C}$. $\mathbb{I}$ is an indication function that outputs value 1 if the condition is true, otherwise 0. (x, y) denotes the spatial position of pixels in the feature map F_j.

The corresponding sub-region X_j^U of the unlabeled image is passed through the network to obtain features map F_j', and the prediction probability map H_j' is further obtained by 1×1 convolutional layer and $softmax$ function. The above calculation formulas are shown in Eq. 4 and Eq. 5:

$$F_j^{'} = f_\theta(X_j^U) \tag{4}$$

$$H_j^{'} = softmax(Conv_{1\times1}(F_j^{'})) \tag{5}$$

Next, calculate the distance map between each spatial position vector and the foreground/background class representation using the cosine similarity measure, as shown in Eq. 6:

$$Dist[k] = \frac{F_j^{'(x,y)} \cdot P_j^k}{\| F_j^{'(x,y)} \| \cdot \| P_j^k \|}, k \in \{f_g, b_g\} \tag{6}$$

By applying a $softmax$ function to the distance map, it will generate a class probability map and then maximizes the probability to obtain a pseudo-label, which are shown in the Eq. 7:

$$Y_j^U = argmax_k(softmax(Disk[k])), k \in \{f_g, b_g\} \tag{7}$$

The unsupervised loss between the unlabeled prediction probability map $H_j^{'}$ and the pseudo-label Y_j^U guided by the labeled class representation is further calculated. The loss function is shown in Eq. 8:

$$L_{unsup} = \frac{1}{n} \sum_{j=1}^{n} Loss(H_j^{'}, Y_j^U) \tag{8}$$

Bidirectional Information Flow Regularization. The bidirectional information flow regularization is inspired by the teacher-student relationship. After obtaining the pseudo-label Y_j^U of the unlabeled images by the pseudo-labels generation method based on prior position, the corresponding mask average pooling of the unlabeled features $F_j^{'}$ is carried out to obtain the class representation of the unlabeled data, as shown in Eq. 9:

$$P_j^{'k} = \frac{1}{B_U} \sum_{i=1}^{B_U} \frac{\sum_{x,y} F_{ij}^{'(x,y)} \mathbb{I}(Y_{ij}^{U(x,y)} = k)}{\sum_{x,y} \mathbb{I}(Y_{ij}^{U(x,y)} = k)}, k \in \{f_g, b_g\} \tag{9}$$

where, B_U represents the batch size. The remaining process is similar to Eq. 6. The regularization loss L_{seg} is calculated by the output soft label and the real label Y_j. The above operations are shown in Eq. 10 and Eq. 11:

$$Disk^{'}[k] = \frac{F_j^{(x,y)} \cdot P_j^{'k}}{\| F_j^{(x,y)} \| \cdot \| P_j^{'k} \|}, k \in \{f_g, b_g\} \tag{10}$$

$$L_{reg} = \frac{1}{n} \sum_{j=1}^{n} Loss(softmax(Disk^{'}[k]), Y_j) \tag{11}$$

3.1 Loss Function

The total loss function consists of supervised loss, unsupervised loss, and regularization loss, As shown in Eq. 12:

$$L_{total} = L_{sup} + \lambda(L_{unsup} + L_{reg}) \tag{12}$$

where, L_{sup}, L_{unsup} and L_{reg} are supervised loss, unsupervised loss and regularization loss, respectively. The weight λ is a time-dependent Gaussian warming up function [24], which balances between the supervised loss and unsupervised consistency loss. L_{unsup} and L_{reg} adopt Dice loss and binary cross entropy loss, respectively. Supervision loss is shown in Eq. 13:

$$L_{sup} = 0.5L_{dice} + 0.5L_{bce} \tag{13}$$

where, L_{dice} and L_{bce} [12] represent Dice loss and binary cross entropy loss, respectively.

4 Experiments and Results

4.1 Datasets

The proposed method was evaluated on the Pancreas-NIH [16] dataset, which is one of the most popular pancreas datasets. It includes 82 subjects at the National Institutes of Health Clinical Center and all volumes were manually annotated by experienced radiologists. CT scan sizes range from $512 \times 512 \times 181$ to $512 \times 512 \times 466$. The 62 samples were used for training and the rest for testing. For image preprocessing, the voxel values are normalized to the range of $[-125, 275]$ Hounsfield Units (HU) and the isotropy resolution of data is further resampling to $1.0 \times 1.0 \times 1.0$ mm.

4.2 Implementation Details

In the 3D V-Net [14] decoder, each layer is concating with the corresponding encoder layer to fuse multi-scale feature information, which helps to improve the accuracy and stability of image segmentation. Therefore, we adopted 3D V-Net [14] as the backbone network. The experiment has a total of 30,000 iterations. The batch size and initial learning rate are 4 and 0.01. The two backbone networks share the same parameters. The 10% or 20% labeled data ratio setting was performed during the training process. In the inference stage, the following evaluations were introduced: Dice, Jaccard, 95HD, Recall, and Precision. The experimental equipment is a GPU NVIDIA RTX 6000/8000.

4.3   Ablation Study

The modules of the proposed method are divided as follows: (1) Baseline(BL): 3D class representation semi-supervised medical image method [5]; (2) PP: pseudo-label generation method based on prior position; (3) RR: bidirectional information flow regularization. The experimental results are shown in Table 1.

Experiments were conducted with 10% labeled data and 20% labeled data. After prior positions were added to the 3D class representation semi-supervised medical image method [5], all evaluation indicators have achieved greater gains. Especially the Dice index increase from 70.87% to 74.16% and from 74.77% to 77.65%, respectively. It proves that applying regional consistency constraints based on prior positions can make more efficient use of unlabeled data. If prior position is not used (when $n = 1$), the model only focuses on the overall images and it is easy to ignore partial details, such as the shape of the pancreatic head and tail, and the narrow region in the middle of the pancreas (see Fig. 4).

In addition, the bidirectional information flow regularization improves the performance of the baseline, especially the Recall is greatly improved. Table 1 shows that the combination of the pseudo-label generation method based on prior position and the bidirectional information flow regularization is superior to adopting a single module. In general, Table 1 proves that each component can improve segmentation performance. Several typical cases are selected for visual comparison, as shown in Fig. 4.

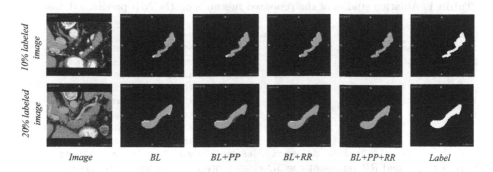

Fig. 4. Comparison results of each component. (The red label indicates the predicted result, while the white label indicates the ground truth.) BL represents the 3D class representation semi-supervised medical image method [5]. BL+PP represents the fusion of prior positions and BL [5]. BL+RR represents the fusion of bidirectional information flow regularization and BL [5]. BL+PP+RR indicates that all modules mentioned are added. (Color figure online)

In order to study the influence of the number of subregions n in the regional constraint on the segmentation results, we selected different values for experiments, and the results are shown in Fig. 5. When the number of n increased from 1 to 4, the Dice improved significantly, but at the time of $n > 4$, the segmentation performance dropped sharply. While $n = 4$, the segmentation results are optimal. The reason is that when the number of subregions is too small, the model only pays attention to global segmentation and lacks the conditions to implement region-level constraints. On the contrary, the model will focus too much on the local area, resulting in too few foreground and background classes, and the corresponding class representation lacks robustness, resulting in a sharp decline in model performance.

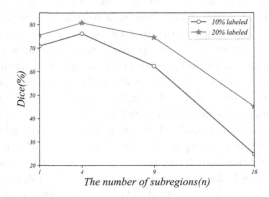

Fig. 5. Influence of the number of subregions for segmentation results.

Table 1. Ablation studies of the proposed methods on the NIH pancreas dataset.

Setting		Structure			Evaluations				
Labeled data	Unlabeled data	BL	PP	RR	Dice(%)↑	Jaccard(%)↑	95HD(mm)↓	Recall(%)↑	Precision(%)↑
6	56	✓			70.87	56.51	21.18	68.38	76.74
			✓		74.16	59.96	**13.44**	72.28	78.77
				✓	72.53	58.26	16.54	70.19	77.95
		✓	✓	✓	**76.23**	**62.29**	13.80	**73.27**	**82.07**
12	50	✓			74.77	60.51	15.09	71.80	81.08
			✓		77.65	64.17	9.27	76.05	80.84
				✓	75.45	61.42	16.82	77.63	74.53
		✓	✓	✓	**80.76**	**68.12**	**6.91**	**81.35**	**81.41**

Note: BL, PP and RR represent the 3D class representation semi-supervised medical image method [5], the pseudo-label generation method based on prior position, and the bidirectional information flow regularization, respectively.

4.4 Comparison Study

The proposed method was conducted under two settings of labeled data (10% and 20%). Following previous works [19–21], we chose V-Net [14] framework as the backbone network and made comparisons with several advanced semi-supervised segmentation methods. Including Mean Teacher model (MT) [18], UA-MT [24], URPC [11] and Deep Adversarial Network (DAN) [25]. In addition, for fair comparison, all experiments cannot rely on any post-processing or use any shape constraints. The experiment results are shown in Table 2. Under

10% and 20% labeled settings, URPC [11] achieved the best performance compared with other methods, but our approach slightly outperforms URPC [11] on most indicators. According to Dice, the proposed method is superior to all comparison methods and has significant improvements in Precision and Jaccard index. Compared with the benchmark V-Net [14], our method improves the Dice index by 10.90% (10% labeled) and 7.09% (20% labeled), respectively. Besides, the upper bound of the proposed method on 20% labeled data approximates the performance of the baseline with 100% labeled data.

Table 2. Comparison of different methods on the NIH pancreas dataset.

Methods	Setting		Evaluations				
	Labeled	Unlabeled	Dice(%)↑	Jaccard(%)↑	95HD(mm)↓	Recall(%)↑	Precision(%)↑
V-Net [14]	6	0	65.33	50.82	14.90	59.29	78.38
DAN [25]	6	56	65.82	50.73	19.87	64.27	71.94
Han [5]			70.87	56.51	21.18	68.38	76.74
MT [18]			69.44	54.69	14.09	65.38	79.12
UA-MT [24]			71.71	56.99	11.18	68.29	78.77
MC-Net [21]			68.76	53.75	26.72	71.47	68.86
MC-Net+ [20]			70.62	56.15	21.97	68.35	75.78
URPC [11]			75.48	61.50	**11.07**	72.40	81.72
Ours			**76.23**	**62.29**	13.80	**73.27**	**82.07**
V-Net [14]	12	0	73.67	59.96	10.53	69.24	83.38
DAN [25]	12	50	77.59	64.13	11.24	79.64	77.88
Han [5]			74.77	60.51	15.09	71.80	81.08
MT [18]			74.95	60.83	13.95	73.16	79.17
UA-MT [24]			76.98	63.22	11.46	77.10	78.02
MC-Net [21]			78.14	64.70	9.52	78.81	78.82
MC-Net+ [20]			79.39	66.28	7.05	78.77	81.39
URPC [11]			80.22	67.34	11.00	**81.77**	79.84
Ours			**80.76**	**68.12**	**6.91**	81.35	**81.41**
V-Net [14]	62	0	81.98	69.90	6.66	86.35	78.89

Compared with UA-MT [24] and MC-Net+ [20], our method shows relatively advanced performance in all indicators. Compared with URPC [11], the proposed method reduces 2.73mm on the 95HD index under 10% labeled data and 0.42% on the Recall index under 20% labeled data. Figure 6 shows 2D and 3D visualizations of different methods. From a qualitative point of view, the segmentation effect of the proposed method has a higher overlap with ground truth. Under the condition of maintaining overall accuracy, our method is much better in terms of details, especially the segmentation boundary of the target region.

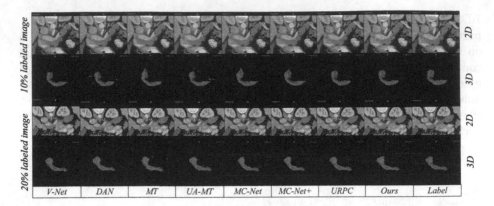

Fig. 6. 2D and 3D segmentation result comparisons of different methods on the NIH pancreas dataset.

5   Conclusion

In this paper, we proposed a semi-supervised pancreas segmentation method based on domain knowledge. By capturing the relationship between prior positions, region-level constraints are applied to guide pseudo label generation. In addition, bidirectional information flow regularization encourages reverse learning and forces the model to align the data distributions. Extensive experiments showed that our method is superior to the latest methods. With the rapid accumulation of unlabeled data, the proposed method has great potential for improving the accuracy and efficiency of segmentation while reducing the cost of labeling. In the future, we will compensate for the shortcomings of several indicators compared with the state-of-the art methods as well as explore some work, including the adjustment of hyperparameters, regional sampling strategies, and network architectures, to provide other perspectives for semi-supervised medical segmentation tasks.

Acknowledgements. This research was supported by the National Natural Science Foundation of China (61976106, 62276116); Jiangsu Six Talent Peak Program (DZXX−122); Jiangsu Graduate Research Innovation Program (KYCX23_3677). Thanks to the open source code from Luo et al.: https://github.com/HiLab-git/SSL4MIS.

References

1. Bai, W., et al.: Semi-supervised Learning for network-based cardiac MR image segmentation. In: Descoteaux, M., Maier-Hein, L., Franz, A., Jannin, P., Collins, D.L., Duchesne, S. (eds.) MICCAI 2017. LNCS, vol. 10434, pp. 253–260. Springer, Cham (2017). https://doi.org/10.1007/978-3-319-66185-8_29

2. Chang, Q., Yan, Z., Lou, Y., Axel, L., Metaxas, D.N.: Soft-label guided semi-supervised learning for bi-ventricle segmentation in cardiac cine MRI. In: 2020 IEEE 17th International Symposium on Biomedical Imaging (ISBI), pp. 1752–1755. IEEE (2020)
3. Duan, J., et al.: Automatic 3D bi-ventricular segmentation of cardiac images by a shape-refined multi-task deep learning approach. IEEE Trans. Med. Imaging **38**(9), 2151–2164 (2019)
4. Enewold, L., et al.: Updated overview of the seer-medicare data: enhanced content and applications. JNCI Monographs **2020**(55), 3–13 (2020)
5. Han, K., et al.: An effective semi-supervised approach for liver CT image segmentation. IEEE J. Biomed. Health Inform. **26**(8), 3999–4007 (2022)
6. Hu, L., et al.: Semi-supervised NPC segmentation with uncertainty and attention guided consistency. Knowl.-Based Syst. **239**, 108021 (2022)
7. Jiao, R., Zhang, Y., Ding, L., Cai, R., Zhang, J.: Learning with limited annotations: a survey on deep semi-supervised learning for medical image segmentation. arXiv preprint arXiv:2207.14191 (2022)
8. Li, K., Zhou, W., Li, H., Anastasio, M.A.: Assessing the impact of deep neural network-based image denoising on binary signal detection tasks. IEEE Trans. Med. Imaging **40**(9), 2295–2305 (2021)
9. Li, X., Yu, L., Chen, H., Fu, C.W., Xing, L., Heng, P.A.: Transformation-consistent self-ensembling model for semisupervised medical image segmentation. IEEE Trans. Neural Netw. Learn. Syst. **32**(2), 523–534 (2020)
10. Liu, T., et al.: Pseudo-3D network for multi-sequence cardiac MR segmentation. In: Pop, M., et al. (eds.) Statistical Atlases and Computational Models of the Heart. Multi-Sequence CMR Segmentation, CRT-EPiggy and LV Full Quantification Challenges: 10th International Workshop, STACOM 2019, Held in Conjunction with MICCAI 2019, Shenzhen, China, October 13, 2019, Revised Selected Papers, pp. 237–245. Springer, Cham (2020). https://doi.org/10.1007/978-3-030-39074-7_25
11. Luo, X., et al.: Semi-supervised medical image segmentation via uncertainty rectified pyramid consistency. Med. Image Anal. **80**, 102517 (2022)
12. Ma, J., et al.: Loss odyssey in medical image segmentation. Med. Image Anal. **71**, 102035 (2021)
13. Masood, S., Sharif, M., Masood, A., Yasmin, M., Raza, M.: A survey on medical image segmentation. Current Med. Imag. **11**(1), 3–14 (2015)
14. Milletari, F., Navab, N., Ahmadi, S.A.: V-net: fully convolutional neural networks for volumetric medical image segmentation. In: 2016 Fourth International Conference on 3D Vision (3DV), pp. 565–571. IEEE (2016)
15. Mizrahi, J.D., Surana, R., Valle, J.W., Shroff, R.T.: Pancreatic cancer. The Lancet **395**(10242), 2008–2020 (2020)
16. Roth, H.R., et al.: DeepOrgan: multi-level deep convolutional networks for automated pancreas segmentation. In: Navab, N., Hornegger, J., Wells, W.M., Frangi, A. (eds.) Medical Image Computing and Computer-Assisted Intervention – MICCAI 2015: 18th International Conference, Munich, Germany, October 5-9, 2015, Proceedings, Part I, pp. 556–564. Springer International Publishing, Cham (2015). https://doi.org/10.1007/978-3-319-24553-9_68
17. Ta, K., Ahn, S.S., Stendahl, J.C., Sinusas, A.J., Duncan, J.S.: A semi-supervised joint network for simultaneous left ventricular motion tracking and segmentation in 4D echocardiography. In: Martel, A.L., et al. (eds.) MICCAI 2020. LNCS, vol. 12266, pp. 468–477. Springer, Cham (2020). https://doi.org/10.1007/978-3-030-59725-2_45

18. Tarvainen, A., Valpola, H.: Mean teachers are better role models: weight-averaged consistency targets improve semi-supervised deep learning results. In: Advances in Neural Information Processing Systems 30 (2017)

19. Wang, K., et al.: Semi-supervised medical image segmentation via a tripled-uncertainty guided mean teacher model with contrastive learning. Med. Image Anal. **79**, 102447 (2022)

20. Wu, Y., et al.: Mutual consistency learning for semi-supervised medical image segmentation. Med. Image Anal. **81**, 102530 (2022)

21. Wu, Y., Xu, M., Ge, Z., Cai, J., Zhang, L.: Semi-supervised left atrium segmentation with mutual consistency training. In: de Bruijne, M., et al. (eds.) Medical Image Computing and Computer Assisted Intervention – MICCAI 2021: 24th International Conference, Strasbourg, France, September 27–October 1, 2021, Proceedings, Part II, pp. 297–306. Springer International Publishing, Cham (2021). https://doi.org/10.1007/978-3-030-87196-3_28

22. Xia, Y., et al.: Uncertainty-aware multi-view co-training for semi-supervised medical image segmentation and domain adaptation. Med. Image Anal. **65**, 101766 (2020)

23. You, C., Zhao, R., Staib, L.H., Duncan, J.S.: Momentum contrastive voxel-wise representation learning for semi-supervised volumetric medical image segmentation. In: Wang, L., Dou, Q., Fletcher, P.T., Speidel, S., Li, S. (eds.) Medical Image Computing and Computer Assisted Intervention – MICCAI 2022: 25th International Conference, Singapore, September 18–22, 2022, Proceedings, Part IV, pp. 639–652. Springer, Cham (2022). https://doi.org/10.1007/978-3-031-16440-8_61

24. Yu, L., Wang, S., Li, X., Fu, C.-W., Heng, P.-A.: Uncertainty-aware self-ensembling model for semi-supervised 3D left atrium segmentation. In: Shen, D., et al. (eds.) MICCAI 2019. LNCS, vol. 11765, pp. 605–613. Springer, Cham (2019). https://doi.org/10.1007/978-3-030-32245-8_67

25. Zhang, Y., Yang, L., Chen, J., Fredericksen, M., Hughes, D.P., Chen, D.Z.: Deep adversarial networks for biomedical image segmentation utilizing unannotated images. In: Descoteaux, M., Maier-Hein, L., Franz, A., Jannin, P., Collins, D.L., Duchesne, S. (eds.) MICCAI 2017. LNCS, vol. 10435, pp. 408–416. Springer, Cham (2017). https://doi.org/10.1007/978-3-319-66179-7_47

26. Zheng, H., et al.: Cartilage segmentation in high-resolution 3D micro-CT images via uncertainty-guided self-training with very sparse annotation. In: Martel, A.L., et al. (eds.) MICCAI 2020. LNCS, vol. 12261, pp. 802–812. Springer, Cham (2020). https://doi.org/10.1007/978-3-030-59710-8_78

27. Zhuang, X.: Multivariate mixture model for myocardial segmentation combining multi-source images. IEEE Trans. Pattern Anal. Mach. Intell. **41**(12), 2933–2946 (2018)

Soybean Genome Clustering Using Quantum-Based Fuzzy C-Means Algorithm

Sai Siddhartha Vivek Dhir Rangoju, Keshav Garg, Rohith Dandi,
Om Prakash Patel$^{(\boxtimes)}$, and Neha Bharill

Department of Computer Science and Engineering, Ecole Centrale School
of Engineering, Mahindra University, Hyderabad, India
{siddhartha20ucse159,keshav20ucse065,rohith20ucse148,omprakash.patel,
neha.bharill}@mahindrauniversity.edu.in

Abstract. Bioinformatics is a new area of research in which many computer scientists are working to extract some useful information from genome sequences in a very less time, whereas traditional methods may take years to fetch this. One of the studies that belongs to the area of Bioinformatics is protein sequence analysis. In this study, we have considered the soybean protein sequence which does not have class information therefore clustering of these sequences is required. As these sequences are very complex and consist of overlapping sequences, therefore Fuzzy C-Means algorithm may work better than crisp clustering. However, the clustering of these sequences is a very time-consuming process also the results are not up to the mark by using existing crisp and fuzzy clustering algorithms. Therefore we propose here a quantum Fuzzy c-Means algorithm that uses the quantum computing concept to represent the dataset in the quantum form. The proposed approach also use the quantum superposition concept which fastens the process and also gives better result than the FCM algorithm.

Keywords: Bioinformatic · Genome Sequence · Soybean · Fuzzy
C-Means · Quantum Computing

1 Introduction

Nowadays, Bioinformatics has emerged as a major field of research among computer scientists. One of the most important studies taken up by computer scientists in the field of Bioinformatics is protein sequence analysis [1]. Proteins are fundamental biological molecules that are responsible for performing a wide range of functions in living organisms. Analysis of the protein sequences helps the researchers in understanding the structure, function, and evolution of proteins. It also involves the identification of patterns and relationships within the protein sequences.

B. Luo et al. (Eds.): ICONIP 2023, LNCS 14450, pp. 83–94, 2024.
https://doi.org/10.1007/978-981-99-8070-3_7

Clustering is one of the most widely used techniques in protein sequence analysis that forms a group of protein sequences based on their similarities which helps in providing valuable insights and relationships between different proteins. The purpose of protein sequence clustering is to organize extensive datasets into significant groups [2]. Several clustering techniques have been applied to protein sequence analysis, including hierarchical clustering, K-means clustering, and Fuzzy C-Means (FCM) clustering. Each of these clustering techniques has its advantages and disadvantages. Many researchers have worked with the K-means clustering algorithm to generate superior-quality sequence clusters [3]. The K-means algorithm measures the distance between data samples with great accuracy. In some cases, where the distance function is not accurately defined, the K-means method may not be able to reveal the sequence-to-structure relationship. Consequently, a significant number of clusters may exhibit inadequate matching of protein sequences. Hence, in some cases, k-means clustering results in poor protein sequence clustering [4]. Thus, the FCM is widely adopted by researchers for protein sequence analysis due to its capability to assign data points to multiple clusters simultaneously [5]. The FCM clustering technique is a widely used clustering algorithm in data analysis. In fuzzy clustering, a single data sample can belong to several clusters with different degrees of membership. The FCM method was introduced by Bezdek [6] and utilizes iterative optimization to minimize an objective function through the use of a similarity measure on the feature space. The advantage of using the FCM can be seen when the clusters are of overlapping nature due to the characteristics of the dataset [7]. The major drawback of the FCM algorithm is that it gets trapped in a problem of local optima.

Recently, nature-inspired metaheuristic approaches have also been employed in clustering applications. Many researchers have worked with metaheuristic-based clustering approaches including ant colony optimization [8], particle swarm optimization (PSO) [9], artificial bee colony algorithm [10], Quantum-inspired optimization approach [11], etc. We have employed the quantum-inspired optimization approach in fuzzy clustering which is a relatively new technique that uses the quantum mechanics principle to cluster data with imprecise boundaries. The objective of our research is to propose a novel quantum-based Fuzzy C-Means clustering approach and to investigate the effectiveness of the proposed approach in the clustering of protein sequences. The effectiveness of the proposed approach is investigated on two well-known cluster validity indexes: the Silhouette index (SI) [12] and the Davies-Bouldin index (DBI) [13] and on several performance measures like NMI, ARI, F-score, respectively.

The rest of the paper is organized as follows: In Sect. 2, we presented the preliminaries related to the proposed work. In Sect. 3, we presented a detailed description of the proposed work. Then in Sect. 4, we presented a detailed discussion of the dataset and the comparative experimental results on various performance measures. Finally, we present the concluding remarks in Sect. 5.

2   Preliminaries

In this section, we presented the preliminaries related to Quantum Fuzzy clustering that includes FCM and quantum computing concepts which are explained briefly subsequently.

2.1   Fuzzy C-Means

One of the most widely used fuzzy clustering methods is the FCM algorithm, proposed by Bezdek [6]. The FCM algorithm attempts to partition a finite collection of n data points $X = \{x_1, x_2, ..., x_n\}$ into c fuzzy clusters with $V = \{v_1, v_2, .., v_c\}$ cluster centroids. The inclusion of data points in a cluster is described by a fuzzy partition matrix $U = [\mu_{ij}]_{n \times c}$ where μ_{ij} defines the degree of membership at which data point x_i belongs to the cluster v_j. The membership degree is allowed to have any values between 0 and 1. The FCM algorithm is based on the minimization of the following objective function, defined as follows:

$$J_m(U, V) = \sum_{i=1}^{c} \sum_{j=1}^{n} (\mu_{ij})^m \|x_j - v_i\|^2, m > 1 \tag{1}$$

where m $(m > 1)$ is a fuzzification parameter that controls the fuzziness of resulting clusters. The FCM algorithm is based on the iterative optimization of the objective function given in Eq. 1 by updating the membership matrix U and cluster centroid V.

2.2   Quantum Computing Concept

The quantum computing concept uses quantum bits (Q_i) for representing the data. In general, the quantum bit (Q_i) is composed of several qubits (q_{ij}) where $j = 1, 2...l$, which is defined as follows:

$$Q_i = (q_{i1}|q_{i2}|......|q_{il}) \tag{2}$$

The classical bit can represent only two possibilities of any event at one time by bit "1" and "0", whereas qubit has the capability to represent a linear superposition of "1" and "0" bits simultaneously using the probability concept [14], which is defined as follows:

$$q_{ij} = \alpha_{ij} \mid 0\rangle + \beta_{ij} \mid 1\rangle = \begin{bmatrix} \alpha \\ \beta \end{bmatrix} \tag{3}$$

where, α and β are the complex numbers representing the probability of qubit in "0" state and "1" state. A probability model is applied here, which represents "0" state by α^2 and "1" state by β^2, defined as follows:

$$\alpha_{ij}^2 + \beta_{ij}^2 = 1; 0 \leq \alpha \leq 1, 0 \leq \beta \leq 1 \tag{4}$$

The quantum bit (Q_i) is formed by a single qubit q_{ij} where $j = 1$ represents two states e.g. "0" or "1" state. Likewise, the quantum bit (Q_i) formed by two-qubits

q_{ij} where $j = 1, 2$ can represent linear superposition with four states i.e. "00", "01", "10", and "11". The quantum bit (Q_i), represents the linear superposition of states due to which it provides better characteristics of population diversity in comparison with other representations [14].

$$Q = \left\langle \begin{array}{c} 1/\sqrt{2}|1/\sqrt{2} \\ 1/\sqrt{2}|1/\sqrt{2} \end{array} \right\rangle \tag{5}$$

$$Q = (1/\sqrt{2} \times 1/\sqrt{2})\langle 00 \rangle + (1/\sqrt{2} \times 1/\sqrt{2})\langle 01 \rangle$$
$$+ (1/\sqrt{2} \times 1/\sqrt{2})\langle 10 \rangle + (1/\sqrt{2} \times 1/\sqrt{2})\langle 11 \rangle \tag{6}$$

A quantum bit (Q) which is made of 3 qubits can represent 2^3, i.e., 8 states as follows:

$$Q = \left\langle \begin{array}{c} 1/\sqrt{2}|1/\sqrt{2}|1/\sqrt{2} \\ 1/\sqrt{2}|1/\sqrt{2}|1/\sqrt{2} \end{array} \right\rangle \tag{7}$$

$$Q = (1/2\sqrt{2})\langle 000 \rangle + (1/2\sqrt{2})\langle 001 \rangle + (1/2\sqrt{2})\langle 010 \rangle + (1/2\sqrt{2})\langle 011 \rangle$$
$$+ (1/2\sqrt{2})\langle 100 \rangle + (1/2\sqrt{2})\langle 101 \rangle + (1/2\sqrt{2})\langle 110 \rangle + (1/2\sqrt{2}))\langle 111 \rangle \tag{8}$$

3   Proposed Work

As discussed in the previous section, the FCM algorithm works on classical data and uses Euclidean distance to measure the similarity between data points with the help of the membership function. In the proposed quantum-based FCM clustering approach, first, we represent the classical data into quantum data. Then we use projections of these vectors and then find the membership of each data point corresponding to clusters.

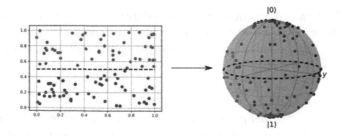

Fig. 1. Classical Data to Quantum Data

3.1   Dataset Preparation

In the proposed Quantum-based FCM clustering approach we need to convert classical data into quantum data in a Hilbert space via a quantum feature map as shown in Fig. 1. It takes a classical data point X and translates it into a quantum state $| X \rangle$.

This process is a crucial part of designing a quantum fuzzy algorithm which helps in reducing computation time and also helps in terms of producing better results. There are different ways to convert classical data into quantum data out of which here we are using the amplitude encoding method to convert classical data into quantum data.

In the amplitude encoding technique, data is encoded in terms of amplitudes of quantum states. A normalized classical N−dimensional data point X is presented by amplitudes of N qubit quantum state. For example, if we have data points as 2 and 3 then it can be represented as $2/\sqrt{(((2)^2 + ((3)^2))}$ and $3/\sqrt{(((2)^2 + ((3)^2))}$ [15]. Once the dataset is converted into quantum data then the proposed Q-FCM algorithm is applied to this data as discussed in the next section.

3.2   Quantum Fuzzy C-Means (QFCM) Clustering Approach

QFCM is an extension of the traditional FCM algorithm that incorporates quantum principles. It aims to cluster data points based on their quantum representations. Firstly dataset X is converted into a quantum data point $| X \rangle = \{| X_1 \rangle, | X_2 \rangle, \ldots, | X_n \rangle \}$. Then we select the number of clusters c, fuzzification parameter m, and termination criteria ϵ as 0.001. We can select the number of clusters (c) in the range $(c_{\min} \leq c \leq c_{\max})$ where $c_{\min} = 2$ and $c_{max} = \sqrt{n}$, where n is the number of data samples in the particular dataset. We have selected different numbers of clusters for different datasets. Once we decided the number of clusters, we select cluster centroids randomly for each cluster (However there are several methods have been proposed to decide cluster centroid but here we have taken as suggested by Bezdek [6]). Let the initial cluster center are $| V \rangle = \{| v_1 \rangle, | v_2 \rangle, \ldots, | v_c \rangle\}$ for a particular cluster. Now we try to find the belongingness of all remaining data points with a particular cluster using the following vector projection rather than Euclidean distance. The distance between two points $| X_i \rangle$ and $| v_j \rangle$ is given by:

$$\text{distance} = \langle X_i.v_j \rangle \tag{9}$$

Using projection, the distance matrix is computed between each quantum data point and the cluster centers. Projection allows for measuring the similarity or dissimilarity between quantum points and cluster centers. The similarity or dissimilarity is calculated based on how much a data point belongs to a particular cluster, this is also represented as the membership degree of the data point with respect to a cluster. The membership degree represents the strength of association between a point and a cluster. Therefore, we calculate the membership matrix corresponding to each data point with respect to each cluster center which is represented as follows:

Compute the cluster membership matrix $\mid U\rangle = \mid \mu_{ij}\rangle$ for $i = 1, 2, \ldots, n$ and $j = 1, 2, \ldots, c$ using Eq. 10:

$$\mid \mu_{ij}\rangle = \frac{\parallel \langle X_i.v_j\rangle \parallel^{\frac{-1}{m-1}}}{\sum_{j=1}^{c} \parallel \langle X_i.v_j\rangle \parallel^{\frac{-1}{m-1}}} \tag{10}$$

The cluster membership matrix $\mid U\rangle$ can be represented as:

$$\mid U\rangle = \begin{bmatrix} \mid u_{11}\rangle & \mid u_{12}\rangle & \ldots & \mid u_{1c}\rangle \\ \mid u_{21}\rangle & \mid u_{22}\rangle & \ldots & \mid u_{2c}\rangle \\ \vdots & \vdots & \ddots & \vdots \\ \mid u_{n1}\rangle & \mid u_{n2}\rangle & \ldots & \mid u_{nc}\rangle \end{bmatrix}$$

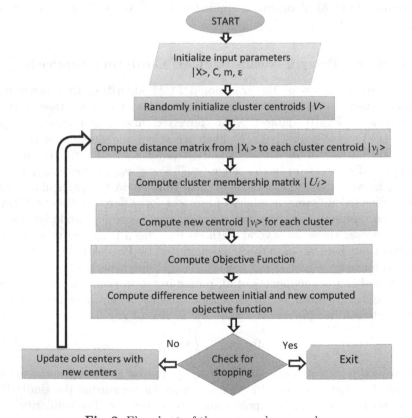

Fig. 2. Flowchart of the proposed approach.

The constraint on the membership matrix U in FCM is given by:

Constraint 1:

$$\sum_{j=1}^{c} \mid u_{ij}\rangle = 1 \quad \forall i = 1, 2, \ldots, n$$

Constraint 2:

$$0 \leq \mid u_{ij}\rangle \leq 1 \quad \forall i = 1, 2, \ldots, n \text{ and } j = 1, 2, \ldots, c$$

Once we get the membership matrix after the first iteration it may or may not be optimal, because we had initialized cluster centers randomly, therefore there is a need of updating cluster centers. The cluster centers are recalculated based on the current memberships of the quantum data points. We compute the updated cluster centroid using the following formulations.

$$\mid v_i \rangle = \frac{\sum_{j=1}^{n} [\mid \mu_{ij}\rangle^m] x_j}{\sum_{j=1}^{n} \mid \mu_{ij}\rangle^m} \tag{11}$$

Once we get the new center, again the distance between quantum data points and the quantum centers is calculated to get the membership matrix. We repeat the process of updating centers until we find no change in cluster center or the difference is very less which is equal to ϵ. The complete above-discussed algorithm is presented in the form of a flowchart as shown in Fig. 2.

4   Experiment and Result

4.1   Experimental Environment

All experiments were carried out on a Dell EMC R720 24-Core(64 threads) HPC Server. The computer was equipped with an Intel(R) Xeon(R) CPU E5-2696 v2 clocked at 2.50 GHz, and had 82 GB of DDR4 RAM. All the codes are written in Python and executed on a version Python version 3.8.5.

4.2   Datasets Description

We have investigated and verified the performance of the proposed (QFCM) approach in comparison with the FCM approach on the following soybean protein datasets. A detailed illustration of these soybean protein datasets used in the experimental investigation is presented in Table 1.

Lee Dataset: The Lee strain, a hybrid between the Chinese lines CNS and S-100 is frequently employed as a parent in several breeding projects in the southern United States and Brazil. The fundamental reason for its widespread usage is its diversified range of desired characteristics, including resistance to certain bacteria from Phytophthora rot, Peanut Mottle Virus, and bacterial pustule [16].

Williams82 Dataset: A soybean cultivar constructed a reference genome sequence known as Williams 82. It was created by reversing the Phytophthora root rot resistance locus from the donor parent kingwa to the recurrent Williams parent [17].

PI483463: The closest wild soybean of Glycine max is called Glycine Soja. Soja is used in breeding projects to identify traits like disease resistance or environmental stress. Glycine swelling accession PI483463 is known to be abnormally tolerable. The genome entry of this specie is sequential and is partly salt tolerance [18].

W05. It is a wild soybean genome that is salt tolerant and primarily designed to serve as a reference genome assembly. Its affiliation is used in various genetic studies of various traits including uncertainty, seed size, number of pods per plant, and seed color [19]. All the protein datasets used in this study are available at the following URL: https://soybase.org/dlpages/.

4.3   Performance Evaluation

Silhouette Index (SI): The Silhouette index is a way to measure how well your data points are clustered. The formula for calculating the Silhouette index is

$$s = (b - a)/max(a, b) \tag{12}$$

where a is the intra-cluster distance and b is the nearest-cluster distance. It ranges from 0-1 to 1. If the value is close to 1 it means good clustering results else the value close to -1 indicates bad clustering results.

Davies-Bouldin Index (DBI): The Davies-Bouldin index is another measure of cluster quality that takes into account both the compactness of each cluster and the separation between clusters. The lower the Davies-Bouldin index, the better the clustering. The formulation of the Davies-Bouldin index is presented as follows:

$$DBI = 1/k * sum(max(R_i + R_j)/d_{ij}), \tag{13}$$

where k is the number of clusters, R_i is the average distance between points within cluster i, R_j is the average distance between points within cluster j, and d_{ij} is the distance between the centroids of clusters i and j.

Normalized Mutual Information (NMI): Normalized mutual information is a measure of similarity between two clusterings of a dataset that takes into

Table 1. Description of Dataset

Parameters	Measures			
	Lee	Williams82	PI483463	W05
Sequences	71358	73320	62102	89477
Size	44 MB	53.4 MB	38.9 MB	50.7 MB

account chance agreement. It ranges from 0 to 1, with higher values indicating greater similarity between the two clusters, the formula of NMI is represented as follows:

$$NMI(X,Y) = [I(X;Y)]/[sqrt(H(X)*H(Y))] \qquad (14)$$

where X and Y are the two sets of data, $I(X;Y)$ is the mutual information between X and Y, and $H(X)$ and $H(Y)$ are the entropy of X and Y, respectively.

Adjusted Rand Index: ARI is the corrected version of the rand index [20] that measures the degree of overlap between two partitions. The Rand index usually suffers from scaling problems therefore ARI is proposed as a corrected version in the form of a formula represented as follows:

$$AdjustedIndex = \frac{Index - ExpectedIndex}{MaxIndex - ExpectedIndex} \qquad (15)$$

F-Score: The F-score helps in evaluating the accuracy of the clustering solution. The F-score for a cluster C_x' with respect to a certain class C_y shows how well a good cluster C_x' describes class C_y by computing the harmonic mean of precision and recall as follows:

$$f(C_x', C_y) = \frac{2 * p_{xy} * r_{xy}}{p_{xy} + r_{xy}} \qquad (16)$$

where p_{xy} represent the precision and r_{xy} represent the recall. The overall f-score is defined as the weighted sum of the maximum f-score for each class C_y, given by the formula defined below:

$$f - score = \sum \frac{n_{xy}}{n} max_x f(C_x', C_y) \qquad (17)$$

where n indicates the total number of samples. The higher value of F-score shows good clustering results.

4.4   Results and Discussion

In this section, we discuss the effectiveness of QFCM in comparison with FCM evaluated on four protein datasets in terms of ARI, NMI, F1-score, SI, and DBI, respectively.

Table 2a highlights the results of the Lee dataset. The FCM algorithm achieved an ARI of 0.0861, NMI of 0.1821, F-score of 0.0105, SI of 0.0182, and DBI of 8.2135. On the other hand, the Quantum-based FCM algorithm obtained an ARI of 0.0461, NMI of 0.1951, F-score of 0.0078, SI of 0.2138, and DBI of 0.8894. The FCM algorithm achieved a SI of 0.0182 and a DBI of 0.82135. On the other hand, the Quantum-based FCM algorithm obtained a higher SI of 0.2138, indicating a better separation between clusters, and a lower DBI of 0.8894, suggesting improved cluster compactness. For the Lee dataset, the time of execution of QFCM and FCM are 173.98 m and 271.36 ms respectively. So we can see here

Table 2. Results of the Proposed (QFCM) and FCM algorithms evaluated on protein datasets on various performance measures.

(a) Results of Lee

Measures	Algorithms	
	FCM	QFCM
ARI	0.0861	0.0461
NMI	0.1821	0.1951
SI	0.0182	0.2138
DBI	0.8213	0.8894
F-score	0.0105	0.0078
Time (m)	271.36	173.98

(b) Results of Williams82

Measures	Algorithms	
	FCM	QFCM
ARI	0.1063	0.0359
NMI	0.1804	0.1956
SI	0.1652	0.2794
DBI	4.3047	1.6469
F-score	0.0032	0.0218
Time (m)	464.4	104.7

(c) Results of PI483463

Measures	Algorithms	
	FCM	QFCM
ARI	0.0407	0.0454
NMI	0.1777	0.1775
SI	0.0509	0.2055
DBI	1.9764	0.9511
F-score	0.0024	0.0064
Time (m)	475.89	162.8

(d) Results of W05

Measures	Algorithms	
	FCM	QFCM
ARI	0.0290	0.0027
NMI	0.0636	0.1107
SI	-0.0522	0.0626
DBI	3.8093	1.963
F-score	0.0025	0.0001
Time (m)	684.7	458.4

that QFCM performs better not only with respect to evaluation parameters but also with respect to time.

Table 2b shows the results of the William82 dataset. The FCM algorithm achieved a SI of 0.1652 and a DBI of 4.3047. In comparison, the Quantum-based FCM algorithm obtained a higher SI of 0.2794, indicating a better separation between clusters, and a lower DBI of 1.6469, suggesting improved cluster compactness. For the William82 dataset, the time of execution of QFCM and FCM is 104.7 m and 464.4 m, respectively. It is clearly shown that QFCM performs 4 times faster than classical FCM and also gives better performance in terms of evaluation parameters discussed here.

In Table, 2c results of the PI483463 dataset are presented. The FCM algorithm obtained SI value of 0.0509 indicating limited cluster distinction, while the Quantum-based FCM algorithm obtained SI value of 0.2055 suggesting better separation and distinctness. The FCM algorithm obtained DBI of 1.9764 suggests some inter-cluster dissimilarity and moderate cluster compactness. In contrast, the Quantum-based FCM algorithm achieved a lower DBI of 0.9511, indicating improved cluster compactness and lower inter-cluster dissimilarity. For PI483463 dataset the time of execution of QFCM and FCM are 162.8 m and 475.89 m, respectively. Again QFCM gives three times faster results than FCM.

Table 2d shows the results of the W05 dataset. The FCM algorithm achieved a negative SI value of -0.0522, indicating poor separation between clusters, and a DBI of 3.8093. In comparison, the Quantum FCM algorithm obtained a positive SI value of 0.0626, suggesting some degree of separation between clusters, and a lower DBI of 1.963, indicating improved cluster compactness. For W05 dataset the time of execution of QFCM and FCM are 458.41 m and 684.7 m, respectively.

Based on the SI and DBI measures, it can be concluded that the Quantum based FCM algorithm outperformed the FCM algorithm in terms of cluster separation and compactness for all the datasets. These results suggest that the Quantum FCM algorithm has the potential to generate more distinct and compact clusters, which can be beneficial for certain clustering tasks. Also, we can observe the time taken by both algorithms, here Q-FCM works much faster than FCM algorithm which is surely beneficial in the case of very large data analysis (big data) problems.

5   Conclusion

Bioinformatics is an area of research where a lot of work still needs to be done for finding new patterns during protein sequence clustering which is very difficult to find through traditional approaches. Here, we are proposing a novel approach to get optimal clusters for Soybean genome sequences. It helps in finding the optimal number of clusters for clustering of genome sequence which can improve crop yield even in extreme weather conditions of drought and rain. To get an optimal number of clusters for genome sequence, we need to cluster genome sequence efficiently and quickly. To get this, we proposed here the Q-FCM algorithm which uses the quantum computing concept to represent classical data into quantum data. It also uses the quantum superposition property to efficiently cluster the dataset. The proposed Q-FCM outperforms with respect to the traditional FCM algorithm in terms of different evaluation parameters and also with respect to time. The proposed work can be further utilized in the future for very large and complex datasets.

References

1. Alawneh, L., Shehab, M.A., Al-Ayyoub, M., Jararweh, Y., Al-Sharif, Z.A.: A scalable multiple pairwise protein sequence alignment acceleration using hybrid CPU-GPU approach. Clust. Comput. **23**, 2677–2688 (2020)
2. de Almeida Paiva, V .: Protein structural bioinformatics: an overview. Comput. Biol. Med., 105695 (2022)
3. Bystrof, C., Thorsson, V., Baker, D.: HMMSTR: a hidden Markov model for local sequence-structure correlations in proteins. J. Mol. Biol. **301**(1), 173–190 (2000)
4. Jha, P., Tiwari, A., Bharill, N., Ratnaparkhe, M., Mounika, M., Nagendra, N.: Apache Spark based kernelized fuzzy clustering framework for single nucleotide polymorphism sequence analysis. Comput. Biol. Chem. **92**, 107454 (2021)

5. Farhangi E., Ghadiri N., Asadi M., Nikbakht M.A., Pitre S.: Fast and scalable protein motif sequence clustering based on Hadoop framework. In: 3th International Conference on Web Research (ICWR), pp. 24–31. IEEE (2017)
6. Bezde J.C.: Fuzzy-Mathematics In Pattern Classification. Cornell University (1973)
7. Pakhira, M.K., Bandyopadhyay, S., Maulik, U.: A study of some fuzzy cluster validity indices, genetic clustering and application to pixel classification. Fuzzy Sets Syst. **155**(2), 191–214 (2005)
8. Shelokar, P.S., Jayaraman, V.K., Kulkarni, B.D.: An ant colony approach for clustering. Anal. Chim. Acta **509**(2), 187–195 (2004)
9. Kao, Y.T., Zahara, E., Kao, I.W.: A hybridized approach to data clustering. Expert Syst. Appli. **34**(3), 1754–1762 (2008)
10. Zhang, C., Ouyang, D., Ning, J.: An artificial bee colony approach for clustering. Expert Syst. Appli. **37**(7), 4761–4767 (2010)
11. Patel O.P., Bharill N., Tiwari A.: A Quantum-inspired fuzzy based evolutionary algorithm for data clustering. In: 2015 IEEE International Conference on Fuzzy Systems (FUZZ-IEEE), Istanbul, Turkey, 2015, pp. 1–8 (2015). https://doi.org/10.1109/FUZZ-IEEE.2015.7337861
12. Bolshakova, N., Azuaje, F.: Cluster validation techniques for genome expression data. Signal Process. **83**(4), 825–833 (2003)
13. Coelho G.P., Barbante C.C., Boccato L., Attux R.R., Oliveira J.R., Von Zuben F.J.: Automatic feature selection for BCI: an analysis using the davies-bouldin index and extreme learning machines. In: The 2012 international joint conference on neural networks (IJCNN), vol. 2012, pp. 1–8. IEEE (2012)
14. Han, H.K., Kim, J.H.: Quantum-inspired evolutionary algorithm for a class of combinatorial optimization. IEEE Trans. Evol. Comput. **6**(6), 580–593 (2002)
15. Peter W.: Quantum machine learning: what quantum computing means to data mining, vol. 2014. Academic Press (2014)
16. Wysmierski, P.T., Vello, N.A.: The genetic base of Brazilian soybean cultivars: evolution over time and breeding implications. Genet. Mol. Biol. **36**, 547–555 (2013)
17. Sedivy, E.J., Wu, F., Hanzawa, Y.: Soybean domestication: the origin, genetic architecture and molecular bases. New Phytol. **214**(2), 539–553 (2017)
18. Lee J.D., Shannon J.G., Vuong T.D., Nguyen H.T.: Inheritance of salt tolerance in wild soybean (Glycine soja Sieb. and Zucc.) accession PI483463. J. Heredity, **100**(6), 798–801 (2009)
19. Xie, M., et al.: A reference-grade wild soybean genome. Nat. Commun. **10**(1), 1216 (2019)
20. Yeung, K.Y., Ruzzo, W.L.: Details of the adjusted rand index and clustering algorithms, supplement to the paper an empirical study on principal component analysis for clustering gene expression data. Bioinformatics **17**(9), 763–774 (2001)

DAMFormer: Enhancing Polyp Segmentation Through Dual Attention Mechanism

Huy Trinh Quang[✉], Mai Nguyen, Quan Nguyen Van, Linh Doan Bao,
Thanh Dang Hong, Thanh Nguyen Tung, and Toan Pham Van

R&D Lab, Sun* Inc., Hanoi, Vietnam
{trinh.quang.huy,nguyen.mai,nguyen.van.quan,doan.bao.linh,
dang.hong.thanh,nguyen.tung.thanh,pham.van.toan}@sun-asterisk.com

Abstract. Polyp segmentation has been a challenging problem for researchers because it does not define a specific shape, color, or size. Traditional deep learning models, based on convolutional neural networks (CNNs), struggle to generalize well on unseen datasets. However, the Transformer architecture has shown promising potential in addressing medical problems by effectively capturing long-range dependencies through self-attention. This paper introduces the DAMFormer model based on Transformer for high accuracy while keeping lightness. The DAMFormer utilizes a Transformer encoder to extract better global information. The Transformer outputs are strategically fed into the ConvBlock and Enhanced Dual Attention Module to effectively capture high-frequency and low-frequency information. These outputs are further processed through the Effective Feature Fusion module to combine global and local features efficiently. In our experiment, five standard benchmark datasets were used Kvasir, CVC-Clinic DB, CVC-ColonDB, CVC-T, and ETIS-Larib.

Keywords: Polyp Segmentation · Transformer

1 Introduction

Colorectal cancer is one of the most commonly diagnosed cancers globally. Early detection and removal of polyps are crucial for the prevention and treatment of colorectal cancer since most of these cancers develop from adenomatous polyps. Colonoscopy is considered the definitive method for identifying and eliminating polyps before progressing into colorectal cancer, establishing it as the gold standard in detection and removal. The process of precisely identifying and segmenting polyps during colonoscopy is challenging due to the wide variation in their shape, size, and texture. This complexity can result in overlooked or misdiagnosed polyps, posing a significant risk to the patient's health. The detection

---

H.T. Quang and M. Nguyen—These authors contributed equally to this work.

© The Author(s), under exclusive license to Springer Nature Singapore Pte Ltd. 2024
B. Luo et al. (Eds.): ICONIP 2023, LNCS 14450, pp. 95–106, 2024.
https://doi.org/10.1007/978-981-99-8070-3_8

of colorectal polyps, along with advancements in endoscopic tools, may also contribute to the advancement of robotic-assisted surgical systems.

The application of machine learning algorithms, specifically CNNs, in medical image segmentation has yielded promising results. These algorithms have been successfully employed for the detection and segmentation of polyps [24]. To enhance the precision and efficiency of polyp segmentation, researchers have introduced several deep-learning architectures employing various techniques. These architectures, including U-Net [27], FCN [20], and their variants like U-Net++ [40], Modified U-Net (mU-Net) [28], ResUNet++ [15], and H-DenseUNet [17], have been developed to address the intricacies of this complex task. Although these methods can deliver accurate segmentation outcomes, their performance might be less consistent when confronted with a diverse array of polyp characteristics.

Recently, there have been notable advancements in computer vision, particularly with the introduction of transformer-based models like ViT [5], Swin-Transformer [19], and Segformer [37]. These models perform significantly better than conventional CNNs models. In this paper, we introduce an improved model to the previous approach. Our model has been equipped with modules to overcome the limitations of the previous model, resulting in significantly improved accuracy while saving computational resources. The main success in our model mainly comes from the components ConvBlock, Enhanced Dual Attention Module, Channel-Wise Scaling, and Effective Feature Fusion.

The contributions of this work can be succinctly summarized as follows:

- We provide a novel model, namely DAMFormer, which can recognize polyps of various sizes due to the powerful information capture capabilities provided by Transformer encoder.
- We designed modules like Enhanced Dual Attention Module to capture global information optimally. Effective Feature Fusion helps the model synthesize information efficiently and amplify or reduce the importance of each channel through the scale factor of Channel-Wise Scaling.
- Our experiments show that our model achieves competitive results compared to other SOTA models while requiring fewer computational resources.

2   Related Work

2.1   Polyp Segmentation

Over the past decade, deep learning has made substantial progress, leading to its practical implementation in various real-world scenarios. Furthermore, it plays an important role in aiding medical diagnosis, specifically in tasks like polyp segmentation. A notable deep-learning model that has demonstrated considerable efficacy in the domain of medical image segmentation is UNet [27], which stands out due to its adoption of a CNN encoder-decoder architecture. Drawing inspiration from UNet, numerous polyp segmentation models have also adopted the encoder-decoder architecture [12,15,40]. These models focus on enhancing the fusion of features at multiple scales in order to achieve improved performance.

PraNet [7], for instance, revolves around a reversed attention mechanism, which facilitates better discrimination between polyps and the surrounding mucosa. In contrast, alternative methods, such as [2,38], combine Transformer-based architectures with CNNs to yield more robust representations for polyps, thus enhancing predictive performance. Moreover, LAPFormer [22] constructs a lightweight CNN decoder atop a Transformer encoder, incorporating appropriate feature connections to achieve a model that is both computationally efficient and capable of producing promising results.

2.2 Attention Mechanism

The concept of attention in the context of machine learning is designed to mimic human cognitive attention. It helps the model to focus on the important part of the input data. Attention mechanisms have proven to be advantageous in numerous visual tasks, such as image classification, object detection, semantic segmentation, and many more. SENet [11] proposed a novel channel-attention network which adaptively enhances important features along channel dimension. ECANet [33] improves SENet [11] channel attention by directly model the correspondence between weight vectors and inputs using 1D Convolution. If channel attention is *what to pay attention to*, then spatial attention is *where to pay attention*. Attention gate [23] focuses on targeted regions while suppresses irrelevant regions using additive attention to obtain the gating signal. Non-local model [34] adopts a self-attention-like mechanism, executing attention across all locations of the input feature maps. This makes it possible for each location to attend to all other locations, capturing their interdependencies. CBAM [36] and BAM [25] utilize the advantages of both channel and spatial attention. Each consists of 2 sub-modules: a channel attention module followed by or parallel with a spatial attention module.

3 Proposed Method

Our network architecture is provided in Fig. 1. The network structure encompasses an encoder and a novelly designed decoder, collectively referred to as DAMFormer. The following sections will provide detailed descriptions of each component.

3.1 Transformer Encoder

MiT (Mix Transformers) [37] are chosen as main backbone. MiT is an encoder based on the hierarchical Transformer architecture that can capture fine-grained details at low resolution and coarse features at high resolution. This capability is of utmost importance in addressing challenges such as polyp segmentation. MHSA (Multi-Head Self-Attention) blocks are often the leading cause of computational bottlenecks. Therefore, MiT addresses this issue by improving the MHSA block into Effective Self-Attention, where the number of attention keys

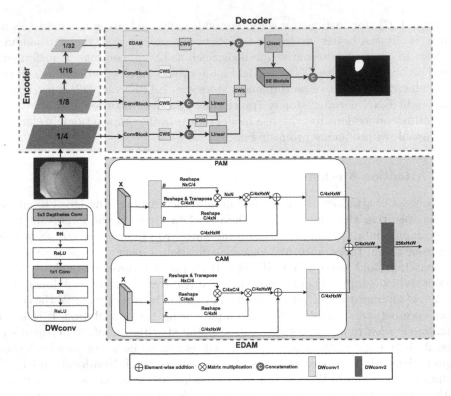

Fig. 1. The architecture of proposed DAMFormer. We chose MiT as the encoder. Features from the first three stages are passed through the ConvBlock block to represent local information better, and EDAM will improve features at the final stage. We propose a Channel-wise scaling method to weight each channel through a learnable scale factor. Then the feature maps will be aggregated using EFF. SE module denotes Squeeze-and-Excitation module.

(K) is reduced to a ratio of R, thereby reducing the computational complexity of self-attention layers. This optimization plays a pivotal role in our selection of the MiT model, as it significantly improves efficiency and alleviates the computational burden associated with MHSA blocks.

3.2   ConvBlock

The encoder's output is commonly passed through lateral connections to aggregate information between stages together. Lateral connections are typically established through the utilization of 1×1 convolution. However, we have observed that this approach may not be optimal [18]. Specifically, the output produced by the first three stages of the transformer encoder consistently tends to capture local information, while the final stage predominantly captures global information. Consequently, employing a uniform 1×1 convolution across all stages fails to fully exploit the transformer encoder's information-rich capabilities.

To address the aforementioned issue, we employed ConvBlock in the first three feature scales, corresponding to $\frac{1}{4}$, $\frac{1}{8}$, and $\frac{1}{16}$ of the total four output stages of the encoder, intending to enhance the representation of local information. ConvBlock consists of a 3×3 convolution layer, a batch normalization layer, and a ReLU activation function.

ConvBlock is represented as a formula:

$$ConvBlock(x) = \sigma(BN(Conv(x)))$$

Where $Conv(.)$ represents 3×3 convolution layer, BN represents batch normalization, and σ is ReLU activation.

3.3 Enhanced Dual Attention Module

One difficulty encountered in Polyps problems is that the foreground and background are often not separate and difficult to distinguish, which requires the model to have a larger Effective receptive field. As demonstrated in [10], it is shown that CNNs exhibit significant ability for texture bias, which is really unhelpful when we want to distinguish boundaries between foreground and background. On the other hand, self-attention again acts as a low-pass filter [26], and it's powerful for shape bias. Furthermore, the output of the final stage typically contains extensive global information. Utilizing a CNN at this stage could lead to an increase in excessive parameters and failing to capitalize on the beneficial information from the last stage. Based on previous research by DANet [9] on Position Attention Module and Channel Attention Module, used to capture long-range contextual information in spatial and channel dimensions, respectively. We have effectively used this technique in our model specifically for the polyps segmentation problem.

CNN has the capability to encode position information through padding [13]. This feature is vital for addressing object positioning challenges like semantic segmentation. Therefore, we have substituted CNN inside the Spatial attention and Channel attention module with Depth-wise Separable Convolution (DWconv). DWconv encodes position information similar to CNN and acts as a low-pass filter [26]. Our method is called EDAM (Enhanced Dual Attention Module), which is illustrated in Fig. 1.

Position Attention Module (PAM): Used to determine the correlation between locations in the feature map with each other. This gives the model the ability to see the entire image, making it easier to identify the area containing the subject. PAM operates on the input feature map $X \in \mathbf{R}^{C \times H \times W}$.

PAM is represented by the following formula:

$$B, C, D = Reshape(DWconv1(X)); \quad B, C, D \in \mathbf{R}^{\frac{C}{4} \times N}$$

$$W_P = Softmax(B \cdot C^T); \quad W_P \in \mathbf{R}^{N \times N}$$

$$A_P = \alpha Reshape(W_P \cdot D) + X; \quad A_P \in \mathbf{R}^{\frac{C}{4} \times H \times W}$$

Where $DWconv1$ produces an output feature map with a size of $\frac{C}{4} \times H \times W$, $N = H \times W$ represents the total number of pixels, and α can be learnable during training initialized as 0.

Channel Attention Module (CAM): Each channel map will capture different semantic information in the image, and there is always an interdependence in the relationship between the channel maps. Therefore, exploiting the relationship between channel maps will help the model be able to represent specific semantic features. Similarly to PAM, CAM takes the feature map $X \in \mathbf{R}^{C \times H \times W}$ as its input.

The following formula represents CAM:

$$O, E, Z = Reshape(DWconv1(X)); \quad O, E, Z \in \mathbf{R}^{\frac{C}{4} \times N}$$

$$W_C = Softmax(O \cdot E^T); \quad W_C \in \mathbf{R}^{\frac{C}{4} \times \frac{C}{4}}$$

$$A_C = \gamma Reshape(W_C \cdot Z) + X; \quad A_C \in \mathbf{R}^{\frac{C}{4} \times H \times W}$$

Where $DWconv1$ gives an output of size $\frac{C}{4} \times H \times W$, $N = H \times W$ represents the total number of pixels, and γ is initialized to 0 as a weight value, which is gradually learned during training.

Finally, the Enhanced Dual Attention Module (EDAM) will be synthesized by performing the sum operation, and the results will continue to be passed through $DWconv2$.

The formula is represented as follows:

$$EDAM(X) = DWconv2(CAM(X) + PAM(X))$$

Where the output feature map generated by $DWconv2$ has dimensions of $256 \times H \times W$.

3.4 Channel-Wise Scaling

The importance of each channel is not uniform. Therefore, amplifying the important channels and minimizing the influence of the unimportant channels is necessary work. This reduces the noise caused by redundant information. SE (Squeeze and Excitation) [11] is an attention mechanism that can solve this problem. However, our objective is to develop an efficient model with a minimal number of parameters. Therefore, we introduce a simple but effective solution while the number of parameters increases insignificantly named Channel-Wise Scaling (CWS) as an alternative to SE module in our problem.

We first set the Scale coefficient $W_S \in \mathbf{R}^{C \times 1 \times 1}$ that can be learned during model training with an initial value of 1, W_S then will be multiplied by the corresponding feature map $X \in \mathbf{R}^{C \times H \times W}$.

The following formula represents it:

$$X_S = X \times W_S; \quad X_S \in \mathbf{R}^{C \times H \times W}$$

3.5   Effective Feature Fusion

LAPFormer [22] has shown that utilizing a Feature pyramid network [18] is unreasonable because of a vast semantic difference between non-adjacent feature scales. Specifically, feature scales $\frac{1}{4}$ and $\frac{1}{32}$ will have a vast difference in semantic correlation, while feature scales $\frac{1}{8}$ and $\frac{1}{16}$ will have a high semantic correlation. Therefore, LAPFormer introduced an approach named Progressive Feature Fusion (PFF) as a potential solution.

However, our experiments indicate that the output of the model depends significantly on the output of the last stage. Meanwhile, the concatenation is progressively from upper scale to lower scale because it will lose a lot of information on high-level features. Therefore, we propose an Effective Feature Fusion (EFF), then the low-level features at the first three stages after Progressive Feature Fusion will be combined with the high-level feature at the last stage to provide more local information without losing global information.

Specifically, our EFF is represented as the following formula:

$$X_{2-i}^{EFF} = \begin{cases} Linear(Cat[x_i, x_{i-1}]) & , i = 2 \\ Linear(Cat[X_{i-1}^{EFF}, x_{i-1}]) & , i = 1 \\ Linear(Cat[X_1^{EFF}, x_3]) & , i = 0 \end{cases}$$

In the equation, i represents the scale level, x_i represents the backbone features obtained from scale i, and x_{i-1} represents the backbone features obtained from scale one level lower than i. X^{EFF} refers to the output after applying a specific process called EFF. The operation $Cat[...]$ denotes the concatenation of the features, and $Linear$ represents a fully-connected layer.

4   Experiments

Dataset and Evaluation Metrics: We conducted experiments on five different datasets for polyp segmentation, namely Kvasir [14], CVC-ClinicDB [1], CVC-ColonDB [30], CVC-T [31], and ETIS-Larib Polyp DB [29]. To ensure consistency, we followed the experimental procedure outlined in PraNet [4] and UACANet [6]. Specifically, we randomly extracted 1450 images from Kvasir and CVC-ClinicDB to construct a training dataset identical to the dataset used in PraNet and UACANet. We then evaluated the performance of our model on the remaining images from Kvasir and CVC-ClinicDB, as well as on the CVC-ColonDB, CVC-T, and ETIS datasets, to assess its generalization capabilities across unseen datasets. For individual dataset evaluation, we used the mean Dice coefficient, while for the combined evaluation of all five datasets, we used the weighted Dice coefficient (wDice). The weight assigned to each dataset was determined based on the number of images it contained, with Kvasir, ClinicDB, ColonDB, CVC-300, and ETIS having weights of $0.1253, 0.0777, 0.4762, 0.0752$, and 0.2456, respectively.

Implementation Details: The implementation of the proposed model is based on PyTorch and utilizes the MMSegmentation [3] toolset. The training process utilizes a single NVIDIA RTX 4090 GPU along with 64GB of RAM. We employ the AdamW optimizer with an initial learning rate of 0.0001. To ensure training stability and reduce variation between runs, we incorporate the SAM [8] optimizer. The input images are resized to a resolution of 352×352 pixels and use multi-scale training with scale values of 0.75, 1.0, and 1.25. Our loss combines weighted IoU loss and weighted focal loss, as in [6]. Our model is trained five times for 50 epochs with a batch size of 8. The reported results are an average of five runs. In particular, no augmentation techniques are employed to demonstrate the true capabilities of the model.

We used the MiT-B1 backbone and train model for 50 epochs averaged over five runs for the ablation study. Table 1 summarizes all of the results.

Table 1. Ablation study on each component

Methods	Kavasir	ClinicDB	ColonDB	CVC-T	ETIS	wDice
LAPFormer	91.1	90.1	78.1	85.9	76.8	80.9
+EDAM	91.4	92.9	78.6	88.2	75.0	81.2 (+0.3)
+CWS	91.4	92.7	79.3	88.3	74.1	81.3 (+0.4)
+EFF	91.8	92.8	79.3	88.2	75.6	**81.7 (+0.8)**

Enhanced Dual Attention Module (EDAM): Implementing the EDAM block demonstrates a notable reduction in parameters count by 5.9% compared to utilizing the ConvBlock. Additionally, its proficiency in capturing global information is substantially enhanced. This improvement is evident in Fig. 2, Where the model detected areas containing polyps even under complex conditions.

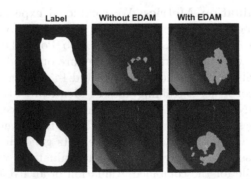

Fig. 2. Impact of EDAM in DAMFormer

Effective Feature Fusion (EFF): By providing more local information without losing global information, the model can accurately identify the entire surface of the polyps. As can be seen in Fig. 3.

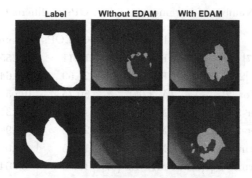

Fig. 3. Impact of EFF in DAMFormer

Our method has been compared to state-of-the-art (SOTA) approaches on five benchmark datasets. The results, presented in Table 2, demonstrate that our model achieves competitive performance while maintaining a relatively lightweight architecture compared to other models. Furthermore, we have compared the number of floating-point operations (FLOPs) and parameters, as depicted in Table 3.

Table 2. Comparison with State-of-the-Arts. Models in a light blue background are proposed in this paper. Our DAMFormer achieves very competitive results while being relatively light.

Methods	Kvasir	ClinicDB	ColonDB	CVC-T	ETIS	wDice
PraNet [7]	89.8	89.9	70.9	87.1	62.8	74.0
Polyp-PVT [4]	91.7	93.7	80.8	90.0	78.7	83.3
SANet [35]	90.4	91.6	75.3	88.8	75.0	79.4
MSNet [39]	90.7	92.1	75.5	86.9	71.9	78.7
TransFuse-L* [38]	92.0	94.2	78.1	89.4	73.7	80.9
CaraNet [21]	91.8	93.6	77.3	90.3	74.7	80.7
SSFormer-S [32]	92.5	91.6	77.2	88.7	76.7	81.0
SSFormer-L [32]	91.7	90.6	80.2	89.5	79.6	83.0
ColonFormer-XS [6]	91.3	92.6	78.4	87.9	75.8	81.2
ColonFormer-S [6]	92.7	93.2	81.1	89.4	78.9	83.6
UACANet-S [16]	90.5	91.6	78.3	90.2	69.4	79.6
UACANet-L [16]	91.2	92.6	75.1	91.0	76.6	80.0
DAMFormer-S (Ours)	91.8	92.8	79.3	88.2	75.6	81.7
DAMFormer-M (Ours)	92.9	93.2	81.7	88.5	78.4	83.7

Table 3. Number of Parameters and FLOPs of different methods

Methods	Backbone	GFLOPs	Params (M)
PraNet [7]	Res2Net [30]	13.11	32.55
CaraNet [21]	Res2Net [30]	21.69	46.64
TransUNet [2]	ViT [5]	60.75	105.5
SSFormer-S [32]	MiT-B2 [37]	17.54	29.31
SSFormer-L [32]	MiT-B4 [37]	28.26	65.96
ColonFormer-XS [6]	MiT-B1 [37]	12.25	22.02
ColonFormer-S [6]	MiT-B2 [37]	16.03	33.04
DAMFormer-S (Ours)	MiT-B1 [37]	8.86	15.04
DAMFormer-M (Ours)	MiT-B2 [37]	11.83	26.08

5 Conclusion

In this research, we propose a unique, lightweight model that retains a high degree of accuracy. The proposed model amalgamates the benefits of both Transformers and CNNs, thereby enabling efficient capture of high-resolution coarse features and low-resolution fine features. This duality enhances its performance in the segmentation tasks. Through various experimental validation, it has been demonstrated that our model produces competitive outcomes in comparison to the current SOTA models. A notable advantage of our model is its reduced demand for computational resources despite delivering comparable results.

References

1. Bernal, J., Sánchez, F.J., Fernández-Esparrach, G., Gil, D., Rodríguez, C., Vilariño, F.: Wm-dova maps for accurate polyp highlighting in colonoscopy: validation vs. saliency maps from physicians. Comput. Med. Imaging Graph. **43**, 99–111 (2015)
2. Chen, J., et al.: Transunet: transformers make strong encoders for medical image segmentation. arXiv preprint arXiv:2102.04306 (2021)
3. Contributors, M.: MMSegmentation: Openmmlab semantic segmentation toolbox and benchmark. https://github.com/open-mmlab/mmsegmentation (2020)
4. Dong, B., Wang, W., Fan, D.P., Li, J., Fu, H., Shao, L.: Polyp-pvt: polyp segmentation with pyramid vision transformers. arXiv preprint arXiv:2108.06932 (2021)
5. Dosovitskiy, A., et al.: An image is worth 16x16 words: transformers for image recognition at scale. arXiv preprint arXiv:2010.11929 (2020)
6. Duc, N.T., Oanh, N.T., Thuy, N.T., Triet, T.M., Dinh, V.S.: Colonformer: an efficient transformer based method for colon polyp segmentation. IEEE Access **10**, 80575–80586 (2022)
7. Fan, D.-P., et al.: PraNet: Parallel Reverse Attention Network for Polyp Segmentation. In: Martel, A.L., et al. (eds.) MICCAI 2020. LNCS, vol. 12266, pp. 263–273. Springer, Cham (2020). https://doi.org/10.1007/978-3-030-59725-2_26

8. Foret, P., Kleiner, A., Mobahi, H., Neyshabur, B.: Sharpness-aware minimization for efficiently improving generalization. arXiv preprint arXiv:2010.01412 (2020)
9. Fu, J., et al.: Dual attention network for scene segmentation. In: Proceedings of the IEEE/CVF Conference on Computer Vision and Pattern Recognition, pp. 3146–3154 (2019)
10. Geirhos, R., Rubisch, P., Michaelis, C., Bethge, M., Wichmann, F.A., Brendel, W.: Imagenet-trained cnns are biased towards texture; increasing shape bias improves accuracy and robustness. arXiv preprint arXiv:1811.12231 (2018)
11. Hu, J., Shen, L., Sun, G.: Squeeze-and-excitation networks. In: Proceedings of the IEEE Conference on Computer Vision and Pattern Recognition, pp. 7132–7141 (2018)
12. Huang, C.H., Wu, H.Y., Lin, Y.L.: Hardnet-mseg: a simple encoder-decoder polyp segmentation neural network that achieves over 0.9 mean dice and 86 fps. arXiv preprint arXiv:2101.07172 (2021)
13. Islam, M.A., Jia, S., Bruce, N.D.: How much position information do convolutional neural networks encode? arXiv preprint arXiv:2001.08248 (2020)
14. Jha, D., et al.: Kvasir-SEG: a segmented polyp dataset. In: Ro, Y.M., et al. (eds.) MMM 2020. LNCS, vol. 11962, pp. 451–462. Springer, Cham (2020). https://doi.org/10.1007/978-3-030-37734-2_37
15. Jha, D., et al.: Resunet++: an advanced architecture for medical image segmentation. In: 2019 IEEE International Symposium on Multimedia (ISM), pp. 225–2255. IEEE (2019)
16. Kim, T., Lee, H., Kim, D.: Uacanet: uncertainty augmented context attention for polyp segmentation. In: Proceedings of the 29th ACM International Conference on Multimedia, pp. 2167–2175 (2021)
17. Li, X., Chen, H., Qi, X., Dou, Q., Fu, C.W., Heng, P.A.: H-denseunet: hybrid densely connected unet for liver and tumor segmentation from ct volumes. IEEE Trans. Med. Imaging 37(12), 2663–2674 (2018)
18. Lin, T.Y., Dollár, P., Girshick, R., He, K., Hariharan, B., Belongie, S.: Feature pyramid networks for object detection. In: Proceedings of the IEEE Conference on Computer Vision and Pattern Recognition, pp. 2117–2125 (2017)
19. Liu, Z., et al.: Swin transformer: hierarchical vision transformer using shifted windows. In: Proceedings of the IEEE/CVF International Conference on Computer Vision, pp. 10012–10022 (2021)
20. Long, J., Shelhamer, E., Darrell, T.: Fully convolutional networks for semantic segmentation. In: Proceedings of the IEEE Conference on Computer Vision and Pattern Recognition, pp. 3431–3440 (2015)
21. Lou, A., Guan, S., Ko, H., Loew, M.H.: Caranet: context axial reverse attention network for segmentation of small medical objects. In: Medical Imaging 2022: Image Processing, vol. 12032, pp. 81–92. SPIE (2022)
22. Nguyen, M., Thanh Bui, T., Van Nguyen, Q., Nguyen, T.T., Van Pham, T.: LAPFormer: A Light and Accurate Polyp Segmentation Transformer https://doi.org/10.48550/arXiv.2210.04393. arXiv e-prints arXiv:2210.04393 (2022)
23. Oktay, O., et al.: Attention u-net: learning where to look for the pancreas. arXiv preprint arXiv:1804.03999 (2018)
24. Pacal, I., Karaboga, D., Basturk, A., Akay, B., Nalbantoglu, U.: A comprehensive review of deep learning in colon cancer. Comput. Biol. Med. 126, 104003 (2020)
25. Park, J., Woo, S., Lee, J.Y., Kweon, I.S.: Bam: bottleneck attention module. arxiv preprint arxiv:1807.06514 (2018)
26. Park, N., Kim, S.: How do vision transformers work? arXiv preprint arXiv:2202.06709 (2022)

27. Ronneberger, O., Fischer, P., Brox, T.: U-Net: convolutional networks for biomedical image segmentation. In: Navab, N., Hornegger, J., Wells, W.M., Frangi, A.F. (eds.) MICCAI 2015. LNCS, vol. 9351, pp. 234–241. Springer, Cham (2015). https://doi.org/10.1007/978-3-319-24574-4_28

28. Seo, H., Huang, C., Bassenne, M., Xiao, R., Xing, L.: Modified u-net (mu-net) with incorporation of object-dependent high level features for improved liver and liver-tumor segmentation in ct images. IEEE Trans. Med. Imaging **39**(5), 1316–1325 (2019)

29. Silva, J., Histace, A., Romain, O., Dray, X., Granado, B.: Toward embedded detection of polyps in wce images for early diagnosis of colorectal cancer. Int. J. Comput. Assist. Radiol. Surg. **9**(2), 283–293 (2014)

30. Tajbakhsh, N., Gurudu, S.R., Liang, J.: Automated polyp detection in colonoscopy videos using shape and context information. IEEE Trans. Med. Imaging **35**(2), 630–644 (2015)

31. Vázquez, D., et al.: A benchmark for endoluminal scene segmentation of colonoscopy images. J. Healthcare Eng. **2017** (2017)

32. Wang, J., Huang, Q., Tang, F., Meng, J., Su, J., Song, S.: Stepwise feature fusion: local guides global. In: Medical Image Computing and Computer Assisted Intervention-MICCAI 2022: 25th International Conference, Singapore, 18–22 September 2022, Proceedings, Part III, pp. 110–120. Springer (2022). https://doi.org/10.1007/978-3-031-16437-8_11

33. Wang, Q., Wu, B., Zhu, P., Li, P., Zuo, W., Hu, Q.: Eca-net: efficient channel attention for deep convolutional neural networks. In: Proceedings of the IEEE/CVF Conference on Computer Vision and Pattern Recognition, pp. 11534–11542 (2020)

34. Wang, X., Girshick, R., Gupta, A., He, K.: Non-local neural networks. In: Proceedings of the IEEE Conference on Computer Vision and Pattern Recognition, pp. 7794–7803 (2018)

35. Wei, J., Hu, Y., Zhang, R., Li, Z., Zhou, S.K., Cui, S.: Shallow attention network for polyp segmentation. In: de Bruijne, M., et al. (eds.) MICCAI 2021. LNCS, vol. 12901, pp. 699–708. Springer, Cham (2021). https://doi.org/10.1007/978-3-030-87193-2_66

36. Woo, S., Park, J., Lee, J.-Y., Kweon, I.S.: CBAM: convolutional block attention module. In: Ferrari, V., Hebert, M., Sminchisescu, C., Weiss, Y. (eds.) ECCV 2018. LNCS, vol. 11211, pp. 3–19. Springer, Cham (2018). https://doi.org/10.1007/978-3-030-01234-2_1

37. Xie, E., Wang, W., Yu, Z., Anandkumar, A., Alvarez, J.M., Luo, P.: Segformer: simple and efficient design for semantic segmentation with transformers. Adv. Neural. Inf. Process. Syst. **34**, 12077–12090 (2021)

38. Zhang, Y., Liu, H., Hu, Q.: TransFuse: fusing transformers and CNNs for medical image segmentation. In: de Bruijne, M., et al. (eds.) MICCAI 2021. LNCS, vol. 12901, pp. 14–24. Springer, Cham (2021). https://doi.org/10.1007/978-3-030-87193-2_2

39. Zhao, X., Zhang, L., Lu, H.: Automatic polyp segmentation via multi-scale subtraction network. In: de Bruijne, M., et al. (eds.) MICCAI 2021. LNCS, vol. 12901, pp. 120–130. Springer, Cham (2021). https://doi.org/10.1007/978-3-030-87193-2_12

40. Zhou, Z., Rahman Siddiquee, M.M., Tajbakhsh, N., Liang, J.: UNet++: A Nested U-Net architecture for medical image segmentation. In: Stoyanov, D., et al. (eds.) DLMIA/ML-CDS -2018. LNCS, vol. 11045, pp. 3–11. Springer, Cham (2018). https://doi.org/10.1007/978-3-030-00889-5_1

BIN: A Bio-Signature Identification Network for Interpretable Liver Cancer Microvascular Invasion Prediction Based on Multi-modal MRIs

Pengyu Zheng, Bo Li, Huilin Lai, and Ye Luo[✉]

School of Software Engineering, Tongji University, Shanghai 201804, China
yeluo@tongji.edu.cn

Abstract. Microvascular invasion (MVI) is a critical factor that affects the postoperative cure of hepatocellular carcinoma (HCC). Precise pre-operative diagnosis of MVI by magnetic resonance imaging (MRI) is crucial for effective treatment of HCC. Compared with traditional methods, deep learning-based MVI diagnostic models have shown significant improvements. However, the black-box nature of deep learning models poses a challenge to their acceptance in medical fields that demand interpretability. To address this issue, this paper proposes an interpretable deep learning model, called Biosignature Identification Network (BIN) based on multi-modal MRI images for the liver cancer MVI prediction task. Inspired by the biological ways to distinguish the species through the biosignatures, our proposed BIN method classifies patients into MVI absence (i.e., Non-MVI or negative) and MVI presence (i.e., positive) by utilizing Non-MVI and MVI biosignatures. The adoption of a transparent decision-making process in BIN ensures interpretability, while the proposed biosignatures overcome the limitations associated with the manual feature extraction. Moreover, a multi-modal MRI based BIN method is also explored to further enhance the diagnostic performance with an attempt to interpretability of multi-modal MRI fusion. Through extensive experiments on the real dataset, it was found that BIN maintains deep model-level performance while providing effective interpretability. Overall, the proposed model offers a promising solution to the challenge of interpreting deep learning-based MVI diagnostic models.

Keywords: Interpretability · Multi-modal MRI · Microvascular Invasion · Liver Cancer

1 Introduction

Hepatocellular carcinoma (HCC) is a major cause of cancer-related deaths worldwide, ranking third in terms of mortality [7]. The grave prognosis of HCC is attributed to its high rate of recurrence after surgical intervention [6]. Microvascular invasion (MVI) is an independent risk factor for HCC recurrence, and

B. Luo et al. (Eds.): ICONIP 2023, LNCS 14450, pp. 107–119, 2024.
https://doi.org/10.1007/978-981-99-8070-3_9

patients diagnosed with MVI exhibit a 4.4-fold increased risk of postoperative recurrence [4]. Accurate diagnosis of MVI before surgery can aid clinicians in selecting optimal treatment strategies and significantly reduce the likelihood of postoperative recurrence in HCC patients. Unfortunately, current diagnostic methods for MVI involve manually counting tumor cells in postoperative biopsy pathological images, which cannot achieve the desired goal of preoperative diagnosis. Therefore, non-invasive technology capable of diagnosing MVI preoperatively would be highly advantageous in clinical settings. Magnetic resonance imaging (MRI) is a promising technique that offers numerous advantages, such as clarity, ease of acquisition, and non-invasiveness, making it a valuable tool for predicting preoperative MVI.

Numerous studies have investigated the potential of MRI images for MVI detection, and have yielded promising initial results. Existing works [12,17] have established a link between various manually extracted biological characteristics and MVI through statistical analysis. Independent factors associated with MVI include enhancement around the tumor, tumor margin, tumor size, etc. However, traditional methods based on statistical analysis have been found to be inadequate in achieving ideal diagnostic performance due to two main reasons. Firstly, the specific causes of MVI have not yet been clearly identified [15], and most existing traditional methods rely on manually extracted biological characteristics. Secondly, the biological characteristics of MVI patients are diverse and result from complex interactions between multiple factors. Therefore, it is imperative to conduct joint research on these characteristics. Deep learning methods offer potential solutions to the shortcomings of statistical analysis and achieved better results [14,19]. Nonetheless, the black box nature of deep learning models limits their interpretability, which is critical in medical fields. Although some previous explainable models exist [18,20], they predominantly fall under post-hoc interpretable models [8], offering limited interpretability, and the interpretation results may be questionable [11].

To address the aforementioned issues, this paper introduces BIN, a transparency interpretable deep model inspired by the biosignature classification methods [9] in biology. It is well known that we usually distinguish species by some key biosignatures (e.g. Wings, beaks and other characteristics are used with various weights to recognize different kinds of birds.). Inspired by this, in this paper, we refer to the unique features of MVI-positive and MVI-negative cases as MVI and Non-MVI biosignatures, respectively. After obtaining two groups of globally learned MVI/Non-MVI biosignatures, BIN measures the similarity between the input MRI image and these MVI and Non-MVI biosignatures respectively, and then the summations of the weighted similarities are as the determinant for the final prediction result. Here, the weight is to identify the importance of each biosignature to the MVI/Non-MVI prediction, and is accordingly learned by the proposed BIN network. As depicted in Fig. 1, given a patient's MRI and the learned MVI/Non-MVI biosignatures (only three of them list for clarity), pixel locations with large similarities to those biosignatures are exhibited on the MRI image, and for each biosignature its similarity to the MRI and its weight are list around it too. The final prediction result is obtained by comparing the

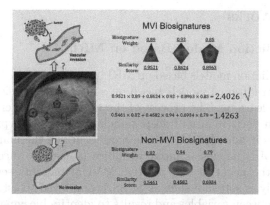

Fig. 1. Illustration of the main idea to predict the liver cancer MVI status by the proposed MVI biosignatures based on the MRI images. Given a testing sample (a MRI image) and the learned MVI/Non-MVI biosignatures (upper/lower half), its MVI status (MVI or Non-MVI, animation figure from [13].) is determined by its similarity to the MVI/Non-MVI biosignatures and also the importance (i.e. weight) of the biosignature to the decision process. The labels on the MRI image show the locations with large similarities to different biosignatures, and for each biosignature its similarity to this MRI as well as its weight are list. It can be observed that the weighted sum of the positive category is greater than that of the negative category.

weighted summations for different MVI classes. By this way, unlike groundless creation, these biosignatures derived from MRI images of patients make our BIN interpretable. Moreover, benefiting from deep models' learning, the MVI biosignatures in BIN are not restricted by any prior knowledge. Last but not least, considering the complex decision process of MVI by MRI images, multi-modal MRI images are as inputs of BIN in the later introduction of this paper, and we implement a multi-modal fusion based interpret-able model. To the best of our knowledge, there is currently no existing research on transparent interpretable models specifically designed for multi-modal fusion especially for medical image analysis. Extensive qualitative and quantitative experiments confirm the effectiveness of BIN. The contributions of this paper can be summarized as follows:

- This paper proposes Biosignature Identification Network (BIN) for liver cancer microvascular invasion prediction, which offers the interpretability while preserving the prediction performance of deep learning models.
- BIN is a model that comprehensively processes multiple modality MRI images, making an exploratory contribution to the interpretability of the multi-modal fusion method.
- Extensive experiments conducted on the reality clinical MVI patient dataset have confirmed the predictive performance and interpretability of BIN.

2   Related Works

2.1   MVI Prediction Models Based on MRIs

Current MVI prediction models based on MRI images can be broadly categorized into two groups: traditional models based on statistical analysis and deep models based on deep learning. The process of the traditional models is transparent, but the prediction performance of deep models is superior. Typical examples of traditional models include [12,17]. These models utilize the manually extracted biosignatures in MRI images as input variables (some also incorporate clinical data [17]), with the negative and positive MVI of patients serving as output results. Data analysis software such as SPSS is employed to establish a statistical relationship between variables and results to identify biosignatures associated with MVI. Although these methods can derive statistical conclusions, they are limited by research methods, leading to a lack of correlation research between biosignatures. Additionally, the input biosignatures must be manually extracted, restricting the study objects to known biosignatures. However, since the mechanism of MVI has yet to be determined [15], the statistical analysis method is ineffective in exploring unknown biosignatures, thereby limiting the prediction performance of statistical analysis methods.

Deep models such as [14,19] can automatically extract features, thus overcoming the limitations of manual extraction and achieving better prediction performance. However, deep models are typically black box models, making it challenging to comprehend the extracted feature content and the decision-making process. Consequently, although deep models offer stronger prediction performance, they are challenging to implement in medical fields that demand reliable interpretation. BIN, being a deep model capable of providing interpretability, undoubtedly offers a highly valuable solution for the MVI prediction task.

2.2   MVI Interpretable Deep Models

Till now, there have been a few methods proposed to interpret and classify the MVI simultaneously, including [18,20]. Among them, [18] use attribution interpretation of the predicted results, which is considered post-hoc interpretable [8]. Post-hoc interpretability only proves the connection between input and output and lacks the model's interpretation itself. TED [20] incorporates decision trees as the final classifier, but it still relies on automatically extracted unknown features. Unlike these models, BIN uses biosignature methods employed in biology for decision-making. Both biosignatures and decisions are interpretable, with a higher degree of interpretability, making BIN a transparency interpretable model. Loosely speaking, BIN can be categorized as a form of case-based learning or prototype-based learning. In addition, BIN represents a preliminary exploration into the interpretability of the multi-modal fusion based methods.

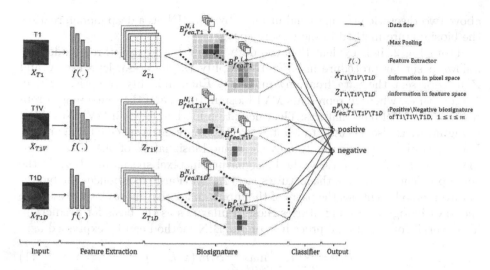

Fig. 2. The network structure of the proposed multi-modal fusion based BIN method. Given three modalities MRI images (e.g. T1, T1V, T1D) of a patient, the feature maps of them are learned by three feature extractors which have the same network architecture, and then their similarities to the biosignatures from every modality are calculated. The classifier applies a fully connected network to achieve modal fusion at the decision-making level.

3   The Proposed Multi-modal Fusion Based BIN Method

Considering that MVI diagnosis based on MRI images is a very challenging problem, we implement a multi-modal fusion based BIN method on T1, T1D, and T1V. The detailed network structure is shown in Fig. 2. Given the biosignatures (including the MVI and Non-MVI biosignatures) learned from T1, T1D, and T1V, BIN calculates the similarity between the feature map of MRI images and each biosignature in the feature space. The different contributions of different biosignatures to the classification result are reflected by the weights of the fully connected layer. The final prediction result is determined by the product of the calculated similarities and the corresponding contributions. The biosignatures employed in BIN are global features obtained through learning, and interpretability is achieved through the center shift to these global features. For the multi-modal model, to maintain the interpretability from end-to-end, we implement the multi-modal fusion strategy at the decision-making level.

Main Architecture of BIN. The deep model can be viewed as a complex mapping function represented by $f(.)$. The deep model is responsible for mapping information from $X \in R^{3 \times h \times w}$ in the pixel space to $Z \in R^{d \times h' \times w'}$ in the feature space through the function $f(.)$. That is $Z = f(X)$. Convolution possesses the property of preserving the relative spatial invariance. Based on the

above two theoretic principles and inspired by [2], BIN, as a deep model, realizes
the biosignature method in the feature space.

From Fig. 2, it is evident that the data processing for the three modalities
follows a similar procedure in BIN. To simplify the notation, let's use X and
Z to represent the data from any modality. Each modality comprises $2 \times m$
biosignatures, where m represents MVI biosignatures, and the remaining m represents Non-MVI biosignatures. The calculation method for each MVI/Non-MVI
biosignature is also consistent, denoted by B to represent a specific biosignature.
For illustration convenience, we only show one biosignature of one modality for
MVI here. Let B_{pix} represent the biosignature in pixel space, and B_{fea} is the
correspondence of B_{pix} in the feature space. As previously mentioned, the biosignature method identifies the parts of the input image that most closely resemble
the MVI biosignatures and utilizes these similarities as the basis for prediction.
The overall computational procedure of our BIN method can be expressed as:

$$Pre = \sum \alpha \max_{z=patch(Z)} Sim\left(z, B_{fea}\right) \tag{1}$$

where Pre represents the output prediction result for MVI positive or
MVI negative, α corresponds to the weight of each biosignature, $z =
patch\left(Z\right)\left(z \in R^{d \times 1 \times 1}\right)$ means that z is the patch of Z, and $Sim\left(\right)$ is the sim-
ilarity calculation function. Since z and B_{fea} are tensors in the feature space,
BIN uses the commonly used L2 distance as the similarity calculation function,
$Sim\left(z, B_{fea}\right) = \log\left(1 + 1 / \left(\|z - B_{fea}\|_2^2 + \epsilon\right)\right)$. In BIN, the maximum similarity between Z and $\hat{B}_{fea}$ is taken as the similarity metric. The contribution of
each biosignature to the result is expressed as the weight α of the fully connected
layer.

Selection of Biosignatures. BIN extract biosignatures automatically, which
leads to better prediction performance and model generalization than traditional
models. BIN achieves automatic extraction of interpretable MVI biosignatures
by using a two-stage processing approach.

In the first stage, BIN initializes a tensor of the same size as B_{fea} and defines
it as $\hat{B}_{fea}$. Similarly to convolutional kernels, $\hat{B}_{fea}$ is learnable and updated
through gradient descent during the training process. Through the optimization
algorithm of the deep model, $\hat{B}_{fea}$ is guided to acquire highly similar content to
the MRI images of MVI-positive patients, analogous to a common characteristic
of all MVI patients. Therefore, during the training phase, center loss is increased
to accelerate the learning of $\hat{B}_{fea}$.

In the second stage, BIN uses center offset operations to achieve interpretability. The calculation formula for center offset is shown here:

$$B_{fea} \leftarrow arg \min_z \left\| \hat{B}_{fea} - z \right\|_2^2 \tag{2}$$

where z is the closest and most similar feature to $\hat{B}_{fea}$. z is the mapping of MRI
information in the feature space, corresponding to x in the pixel space, therefore

B_{fea} has interpretability after the center offset, that is $B_{fea} = z$. The center offset operation may cause some performance loss, but the effectiveness of this method has been theoretically proven [2].

As mentioned earlier, B_{fea} and B_{pix} have the same relative spatial position. So after spatial alignment through upsampling, the spatial position of B_{pix} in X corresponds to B_{fea} in Z, which can also be referred to as the visualization of B_{fea}, as shown in Fig. 3.

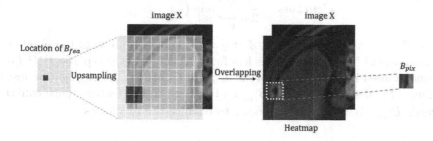

Fig. 3. The figure on the left shows the location of B_{fea}. Aligning feature space and pixel space through upsampling. The third image is the heatmap obtained by overlapping, which shows the location of B_{pix}. On the right is the B_{pix} cropped from X, which can also be considered as the visualization of B_{fea}.

Multi-modal Fusion. Like [16], BIN combines T1, T1D and T1V MRI images to boost the performance. Multi-modal fusion methods can be divided into three main categories: early fusion, feature fusion, and decision fusion [1]. Both early fusion and feature fusion methods involve integrating information from multiple modalities into Z, which disrupt the single mapping relationship between X and Z. As a consequence, Z loses its interpretability. Since $z = patch\,(Z)$ lacks interpretability, even after the center shift operation, B_{fea} loses its interpretability. To maintain the interpretability of biosignatures, BIN uses decision fusion methods, as shown in Fig. 2. Each modal data is not fused before similarity calculation but only fused at the final fully connected layer, which is the decision-making layer, to determine the final prediction result.

Training Strategy. During the training process, there is prior knowledge that MVI biosignatures should have a primary positive impact on positive prediction, while Non-MVI biosignatures should not affect the judgment of positive results. This is because the biosignature method determines the result only based on similarity instead of dissimilarity. To achieve this goal, BIN uses prior knowledge during training. Specifically, the weight of the last fully connected layer is set to 0/1, where the weight of MVI features (with $y = 1$ as the label of the MVI positive) and positive classification results is set to 1, and the weight of negative classification results is set to 0. Non-MVI features(with $y = 0$ as the label of the MVI negative) are mirrored. First, fully connected layer weights are fixed, and only train features and deep networks. After training to some extent, the deep network and features are fixed, and only the fully connected layer is trained to learn the importance of biosignatures in predicting results. The training is completed by alternating between these two steps. The Loss formula used in training is as follows:

$$Loss = BCELoss + \lambda_1 CentLoss, \tag{3}$$

$$CentLoss = \frac{1}{2m} \sum_{i=1}^{2m} \min_{z \in P} (\|z - B_{fea}^i\|), \tag{4}$$

where $P = \{z| \forall z = patch\,(Z), \forall Z\ if\ y = 1\}$, and m represents the number of MVI/Non-MVI biosignatures. The loss function comprises two parts: $BCELoss$, commonly used in classification problems, and $CentLoss$, an aggregation loss to minimize the distance between B_{fea}^i and z to guide the feature space converge towards B_{fea}^i. λ_1 is a hyperparameter to balance the two losses.

4   Experiment and Analysis

Dataset. The proposed model was evaluated on a dataset of 160 HCC patients collected by Zhongshan Hospital in 2015, with 110 patients diagnosed with MVI through pathological imaging. Each patient had three MRI images for each of the three modalities (T1, T1V, T1D), resulting in 1440 MRI images. To verify the generalization performance of our proposed method, we conducted 5-fold cross-validation on the dataset.

Experimental Details. In the experiment, Adam optimizer was selected. The learning rate was set to 1e-4, and λ_1 of the loss was set to 0.8. For each modality, the number of MVI biosignatures and Non-MVI biosignatures was set to 10. All experiments were conducted on a 1080ti graphics card.

Evaluation Metric. To evaluate the performance of our proposed method, we used several widely-used metrics for medical classification tasks, including AUC, Accuracy, Sensitivity, Precision, and F1 score. These metrics enable us to comprehensively evaluate the model's ability.

4.1   Performance Comparisons

Transparent interpretable models with high interpretability typically result in a loss of model performance. The quantitative experiments aim to verify whether the BIN model can improve interpretability while maintaining deep model performance. The baseline in the experiment erases the calculation process of biosignatures in BIN, directly pools Z and unfolds them using fully connected layers and predictive outputs. The network structure of the baseline is very similar to

Table 1. Compares various evaluation indicators using interpretable networks BIN and black box deep networks as baselines, with red values indicating increased indicator values and green values representing decreased ones.

backbone		AUC	Accuracy	Sensitivity	Precision	F1 Score
	Sun et al [16].	95	92.11	83	100	90.91
ResNet18	baseline	92.27	84.93	85.93	94.29	89.88
	BIN	93.77(+1.5)	87.54(+2.61)	91.11(+5.18)	92.93(-1.36)	91.92(+2.04)
ResNet34	baseline	89.24	81.74	82.96	93.17	87.61
	BIN	92.25(+3.01)	86.96(+5.22)	87.78(+4.82)	95.39(+2.22)	91.3(+3.69)
VGG11	baseline	91.85	85.22	88.89	92.02	90.4
	BIN	90.7(-1.15)	81.74(-3.48)	97.78(+8.89)	82.31(-9.71)	89.35(-1.05)
VGG13	baseline	87.44	81.45	82.96	92.65	87.39
	BIN	92.16(+4.72)	85.8(+4.35)	89.63(+6.67)	92.17(-0.48)	90.8(+3.41)
VGG16	baseline	89.11	83.19	82.22	95.74	88.43
	BIN	90.64(+1.53)	80(-3.19)	87.04(+4.82)	87.56(-8.18)	87.23(-1.2)
VGG19	baseline	88.2	80.29	77.78	96.65	85.65
	BIN	89.3(+1.1)	83.77(+3.48)	90(+12.22)	89.46(-7.19)	89.62(+3.97)

the classification network corresponding to the backbone. To verify universality, scientific controls were performed under several typical deep convolutional networks used as backbones. We also compared with a latest multi-modal fusion deep model that applied feature fusion [16].

The experimental results are shown in Table 1. It can be seen that under commonly used ResNet and VGG as the backbone settings, the proposed model not only did not show a decrease in performance on most indicators but also demonstrated better results, with an increase of 2 to 5% points. There was a significant decrease in Precision, which may be related to BIN's less attention to Non-MVI. When comparing BIN to [16], both of which utilize ResNet18 as the backbone, BIN demonstrates either higher or lower scores across the five evaluation indicators. The utilization of decision fusion in BIN, as opposed to feature fusion like [16], could potentially account for the decrease in accuracy. However, when evaluating sensitivity, BIN outperforms [16] significantly. It is worth noting that sensitivity holds paramount importance as an evaluation indicator within the medical field. In addition, BIN offers transparent interpretability, allowing for a better understanding of its decision-making processes, whereas [16] remains a black box. The quantitative experimental results demonstrate that BIN maintains the performance of the deep model while offering high interpretability.

4.2 Qualitative Experiment

Interpretability of MVI Classification. A qualitative experiment was designed to evaluate the interpretability of BIN, using the biomarker method. In Fig. 4, we present the visualization results of several easily observable biosignatures: incomplete capsule, unsmooth edge and tumor size, which align with the research results attained through traditional models [3,5,10]. Figure 4 also shows the specific features that BIN emphasizes, but not been validated by traditional methods. These findings present promising avenues for further exploration in medical research.

Interpretability of Multi-modal Fusion. We performed qualitative experiments to assess the significance of multi-modal fusion in BIN. The outcomes of these experiments are presented in Fig. 5. We computed the similarity of biosignatures from three modalities separately for the test cases and determined the contribution values of MVI biosignatures and Non-MVI biosignatures from each modality towards the final prediction results. Figure 5 illustrates a specific test case where the T1V modality plays a pivotal role. It can be observed that all three modalities contribute to the prediction results, but one of them assumes a decisive role. Furthermore, the modalities that exhibit crucial roles vary across different test cases. On the right side of Fig. 5, the pie chart showcases the

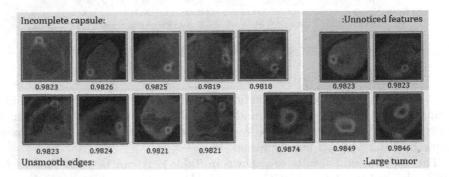

Fig. 4. The visualization results of our biosignatures (i.e. heat maps) overlapped on the tumor regions with the hot color indicating the large value. High correlations are found between the incomplete capsule, large tumor size, unsmoothed tumor contour, etc. and our visulized biosignatures. The numerical value displayed below each visual biosignatures indicates the degree of impact on the predicted results. Blue, green and red rectangles are corresponding to T1, T1V and T1D modality. (Color figure online)

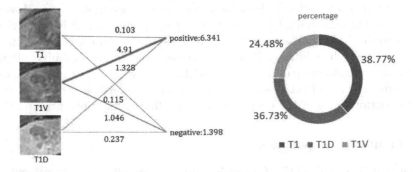

Fig. 5. Illustration of our method's interpretability onto the importance of different MRI modalities to the MVI prediction. Left figure shows a patient was diagnosed into a MVI positive, because among the three modalities, biosignatures from T1V are the most important clues. While the right figure shows the percentage of each modality which is the most important clue to do the diagnosis, and we can see that T1 and T1D have relatively more importance to T1V on our dataset.

proportion of modal importance in all test cases. Notably, on the test dataset, the T1 and T1D modal data are more influential than the T1 modality.

Effect of Different Number of Biosignatures. The paper conducted ablation experiments on the number of biosignatures. Figure 6 displays the results of the ablation experiment on the number of biosignatures. We conducted ablation experiments on the number of biosignatures to determine their impact on the model's performance. The results of the ablation experiments are presented in Fig. 6. We observed that the best overall performance was achieved when the number of biosignatures was set to 10. When the number of biosignatures was less than 10 or greater than 10, there was a certain degree of performance degradation. This may be due to the fact that when the number of biosignatures is too small, it is difficult to establish a set of all MVI biosignatures. When the number of biosignatures is too large, it will be affected by noise and affect the performance of the BIN.

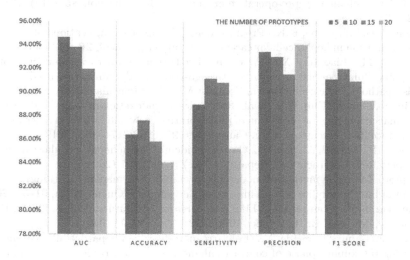

Fig. 6. Performance comparisons by different numbers of biosignatures on our collected dataset. The results indicate that the best overall performance was achieved when the number of biosignatures is set to 10.

5   Conclusion

Accurate and reliable preoperative prediction of Microvascular Invasion (MVI) is a crucial matter that requires immediate attention for patients with Hepatocellular Carcinoma (HCC). In this article, we propose BIN, a multi-modal deep model with transparent interpretability, as a potential solution. Inspired by biosignature method, BIN enables transparent decision-making by comparing biosignatures. Through comprehensive experiments, the MVI prediction performance and interpretability of BIN have been thoroughly validated. Muli-modal and interpretability pose significant challenges in the application of deep learning within the medical field and BIN just represents a preliminary exploratory research.

Acknowledgements. This work was partially supported by the General Program of National Natural Science Foundation of China (NSFC) under Grant 62276189, and the Fundamental Research Funds for the Central Universities No. 22120220583.

References

1. Atrey, P.K., Hossain, M.A., El Saddik, A., et al.: Multimodal fusion for multimedia analysis: a survey. Multimedia Syst. **16**, 345–379 (2010)
2. Chen, C., Li, O., Tao, D., et al.: This looks like that: deep learning for interpretable image recognition. In: Advances in Neural Information Processing Systems 32 (2019)
3. Chou, C., Chen, R., Lee, C., et al.: Prediction of microvascular invasion of hepatocellular carcinoma by pre-operative ct imaging. Br. J. Radiol. **85**(1014), 778–783 (2012)
4. Erstad, D.J., Tanabe, K.K.: Prognostic and therapeutic implications of microvascular invasion in hepatocellular carcinoma. Ann. Surg. Oncol. **26**, 1474–1493 (2019)
5. Huang, M., Liao, B., Xu, P., et al.: Prediction of microvascular invasion in hepatocellular carcinoma: preoperative gd-eob-dtpa-dynamic enhanced mri and histopathological correlation. In: Contrast Media & Molecular Imaging 2018 (2018)
6. Kim, S., Shin, J., Kim, D.Y., et al.: Radiomics on gadoxetic acid-enhanced magnetic resonance imaging for prediction of postoperative early and late recurrence of single hepatocellular carcinoma. Clin. Cancer Res. **25**(13), 3847–3855 (2019)
7. Konyn, P., Ahmed, A., Kim, D.: Current epidemiology in hepatocellular carcinoma. Expert Rev. Gastroenterol. Hepatol. **15**(11), 1295–1307 (2021)
8. Lipton, Z.C.: The mythos of model interpretability: in machine learning, the concept of interpretability is both important and slippery. Queue **16**(3), 31–57 (2018)
9. Nealson, K.H., Conrad, P.G.: Life: past, present and future. Philos. Trans. Royal Soc. London. Ser. B: Biolog. Sci. **354**(1392), 1923–1939 (1999)
10. Nuta, J., Shingaki, N., Ida, Y., et al.: Irregular defects in hepatocellular carcinomas during the kupffer phase of contrast-enhanced ultrasonography with perfluorobutane microbubbles: pathological features and metastatic recurrence after surgical resection. Ultrasound Med. Biol. **43**(9), 1829–1836 (2017)
11. Rudin, C.: Stop explaining black box machine learning models for high stakes decisions and use interpretable models instead. Nat. Mach. Intell. **1**(5), 206–215 (2019)
12. Ryu, T., Takami, Y., Wada, Y., et al.: A clinical scoring system for predicting microvascular invasion in patients with hepatocellular carcinoma within the milan criteria. J. Gastrointest. Surg. **23**, 779–787 (2019)
13. Sharif, G.M., Schmidt, M.O., Yi, C., et al.: Cell growth density modulates cancer cell vascular invasion via hippo pathway activity and cxcr2 signaling. Oncogene **34**(48), 5879–5889 (2015)
14. Song, D., Wang, Y., Wang, W., et al.: Using deep learning to predict microvascular invasion in hepatocellular carcinoma based on dynamic contrast-enhanced mri combined with clinical parameters. J. Cancer Res. Clinical Oncol., 1–11 (2021)

15. Sumie, S., Kuromatsu, R., Okuda, K., et al.: Microvascular invasion in patients with hepatocellular carcinoma and its predictable clinicopathological factors. Ann. Surg. Oncol. **15**, 1375–1382 (2008)
16. Sun, B.Y., Gu, P.Y., Guan, R.Y., et al.: Deep-learning-based analysis of preoperative mri predicts microvascular invasion and outcome in hepatocellular carcinoma. World J. Surgical Oncol. **20**(1), 1–13 (2022)
17. Witjes, C.D., Willemssen, F.E., Verheij, J., et al.: Histological differentiation grade and microvascular invasion of hepatocellular carcinoma predicted by dynamic contrast-enhanced mri. J. Magn. Reson. Imaging **36**(3), 641–647 (2012)
18. Xiao, H., Guo, Y., Zhou, Q., et al.: Prediction of microvascular invasion in hepatocellular carcinoma with expert-inspiration and skeleton sharing deep learning. Liver Int. **42**(6), 1423–1431 (2022)
19. Zhou, W., Jian, W., Cen, X., et al.: Prediction of microvascular invasion of hepatocellular carcinoma based on contrast-enhanced mr and 3d convolutional neural networks. Front. Oncol. **11**, 588010 (2021)
20. Zhou, Y., Sun, S.W., Liu, Q.P., et al.: Ted: Two-stage expert-guided interpretable diagnosis framework for microvascular invasion in hepatocellular carcinoma. Med. Image Anal. **82**, 102575 (2022)

Human-to-Human Interaction Detection

Zhenhua Wang[1]($\boxtimes$) (iD), Kaining Ying[2] (iD), Jiajun Meng[2], and Jifeng Ning[1]

[1] Intelligent Media Processing Group, College of Information Engineering,
Northwest A&F University, Yangling 712100, Shanxi, China
`zhenhuawang@nwafu.edu.cn`
[2] College of Computer Science and Technology, Zhejiang University of Technology,
Hangzhou 310023, China

Abstract. Different from conventional human interaction recognition, which uses choreographed videos as inputs, neglects concurrent interactive groups, and performs detection and recognition in separate stages, we introduce a new task named **human-to-human interaction detection (HID)**. HID devotes to detecting subjects, recognizing person-wise actions, and grouping people according to their interactive relations, in one model. First, based on the popular AVA dataset created for action detection, we establish a new HID benchmark, termed **AVA-Interaction (AVA-I)**, by adding annotations on interactive relations in a frame-by-frame manner. AVA-I consists of 85,254 frames and 86,338 interactive groups, and each image includes up to 4 concurrent interactive groups. Second, we present a novel baseline approach **SaMFormer** for HID, containing a visual feature extractor, a split stage which leverages a Transformer-based model to decode action instances and interactive groups, and a merging stage which reconstructs the relationship between instances and groups. All SaMFormer components are jointly trained in an end-to-end manner. Extensive experiments on AVA-I validate the superiority of SaMFormer over representative methods.

Keywords: Human-to-human interaction detection · benchmark

1 Introduction

Understanding human interactions in video streams is pivotal to downstream applications, such as security surveillance, human-centered video analysis, key event retrieval and sociological investigation. In this paper, we introduce a novel task termed human-to-human interaction detection (HID) in videos, which aims to predict all **I**nteractive **C**lusters (ICs) per video frame. As defined in F-formation [5] in sociology studies, we assume that different ICs in the same scene are non-overlapping, *i.e.* each person belongs to only one IC. Each IC prediction consists of the bounding boxes and the action labels of all its participants. For example, there are two ICs in Fig. 1(b), each of which forms a conversational group.

The HID task is inspired by the well-studied task of *human-object interaction detection* (**HOID**) [8,14,19], which takes into consideration the target

B. Luo et al. (Eds.): ICONIP 2023, LNCS 14450, pp. 120–132, 2024.
https://doi.org/10.1007/978-981-99-8070-3_10

Fig. 1. Human interaction understanding (HIU) with existing methods (left) and the proposed human-to-human interaction detection (HID) model (right).

detection, target classification and the pairwise interactive relation recognition in the form of ⟨*person ID, object ID, interactive category*⟩. HID differs from HOID on a particular concentration on human-to-human interactions and the inclusion of high-order interactive relations, which commonly appear in practice. Based on AVA [6], which is a benchmark for action detection, we build a large-scale dataset (namely, AVA-I) for HID, in order to mitigate the manifest gap between existing datasets and practical human interactions. We use AVA because it includes abundant daily interactions such as *fight, grab, serve, talk and sing to people, etc.*. More importantly, it contains many challenging cases, *e.g.*, *concurrent interactions, heavy occlusions* and *cluttered backgrounds*, which are missing in existing benchmarks. To create AVA-I, we developed a toolbox, which loads and visualizes AVA videos and annotations (including per-person bounding boxes and action labels), and offers the function of adding annotation on interactive relations. In a nutshell, AVA-I includes 85,254 annotated frames, 17 interactive categories and 86,338 interactive groups in total, which significantly surpass existing datasets available to HID (as illustrated by Table 1).

To tackle the proposed HID task, we propose a novel one-stage Transformer-based framework, termed *SaMFormer* (shown in Fig. 2), which predicts human bounding boxes, per-person action labels and interactive relations jointly in a *Split-and-Merging* manner. Specifically, we design two Siamese decoders (sharing a global encoder), *i.e.*, an instance decoder and a group decoder to detect people and interactive groups, respectively (the *split*). Then the model incorporates both spatial cues and semantic representations to associate each human instance to a particular interactive group (the *merging*). We show that SaMFormer is a strong baseline compared with other alternative models.

2   Related Work

Closely Related Tasks. Action detection (**AD**) extends the well-explored object detection task [2,12,15] to the localization of human actions in videos [4]. The key difference to our HID task is that AD and its associated models are interactive-relation agnostic. Human interaction understanding (**HIU**) assumes that bounding boxes of human bodies have been obtained beforehand, and performs per-person action recognition and *pairwise* interactive relation estimation simultaneously [17]. In comparison, HID takes the prediction of both bounding boxes and *high-order* interactive relations into consideration. We empirically demonstrate that our proposed one-stage framework outperforms recent two-stage HIU models significantly (revealed in Table 2). Compared with HOID, HID focuses specifically on detecting human-to-human interactions. Another key difference is that people might have different personal actions (*e.g. pushing vs. falling*) in HID, even if they belong to the identical interactive group (*e.g. fighting*).

Available Datasets for HID. Training HID models requires comprehensive annotations on per-person bounding boxes, per-person actions as well as the interactive relations among people. Available benchmarks, including TVHI [11], BIT [9] and UT [13], are originally crafted for the task of image/video-level interaction classification. Unfortunately, these datasets are of small-scale in terms of the number of interactive groups (see Table 1 for statistics), and only include artificial human interactions performed by a few amateurish actors. Moreover, none of them involve concurrent human interactions (*i.e.* multiple interactions performed by different groups). AVA [6] is a recently created dataset for spatio-temporal action detection. It is a YouTube-sourced dataset including abundant interactive groups and richer action classes. Also, AVA videos usually include crowded foregrounds, cluttered backgrounds and concurrent interactions. Our AVA-I utilizes all videos and annotations of AVA that contain interesting interactions and adds high-quality annotations on interactive relations among people.

Vision Transformers for Set Prediction. In the vision, Transformers have been taken to accomplish the so-called *set prediction* tasks, such as object detection [1], instance segmentation [18] and HOID [8,14,19], where the final prediction comprises of a set stand-alone components. The most closely related work to ours is HOTR [8] for HOID, which first predicts sets of instances and interactions, then reconstructs human-object interactions with *human-object Pointers*. Nevertheless, each such pointer is designed to recompose the interactive relation between a person and an object, which is incapable of processing high-order interactions among multiple targets in HID. In comparison, SaMFormer introduces a *split-and-merging* way to associate instances with interactive groups, which enables the prediction of interactions with arbitrary orders.

3 HID Task

3.1 Problem Definition

Given an input video, HID aims to make each frame a joint prediction on *per-person bounding boxes* $\boldsymbol{b} \in \mathbb{R}^4$, *per-person interactive actions* $\boldsymbol{c} \in \{0,1\}^K$ where K is the number of interested interactive actions, and the *interactive relations* among people. Here $\boldsymbol{c}$ is a sequence containing K zero-one values, in which 1 indicates that the corresponding action is activated. Note that $\boldsymbol{c}$ can take multiple 1 s in order to allow the prediction of plural actions performed by the same person (*e.g. watching* and *talking*). The interactive relations are denoted by $\mathcal{G} = [g_i]_{i=1}^M$,

Table 1. Compare AVA-I against existing datasets that are available to HID.

	BIT	UT	TVHI	AVA-I
# Interaction categories	9	6	5	**17**
Max # group per frame	1	1	1	**4**
Max # people per group	2	2	2	**13**
Mean # people per group	2	2	2	**2.5**
# Annotated frames	12,896	9,228	2,815	**85,254**
# Groups in total	12,896	9,228	2,815	**86,338**

where $g_i \subseteq \{1, \ldots, N\}$, is a grouping of N people (proposals) into M groups such that $g_i \cap g_j = \emptyset$, $\forall i, j \in \{1, 2, \ldots, M\}$, and $i \neq j$. In other words, a proposal can belong to only one group. Moreover, we predict a foreground confidence score s_i^P for each person i, and a group confidence score s_k^G for each group k. To sum up, each HID prediction takes a form: $\boldsymbol{y} = [B, C, \mathcal{G}, \boldsymbol{s}^P, \boldsymbol{s}^G]$, where $B = [\boldsymbol{b}_i]$, $C = [\boldsymbol{c}_i]$, $\boldsymbol{s}^P = [s_i^P]$, and $\boldsymbol{s}^G = [s_k^G]$. Ideally, individuals in a predicted group g have interactive relations with each other (*e.g.* the handshaking and talking scenes in Fig. 1). Note that they might have different personal actions, *e.g.* talking *vs.* listening. Nevertheless, their actions are closely and strongly correlated with each other, thus forming a semantic group.

3.2 Evaluation Metrics

We use two evaluation metrics for HID. For per-person interactive action detection (a mixture of human bounding box detection and action classification), we use mean average precision (mAP), which is a popular protocol in AD [4,6]. For the prediction of interactive relations (*i.e.* $\mathcal{G}$), we design a metric termed *group average precision* (AP^G). To this end, we introduce the group-level intersection-over-union IoU^G between a ground-truth group g and a predicted group g'. Let $U = |g|$ and $V = |g'|$. First, we obtain an *action-agnostic* matching between these two groups with the *optimal bipartite matching* [10], which relies the cost matrix $O = [cost_{ij}]_{U \times V}$, with each $cost_{ij}$ representing the matching cost between two targets $i \in g$ and $j \in g'$:

$$cost_{ij} = \begin{cases} -\text{IoU}(\boldsymbol{b}^i, \boldsymbol{b}^j) & \text{if } \text{IoU}(\boldsymbol{b}^i, \boldsymbol{b}^j) \geq 0.5, \\ \epsilon & \text{otherwise.} \end{cases} \tag{1}$$

Here ϵ is a big positive value used to reserve matches taking high IoU values. Second, taking as input the cost matrix O, a matching candidate between g and g' is computed efficiently utilizing hungarian algorithm [10]. The final matching is obtained by ruling out bad matches having IoU < 0.5. Finally, the IoU^G is given by $\text{IoU}^G = \frac{R}{U+V-R}$, where R is number of matched elements in g'. Like the IoU in object detection, IoU^G measures the rightness of associating the ground-truth g with the prediction g'. The ideal case is that g and g' admit a perfect match (*i.e.* $U = V = R$) such that $\text{IoU}^G = 1$. To calculate AP^G, we use six IoU^G thresholds (denoted by δ) ranging from 50% to 100% with a resolution of 10%. For each δ, the *average precision* (AP_δ^G) is computed under a retrieval scheme of the top-K predicted groups based on IoU^G values and $\mathbf{s}^G$ confidences. Finally, AP^G is computed by averaging all AP_δ^G values.

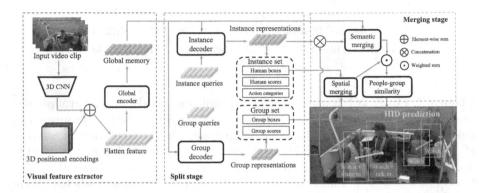

Fig. 2. The framework of SaMFormer.

3.3   The AVA-Interaction Dataset

As mentioned in Sect. 2, available datasets for HID fail to match practical human interactions in terms of scale and complexity. To fix this, we present a large-scale benchmark. Instead of building the benchmark from scratch, the proposed benchmark takes videos from AVA [6], which is a large-scale action detection dataset consisting of 437 YouTube videos and 80 atomic daily actions. To make it available to HID, we upgrade the original annotation of AVA by adding frame-by-frame interactive relations among people. Due to this, we call this new version *AVA-Interaction* (AVA-I).

To shape AVA (created for action detection) for HID, first we discard videos without any human-to-human interactions. Consequently, we obtain 298 videos containing 17 interactive action categories. Since AVA does not cover interactive relations among people, we next annotate this information in the frequency of 1 frame per second, which aligns with the original AVA annotation. To this end, we develop an annotation toolbox, which first loads and visualizes an AVA video and its original annotations of bounding boxes and action labels, and then enables the user to browse and annotate the interactive relations in a frame-by-frame manner. It considers three scenarios for efficiency: 1) For frames that only include one person, they are simply discarded as any interaction must include multiple participants; 2) For frames that exactly include two interactive people (the frame might take non-interactive people as well), the interactive relations are automatically recorded; 3) For all rest scenarios, two groups of human annotators are employed to carefully and independently draft the annotation. If the two versions are identical, they are believed to be trustworthy and are kept.

Otherwise, they are double-checked by a senior researcher in this field to finalize the annotation. In this way, the quality of the annotated interactive relations can be guaranteed. The comparison of AVA-I against existing datasets is shown in Table 1.

4   SaMFormer

The architecture of our proposed SaMFormer is shown in Fig. 2. We describe each component below.

4.1   Visual Feature Extractor

The VFE combines a 3D CNN backbone [4] and a Transformer encoder. Feeding with an input video clip $V \in \mathbb{R}^{T \times H_0 \times W_0 \times 3}$, the backbone generates a feature map as a tensor, taking the shape (T, H, W, D), where T, H, W, D denote the number of frames, the height, the width and the depth of the feature map respectively. Then the depth of the feature map is reduced to d (e.g. 256) by applying an $1 \times 1 \times 1$ filter. Next, a flatten operation is taken to get the feature $x \in \mathbb{R}^{(T \cdot H \cdot W) \times d}$ by collapsing the temporal and spatial dimensions. Since the Transformer architecture is permutation-invariant, we supplement x with fixed 3D positional encodings $e \in \mathbb{R}^{(T \cdot H \cdot W) \times d}$, and feed them into a Transformer encoder. Benefiting from the attention mechanism [16], the encoder injects rich contextual information into x, and the resulting representations $x_g \in \mathbb{R}^{(T \cdot H \cdot W) \times d}$ are stored into a global memory for subsequent usage.

4.2   The Split Stage

This stage is designed to predict the instance and group sets in parallel. To this end, we first randomly initialize two sets of learnable queries, i.e., $Q^P \in \mathbb{R}^{u \times d}$ as instance queries, and $Q^G \in \mathbb{R}^{v \times d}$ as group queries, where u and v denote the allowed largest numbers of instances and groups. Then these queries, together with x_g (served as key and value in cross-attention), are fed into the instance decoder and the group decoder as illustrated by Fig. 2. The instance decoder transforms the instance queries into instance representations $\{r_i^P\}_{i=1}^u$ for the detection of person proposals and the recognition of per-person actions, while the group decoder leverages the group queries to extract group representations $\{r_j^G\}_{j=1}^v$ for the detection of interactive groups. We then apply three shared MLPs to each instance representation r_i^P, which respectively predict a confidence score s_i^P, an interactive action category c_i and a bounding box b_i^P for the i-th instance query. Likewise, we apply two extra FFNs to each group representation r_j^G, in order to predict a group confidence score s_j^G and a group bounding box b_j^G (i.e. a bounding box enclosing all interactive people in this group). Consequently, we obtain an estimated instance set $\{s_i^P, c_i, b_i^P\}_{i=1}^u$ using the instance decoder, and an estimated group set $\{s_j^G, b_j^G\}_{j=1}^v$ with the group decoder.

4.3   The Merging Stage

The goal of this merging stage is to merge the output from the previous split stage to predict the interactive relationships. Specifically, we introduce two forms of merging: semantic merging and spatial merging. For the former, we take the instance representations $\{r_i^P\}_{i=1}^u$ and group representations $\{r_j^G\}_{j=1}^v$ output by the decoders to compute the people-group similarity $\Theta = [\theta_{ij}]_{u \times v}$ (*i.e.* the *semantic similarity*), with $\theta_{ij} = 1$ indicating that person i definitely belongs to group j. We implement this with three alternatives.

Inner Product performs the inner (dot) product between linearly transformed instance and group representations:

$$\theta_{ij} = \frac{\langle f^P(r_i^P), f^G(r_j^G) \rangle}{\|f^P(r_i^P)\| \cdot \|f^G(r_j^G)\|}, \tag{2}$$

where $f^P(\cdot)$ and $f^G(\cdot)$ are linear transformations, $\langle \cdot, \cdot \rangle$ denotes inner product, and $\|\cdot\|$ is $L2$ normalization.

Linear Transformation first concatenates instance and group representations. Then it applies a linear transformation to the concatenated feature:

$$\theta_{ij} = \text{sigmoid}\big(f^C(r_i^P \otimes r_j^G)\big), \tag{3}$$

where $\otimes$ denotes vector concatenation.

Semantic Decoder. Both the *inner product* and the *linear transformation* have omitted the global memory, which might contain key contextual information beyond the scope of local features in terms of discriminating interactive relations. Motivated by this, we design an additional Transformer decoder named *semantic decoder*, which takes $[f^R(r_i^P \| r_j^G)]_{i=1,\ldots,u,j=1,\ldots,v}$ as queries, and x_g as keys and values. Here $f^R(\cdot)$ maps each concatenated vector to a new vector in $\mathbb{R}^d$. Let $r_{ij}^{PG} \in \mathbb{R}^d$ be the output of the decoder for the person i and the group j. Then a FFN (shared by all r_{ij}^{PG} vectors) is taken to compute the semantic similarity θ_{ij}. Although the *semantic merging* block is capable of discriminating most interactive relations, it could fail under some tricky circumstances. To alleviate this, we introduce a spatial prior, which is encoded by another $\Theta' = [\theta'_{ij}]_{u \times v}$ matrix with entries given by $\theta'_{ij} = \text{IoF}(b_i^P, b_j^G)$, where $\text{IoF}(\cdot, \cdot)$ computes *intersection-over-foreground*, and the person-box is used as the foreground. Specifically, if the person-box i is thoroughly enclosed by the group-box j, then $\theta'_{ij} = 1$. We name grouping with θ' the *spatial merging*. We get the final similarity by blending spatial and semantic similarities: $\hat{\theta}_{ij} = \alpha \theta'_{ij} + (1 - \alpha)\theta_{ij}$, $\alpha \in [0, 1]$. Here, α can be pre-designated empirically or learned from data.

4.4   Training and Inference

Training. Following the training protocol of DETR [1], we first match each instance or group ground truth with its best-matching prediction with the Hungarian algorithm. These matches are then utilized to calculate losses for backpropagation. The total loss comprises of four terms: $L = L_{detr}^P + L_{detr}^G + \lambda_a^P L_a^P +$

$\lambda^{PG}L^{PG}$, where L^{P}_{detr} and L^{G}_{detr} are of DETR losses (we follow the loss items and loss weights in DETR, please refer to [1]), used to supervise foreground-background classification and bounding box regression of people and groups respectively. L^{P}_{a} and L^{PG} are cross-entropy losses to supervise per-person action classification and instance-group matching, and λ weights different terms.

Inference is to recognize all instances and associate them to form proper groups. We first get all instances from the instance decoder. Next, given the similarity matrix $\hat{\Theta} = [\hat{\theta}_{ij}]_{u \times v}$, we compute the group ID via $\text{argmax}_{k \in \{1,...,v\}} \hat{\theta}_{ik}$, for any instance i. This operation guarantees that each instance is assigned a unique group ID to meet our grouping assumption.

5 Experiments

Table 2. Comparing SaMFormers with other baselines on AVA-I. E2E means end-to-end trainable. $T \times \tau$ tells the number of frames (T) and the sampling-rate (τ) on each input video. For all two-stage methods, Faster R-CNN [12] is taken to detect human bodies as RoIs. * indicate that the annotated bounding boxes are used instead of the detected ones.

Type	Method	Backbone	E2E	$T \times \tau$	AP^{G}	AP^{G}_{50}	AP^{G}_{80}	AP^{P}_{50}
Action Only	SlowOnly-R50	SlowOnly-R50	✗	4×16	–	–	–	17.05
	SlowFast-R50	SlowFast-R50	✗	8×8	–	–	–	18.06
Two-Stage	SCG	SlowOnly-R50	✗	4×16	59.45	73.68	51.34	17.12
	SCG*	SlowOnly-R50	✗	4×16	73.88	84.19	67.19	19.52
	CAGNet	SlowOnly-R50	✗	4×16	57.11	70.77	50.85	15.08
	CAGNet*	SlowOnly-R50	✗	4×16	**76.62**	86.2	**72.2**	18.36
One-Stage	QPIC [14]	SlowOnly-R50	✗	4×16	63.74	84.23	53.91	16.71
	CDN [19]	SlowOnly-R50	✗	4×16	63.81	85.26	53.27	17.59
	SaMFormer (ours)							
	Spatial	SlowOnly-R50	✓	4×16	56.1	79.38	47.29	17.88
	Semantic	SlowOnly-R50	✓	4×16	67.14	87.53	60.13	18.01
	Both	SlowOnly-R50	✓	4×16	72.52	89.11	66.09	18.46
	Both	SlowFast-R50	✓	8×8	74.05	**89.12**	68.58	**19.97**

Implementation Details. We use AdamW with a weight decay of 0.0001 as the optimizer for all experiments. The mini-batch consists of 24 video clip. We train the network in 20 epochs with an initial learning rate of 0.0001. The decay factor is 0.1 applied at epoch 10 and 16, respectively. The backbone is initialized with the pre-trained weights on Kinetics [7] and the rest layers are initialized with Xavier. We resize each frame to 256×256 for all methods in both training and evaluation. For the loss weights, we use $\lambda^{P}_{a} = 1$ and $\lambda^{PG} = 5$, without meticulously tuning.

Dataset and Evaluation Metrics. Following AVA [6], we split AVA-I into training and testing sets, which contain 234 (66,285 frames) and 64 (18,969 frames) videos respectively. We use AP^G, AP^G_{50}, AP^G_{80} to evaluate the performance on interactive relation detection. For human detection and action classification, we take the mean average precision denoted by AP^P_{50} (where the IoU threshold is 50% as suggested in [6]).

5.1   Main Results on AVA-I

As HID is new, we absorb ideas from relevant tasks [3,4,6,17] to create several baselines, which tackle HID in either two-stage or one-stage. The former detects human bodies in the first stage, and then performs interaction recognition in the second stage, while the latter accomplishes both sub-tasks in a unique model. For the two-stage methods, we choose the CAGNet [17] for HIU and SCG [20] for HOID as our competitors. For the one-stage approaches, we choose the classic QPIC [14] and CDN [19] for HOID.

Table 3. Compare the performance of learned α with that of pre-designated α (*i.e.* $\alpha = 0.3$).

Type	α	AP^G	AP^G_{50}	AP^G_{80}	AP^P_{50}
Pre-designated	0.3	72.52	89.11	66.09	**18.46**
Learned	0.3723	**73.04**	**89.44**	**66.81**	18.42

Table 4. Performance comparison on *semantic merging* variants.

Methods	AP^G	AP^G_{50}	AP^G_{80}	AP^P_{50}
Linear Transformation	70.83	87.44	64.97	17.82
Inner Product	70.92	87.15	65.15	17.79
Semantic Decoder	**72.52**	**89.11**	**66.09**	**18.46**

Furthermore, we also apply SlowFast/SlowOnly [4] to detect person-wise actions on AVA-I, which serves as a strong baseline for both human detection and per-person action recognition. Table 2 gives the main quantitative results. Thanks to the self-attention modelling and the effective multi-task learning, SaMFormer surpasses the baseline of AD (Action Only), by 1.41 points under SlowOnly and 1.91 points under SlowFast on AP^P_{50}. For interaction detection, SaMFormer (Spatial + Semantic) outperforms both one-stage and two-stage baselines by large margins (except for models using *ground-truth bounding boxes*) under all metrics. When utilizing an out-of-box detector [12] to provide RoIs, the performance of the two-stage models is severely degraded (-14.43 and -19.51 points in term of AP^G for SCG and CAGNet), which indicates that these models typically suffer from the quality of detection. One-stage models jointly optimize three sub-tasks, including human detection, per-person action recognition and interactive relation estimation, and their results are generally better than two-stage methods, which is consistent with the observations in HOID [19]. Apart from the one-stage design, SaMFormer also leverages a simple yet effective Split-and-Merging paradigm to achieve detection and grouping, which in turn removes the heuristic post-processing procedure (*i.e.* the *greedy search* to form groups).

With this, SaMFormer surpasses the second best CDN [19] by 8.71 points (72.52 vs. 63.81).

Furthermore, we investigate the effectiveness of our merging design. The necessity of combing the *semantic* and the *spatial* merging could be justified by that the conjunctive usage of them improves the result by 5.38 points (last four rows in Table 2), which suggests that both the semantic feature and the spatial prior are essential to the estimation of interactive relations. In addition, semantic merging is more reliable than spatial merging, as spatial cues are typically non-robust against dynamic human interactions. Replacing SlowOnly with SlowFast, a more powerful backbone for video-level representation, the result is further improved by 1.53, which demonstrates that SaMFormer largely benefits from our *split* and *merging* designs.

5.2 Ablation Study

We take an ablation to validate our model design. Here all methods use SlowOnly-R50 as the backbone for fairness and effectiveness.

Table 5. Comparison with CAGNet [17] on BIT [9] and UT [13].

Model	Detector	BIT [9]	UT [13]
CAGNet	YOLO v5	59.39 (+17.22)	75.63 (+10.5)
CAGNet	FCOS	68.62 (+7.99)	81.85 (+4.28)
CAGNet	Annotated	**81.73** (−5.12)	86.03 (+0.1)
SaMFormer	–	76.61	**86.13**

Spatial *vs.* Semantic. SaMFormer incorporates both spatial prior and learned semantic representations to discriminate the interactive relations in the *merging* stage. To ablate such design, we evaluate SaMFormer with both pre-designated and learned α on Table 3.

Variants of Semantic Merging. Section 4.3 presents three alternatives for semantic merging, which are compared in Table 4. *Semantic Decoder* is notably better than *Linear Transformation* and *Inner Product*. Indeed, the decoder allows to attentively learn informative local and contextual representations for each *person-group* query, such that the interactive relations among people could be determined under a group perspective. In comparison, the other merging methods compute *person-group* similarities under a local perspective, which is insufficient in addressing complicated human interactions.

5.3 Qualitative Results

To provide a qualitative analysis of different models, we visualize a few predictions in Fig. 3. Here we select CAGNet and QPIC as the competitors. The three leftmost examples show that CAGNet and QPIC could give defective HID results, either caused by missing detection (the two leftmost columns) or incorrect grouping (the third column from left to right), while our proposed SaMFormer is able to generate correct HID predictions. We attribute this success to the split-and-merging design of SaMFormer, which enables the model to interpret the scene in both micro (*i.e.* instances) and macro (*i.e.* groups) perspectives. However, for the rightmost example, where severe occlusion appears, CAGNet performs much better than QPIC and our SaMFormer. Specifically, QPIC and SaMFormer mistakenly put all targets into an identical group, probably caused by the partial shared representations and the proximity between occluded people.

5.4   Evaluation on BIT and UT

Table 5 compares SaMFormer with CAGNet on BIT and UT. Since CAGNet takes as inputs the bounding boxes, for fair comparison, we train three CAGNet models using YOLOv5 detections, FCOS detections and annotated boxes (testing uses the identical source of boxes). On BIT, SaMFormer outperforms by 17.22 and 7.99 points in terms of mAP against CAGNet with YOLOv5 and FCOS detections respectively. On UT, SaMFormer surpasses CAGNet in all settings, and to our surprise, it is even marginally better than CAGNet with *annotated* boxes.

Fig. 3. Visualize HID results on AVA-I. Red highlights incorrect grouping results. (Color figure online)

6   Conclusion

We presented the HID task, which devotes to locate people and predict their actions and interactive relations in videos. Along with HID, a large-scale, realistic and challenging dataset was proposed to benchmark HID, on which we observed salient performance gaps against available datasets with strong baselines. To alleviate this, we proposed SaMFormer based on the encoding-decoding mechanism of Transformer, which allows tackling HID in an end-to-end *Split-and-Merging* manner. Without bells and whistles, SaMFormer outperforms strong baselines by clear margins on both existing and new benchmarks. Proper ablation experiments are also conducted, showing that a proper blending of semantic and spatial cues is vital to HID.

Acknowledgement. This work was supported by Zhejiang Provincial Natural Science Foundation of China (No. LY21F020024), National Natural Science Foundation of China (No. 62272395), and Qin Chuangyuan Innovation and Entrepreneurship Talent Project (No. QCYRCXM-2022-359).

References

1. Carion, N., Massa, F., Synnaeve, G., Usunier, N., Kirillov, A., Zagoruyko, S.: End-to-end object detection with transformers. In: Vedaldi, A., Bischof, H., Brox, T., Frahm, J.-M. (eds.) ECCV 2020. LNCS, vol. 12346, pp. 213–229. Springer, Cham (2020). https://doi.org/10.1007/978-3-030-58452-8_13
2. Chan, S., Tao, J., Zhou, X., Bai, C., Zhang, X.: Siamese implicit region proposal network with compound attention for visual tracking. IEEE Trans. Image Process. **31**, 1882–1894 (2022)
3. Chao, Y., Liu, Y., Liu, X., Zeng, H., Deng, J.: Learning to detect human-object interactions. In: IEEE Winter Conference on Applications of Computer Vision (WACV) (2018)
4. Feichtenhofer, C., Fan, H., Malik, J., He, K.: SlowFast networks for video recognition. In: International Conference on Computer Vision (ICCV) (2019)
5. Gan, T., Wong, Y., Zhang, D., Kankanhalli, M.: Temporal encoded F-formation system for social interaction detection. In: ACM International Conference on Multimedia (ACM MM) (2013)
6. Gu, C., et al.: AVA: a video dataset of spatio-temporally localized atomic visual actions. In: IEEE/CVF Conference on Computer Vision and Pattern Recognition (CVPR) (2018)
7. Kay, W., et al.: The kinetics human action video dataset. arXiv preprint arXiv:1705.06950 (2017)
8. Kim, B., Lee, J., Kang, J., Kim, E., Kim, H.: HOTR: end-to-end human-object interaction detection with transformers. In: IEEE/CVF Conference on Computer Vision and Pattern Recognition (CVPR) (2021)
9. Kong, Yu., Jia, Y., Fu, Y.: Learning human interaction by interactive phrases. In: Fitzgibbon, A., Lazebnik, S., Perona, P., Sato, Y., Schmid, C. (eds.) ECCV 2012. LNCS, vol. 7572, pp. 300–313. Springer, Heidelberg (2012). https://doi.org/10.1007/978-3-642-33718-5_22
10. Kuhn, H.: The Hungarian method for the assignment problem. Naval Res. Logistics Q. **2**(1–2), 83–97 (1955)
11. Patron-Perez, A., Marszalek, M., Zisserman, A., Reid, I.: High five: recognising human interactions in TV shows. In: British Machine Vision Conference (BMCV) (2010)
12. Ren, S., He, K., Girshick, R., Sun, J.: Faster R-CNN: towards real-time object detection with region proposal networks. In: Advances in Neural Information Processing Systems (NeurIPS) (2015)
13. Ryoo, M., Aggarwal, J.: UT-Interaction Dataset, ICPR contest on Semantic Description of Human Activities (SDHA) (2010). http://cvrc.ece.utexas.edu/SDHA2010/Human_Interaction.html

14. Tamura, M., Ohashi, H., Yoshinaga, T.: QPIC: query-based pairwise human-object interaction detection with image-wide contextual information. In: IEEE/CVF Conference on Computer Vision and Pattern Recognition (CVPR) (2021)
15. Tian, Z., Shen, C., Chen, H., He, T.: FCOS: fully convolutional one-stage object detection. In: International Conference on Computer Vision (ICCV) (2019)
16. Vaswani, A., et al.: Attention is all you need. In: Advances in Neural Information Processing Systems (NeurIPS) (2017)
17. Wang, Z., Meng, J., Guo, D., Zhang, J., Shi, J., Chen, S.: Consistency-aware graph network for human interaction understanding. In: Proceedings of the IEEE/CVF International Conference on Computer Vision (ICCV) (2021)
18. Ying, K., Wang, Z., Bai, C., Zhou, P.: ISDA: position-aware instance segmentation with deformable attention. In: International Conference on Acoustics, Speech and Signal Processing (ICASSP) (2022)
19. Zhang, A., et al.: Mining the benefits of two-stage and one-stage HOI detection. In: Advances in Neural Information Processing Systems (NeurIPS) (2021)
20. Zhang, F., Campbell, D., Gould, S.: Spatially conditioned graphs for detecting human-object interactions. In: IEEE/CVF International Conference on Computer Vision (ICCV) (2021)

Reconstructing Challenging Hand Posture from Multi-modal Input

Xi Luo[1,2,3(✉)], Yuwei Li[1,2,3], and Jingyi Yu[1]

[1] School of Information Science and Technology, ShanghaiTech University, Shanghai, China
sunshine.just@outlook.com
[2] University of Chinese Academy of Sciences, Beijing, China
[3] Shanghai Institute of Microsystem and Information Technology, Shanghai, China

Abstract. 3D Hand reconstruction is critical for immersive VR/AR, action understanding or human healthcare. Without considering actual skin or texture details, existing solutions have concentrated on recovering hand pose and shape using parametric models or learning techniques. In this study, we introduce a challenging hand dataset, CHANDS, which is composed of articulated precise 3D geometry corresponding to previously unheard-of challenging gestures performed by real hands. Specifically, we construct a multi-view camera setup to acquire multi-view images for initial 3D reconstructions and use a hand tracker to separately capture the skeleton. Then, we present a robust method for reconstructing an articulated geometry and matching the skeleton to the geometry using a template. In addition, we build a hand pose model from CHANDS that covers a wider range of poses and is particularly helpful for difficult poses.

Keywords: 3D Hand Reconstruction · Mesh Models · Pose Space

1 Introduction

Reconstructing hand shapes and recovering their motions have practical applications in human-computer interaction, gesture recognition, and virtual/augmented reality. However, the problem is challenging owing to the many degrees of freedom (DoF) of hand motions: each finger can flex and extend, abduct and adduct, and also circumduct; and all fingers can move independently as well as coordinately to form specific gestures. Such high DoF causes complex occlusions and hence imposes significant difficulties in 3D reconstruction, skeleton extraction, etc. Directly using state-of-the-art techniques such as skeleton estimation [10] and multi-view stereo (MVS) [9] cannot produce satisfactory results.

More recent deep learning techniques manage to greatly improve the performance of hand pose and shape estimation, especially in gesture recognition and finger tracking tasks. Learning-based methods by far unanimously rely on training on massive labeled datasets. For instance, ground-breaking method [11] trained their hand pose estimation network with 30K annotated images, recent methods [14] are mostly trained on Freihand [15] which contains 130K images. In contrast, very few datasets [6] provide hand

---

X. Luo and Y. Li—Contributed equally to the paper.

B. Luo et al. (Eds.): ICONIP 2023, LNCS 14450, pp. 133–145, 2024.
https://doi.org/10.1007/978-981-99-8070-3_11

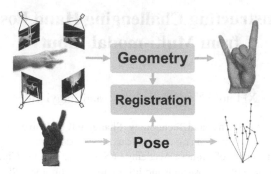

Fig. 1. The data acquisition setup for producing CHANDS. We use the NANSENSE [1] glove for pose capture and have the same gesture captured in a small dome system composed of 24 synchronized cameras.

images along with their ground truth skeleton/poses together with detailed 3D geometry. In fact, even for humans, it would be very difficult to manually label the hand skeleton of complex gestures at high precision, largely due to ambiguity caused by occlusions. Using multiple cameras only marginally benefits the skeleton labeling task since one can easily perform gestures with finger(s) hidden from any perspective. The goal of this paper is to fill in this gap by reconstructing articulated 3D hands under extremely challenging gestures and constructing a challenging hand dataset which we call CHANDS. The dataset includes multi-view images, 3D skeletons, and high-precision 3D textured geometry of hands under respective gestures.

Existing hand datasets have focused on one of three levels: low-level imagery data suitable for gesture recognition [13], middle-level pose/skeleton data applicable to high-precision human-computer interactions [6], and high-level detailed 3D geometry for producing virtual avatars [8]. Our CHANDS dataset, in contrast, captures hands at all three levels for each gesture. Figure 1 shows the data capture process of CHANDS. We first construct a multi-camera dome system composed of 24 synchronized cameras for 3D geometry capture. For a specific gesture, we use the dome system to capture multi-view images and subsequently conduct MVS for 3D reconstruction. We have 10 subjects performing a set of 50 challenging gestures. Some complex gestures included in our dataset are shown in Fig. 2. The raw 3D reconstruction cannot recover occluded fingers and may contain holes and strong noise. We then have the subjects wear the NANSENSE [1] hand tracker to perform the same set of gestures to obtain the 3D skeletons as an initialization. Although a human cannot recreate a pose precisely in reality, precision is sufficient here for initializing registration. Also, we observe that the skeletons deviate from the ground truth due to noise on the inertial sensor and differences between the template hand and the real ones. We, therefore, present an additional alignment algorithm that refines the skeleton using the acquired 3D data. Specifically, for each subject, we first capture the 3D geometry of the hand at the rest pose and then deform the template to align with recovered 3D point clouds as a personalized template hand. For the remaining gestures, we perform skeleton animation and non-rigid

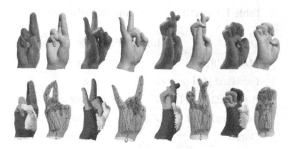

Fig. 2. Sample poses in CHANDS. The top row shows the dense point cloud and the corresponding registered template mesh. The bottom row shows the semantic point cloud and the template mesh overlaid with its corresponding 3D skeleton.

deformation to align the personalized template to the recovered point cloud through skeleton/pose optimization and non-rigid deformation.

To summarize, our main technical contributions include:

– We propose a novel multi-modality hand data acquisition pipeline that separates pose and geometry to improve accuracy for challenging poses.
– We present the first challenging hand dataset, CHANDS, which includes multi-view imagery data, pose/skeleton of fingers, and high-precision 3D geometry with textures.
– We present a hand pose model learned from CHANDS that spans a larger pose space and is especially useful for complex poses.

2   Related Work

Recently, Remero [8] proposed a parametric hand model MANO [8]. They released a training dataset that contains 1,554 high-quality hand scans and registered MANO meshes, the corresponding 3D skeleton can be acquired using the joint regression matrix of MANO. However, the dataset does not provide natural hand images or model textures. Its corresponding gestures are also relatively simple and do not cover the diverse needs of movements in sign languages. FreiHand [15] proposed by Zimmermann is a large-scale, multi-view hand dataset that is accompanied by both 3D hand pose and shape annotations. It contains 37K frames of RGB image, 3D hand skeleton, and geometry. But their hand geometries are generated through MANO fitting, so it is not accurate enough for mesh registration evaluation. BigHand2.2M [13] contains 290K depth images and corresponding 3D hand skeletons, as well as fitted MANO geometry. Inter-Hand2.6M [6] is the largest two-hand interaction dataset with 2.6M labeled single and interacting hand frames from multi-view cameras, annotated 3D skeleton, and MANO geometry under different poses from multiple subjects. We briefly compare these 4 state-of-the-art datasets with ours and summarize them in Table 1.

Table 1. CHANDS vs. existing hand datasets.

Dataset	Image	Pose	Geometry
MANO [8]	×	✓	coarse
FreiHand [15]	✓	✓	×
BigHand2.2M [13]	✓	✓	coarse
InterHand2.6M [6]	✓	✓	coarse
Ours	✓	✓	detail

3   Capture

We capture the data in two separate types: pose and geometry. We first use a commercial motion capture glove to extract the pose information. We later build a multi-camera system to capture high-resolution images and then apply multiview reconstruction methods to recover the dense point clouds.

Capture Pose. We collect the hand pose by having the subjects wear the motion capture glove (NANSENSE BIOMED gloves [1]). The glove contains 16 IMU sensors on each hand, structurally distributed on finger segments. It continuously records the joint rotation between 2 connected IMUs during the capture. The accompanied glove software then processes the IMU data to obtain relative bone rotations from each bone segment to its parent bone. We fix the global position of the wrist since only finger motions are considered. In practice, we align the middle finger metacarpal of all poses and keep the rotation angle for the rest four metacarpals. By this scheme, we can record all types of motion. While the MANO model and dataset [8] lacks carpometacarpal joints, which might negatively influence accuracy for nuanced poses.

The poses we captured fall into 3 categories: (1) Sign language: American sign language fingerspelling and British sign language fingerspelling as well as number gestures. (2) Articulation poses, which cover the range of motion of each finger. (3) Hand ability rehabilitation poses, including finger pinching and fist clenching.

The poses are recorded in sequence, so the intermediate frames provide more continuous data. In addition, we mirror the action that is done by the left hand to the right. The glove processes the data at 30 fps, and there are 12K frames of pose data per subject in total.

Pose Space Generation. Our hand model contains 20 bones and 20 joints (as shown in Fig. 3). We use a 3-dimensional axis-angle to represent the rotation for each bone, so we have parameters in 60 dimensions. However, ElKoura [4] show that the effective dimensionality needed for hand poses is much less than 60. We employ Principal Component Analysis (PCA) to pose parameters for dimension reduction due to the parameter redundancy. We evaluate dimensional reduction with a different number of principal components, as reported in Fig. 7, and we find that more than 98% of the motions in our dataset are still represented when we reduce the dimension to 30.

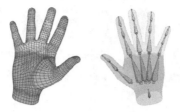

Fig. 3. Our template hand model is composed of a quad mesh with 3,817 vertices and 7,602 faces, 6 semantic components, and a kinematic rig.

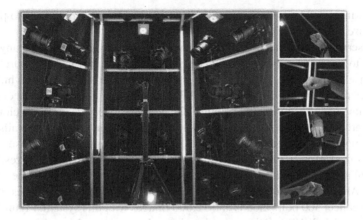

Fig. 4. Our multi-view hand capture system uses 24 calibrated and synchronized Nikon DSLR cameras. We use the commercial structure-from-motion software CaptureReality for recovering the initial point cloud.

Capture 3D Geometry. Figure 4 shows our multi-camera acquisition system, which consists of 24 calibrated and synchronized Nikon D750 DSLR cameras with 24-120MM F/4G ED VR Lens. Cameras are mounted on a 1.6 m height and 1.3 m diameter hexagonal cylinder. The furthest distance from the camera to the center is 1.0 m and the closest is 0.6 m. We ask the participants to lay their wrists on a fixed frame to reduce hand motions and calibrate their hand positions. We also require that their palms face down during the captures to ensure similar pose orientations. We use the position of the frame as an anchor point to crop the hand point cloud out of the background to reduce reconstruction noise.

For the camera arrangement, we mount 4 cameras on the top to directly look at the back of the hand. And we uniformly place 18 cameras for 3 lower levels and keep the angles between every neighboring camera to be 20°. These cameras are mainly used for capturing finger motion from different viewpoints. We only place 2 auxiliary cameras on the left and right with a horizontal viewing direction. That's because the information for the reconstruction is limited on the horizontal level due to heavy self-occlusion. We use full resolution (4016×6016) for multi-view acquisition. We adjust the focal length for each camera individually, so the hand occupies about 2500×2500 pixels and the entire capture volume covers $0.6\,\text{m} \times 0.6\,\text{m} \times 0.6\,\text{m}$. To avoid high specular reflections

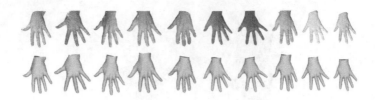

Fig. 5. Captured rest pose and registered personalized hand templates from ten human subjects. The top row shows the initial point clouds and the bottom row shows aligned meshes.

from the skin, we add 12 light units (6 LED strips and 6 LED spotlights) to provide a fairly diffuse lighting environment.

To preserve the realistic appearance of hands, we do not use markers or any speckle projectors to assist 3D reconstruction. The point cloud reconstruction is purely image based. Since the full-resolution image preserves very fine skin details, which benefits correspondence matching and the final 3D reconstruction, we run the commercial photogrammetry software Reality Capture [3] and obtain very dense point clouds. The points are around 400K on average. Downsampling wipes out most fine details and the number of points is reduced more dramatically than the image pixel decrease.

We recruited 10 participants (5 males and 5 females). We first capture the rest pose for both the left and right hands of the participants. The rest poses are used for personalized templates. After that, participants perform 50 gestures listed in the categories in Sect. 3. Hence, we collect 500 point clouds and 12,000 images with this system. Figure 5 shows hand variance for all 10 participants and registered personalized templates.

4   Skeleton-Shape Alignment

In this section, we develop an approach similar to [8]. We first compute a personalized hand template (PHT) $\overline{T}$ for each subject in the rest pose. We then align $\overline{T}$ to more complex poses and perform hole inpainting of point clouds after fitting. In this way, we are able to compensate for pose misalignment caused by the different hand shapes of each subject.

Registration. Figure 6 shows our registration process. We treat the registration process as an optimization problem. Our goal is to minimize the distance between the point cloud and the template. The registration can be separated into three parts: personalized hand template (PHT) calibration, pose/skeleton deformation, and non-rigid shape deformation.

Personalized Hand Template (PHT) Calibration. We start with an initial global hand template (GHT) provided by Wrap3D [2]. The GHT has 3,817 vertices and 7,602 faces. This model has several favorable qualities. First, it has a quad-based topology which is preferred in computer animation. Second, the edges in this model follow the palm print and fingernails, which is more realistic and better for pose fitting. To label the semantic parts, we asked an artist to rig this model and assign vertex labels based on the largest bone weights. The GHT comes in the rest pose.

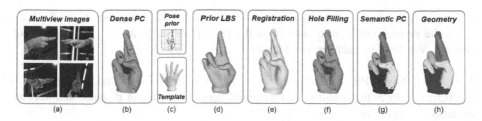

Fig. 6. From left to right: (a) shows sample images captured by our multi-camera system. (b) shows the recovered dense point cloud. (c) Pose prior from the glove and the Personalized template from the rest pose. (d) shows the warped results using the template (c). (e) shows the registered template. (f) shows the point cloud after we conduct semantic Poisson meshing (red). (g) shows the final semantically labeled point cloud. (h) shows the final mesh geometry. (Color figure online)

To generate PHT for each participant, we first capture their hand in the rest pose. Then we register the GHT to the point cloud to generate the PHT. Since the rest pose point cloud is mostly clean and complete, we are able to register PHT using Wrap3D [2] with minimal human supervision, the process for one subject is under 10 min.

Pose/Skeleton Deformation. Then for other poses, we perform pose/skeleton deformation. The optimization variables are the pose parameters Θ in our pose space. Although we fixed wrist position and orientation during capture, there still exists a small misalignment caused by human movements. Therefore, we add three more dimensions at the beginning of the pose parameter for global rotation.

Inspired by the tree-like structure of human hands, we estimate the deformation in a hierarchical manner. In practice, we optimize the first three dimensions of pose parameters with the rest parameters fixed. We run the optimization once to get the global rotation. Next, since we have semantic labels on $\overline{T}$, we optimize the palm and then the four fingers separately in multiple passes. More specifically, instead of finding correspondence between $\overline{T}$ and point set S, in each pass, we search within a subset $\overline{T}' \in \overline{T}$, which is a semantic part of $\overline{T}$ (palm, thumb, index, middle, and pinky finger). In addition to reducing computational costs, our experiments show that such hierarchical optimization also improves accuracy by a small amount, see Sect. 5 for a discussion on this method.

Non-rigid Deformation. Following the method from [12], we also adopt an embedded deformation (ED) graph for non-rigid shape deformation. We sample control nodes $\mathcal{N}$ by interval ϵ between nodes on the PHT $\overline{T}$. Then we construct a transformation graph from these nodes, where each node defines a local warp field. Each node affects vertices within a geodesic distance of 3ϵ. We follow the method in [12] to update each vertex according to node transformations.

Optimization. To solve for the deformation variables, we set out to minimize template pose parameters Θ and model control nodes deformation $\mathcal{N}$ on the template against point set S:

$$E(\Theta, \mathcal{N}; S) = \lambda_d E_d + \lambda_r E_r + \lambda_p E_p, \qquad (1)$$

where E_d is the data term evaluating the differences between point set S and PHT $\overline{T}$, E_r is the regularization term, and E_p is the pose prior term. λ_d, λ_r, and λ_p are the weights. We describe each term as follows.

Data Term. Given PHT $\overline{T}$ and point set S, for each $v_i \in \overline{T}$, we find its corresponding vertex v_j in S that minimizes a weighted sum of Euclidean distance and normal deviation, denoted as:

$$\mathcal{C}_S(v_i) = \arg \min_{v_j} \alpha e^{||v_i - v_j||_2^2/\tilde{d}} + (1 - \alpha)e^{1 - n_j n_i^T/\tilde{\theta}} \tag{2}$$

where n_j is the normal of v_j. We exclude point pairs with distance over $\tilde{d}$ and normal dot product lower than $\tilde{\theta}$. We set $\tilde{d}$ to 2 cm and $\tilde{\theta}$ to $cos(20°)$. For $v_j \in S$, we also find its corresponding point v_i in template $\overline{T}$ and denote it as $\mathcal{C}_T(v_j)$. We then minimize the Chamfer Distance between $\overline{T}$ and S:

$$E_d = \frac{1}{k} \sum_{i=1}^{k} ||v_i - \mathcal{C}_S(v_i)||_2^2 + \frac{1}{m} \sum_{j=1}^{m} ||v_j - \mathcal{C}_{\overline{T}}(v_j)||_2^2, \tag{3}$$

where $k = ||\overline{T}||$ and $m = ||S||$.

Regularization Term. This term prevents the transformations between adjacent nodes (of the non-rigid deformation) to deviate too far. It is as follows:

$$E_r = \sum_{N_i \in \mathcal{N}} \sum_{N_j \in \mathcal{N}} w_{ij}(\mathbf{M}(N_i)N_j - \mathbf{M}(N_j)N_i), \tag{4}$$

$\mathcal{N}$ is the set of control nodes. $\mathbf{M}(\cdot)$ denotes the rigid transformation matrix of the corresponding node to its position solved by the skeleton deformation registration. $w_{ij} = e^{-||N_i - N_j||_2^2/2\sigma^2}$ is the weight between adjacent nodes. Higher weights correspond to closer nodes. Here we use 2 for σ.

Pose Prior Term. We consider the pose parameters obtained from the data glove as the *pose prior*. They are utilized in two ways. First, they are used to initialize the pose parameters Θ of the skeleton deformation. Second, we add the term $E_p = ||\Theta - \overline{\Theta}||^2$ to the objective function to prevent the solved pose parameters from deviating too much from the pose prior. $\overline{\Theta}$ are the pose parameters from the data glove.

Implementation Details. For the initial PHT calibration, we only consider the first 10 dimensions in pose parameters found by PCA to improve robustness. Then all the principal components are used for pose registration. We conduct an experiment with a different number of principal components. See Sect. 5 for more detail.

We use Limited-memory BFGS (L-BFGS) optimizer to minimize the objective functions. Gradients are computed on GPU by automatic differentiation using PyTorch [7]. The optimization takes approximately 4 min with an NVIDIA 1080 Ti GPU.

Mesh Generation. The template registration enables efficient hole filling in the point cloud S. To do so, we first subdivide the registered PHT, $\overline{T}$, to achieve a roughly similar density between S and $\overline{T}$. Afterward, we add vertices of the subdivided $\overline{T}$ that do not have enough close neighbors in S (i.e., there were holes in S near such vertices) as points to S. See Fig. 6 for an example. The red points are copied from $\overline{T}$ to fill in holes in the palm and ring finger regions.

For each point in S, we use Eq. (2) to find the correspondence point in $\overline{T}$ and copy its semantic label. Directly applying Poisson surface reconstruction [5] on those points still causes topology errors like stitched fingers, because points between adjacent fingers are too close to be separated. To solve this artifact, we develop a semantic-based Poisson algorithm by considering the part labels. For normal estimation on the finger point, we only select close points marked as the same finger label as that point to form the local patch. We also add semantic labels as weights to the volume when solving the Poisson equation, and finally, use this semantic volume to extract surface mesh. One recovered geometry is shown in Fig. 6(h).

5   Data Evaluation and Applications

The main contribution of our work is a new hand dataset. To evaluate its usefulness, we first evaluate the compactness of its pose space and measure how pose prior and hierarchy optimization improve registration accuracy and reliability. Finally, we discuss potential evaluations. The registration parameters are set as follows: $\lambda_d = 0.8, \lambda_r = 0.1, \lambda_p = 0.2, \alpha = 0.5$.

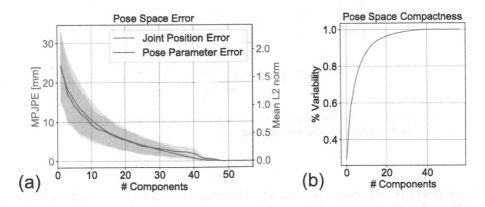

Fig. 7. Pose space evaluation. (a) plots the Mean Per Joint Projection error (blue) and parameter distance (red) using different numbers of principal components in the pose space. The vertical span depicts the mean and standard deviation of the error. (b) plots the compactness measure in terms of the variance vs. the number of principal components. (Color figure online)

Pose Space Compactness. Figure 7(a) shows the mean L2 norm of the pose parameters using the complete pose dimension vs. partial pose dimension. We further use the recovered parameters to deform the skeleton and then compute the Mean Per Joint Projection Error (MPJPE), obtained as the averaged L2 distance with respect to the ground truth over all joints. MPJPE of the first 24 principal components is 3.97 mm and the parameter error is 0.30. Figure 7(b) shows the compactness plot of the pose space. The plot shows that the first 5, 15, and 30 components manage to cover 68%, 92%, and 98% of the complete space. We thus use the first 30 principal components in our optimization procedure for robustness and efficiency.

Pose Prior. We have also conducted experiments to evaluate the usefulness of pose prior with varying principal components, as shown in Fig. 8. The blue line shows the registration error without pose prior whereas the green shows the one with pose prior. Here we use Eq. (1) for evaluation since the ground truth pose is not available. It can be seen that, using pose prior performs better. As when optimization of Eq. (1) starts from far away from the desired minimum, it can be easily trapped in local minima, however, this can be avoided by using the pose prior to starting the optimization from a point that is intuitively much closer to the desired minimum.

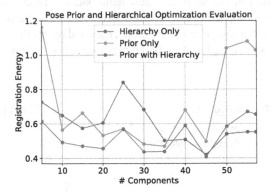

Fig. 8. Pose prior and hierarchical optimization evaluation. We compare the alignment quality using prior only, hierarchy optimization only, and both.

Hierarchical Optimization. We conduct additional experiments to evaluate the effect of hierarchical optimization for pose deformation parameters. Figure 8 plots the registration error in Eq. (1) with (green) and without (yellow) hierarchical optimization. With hierarchical optimization, our deformation module is more stable with sufficient numbers of principal components, and the average error improves by 0.1 compared to direct optimization with 45 principal components. The plot also shows that the registration procedure produces better results with more principal components, although, as aforementioned, the first 30 principal components suffice the need in nearly all cases, even challenging gestures.

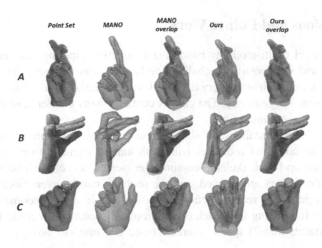

Fig. 9. Qualitative comparisons on shape registration using MANO vs. CHANDS. From left to right, the reconstructed point clouds, MANO's pose-fitting results, and CHANDS's pose-fitting results. CHANDS produces a much tighter fit to the recovered 3D geometry.

CHANDS vs. MANO. We have conducted point cloud registration on both CHANDS and MANO dataset [8]. Specifically, we register MANO to our acquired point clouds. Recall MANO was generated without personalized hand templates as CHANDS. Nor does it conduct non-rigid shape deformations for geometry. We compare three hand gestures of the same person as shown in Fig. 9. For MANO, we fit the shape parameters to the rest pose for template personalization. We report the results in Table 2. Here our model denotes the pose space learned from CHANDS with the artist-rigged template. It can be seen that the fitting results using CHANDS outperform MANO even with low (10) to moderate (30) numbers of principal components. Figure 9 further illustrates that the recovered pose is more accurate and the final 3D model can tightly fit our recovered point cloud where MANO produces significant gaps. This is largely attributed to that MANO does not contain complex poses as in CHANDS for training.

Table 2. Evaluations of our model and MANO on challenging poses. Errors are illustrated using the registration energy with 10 and 30 principal components respectively.

Pose	Ours-10	Ours-30	MANO-10	MANO-30
A	0.50	0.45	0.64	0.69
B	0.41	0.30	0.75	0.58
C	0.54	0.53	0.67	0.63

6 Conclusions and Future Work

We present a novel multi-modality hand data acquisition pipeline that separates pose and geometry, and we construct a challenging hand gesture dataset, CHANDS, comprising low-level multi-view imagery data, mid-level pose/skeleton of fingers and palm, and high-precision 3D geometry. The dataset contains many challenging gestures such as a closed fist, crossed fingers, etc.

So far CHANDS contains only single hand gestures. One interesting venue for future work is to construct a dataset with both hands. There are several challenges to do so. For mocap-based skeleton capture, the inertial sensors exhibit strong interference when the fingers are crossed, and the problem can be even more severe with both hands. For geometry reconstruction, occlusions are likely to become more severe with both hands interacting with each other. Promising solutions include 1) the use of semantic constraints and 2) special markers (e.g., UV inks) to assist in tracking and reconstruction.

Acknowledgement. This work was supported by NSFC programs (61976138, 61977047), the National Key Research and Development Program (2018YFB2100500), STCSM (2015F0203-000-06), and SHMEC (2019-01-07-00-01-E00003).

References

1. Nansense. https://www.nansense.com/gloves/
2. R3ds. https://www.russian3dscanner.com/
3. Realitycapture. https://www.capturingreality.com/
4. ElKoura, G., Singh, K.: Handrix: animating the human hand. In: Proceedings of the 2003 ACM SIGGRAPH/Eurographics Symposium on Computer Animation, pp. 110–119 (2003)
5. Kazhdan, M., Hoppe, H.: Screened Poisson surface reconstruction. ACM Trans. Graph. (ToG) 32(3), 29 (2013)
6. Moon, G., Yu, S.-I., Wen, H., Shiratori, T., Lee, K.M.: InterHand2.6M: a dataset and baseline for 3D interacting hand pose estimation from a single RGB image. In: Vedaldi, A., Bischof, H., Brox, T., Frahm, J.-M. (eds.) ECCV 2020. LNCS, vol. 12365, pp. 548–564. Springer, Cham (2020). https://doi.org/10.1007/978-3-030-58565-5_33
7. Paszke, A., et al.: Automatic differentiation in PyTorch (2017)
8. Romero, J., Tzionas, D., Black, M.J.: Embodied hands: modeling and capturing hands and bodies together. ACM Trans. Graph. (TOG) 36(6), 245 (2017)
9. Schönberger, J.L., Frahm, J.M.: Structure-from-motion revisited. In: Proceedings of the IEEE Conference on Computer Vision and Pattern Recognition, pp. 4104–4113 (2016)
10. Shotton, J., et al.: Real-time human pose recognition in parts from single depth images. In: Proceedings of the IEEE Conference on Computer Vision and Pattern Recognition, pp. 1297–1304 (2011)
11. Simon, T., Joo, H., Matthews, I., Sheikh, Y.: Hand keypoint detection in single images using multiview bootstrapping. In: Proceedings of the IEEE Conference on Computer Vision and Pattern Recognition, pp. 1145–1153 (2017)
12. Sumner, R.W., Schmid, J., Pauly, M.: Embedded deformation for shape manipulation. ACM Trans. Graph. (TOG) 26(3), 80 (2007)
13. Yuan, S., Ye, Q., Stenger, B., Jain, S., Kim, T.K.: BigHand2.2M benchmark: hand pose dataset and state of the art analysis. In: Proceedings of the IEEE Conference on Computer Vision and Pattern Recognition, pp. 4866–4874 (2017)

14. Zhou, Y., Habermann, M., Xu, W., Habibie, I., Theobalt, C., Xu, F.: Monocular real-time hand shape and motion capture using multi-modal data. In: Proceedings of the IEEE/CVF Conference on Computer Vision and Pattern Recognition, pp. 5346–5355 (2020)
15. Zimmermann, C., Ceylan, D., Yang, J., Russell, B., Argus, M., Brox, T.: FreiHAND: a dataset for markerless capture of hand pose and shape from single RGB images. In: Proceedings of the IEEE/CVF International Conference on Computer Vision, pp. 813–822 (2019)

A Compliant Elbow Exoskeleton
with an SEA at Interaction Port

Xiuze Xia[1,2], Lijun Han[1,2], Houcheng Li[1,2], Yu Zhang[1,2], Zeyu Liu[1,2],
and Long Cheng[1,2(✉)]

[1] State Key Laboratory of Multimodal Artificial Intelligence Systems, Institute of
Automation, Chinese Academy of Sciences, Beijing, China
[2] School of Artificial Intelligence, University of Chinese Academy of Sciences, Beijing,
China
long.cheng@ia.ac.cn

Abstract. In recent years, various series elastic actuators (SEAs) have
been proposed to enhance the flexibility and safety of wearable exoskeletons. This paper proposes an SEA composed of wave springs and installs
it at human-robot interaction port. Considering the hysteresis nonlinear
characteristics of the SEA, displacement-force models of the SEA are
established based on long short-term memory (LSTM) model and T-S
fuzzy model in a nonlinear auto-regression moving average with exogenous input (NARMAX) structure. Based on the established models, the
SEA can effectively serve as an interaction force sensor. Subsequently, the
SEA is integrated into an elbow exoskeleton, and a compliant admittance
controller is designed based on the displacement-force model. Experimental results demonstrate that the proposed approach effectively enhances
the flexibility of human-robot interaction.

Keywords: Series elastic actuators · Exoskeleton · LSTM model · T-S
fuzzy model

1 Introduction

Robots play an increasingly important role in our daily lives. With the increasing intelligence of robots [1], safe, comfortable, and efficient human-robot collaboration and interaction become the hot topic of the robot research [2,3].
Exoskeletons are typical human-robot interaction devices, and the research of
exoskeletons can be traced back to the 1960s. Exoskeletons can be used in various
applications such as rehabilitation [4,5], assistance in heavy lifting [6], industrial
scene [7,8], and so on.

In this paper, we focus on the design and application of elbow exoskeletons.
Elbow exoskeletons can be classified into two categories based on the mechanical structure, rigid exoskeleton and soft exoskeleton. For the first category, the
rigid structure helps to maintain the desired elbow joint angle during movement, reducing the risk of misalignment or instability [9]. Rigid robots allow for

B. Luo et al. (Eds.): ICONIP 2023, LNCS 14450, pp. 146–157, 2024.
https://doi.org/10.1007/978-981-99-8070-3_12

precise control and record of the joint angles [10]. They can be designed with specific mechanical constraints and limitations to guide and restrict the range of motion [11]. Moreover, rigid exoskeletons often have simpler control systems, which can result in easier maintenance and operation, making them more accessible and user-friendly [12]. It's important to note that while rigid exoskeletons offer these advantages, they may also have limitations, such as higher weight and inertia, reduced flexibility, potential discomfort due to rigid structures, and limited adaptability to individual anatomies.

Soft elbow exoskeletons are made of flexible and elastic materials, such as fabric or elastomers [13]. This allows them to conform to the user's arm shape and size, providing a customized and comfortable fit. The flexibility of the materials also enables a greater range of motion, mimicking the natural movement of the elbow joint. Soft exoskeletons are generally lighter and more compact compared to rigid exoskeletons [14, 15]. The compact design also makes them easier to wear and transport. Moreover, soft exoskeletons are inherently safer due to their compliance and flexibility. The soft and flexible materials absorb and distribute forces, reducing the risk of injury to both the user and the surrounding environment. However, it is also necessary to point out the limitations of flexible exoskeletons, such as reduced strength augmentation, lower control accuracy, and lower mechanical stability compared to rigid exoskeletons.

To combine the advantages of both rigid and soft elbow exoskeletons, rigid elbow exoskeletons with Series Elastic Actuators (SEAs) are proposed [16]. SEAs integrated into rigid exoskeletons allow for force sensing and control capabilities. The elastic element in the SEA can measure the interaction force between the exoskeleton and the user's arm, providing valuable feedback for force control strategies [17,18]. This enables more precise and adaptive force assistance or resistance based on the user's needs. The elastic element in the SEA provides passive compliance, allowing the exoskeleton to absorb and dampen external forces [19]. This can improve the user's comfort and enhance the overall safety of the interaction. Additionally, the SEA can adapt to variations in user anatomy and arm dimensions, making the exoskeleton more versatile and suitable for a wider range of users [20, 21]. We also need to be aware of the limitations of the SEAs, such as increased complexity in design and control [22–24]. Therefore, it is necessary to simplify the design and control of rigid exoskeletons with SEA on the basis of retaining the original function.

2 Mechanical Design

2.1 Exoskeleton Design

This paper proposes an elbow exoskeleton with SEA, as shown in Fig. 1a. The motor shaft is directly aligned with the human elbow joint. To prevent the exoskeleton from sliding and causing joint misalignment, we designed a back fixation device to secure the elbow exoskeleton, as shown in Fig. 1b.

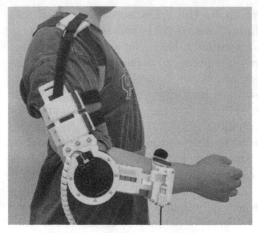

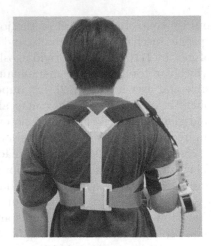

(a) Side view of the exoskeleton.   (b) Back fixation device.

Fig. 1. The elbow exoskeleton.

2.2 SEA Analysis

The design goal of SEA in this paper is to minimize space occupation and better serve the controller design. The proposed SEA consists of two symmetrical wave springs, and the compression of the SEA is measured by a displacement sensor, as illustrated in Fig. 2. The compact structure of the wave springs can reduce the space occupied by the SEA. Additionally, the high linear stiffness of the wave springs is beneficial for SEA modeling. Furthermore, in contrast to other research, we positioned the SEA at the human-robot interaction port, as shown in Fig. 2. The interaction port typically refers to the area where human and robot make contact, as described in [25]. This installation position offers the following two advantages.

Firstly, after establishing the displacement-force model of the SEA, it can be utilized as a force sensor. Accurate measurement of human-robot interaction force is a focus in robotics research [26]. The SEA installed at the human-robot interaction port can directly measure the interaction force, whereas the force measured by the SEA installed at the joint comprises a mixture of interaction force, exoskeleton gravity, inertial force, and other factors. The accurate measurement of pure interaction force is crucial for the design of human-robot interaction control. We illustrate this with an example of the admittance controller design for the elbow exoskeleton. The expression of the admittance controller is as follows.

$$\ddot{q}_e = M^{-1}(\tau_i - B\dot{q}_e - Kq_e)$$
$$\dot{q}_e = \dot{q}_e + \ddot{q}_e \Delta T$$
$$\dot{q}_c = \dot{q}_d + \dot{q}_e \tag{1}$$

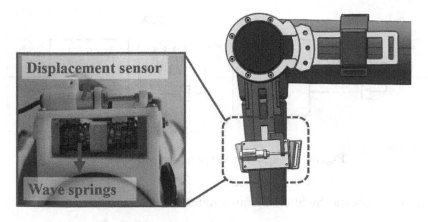

Fig. 2. SEA installed at the interaction port.

where q_e is the difference between the desired and the actual joint angle, M is the inertial parameter, τ_i is the human-robot interaction force, B is the desired damping coefficient, K is the desired stiffness coefficient, ΔT is the control period, $\dot{q}_c$ is the actual velocity control command, and $\dot{q}_d$ is the reference velocity. If the SEA is installed at the human-robot interaction port, the measured torque represents the interaction torque and can be directly substituted as τ_i into Eq. (1). However, if the SEA is installed at the elbow joint, the measured torque is the resultant torque of interaction torque, exoskeleton link gravity, inertial torque, and so on. It requires estimating the interaction torque based on the measured torque. This increases the design difficulty of the controller and decreases the measurement accuracy of the interaction force.

Secondly, the SEA installed at the human-robot interaction port can filter out external high-frequency interference, which contributes to the stability of the system [25]. For the sake of convenience in presentation, we convert the motor motion into translational motion for analysis, as shown in Fig. 3. The subscript "m" represents the motor, and the subscript "s" represents the SEA. F_e represents the interaction force, V_e represents the interaction velocity, F_f is the filtered interaction force, and V_f is the filtered interaction velocity. Then we can establish the following mapping relationship.

$$\begin{bmatrix} V_e \\ F_f \end{bmatrix} = \begin{bmatrix} Y_{ee} & T_{ef} \\ T_{fe} & Z_{ff} \end{bmatrix} \begin{bmatrix} F_e \\ V_f \end{bmatrix} \qquad (2)$$

$$T_{ef} = \frac{B_s s + K_s}{M_s s^2 + B_s s + K_s}, \quad T_{fe} = -\frac{B_s s + K_s}{M_s s^2 + B_s s + K_s},$$

$$Y_{ee} = \frac{s}{M_s s^2 + B_s s + K_s}, \quad Z_{ff} = \frac{M_s B_s s^2 + M_s K_s}{M_s s^2 + B_s s + K_s}. \qquad (3)$$

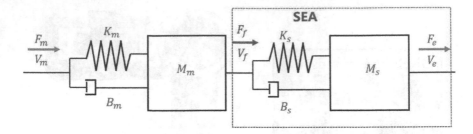

Fig. 3. The model of the proposed exoskeleton.

Then we can obtain the filtered human-robot interaction impedance.

$$Z_f = \frac{F_f}{V_f} = Z_{ff} + Z_{ee}\frac{T_{ef}T_{fe}}{1 - Y_{ee}Z_e} = \frac{M_s s(B_s s + K_s)}{M_s s^2 + B_s s + K_s}$$

$$+ Z_e\frac{(B_s s + K_s)^2}{(M_s s^2 + B_s s + K_s)}\frac{1}{(M_s s^2 + B_s s + K_s + sZ_e)} \qquad (4)$$

where Z_e is the interaction impedance. From the Eq. (4), it can be observed that for high-frequency interference, the filtered interaction impedance exhibits damping characteristics, and $Z_f \approx B_s$. Therefore, the SEA installed at the human-robot interaction port can effectively filter out the high-frequency interference.

3   SEA Modeling

The use of wave springs in the SEA allows for higher linearity in static displacement-force characteristics. However, due to the flexible properties and friction, there is a nonlinear hysteresis phenomenon in the dynamic displacement-force relationship. To fulfill the requirements for measuring interaction force, we separately established the SEA's T-S fuzzy model and long short term memory (LSTM) model. To enhance the hysteresis modeling capability of these models, they are designed in a nonlinear auto-regression moving average with exogenous input (NARMAX) structure.

3.1   NARMAX Model

The structure of the NARMAX model is shown as follows

$$f(k) = g[f(k-1), \cdots, f(k-n_f), d(k), \cdots, d(k-n_d+1)] \qquad (5)$$

where k is the current time, $f(k)$ and $d(k)$ represent the SEA's output force and displacement at time k, $g(\cdot)$ represents the mapping between the model's input and output variables, n_f and n_d represent the time delays of input and output variables respectively. Since $g(\cdot)$ contains historical information, it is suitable for nonlinear hysteresis systems modeling [27–29].

3.2   T-S Fuzzy Model

This section adopts the T-S fuzzy model as a specific expression of $g(\cdot)$. According to the model structure provided in Eq. (5), the fuzzy T-S model is represented as follows:

$$R_i : \text{If } X(k) \text{ is } A_i, \text{ then } f_i(k) = p_i(X(k))(i = 1, 2, \cdots, K) \qquad (6)$$

where R_i represents the i-th fuzzy rule of the T-S fuzzy model, A_i denotes the i-th fuzzy set, $X(k) = [f(k-1), \cdots, f(k-n_f), d(k), \cdots, d(k-n_d+1)] \in \Re^{n_f+n_d}$ represents the input vector, K is the total number of fuzzy rules, $f_i(k)$ represents the output of the i-th sub-model. $p_i(\cdot)$ represents the mapping in the i-th sub-model at time k which is shown in Eq. (7).

$$p_i(X(k)) = \sum_{j=1}^{n_f}(-a_{ij}f(k-j)) + \sum_{l=1}^{n_d} b_{il}d(k-l+1) \qquad (7)$$

where a_{ij} and b_{il} are the i-th sub-model's coefficients. The final output of the T-S fuzzy model is obtained by weighting the outputs of each sub-model. The weights are determined by the degree of matching between the current state and the fuzzy sets. The Gaussian membership functions are employed to calculate sub-models' matching degrees $\beta_i(X(k))$ $(i = 1, 2, \cdots, K)$, that is, the degree $X(k)$ belongs to A_i. The expression of the model's output is shown in Eq. (8).

$$f(k) = \frac{\sum\limits_{i=1}^{K} \beta_i(X(k)) f_i(k)}{\sum\limits_{i=1}^{K} \beta_i(X(k))} \qquad (8)$$

3.3   LSTM Model

In addition to the T-S fuzzy model, we also established an LSTM model for the SEA. The LSTM model is a type of gated recurrent neural network that introduces gate mechanisms to address the issue of long-term dependencies in traditional recurrent neural networks [30, 31]. The proposed LSTM model consists of 5 layers: input layer, two LSTM layers, fully connected layer, and output layer. The input and output of the LSTM model follow the NARMAX structure in Eq. (5) as well.

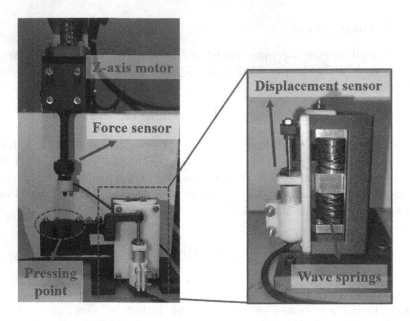

Fig. 4. Model training data acquisition platform.

3.4   Model Training

Both the T-S fuzzy model and LSTM model are data-driven models that require the collection of displacement and force data. We designed a data acquisition device as shown in Fig. 4. A stepper motor is used to achieve Z-axis motion, pressing the SEA at different frequencies and amplitudes. A force sensor is used to record the interaction force, and a displacement sensor is used to record the pressing displacement.

To obtain the specific expression of fuzzy rules in Eq. (6), the collected experimental data is subjected to fuzzy clustering to calculate the corresponding fuzzy partition matrix. Fuzzy sets A_i are obtained from the fuzzy partition matrix. In each cluster, the parameters of the sub-model Eq. (6) are obtained through the method of least squares identification. The parameters of the LSTM model are trained using the "trainNetwork" function in MATLAB. Table 1 shows the hyperparameters for the two models.

3.5   Model Validation

The effectiveness of the proposed models is validated during periodic motions with frequencies of 1 Hz, 0.5 Hz, 0.25 Hz, and 0.125 Hz. The models' outputs in 1 Hz and 0.125 Hz are shown in Fig. 5. It is obvious that with the frequency increasing, the nonlinear hysteresis phenomenon of the SEA becomes more pronounced. Both models effectively capture the nonlinear hysteresis characteristics of the SEA, and the mean absolute errors (MAE) of the models are shown in

Table 1. Hyperparameters of the models.

	Hyperparameter	Value
LSTM model	The number of LSTM layers	2
	The number of neurons in LSTM layers	20
T-S fuzzy model	n_f	3
	n_d	3
	K	3

Table 2. Model prediction MAE in different frequencies.

MAE﹨FERQ Model	0.125 Hz	0.250 Hz	0.500 Hz	1.000 Hz
T-S fuzzy model	0.095 N	0.125 N	0.204 N	0.337 N
LSTM model	0.087 N	0.110 N	0.185 N	0.310 N

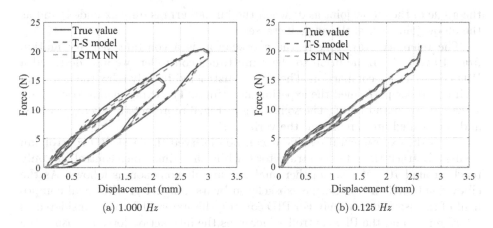

(a) 1.000 Hz   (b) 0.125 Hz

Fig. 5. Validation of the T-S fuzzy model and the LSTM model.

Table 2. It can be observed that both models achieve comparable prediction accuracy, with slightly higher accuracy for the LSTM model. And the prediction error increases with higher frequency.

4   Exoskeleton Flexible Control

To validate the effectiveness of the proposed exoskeleton, we utilized the SEA as a human-robot interaction force sensor and designed the admittance control for the exoskeleton. The expression of the controller is shown in Eq. (1). We defined

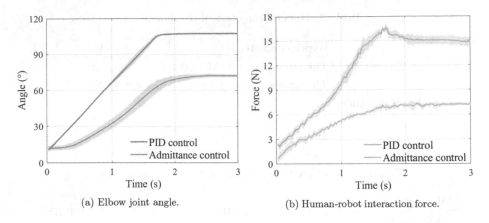

(a) Elbow joint angle.

(b) Human-robot interaction force.

Fig. 6. Validation of the controller compliance performance.

the angle of the elbow joint as 0° when the human arm is fully extended, and as the elbow joint flexes, the angle increases.

The reference trajectory of the exoskeleton is a constant velocity motion from 10 to 110°. In addition to the admittance controller, we also designed a PID controller for comparison. The user is instructed to fully relax the arm, and each controller performed the experiment 5 times. The experimental results are shown in Fig. 6 and Fig. 7. The solid lines in Fig. 6 represent the average values, and the shaded areas represent the variance.

From the experimental results, it can be observed that the PID control can accurately track the reference trajectory, while the admittance control can only reach around 70°. We can further analyze this phenomenon as follows. As the elbow joint angle increases, the exoskeleton bears a higher gravitational component of the user's upper limb. For PID control, this external force is considered as interference, and the PID controller increases the interaction force to ensure the trajectory tracking accuracy. In contrast, admittance control adjusts the reference trajectory based on the interaction force. As the interaction force measured by the SEA increases, admittance controller compromises with the user's gravity, sacrificing the tracking accuracy in exchange for lower human-robot interaction force. This compromise is crucial for ensuring the safety and comfort of user during the human-robot interaction.

(a) Performance of PID controller.

(b) Performance of admittance controller.

Fig. 7. Snapshots of controllers' performance.

5 Conclusion

This paper proposes an elbow exoskeleton with an SEA. Based on the NARMAX structure, T-S fuzzy model and LSTM model for the SEA are designed. Experiments show that the proposed models effectively capture the nonlinear hysteresis characteristics of the SEA. By directly installing the SEA at the human-robot interaction port, the measurement accuracy of the interaction force is improved. Using the established displacement-force model of the SEA, an admittance controller is designed. Comparative experiments demonstrate that the designed admittance controller provides excellent compliance for human-robot interaction.

Acknowledgements. This work is supported by National Key Research & Development Program (Grant No. 2022YFB4703204) and National Natural Science Foundation of China (Grant Nos. 62025307 and 62311530097).

References

1. Qiao, H., Wu, Y.X., Zhong, S.L., Yin, P.J., Chen, J.H.: Brain-inspired intelligent robotics: theoretical analysis and systematic application. Mach. Intell. Res. **20**(1), 1–18 (2023)
2. Weiss, A., Wortmeier, A.K., Kubicek, B.: Robots in industry 4.0: a roadmap for future practice studies on human-robot collaboration. IEEE Trans. Hum. Mach. Syst. **51**(4), 335–345 (2021)
3. Cao, R., Cheng, L., Li, H.: Passive model predictive impedance control for safe physical human-robot interaction. IEEE Trans. Cogn. Dev. Syst. (2023). https://doi.org/10.1109/TCDs.2023.3275217

4. Cheng, L., Xia, X.: A survey of intelligent control of upper limb rehabilitation exoskeleton. Robot **44**(6), 750–768 (2022)
5. Qian, W., et al.: CURER: a lightweight cable-driven compliant upper limb rehabilitation exoskeleton robot. IEEE/ASME Trans. Mechatron. **28**(3), 1730–1741 (2023)
6. Liang, J., Zhang, Q., Liu, Y., Wang, T., Wan, G.: A review of the design of load-carrying exoskeletons. Sci. China Technol. Sci. **65**(9), 2051–2067 (2022)
7. Samper-Escudero, J.L., Coloma, S., Olivares-Mendez, M.A., Gonzalez, M.A.S.U., Ferre, M.: A compact and portable exoskeleton for shoulder and elbow assistance for workers and prospective use in space. IEEE Trans. Hum. Mach. Syst. **53**(4), 668–677 (2022)
8. Grazi, L., Trigili, E., Proface, G., Giovacchini, F., Crea, S., Vitiello, N.: Design and experimental evaluation of a semi-passive upper-limb exoskeleton for workers with motorized tuning of assistance. IEEE Trans. Neural Syst. Rehabil. Eng. **28**(10), 2276–2285 (2020)
9. Zimmermann, Y., et al.: Digital Guinea Pig: merits and methods of human-in-the-loop simulation for upper-limb exoskeletons. In: 2022 International Conference on Rehabilitation Robotics, Rotterdam, Netherlands, pp. 1–6, IEEE (2022)
10. Zhang, Y., Cheng, L., Cao, R., Li, H., Yang, C.: A neural network based framework for variable impedance skills learning from demonstrations. Robot. Auton. Syst. **160**, 104312 (2023)
11. Li, J.F., Cao, Q., Dong, M.J., Zhang, C.: Compatibility evaluation of a 4-DOF ergonomic exoskeleton for upper limb rehabilitation. Mech. Mach. Theor. **156**, 104146 (2021)
12. He, C., Xiong, C.H., Chen, Z.J., Fan, W., Huang, X.L., Fu, C.L.: Preliminary assessment of a postural synergy-based exoskeleton for post-stroke upper limb rehabilitation. IEEE Trans. Neural Syst. Rehabil. Eng. **29**, 1795–1805 (2021)
13. Ang, B.W.K., Yeow, C.H.: Design and modeling of a high force soft actuator for assisted elbow flexion. IEEE Robot. Autom. Lett. **5**(2), 3731–3736 (2020)
14. Jarrett, C., McDaid, A.J.: Robust control of a cable-driven soft exoskeleton joint for intrinsic human-robot interaction. IEEE Trans. Neural Syst. Rehabil. Eng. **25**(7), 976–986 (2017)
15. Gao, G., Liang, J., Liarokapis, M.: Mechanically programmable jamming based on articulated mesh structures for variable stiffness robots. In: 2022 IEEE/RSJ International Conference on Intelligent Robots and Systems, Kyoto, Japan, pp. 11586–11593. IEEE (2022)
16. Trigili, E., et al.: Design and experimental characterization of a shoulder elbow exoskeleton with compliant joints for post-stroke rehabilitation. IEEE/ASME Trans. Mechatron. **24**(4), 1485–1496 (2019)
17. Li, J., Li, S.Q., Tian, G.H., Shang, H.C.: Muscle tension training method for series elastic actuator (SEA) based on gain-scheduled method. Robot. Auton. Syst. **121**, 103253 (2019)
18. Li, S.H., Shi, Y., Hu, L.N., Sun, Z.: A generalized model predictive control method for series elastic actuator driven exoskeleton robots. Comput. Electr. Eng. **94**, 107328 (2021)
19. Sun, N., Cheng, L., Xia, X.: Design and hysteresis modeling of a miniaturized elastomer-based clutched torque sensor. IEEE Trans. Instrum. Measur. **71**, 7501409 (2022)
20. Lin, Y.J., Chen, Z., Yao, B.: Decoupled torque control of series elastic actuator with adaptive robust compensation of time-varying load-side dynamics. IEEE Trans. Industr. Electron. **67**(7), 5604–5614 (2019)

21. Aguirre-Ollinger, G., Yu, H.Y.: Lower-limb exoskeleton with variable-structure series elastic actuators: phase-synchronized force control for gait asymmetry correction. IEEE Trans. Rob. **37**(3), 763–779 (2020)
22. Pan, J., et al.: NESM-γ: an upper-limb exoskeleton with compliant actuators for clinical deployment. IEEE Robot. Autom. Lett. **7**(3), 7708–7715 (2022)
23. Chen, T., Casas, R., Lum, P.S.: An elbow exoskeleton for upper limb rehabilitation with series elastic actuator and cable-driven differential. IEEE Trans. Rob. **35**(6), 1464–1474 (2019)
24. Wu, K.Y., Su, Y.Y., Yu, Y.L., Lin, C.H., Lan, C.C.: A 5-degrees-of-freedom lightweight elbow-wrist exoskeleton for forearm fine-motion rehabilitation. IEEE/ASME Trans. Mechatron. **24**(6), 2684–2695 (2019)
25. Buerger, S.P., Hogan, N.: Complementary stability and loop shaping for improved human-robot interaction. IEEE Trans. Rob. **23**(2), 232–244 (2007)
26. Zou, Y., Cheng, L., Li, Z.: A multimodal fusion model for estimating human hand force: comparing surface electromyography and ultrasound signals. IEEE Robot. Autom. Mag. **29**(4), 10–24 (2022)
27. Chen, S., Billings, S.A.: Representation of non-linear systems: the NARMAX model. Int. J. Control **49**(3), 1012–1032 (1999)
28. Liu, W., Cheng, L., Hou, Z.G., Yu, J., Tan, M.: An inversion-free predictive controller for piezoelectric actuators based on a dynamic linearized neural network model. IEEE/ASME Trans. Mechatron. **21**(1), 214–226 (2016)
29. Xia, X.Z., Cheng, L.: Adaptive Takagi-Sugeno fuzzy model and model predictive control of pneumatic artificial muscles. Sci. China Technol. Sci. **64**(10), 2272–2280 (2021). https://doi.org/10.1007/s11431-021-1887-6
30. Hochreiter, S., Schmidhuber, J.: Long short-term memory. Neural Comput. **9**(8), 1735–1780 (1997)
31. Gao, H., et al.: Trajectory prediction of cyclist based on dynamic Bayesian network and long short-term memory model at unsignalized intersections. Sci. China Inf. Sci. **64**(7), 1–13 (2021). https://doi.org/10.1007/s11432-020-3071-8

Applications

Differential Fault Analysis Against AES Based on a Hybrid Fault Model

Xusen Wan(ID), Jinbao Zhang(ID), Weixiang Wu(ID), Shi Cheng(ID), and Jiehua Wang(✉)(ID)

School of Information Science and Technology, Nantong University, Nantong 226019, China
wang.jh@ntu.edu.cn

Abstract. In this paper, a differential fault analysis based on a hybrid fault model is proposed. The hybrid fault model is comprising a one-byte and multi-byte by injecting faults in the state. Through both theory and simulations, which successfully derived the key of AES-128, 192, and 256 with two, three, and four pairs of faulty ciphertexts (pairs of faulty ciphertext refers to the correct ciphertext and the responding faulty ciphertext after injecting faults) without exhaustive search, respectively. Compared with the latest methods, the method proposed only requires the fault injected in a single round, thus it is easier to carry out to an attacker. When considering AES-192, fewer faulty ciphertexts are needed. In addition, for both AES-192 and 256, our method requires fewer depths of induced fault (the entire key can be retrieved only need to induce fault in the T-2 round). Thus, the DFA proposed in this article is more efficient.

Keywords: AES · DFA · block cipher · hybrid fault model

1 Introduction

In 2001, the National Institute of Standards and Technology (NIST) adopted the Advanced Encryption Standard (AES), as a symmetric block cryptographic standard [1]. It has a wide spectrum of applications in SIM cards, WIFI routing, and encrypted transmission of vital information. For public resources, the protection of data is more important, such as climate measurement, environmental monitoring, and other data, and there is no room for error [2, 3]. AES supports three key sizes of 128, 192, and 256, and 128 with block sizes.

Block ciphers are subjected to Differential Fault Analysis (DFA), such as AES, Blowfish, MacGuffin, RC6, and SMS4 by analyzing the differential information of the correct and faulty ciphertext and have posed a significant threat to information security. The application of DFA on AES is discussed in the paper. In 1997, Biham and Shamir first proposed DFA and applied it in DES [4]. Subsequently, DFA has been employed to crack RC4 [5, 6], RSA [7, 8], LEA [9], SIMON [10, 11], and PRESENT [12].

When considering AES, DFA is divided into two categories depending on the fault location. One is that fault is injected into the key schedule [13–17]. Another in which the fault is induced at the state [18–24]. Besides, according to the fault model, DFA is also be classed into one-byte fault models and multi-byte fault models.

B. Luo et al. (Eds.): ICONIP 2023, LNCS 14450, pp. 161–171, 2024.
https://doi.org/10.1007/978-981-99-8070-3_13

In 2009, a one-byte fault was injected in the key schedule for retrieving the key of AES-192 with 16 pairs of faulty ciphertexts [16]. Later, in 2010, Barenghi *et al.* [22] described a method based on a one-byte fault to crack AES-192 and 256 with the same amount of 16 pairs of faulty ciphertexts. In 2011, Tunstall *et al.* [21] utilized a fault model of a one-byte fault and succeed to retrieve the key of AES-128 with one pair of faulty ciphertext and 2^8 exhaustive searches. Whereafter, Kim [14] showed another one-byte fault model to retrieve the key of AES-128, 192, and 256 with two, four, and four pairs of faulty ciphertexts, respectively.

Regarding multi-byte fault models, in 2009, the authors conducted a DFA based on regular four-byte fault models against AES-128 with two pairs of faulty ciphertexts [19]. In 2011, another regular eight-byte fault model was presented by Kim [20] to crack AES-192 and 256, which required two pairs of faulty ciphertexts with 2^8 exhaustive searches and four pairs of faulty ciphertexts, respectively [20]. In 2019, Zhang *et al.* [15] mentioned a two-byte of discontinuous rows fault model to crack AES, in which the key of AES-128, 192, and 256 could be found with two, four, and four pairs of faulty ciphertexts without exhaustive search, respectively.

By establishing a hybrid model, which makes use of both one-byte and multi-byte fault models, an efficient DFA is proposed in this paper, and the innovation points are as follows:

- The efficiency of cracking AES has been improved. The method proposed to crack AES-128, 192, and 256 with two, three, and four pairs of faulty ciphertexts without exhaustive search, respectively. In [25], Liu proposed a hybrid fault model, using six pairs of faulty ciphertexts to crack AES-128. Kim [14] used two and four pairs of faulty ciphertexts to crack AES-128 and 256, which is the same number of pairs compared with our method. However, our method has a lower depth of injected fault. Additionally, in cracking AES-192, the hybrid model proposed reduces the times of inducing faulty due to the same location of the induced fault.

  The DFA proposed in [20] could crack AES-192 using two pairs of faulty ciphertexts and 2^8 exhaustive searches. Although our method needs three pairs, the AES-192 key can be recovered without exhaustive search.

- The faults are injected in a single round, thus, in the process of injecting a fault, the clock or voltage glitches are invariable. Specifically, the proposed method needs to induce fault twice at S^8 of AES-128, three times at S^{10} of AES-192, and four times at S^{12} of AES-256, respectively. Whereas, in [25], another hybrid model requires inducing faults both in the key schedule and state.

  Compared with these existing DFAs on AES, the hybrid model is used in our method, and the faults are injected in a single round. Thus, the condition of inducing fault is changeless.

- No need to know the plaintext. As we conduct the procedure of recovering the AES key without exhaustive search, therefore, it is unnecessary to figure out the plaintext to an adversary.

2   DFA on AES State

AES supports three key sizes of 128, 192, and 256, and 128 with block sizes, responding to the number of encryption rounds (R) is 10, 12, and 14, respectively. The process of AES encryption is shown in Fig. 1. The 128 bits of input data are calculated through the R round to get the 128 bits of output data, and each round will have a corresponding round key involved.

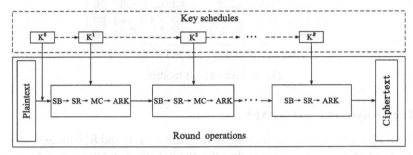

Fig. 1. The process of AES encryption.

If the plaintext (P) and key (K) are constant throughout the attack, i.e., a fault is injected at state, and obtain faulty ciphertexts (C^*) with the same plaintext and key. In the below analysis, supposing that the fault can be instantaneously injected at the state.

According to previous research work, there are various fault models. For one-byte fault models, there have been many work studies, and multi-byte fault models are also widely used. However, both the one-byte fault model and multi-byte fault model need to change the depth of fault induction when cracking AES-192 and AES-256, which will make fault induction more difficult. From the point of view of the number of fault induction and the number of induction rounds, it is necessary to reduce the number of fault induction and minimize the need to change the fault induction depth when cracking AES-192 and AES-256. Therefore, we proposed another fault model, that is, the hybrid fault model.

2.1   Proposed Fault Model

Both one-byte and multi-byte models are exploited to derive the hybrid fault model in the paper. The one-byte fault is induced at S^{R-2} and the propagation of the fault is shown in (a) of Fig. 2. The propagation of the two-byte faults is presented in (b) and (c). The propagation of the three-byte fault is shown in (d).

The analysis of cracking three variants of AES by utilizing the hybrid model will be discussed. From the perspective of fault induction depth, all ciphertexts will be affected in the end, and more differential fault equations can be obtained, which will reveal more key information, which is very useful for attackers.

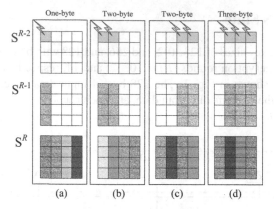

Fig. 2. Hybrid fault models.

2.2 The Analysis of Cracking AES

For AES-128, assuming that the location of the fault is at S_1^8 and S_2^8 (this assumes only for the sake of analysis and does not affect the results). The key schedule of AES-128 is shown in Fig. 3, and the green part is the round key that needs to be obtained, which is K^{10}.

Firstly, the fault is induced at S_1^8. Secondly, induced both at S_1^8 and S_2^8. The attacker unknows the fault value, but the location of faults is known.

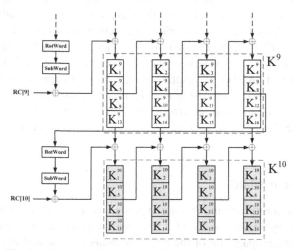

Fig. 3. The key schedule of AES-128. (Color figure online)

Figure 3 only shows the process of the last two round-key of AES-128, for all this, it contains all information of the initial key, therefore, the initial key can be recovered according to the key schedule.

For AES-192, the location of the induced fault is at S_1^{10}, S_2^{10}, S_3^{10} and S_4^{10}. If the K^{12} and the right half of K^{11} are obtained, the initial key could be found.

First of all, the fault is induced at S_1^{10}. In the second place, induced at S_1^{10} and S_2^{10}. Finally, induced at S_3^{10} and S_4^{10}. The key schedule of AES-192 is shown in Fig. 4, it only shows the process of the last two rounds' key schedule of AES-192, the key is 192 bits, and it contains all information of the initial key, therefore, the initial key can be recovered according to the last two round-key.

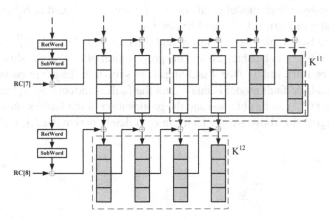

Fig. 4. The key schedule of AES-192.

For AES-256, the location of the fault is at S_1^{12}, S_2^{12}, S_3^{12} and S_4^{12}. To crack AES-256, K^{14} and K^{13} should be recovered.

The key of AES-256 can be cracked by inducing four times of fault. For the first time, the fault is induced S_1^{12}. The second time, induced at S_1^{12} and S_2^{12}. The third time, induced at S_3^{12} and S_4^{12}. For the fourth time, induced at S_2^{12}, S_3^{12} and S_4^{12}.

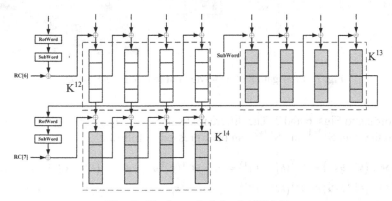

Fig. 5. The key schedule of AES-256.

The key schedule of AES-256 is shown in Fig. 5, and it only shows the process of the last round key schedule of AES-256, the key is 256 bits, and it contains all information of the initial key, therefore, the initial key can be recovered according to the last two round-key.

2.3 The Process of Cracking AES

Since the key of AES-192 and 256 require the last two round keys, AES-128 only recovers the last round key. Thus, the process of cracking AES-192 could be a representative, and the process is divided into the following steps:

We conduct encryption without fault, and fault-free ciphertext (C) is obtained.

Step 1: One-byte fault model is taken, and the fault is induced at S_1^{10}, as shown in Fig. 6. The faulty ciphertext C^{*1} could be obtained.

Step 2: Two-byte fault model is taken, and the fault is induced at S_1^{10} and S_2^{10}. The faulty ciphertext C^{*2} can be obtained and the propagation of the fault is shown in Fig. 7.

K^{12} could be retrieved by Step 1 and Step 2, as shown in the green parts of Fig. 4.

Step 3: Two-byte fault model is taken, and the fault is induced at S_3^{10} and S_4^{10}. The faulty ciphertext C^{*3} can be obtained and the propagation of the fault is shown in Fig. 8.

Final: Merging the above steps which can completely derive the AES-192 key.

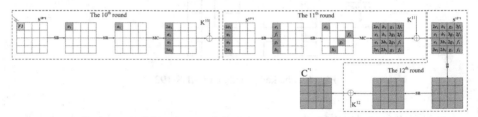

Fig. 6. Propagation of one-byte fault induced at AES-192.

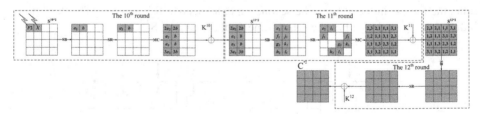

Fig. 7. Propagation of two-byte fault induced at AES-192.

According to Figs. 6 and 7. The differential values $\{e_1, h_1, g_1, f_1, e_2, h_2, g_2, f_2, l_1, i_1, j_1, k_1\}$ in the state S^{12*1} and S^{12*2} can be expressed as follows:

$$\Delta S_1^{12} = SB^{-1}\left(K_1^{12} \oplus C_1\right) \oplus SB^{-1}\left(K_1^{12} \oplus C_1^{*1}\right) = 2e_1 = SB^{-1}\left(K_1^{12} \oplus C_1\right) \oplus SB^{-1}\left(K_1^{12} \oplus C_1^{*2}\right) = 2e_2 \oplus 3j_1$$

$$\Delta S_5^{12} = SB^{-1}\left(K_8^{12} \oplus C_8\right) \oplus SB^{-1}\left(K_8^{12} \oplus C_8^{*1}\right) = e_1 = SB^{-1}\left(K_8^{12} \oplus C_8\right) \oplus SB^{-1}\left(K_8^{12} \oplus C_8^{*2}\right) = e_2 \oplus 2j_1$$

$$\Delta S_9^{12} = SB^{-1}\left(K_{11}^{12} \oplus C_{11}\right) \oplus SB^{-1}\left(K_{11}^{12} \oplus C_{11}^{*1}\right) = e_1 = SB^{-1}\left(K_{11}^{12} \oplus C_{11}\right) \oplus SB^{-1}\left(K_{11}^{12} \oplus C_{11}^{*2}\right) = e_2 \oplus j_1 \tag{1}$$

$$\Delta S_{13}^{12} = SB^{-1}\left(K_{14}^{12} \oplus C_{14}\right) \oplus SB^{-1}\left(K_{14}^{12} \oplus C_{14}^{*1}\right) = 3e_1 = SB^{-1}\left(K_{14}^{12} \oplus C_{14}\right) \oplus SB^{-1}\left(K_{14}^{12} \oplus C_{14}^{*2}\right) = 3e_2 \oplus j_1$$

$$\Delta S_2^{12} = SB^{-1}\left(K_2^{12} \oplus C_2\right) \oplus SB^{-1}\left(K_2^{12} \oplus C_2^{*1}\right) = h_1 = SB^{-1}\left(K_2^{12} \oplus C_2\right) \oplus SB^{-1}\left(K_2^{12} \oplus C_2^{*2}\right) = h_2 \oplus 2i_1$$

$$\Delta S_6^{12} = SB^{-1}\left(K_5^{12} \oplus C_5\right) \oplus SB^{-1}\left(K_5^{12} \oplus C_5^{*1}\right) = h_1 = SB^{-1}\left(K_5^{12} \oplus C_5\right) \oplus SB^{-1}\left(K_5^{12} \oplus C_5^{*2}\right) = h_2 \oplus i_1$$

$$\Delta S_{10}^{12} = SB^{-1}\left(K_{12}^{12} \oplus C_{12}\right) \oplus SB^{-1}\left(K_{12}^{12} \oplus C_{12}^{*1}\right) = 3h_1 = SB^{-1}\left(K_{12}^{12} \oplus C_{12}\right) \oplus SB^{-1}\left(K_{12}^{12} \oplus C_{12}^{*2}\right) = 3h_2 \oplus i_1 \tag{2}$$

$$\Delta S_{14}^{12} = SB^{-1}\left(K_{15}^{12} \oplus C_{15}\right) \oplus SB^{-1}\left(K_{15}^{12} \oplus C_{15}^{*1}\right) = 2h_1 = SB^{-1}\left(K_{15}^{12} \oplus C_{15}\right) \oplus SB^{-1}\left(K_{15}^{12} \oplus C_{15}^{*2}\right) = 2h_2 \oplus 3i_1$$

$$\Delta S_3^{12} = SB^{-1}\left(K_3^{12} \oplus C_3\right) \oplus SB^{-1}\left(K_3^{12} \oplus C_3^{*1}\right) = g_1 = SB^{-1}\left(K_3^{12} \oplus C_3\right) \oplus SB^{-1}\left(K_3^{12} \oplus C_3^{*2}\right) = g_2 \oplus l_1$$

$$\Delta S_7^{12} = SB^{-1}\left(K_6^{12} \oplus C_6\right) \oplus SB^{-1}\left(K_6^{12} \oplus C_6^{*1}\right) = 3g_1 = SB^{-1}\left(K_6^{12} \oplus C_6\right) \oplus SB^{-1}\left(K_6^{12} \oplus C_6^{*2}\right) = 3g_2 \oplus l_1$$

$$\Delta S_{11}^{12} = SB^{-1}\left(K_9^{12} \oplus C_9\right) \oplus SB^{-1}\left(K_9^{12} \oplus C_9^{*1}\right) = 2g_1 = SB^{-1}\left(K_9^{12} \oplus C_9\right) \oplus SB^{-1}\left(K_9^{12} \oplus C_9^{*2}\right) = 2g_2 \oplus 3l_1 \tag{3}$$

$$\Delta S_{15}^{12} = SB^{-1}\left(K_{16}^{12} \oplus C_{16}\right) \oplus SB^{-1}\left(K_{16}^{12} \oplus C_{16}^{*1}\right) = g_1 = SB^{-1}\left(K_{16}^{12} \oplus C_{16}\right) \oplus SB^{-1}\left(K_{16}^{12} \oplus C_{16}^{*2}\right) = g_2 \oplus 2l_1$$

$$\Delta S_4^{12} = SB^{-1}\left(K_4^{12} \oplus C_4\right) \oplus SB^{-1}\left(K_4^{12} \oplus C_4^{*1}\right) = 3f_1 = SB^{-1}\left(K_4^{12} \oplus C_4\right) \oplus SB^{-1}\left(K_4^{12} \oplus C_4^{*2}\right) = 3f_2 \oplus k_1$$

$$\Delta S_8^{12} = SB^{-1}\left(K_7^{12} \oplus C_7\right) \oplus SB^{-1}\left(K_7^{12} \oplus C_7^{*1}\right) = 2f_1 = SB^{-1}\left(K_7^{12} \oplus C_7\right) \oplus SB^{-1}\left(K_7^{12} \oplus C_7^{*2}\right) = 2f_2 \oplus 3k_1$$

$$\Delta S_{12}^{12} = SB^{-1}\left(K_{10}^{12} \oplus C_{10}\right) \oplus SB^{-1}\left(K_{10}^{12} \oplus C_{10}^{*1}\right) = f_1 = SB^{-1}\left(K_{10}^{12} \oplus C_{10}\right) \oplus SB^{-1}\left(K_{10}^{12} \oplus C_{10}^{*2}\right) = f_2 \oplus 2k_1 \tag{4}$$

$$\Delta S_{16}^{12} = SB^{-1}\left(K_{13}^{12} \oplus C_{13}\right) \oplus SB^{-1}\left(K_{13}^{12} \oplus C_{13}^{*1}\right) = f = SB^{-1}\left(K_{13}^{12} \oplus C_{13}\right) \oplus SB^{-1}\left(K_{13}^{12} \oplus C_{13}^{*2}\right) = f_2 \oplus k_1$$

According to sets of (1)–(4), the last round key can be deduced, and the left part of K^{11} could be found with key schedule, thus, K_1^{11}, K_2^{11}, K_5^{11}, K_6^{11}, K_9^{11}, K_{10}^{11}, K_{13}^{11} and K_{14}^{11} is known to the attacker. The result of the penultimate round is figured out as Eq. (5).

$$S^{12} = SB^{-1}(SR^{-1}(K^{12} \oplus C))$$
$$S^{12*2} = SB^{-1}(SR^{-1}(K^{12} \oplus C^{*2})) \tag{5}$$
$$S^{12*3} = SB^{-1}(SR^{-1}(K^{12} \oplus C^{*3}))$$

The differential values $\{a_2, b\}$ in the state S^{11} of Fig. 7 can be written as follows:

$$\Delta S_1^{11} = SB^{-1}\left(MC^{-1}\left(K_1^{11} \oplus S_1^{12}\right)\right) \oplus SB^{-1}\left(MC^{-1}\left(K_1^{11} \oplus S_1^{12*2}\right)\right) = 2a_2$$

$$\Delta S_2^{11} = SB^{-1}\left(MC^{-1}\left(K_2^{11} \oplus S_2^{12}\right)\right) \oplus SB^{-1}\left(MC^{-1}\left(K_2^{11} \oplus S_2^{12*2}\right)\right) = 2b$$

$$\Delta S_6^{11} = SB^{-1}\left(MC^{-1}\left(K_5^{11} \oplus S_5^{12}\right)\right) \oplus SB^{-1}\left(MC^{-1}\left(K_5^{11} \oplus S_5^{12*2}\right)\right) = b \tag{6}$$

$$\Delta S_{13}^{11} = SB^{-1}\left(MC^{-1}\left(K_{14}^{11} \oplus S_{14}^{12}\right)\right) \oplus SB^{-1}\left(MC^{-1}\left(K_{14}^{11} \oplus S_{14}^{12*2}\right)\right) = 2a_2$$

In the above equation, K_1^{11}, K_2^{11}, K_5^{11} and K_{14}^{11} is obtained, therefore, b and a_2 could be calculated.

$$\Delta S_5^{11} = SB^{-1}\left(MC^{-1}\left(K_8^{11} \oplus S_8^{12}\right)\right) \oplus SB^{-1}\left(MC^{-1}\left(K_8^{11} \oplus S_8^{12*2}\right)\right) = a_2$$

$$\Delta S_9^{11} = SB^{-1}\left(MC^{-1}\left(K_{11}^{11} \oplus S_{11}^{12}\right)\right) \oplus SB^{-1}\left(MC^{-1}\left(K_{11}^{11} \oplus S_{11}^{12*2}\right)\right) = a_2$$

$$\Delta S_{10}^{11} = SB^{-1}\left(MC^{-1}\left(K_{12}^{11} \oplus S_{12}^{12}\right)\right) \oplus SB^{-1}\left(MC^{-1}\left(K_{12}^{11} \oplus S_{12}^{12*2}\right)\right) = b \tag{7}$$

$$\Delta S_{14}^{11} = SB^{-1}\left(MC^{-1}\left(K_{15}^{11} \oplus S_{15}^{12}\right)\right) \oplus SB^{-1}\left(MC^{-1}\left(K_{15}^{11} \oplus S_{15}^{12*2}\right)\right) = 3b$$

Repeating the above procedure, plugging b and a_2 in the Eq. (7), K_8^{11}, K_{11}^{11}, K_{12}^{11} and K_{15}^{11} can be recovered.

Under the condition of K^{12} is known. The correct and faulty states of the penultimate round of encryption can be obtained.

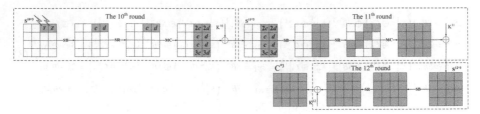

Fig. 8. Propagation of two-byte fault induced at AES-192.

The differential values $\{c,d\}$ in the state S^{11} of Fig. 8 can be described as follows:

$$\Delta S_7^{11} = SB^{-1}\left(MC^{-1}\left(K_6^{11} \oplus S_6^{12}\right)\right) \oplus SB^{-1}\left(MC^{-1}\left(K_6^{11} \oplus S_6^{12*3}\right)\right) = c$$

$$\Delta S_{11}^{11} = SB^{-1}\left(MC^{-1}\left(K_9^{11} \oplus S_9^{12}\right)\right) \oplus SB^{-1}\left(MC^{-1}\left(K_9^{11} \oplus S_9^{12*3}\right)\right) = c$$

$$\Delta S_{12}^{11} = SB^{-1}\left(MC^{-1}\left(K_{10}^{11} \oplus S_{10}^{12}\right)\right) \oplus SB^{-1}\left(MC^{-1}\left(K_{10}^{11} \oplus S_{10}^{12*3}\right)\right) = d$$

$$\Delta S_{16}^{11} = SB^{-1}\left(MC^{-1}\left(K_{13}^{11} \oplus S_{13}^{12}\right)\right) \oplus SB^{-1}\left(MC^{-1}\left(K_{13}^{11} \oplus S_{13}^{12*3}\right)\right) = 3d$$

$$(8)$$

In the above equation, K_6^{11}, K_9^{11}, K_{10}^{11} and K_{13}^{11} is obtained, therefore, c and d could be calculated.

$$\Delta S_3^{11} = SB^{-1}\left(MC^{-1}\left(K_3^{11} \oplus S_3^{12}\right)\right) \oplus SB^{-1}\left(MC^{-1}\left(K_3^{11} \oplus S_3^{12*3}\right)\right) = 2c$$

$$\Delta S_4^{11} = SB^{-1}\left(MC^{-1}\left(K_4^{11} \oplus S_4^{12}\right)\right) \oplus SB^{-1}\left(MC^{-1}\left(K_4^{11} \oplus S_4^{12*3}\right)\right) = 2d$$

$$\Delta S_8^{11} = SB^{-1}\left(MC^{-1}\left(K_7^{11} \oplus S_7^{12}\right)\right) \oplus SB^{-1}\left(MC^{-1}\left(K_7^{11} \oplus S_7^{12*3}\right)\right) = d$$

$$\Delta S_{15}^{11} = SB^{-1}\left(MC^{-1}\left(K_{16}^{11} \oplus S_{16}^{12}\right)\right) \oplus SB^{-1}\left(MC^{-1}\left(K_{16}^{11} \oplus S_{16}^{12*3}\right)\right) = 2c$$

$$(9)$$

Then, K_3^{11}, K_4^{11}, K_7^{11} and K_{16}^{11} is known, thus, the penultimate round key is deduced. The process of deducing AES-192 key thus far overall. For AES-128, the last round key is required, and the penultimate round key is additionally needed for cracking AES-256.

3   Experimental Results and Comparisons

By exploiting DFA based on the hybrid model proposed in this paper, AES can be cracked under the plaintext unknown to an attacker. The Vivado software is used for simulation. The simulation results reveal that the method taken improves the cracking efficiency of AES and reduce the difficulty of inducing faults.

The existing DFAs on AES state are listed in Tables 1, 2, and 3. In Table 1, the location of injecting fault is not in the single round, the induced fault is relatively tricky, and two pairs of faulty ciphertext are required [19]. Although one pair of faulty ciphertext is used, it requires a 2^8 exhaustive search [21]. In Tables 1, 2, and 3, a multi-byte fault model was used to crack AES-128 and 192 with less faulty ciphertext, but exhaustive searches are required [20]. The one-byte fault model proposed in the paper [23] needs two pairs of faulty ciphertexts but also requires a 2^{32} exhaustive search. 16 pairs of faulty ciphertexts are used to crack AES-192 [22]. Paper [25] shows a hybrid model that induces faults both in the state and in the key schedule, while our hybrid model is only aimed at injecting faults in the state S^{R-2} by exploiting one-byte and multi-byte fault models and requires less faulty ciphertext. In Table 1, M2 expresses a random non-zero fault that occurs across three diagonals.

This hybrid model we proposed has two merits: On one hand, a one-multi-byte hybrid fault model is employed, which reduces the difficulty of inducing faults. Compared with other models, there is no need to induce fault in the previous round or the next round. On the other hand, the second point is to reduce the number of faulty ciphertexts required to crack AES-192, and the method proposed in this paper only required three pairs of faulty ciphertexts without exhaustive search.

Table 1. Existing DFAs on the AES-128 state.

Ref.	Fault Type	No. of Faults	Fault Round	Exhaustive Search
[20] M2	Multi-byte	4	8th and 7th	1
[20]	Multi-byte	2	8th	2^8
[21]	One byte	1	9th	2^8
[19]	One byte	2	8th and 7th	1
This work	Hybrid byte	2	8th	1

Table 2. Existing DFAs on the AES-192 state.

Ref.	Fault Type	No. of Faults	Fault Round	Exhaustive Search
[20]	Multi-byte	2	11th and 10th	2^8
[22]	One byte	16	10th and 9th	1
[23] A1	One byte	2	10th and 9th	2^{32}
[20]	One byte	2	10th and 9th	1
This work	Hybrid byte	3	10th	1

4   Conclusions

We implement the proposal of an efficient DFA against AES based on the hybrid fault model presented in this paper. The induction of faults in the R-2 round allows for the retrieval of the AES-128, 192, and 256 keys with only two, three, and four pairs of faulty ciphertexts, respectively, without exhaustive search when the plaintext is unknown. This method improves the efficiency of cracking AES and reduces the difficulty of implementation compared to existing DFAs. It is also noteworthy that the single round

Table 3. Existing DFAs on the AES-256 state.

Ref.	Fault Type	No. of Faults	Fault Round	Exhaustive Search
[20]	Multi-byte	4	12th and 11th	1
[22]	One byte	16	12th and 11th	1
[20]	One byte	3	12th and 11th	1
[24]	One byte	4	11th	2^{13}
This work	Hybrid byte	4	11th	1

can be targeted for injecting faults in AES-192 and 256, making the induction of faults easier. Although the fault model proposed in this paper is specifically applied to cracking AES, it can be further explored for other block ciphers with SPN structures such as RC6 and SM4, which is also an area of interest for our future research.

Acknowledgments. The authors would like to thank the anonymous reviewers for their valuable suggestions and comments that improved the quality of this paper. This work was supported in part by the Natural Science Research in Colleges of Jiangsu Province under Grant 21KJB520040; the Basic Science Research Project of Nantong under Grant JC2020143.

References

1. Biham, E.: Advanced encryption standard. In: Biham, E. (eds) Fast Software Encryption, FSE 1997. LNCS, vol. 1267, pp. 83–87. Springer, Heidelberg (1997). https://doi.org/10.1007/BFb 0052336
2. Ramani, S., Jhaveri, R.H.: ML-based delay attack detection and isolation for fault-tolerant software-defined industrial networks. Sensors **22**(18), 6958 2022
3. Azam, S., Bibi, M., Riaz, R., et al.: Collaborative learning based sybil attack detection in Vehicular AD-HOC Networks (VANETS). Sensors **22**(18), 6934 (2022)
4. Biham, E., Shamir, A.: Differential fault analysis of secret key cryptosystems. In: Kaliski, B.S. (eds.) Advances in Cryptology — CRYPTO 1997, CRYPTO 1997. LNCS, vol. 1294, pp. 513–525. Springer, Heidelberg (1997). https://doi.org/10.1007/BFb0052259
5. Biham, E., Granboulan, L., Nguyễn, P.Q.: Impossible fault analysis of RC4 and differential fault analysis of RC4. In: Gilbert, H., Handschuh, H. (eds.) FSE 2005. LNCS, vol. 3557, pp. 359–367. Springer, Heidelberg (2005). https://doi.org/10.1007/11502760_24
6. Hoch, J.J., Shamir, A.: Fault analysis of stream ciphers. In: Joye, M., Quisquater, J.-J. (eds.) CHES 2004. LNCS, vol. 3156, pp. 240–253. Springer, Heidelberg (2004). https://doi.org/10.1007/978-3-540-28632-5_18
7. Lin, I.C., Chang, C.C.: Security enhancement for digital signature schemes with fault tolerance in RSA. Inf. Sci. **177**(19), 4031–4039 (2007)
8. Trichina, E., Korkikyan, R.: Multi fault laser attacks on protected CRT-RSA. In: 2010 Workshop on Fault Diagnosis and Tolerance in Cryptography, pp. 75–86. IEEE, Santa Barbara, CA, USA (2010)
9. Jap, D., Breier, J.: Differential fault attack on LEA. In: Khalil, I., Neuhold, E., Tjoa, A.M., Da Xu, L., You, I. (eds.) CONFENIS/ICT-EurAsia-2015. LNCS, vol. 9357, pp. 265–274. Springer, Cham (2015). https://doi.org/10.1007/978-3-319-24315-3_27

10. Jeong, K., Lee, Y., Sung, J., et al.: Improved differential fault analysis on PRESENT-80/128. Int. J. Comput. Math. **90**(12), 2553–2563 (2013)
11. Tupsamudre, H., Bisht, S., Mukhopadhyay, D.: Differential fault analysis on the families of SIMON and SPECK ciphers. In: 2014 Workshop on Fault Diagnosis and Tolerance in Cryptography, pp. 40–48. IEEE, Busan, Korea (South) (2014)
12. Zhang, J.B., Wang, J.H., Bin, G., et al.: An efficient differential fault attack against SIMON key schedule. J. Inf. Secur. Appl. **66**, 103155 (2022)
13. Kim, C.H., Quisquater, J.-J.: New differential fault analysis on AES key schedule: two faults are enough. In: Grimaud, G., Standaert, F.-X. (eds.) CARDIS 2008. LNCS, vol. 5189, pp. 48–60. Springer, Heidelberg (2008). https://doi.org/10.1007/978-3-540-85893-5_4
14. Kim, C.H.: Improved differential fault analysis on AES key schedule. IEEE Trans. Inf. Forensics Secur. **7**(1), 41–50 (2012)
15. Jinbao, Z., et al.: A novel differential fault analysis using two-byte fault model on AES key schedule. IET Circuits Devices Syst. **13**(5), 661–666 (2019)
16. Fukunaga, T., Takahashi, J.: Practical fault attack on a cryptographic LSI with IOS/IEC 18033-3 block ciphers. In: 6th International Workshop on Fault Diagnosis and Tolerance in Cryptography, FDTC 2009, pp. 84–92. IEEE Computer Society, Lusanne, Switzerland (2009)
17. Floissac, N., L'Hyver, Y.: From AES-128 to AES-192 and AES-256, how to adapt differential fault analysis attacks. In: 2011 Workshop on Fault Diagnosis and Tolerance in Cryptography, pp. 43–53, Nara, Japan (2011)
18. Piret, G., Quisquater, J.-J.: A differential fault attack technique against SPN structures, with application to the AES and Khazad. In: Walter, C.D., Koç, Ç.K., Paar, C. (eds.) Cryptographic Hardware and Embedded Systems - CHES 2003. LNCS, vol. 2779, pp. 77–88. Springer, Heidelberg (2003). https://doi.org/10.1007/978-3-540-45238-6_7
19. Saha, D., Mukhopadhyay, D., Chowdhury, D.R.: A diagonal fault attack on the advanced encryption standard. Cryptology eprint archive (2009)
20. Kim, C.H.: Differential fault analysis of AES: toward reducing number of faults. Inf. Sci. **199**, 43–57 (2011)
21. Tunstall, M., Mukhopadhyay, D., Ali, S.: Differential fault analysis of the advanced encryption standard using a single fault. Community Mental Health J. **49**(6), 658–667 (2011)
22. Barenghi, A., Bertoni, G.M., Breveglieri, L., et al.: Low voltage fault attacks to AES. In: 2010 IEEE International Symposium on Hardware-Oriented Security and Trust (HOST), pp. 7–12. IEEE, Anaheim, CA, USA (2010)
23. Kim, C.H.: Differential fault analysis against AES-192 and AES-256 with minimal faults. In: 2010 Workshop on Fault Diagnosis and Tolerance in Cryptography, pp. 3–9. IEEE, Santa Barbara, CA, USA (2010)
24. Takahashi, J., Fukunaga, T.: Differential fault analysis on AES with 192 and 256-bit keys. IACR eprint archive, 023 (2010)
25. Liu, Y., Cui, X., Cao, J., et al.: A hybrid fault model for differential fault attack on AES. In: 2017 IEEE 12th International Conference on ASIC, pp. 784–787. IEEE, Guiyang, China (2017)

Towards Undetectable Adversarial Examples: A Steganographic Perspective

Hui Zeng[1,2]([✉]) [ID], Biwei Chen[3], Rongsong Yang[1], Chenggang Li[1],
and Anjie Peng[1,2]

[1] School of Computer Science and Technology, Southwest University of Science
and Technology, Mianyang, China
zengh5@mail2.sysu.edu.cn
[2] Guangdong Provincial Key Laboratory of Information Security Technology,
Guangzhou, China
[3] Beijing Normal University, Beijing, China
bchen@colby.edu

Abstract. Over the past decade, adversarial examples have demonstrated an enhancing ability to fool neural networks. However, most adversarial examples can be easily detected, especially under statistical analysis. Ensuring undetectability is crucial for the success of adversarial examples in practice. In this paper, we borrow the idea of the embedding suitability map from steganography and employ it to modulate the adversarial perturbation. In this way, the adversarial perturbations are concentrated in the hard-to-detect areas and are attenuated in predictable regions. Extensive experiments show that the proposed scheme is compatible with various existing attacks and can significantly boost the undetectability of adversarial examples against both human inspection and statistical analysis of the same attack ability. The code is available at github.com/zengh5/Undetectable-attack.

Keywords: Adversarial examples · statistical analysis · embedding suitability map · steganography

1 Introduction

Adversarial examples (AE) [1] have attracted extensive attention in artificial intelligence and information security. Recent works show that carefully crafted AEs can fool the source model and transfer to unknown models, even with significantly different structures [2,3]. Moreover, under certain circumstances, the attacker can add image-agnostic perturbation to an arbitrary image and mislead a convolutional neural network (CNN) model to output an arbitrarily specified classification [4].

In the meantime, there are also studies on how to defend the CNN models from being fooled by AE. A typical defensive strategy is to enhance the robustness of the CNNs [5,6]. Unfortunately, these defenses are vulnerable to a

B. Luo et al. (Eds.): ICONIP 2023, LNCS 14450, pp. 172–183, 2024.
https://doi.org/10.1007/978-981-99-8070-3_14

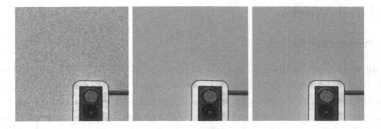

Fig. 1. Adversarial examples generated by the baseline IFGSM (left), the proposed W-IFGSM (middle) and WA-IFGSM (right).

secondary attack [7]. To our knowledge, adversarial training is the most efficient way to boost CNNs' robustness [8]. However, it suffers from huge computational demand and non-negligible accuracy loss on clean images. Facing the great challenge of designing a robust CNN, some researchers try to detect AEs before feeding them into the CNN model. For instance, Hendrycks [9] and Li [10] found the discrepancy between benign and adversarial images in their principal components. Liang [11] and Deng [12] leveraged the spatial instability nature of the AEs to detect them. Liu reformulated the AE detection as a steganalysis problem [13] and proposed a spatial rich model (SRM) [14] -based detector. Peng utilized the dependencies among three color channels of images to detect AEs [15]. Among such detect-and-reject defenses, those based on statistical analysis [13,15] have shown the most promising accuracies [16].

Emerging detectors suggest that a successful AE in practice should be able to circumvent common detectors first. The most intuitive strategy to escape statistical analysis is decreasing the attacking strength. However, this will inevitably sacrifice the attack ability. Developing undetectable AEs while preserving attack ability remains a challenge for AE designers. To fill this gap, this paper leverages the suitability map from steganography to adjust the added adversarial perturbation. As shown in Fig. 1, our proposed method encourages more substantial perturbations in unpredictable regions (e.g., the traffic light panel) and weaker perturbations in predictable regions (e.g., the blue sky). AEs crafted like this are hard to spot by human eyes and statistical detectors. Our contributions can be summarized as follows: 1) We show that constraining the adversarial perturbation to statistically unpredictable areas can reduce image quality degradation while preserving attack ability. 2) We propose to modulate adversarial perturbation with embedding suitability, making the AEs hard to be revealed by steganalysis analysis. 3) Extensive experiments verify that the proposed weighting scheme is compatible with various attacks and can boost AEs' undetectability under the same attack ability.

This paper is organized as follows. Section 2 reviews related works. In Sect. 3, we discuss our motivation and propose our undetectable attack scheme. Experiments are presented and analyzed in Sect. 4. Section 5 concludes.

2   Related Works

2.1   Adversarial Attack

According to the perturbation limitation in generating AEs, existing attacks can be roughly divided into perturbation-constrained and perturbation-optimized. Perturbation-constrained attacks aim to maximize the adversarial loss of a target classifier under given perturbation restrictions, usually measured with L_p distance. The representative algorithms of this class are the projected gradient descent (PGD) [17], iterative FGSM (IFGSM) [18], and momentum IFGSM (MI) [19], to name a few. On the other hand, perturbation-optimized attacks [1,20,21] minimize the perturbation under certain criteria, e.g., the attack confidence exceeds a predefined threshold. In this paper, we use the perturbation-constrained attacks as the baselines. However, the proposed scheme can also integrate with perturbation-optimized attacks.

As a common baseline, IFGSM iteratively modifies a benign image I:

$$I'_{N+1} = clip_{I,\epsilon}\{I'_N + \alpha sign(\nabla_{I'_N} J(I'_N, y_o))\}, I'_0 = I \tag{1}$$

where α is the step size, y_o is the original label, and $\nabla_{I'_N} J()$ denotes the gradient of the loss function $J()$ to I'_N. The accumulated perturbation for each pixel is restricted to $[-\epsilon, \epsilon]$. There are many variants of IFGSM, e.g., MI attack stabilizes the update direction with a momentum term, DI attack [22], Admix attack [23], and TI attack [24] utilize data augmentation to enhance the transferability of AE. Moreover, these enhanced schemes can be integrated for even better transferability, e.g., Translation-invariant-diverse-inputs IFGSM (TDI).

Despite the success of the attacks mentioned above, they mostly treat all the pixels in one image equally, which may not be optimal because a CNN model focuses only on its region of interest in classification. To make the attack more efficient, [25] proposes an attentional attack scheme. They first calculate the attentional map from a benign image with the Class Activation Mapping (CAM) method [26]. Then, the adversarial perturbation is constrained in the regions of high attention value. Since the modified regions are reduced, and the high-attention regions often contain more textures, a byproduct of the attentional attack is the obtained AEs are hard to detect under statistical analysis. Another recent work of improving AE's undetectability is [27], in which microscopical regularization is used to encourage the high-pass residues of an AE to be similar to that of the corresponding benign image. Our method shares the same view with [27] that adversarial perturbations should be proportional to the texture richness of the image, which the embedding suitability from steganography can measure. However, significant differences exist between the two. First, the embedding suitability map is used to weight the regularization term of the loss function in [27], whereas we leverage it to modulate the gradients. Second, we do not require the resulting AE to display similar statistical features as the corresponding benign image, as done in [27], i.e., we encourage the obtained AE like 'a' benign image, whereas [27] encourages it like 'the' benign one in the feature domain. In the following, we denote [25] with a prefix 'A-' and [27] with a prefix 'MR-' to a baseline attack, e.g., A-IFGSM or MR-IFGSM.

2.2 Embedding Suitability Map

Steganography is the art of covert communications that aims to make the steganographic images unperceivable to the human eye and undetectable to its counterpart: steganalysis. Contemporary steganography schemes proceeds two steps: 1) calculating an embedding suitability map according to a heuristically designed distortion function; 2) embedding secret messages into the cover image guided by this suitability map. Representative distortion functions include WOW [28], S-UNIWARD [29], HILL [30], and MiPOD [31], to name a few. Figure 2(b–d) show the embedding suitability maps for the image in Fig. 2(a) using three distortion functions. These functions have one thing in common: they encourage embedding in textural areas. This is because such areas are difficult to model by steganalysis, i.e., suitable for message embedding. We realize that generating undetectable AE is similar to steganography, inspiring us to leverage the embedding suitability map for crafting undetectable AEs.

3 Proposed Scheme

3.1 Motivation

Our method is motivated by the observation that adding adversarial perturbation to the textural area is less perceptible to human eyes than to the smooth area while preserving the attack ability. An intuitive example is provided in Fig. 3. The white area of Fig. 3(a) corresponds to the smoothest quarter of Fig. 2(a), whereas that of Fig. 3(c) corresponds to the most complex quarter. By retaining only the corresponding region of the adversarial noise (IFGSM, $\epsilon = 16$), we obtain the modified image as shown in Fig. 3(b) and (d). The visual quality of Fig. 3(d) is better than that of Fig. 3(b) in terms of less perceptible noise found in the blue sky of Fig. 3(d). Moreover, the classification confidence to the original label 'traffic light' of Fig. 3(d) is 1e-6, much lower than that of Fig. 3(b) (0.5100), which means attacking the textural area is more efficient. The visually appealing result of Fig. 3(d) suggests it may also be undetectable against statistical detectors. To verify this hypothesis, we mount the IFGSM attack on the ImageNet-compatible dataset [32] with $\epsilon = 3$. Then, we retain adversarial perturbation in the smoothest, sub-smooth, sub-complex or most complex quarter for each attacked image, similar to that of Fig. 3. Finally, we use a

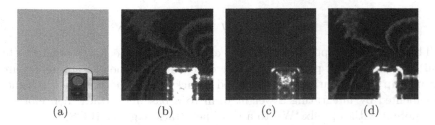

(a)   (b)   (c)   (d)

Fig. 2. Embedding suitability maps using WOW (b), S-UNIWARD (c), and MiPOD (d) for the image shown in (a). White color indicates suitable area for embedding.

| (a) | (b) | (c) | (d) |

Fig. 3. Retaining adversarial noise of different regions in Fig. 2(a). (a) the smoothest quarter, (b) the perturbation in the smoothest region is retained, the classification confidence to '*traffic light*' is 0.51, (c) the most complex quarter, (d) the perturbation in the most complex region is retained, the classification confidence to '*traffic light*' is 1e-6. Lower confidence to the original label means stronger attack ability.

Table 1. Undetectability against statistical analysis [13] when adversarial perturbations of different types of regions are retained.

	smoothest	sub-smooth	sub-complex	most complex
P_E	0.040	0.074	0.149	0.261

steganalysis-based detector [13] to distinguish them from benign images. As shown in Table 1, retaining adversarial perturbations of different regions results in strikingly different undetectability. For instance, the detector's error probability (P_E) is 0.04 in the case of retaining perturbation of the smoothest region versus 0.261 in the case of retaining perturbation of the most complex region. To summarize, attacking the textural area is more imperceptible and effective.

3.2 Embedding Suitability Map-Weighted Attack

Based on the analysis above, we propose our undetectable attack, in which a steganographic embedding suitability map weights the adversarial perturbation. Starting from a benign image I, we first calculate its embedding suitability map ξ_I with S-UNIWARD [29]. Next, ξ_I is normalized by its maximum value.

$$N\xi_I = \xi_I/max(\xi_I) \qquad (2)$$

In each attack iteration, the signed gradient in (1) is modulated by $N\xi_I$.

$$I'_{N+1} = clip_{I,\epsilon}\{I'_N + N\xi_I \cdot \alpha sign(\nabla_{I'_N} J(I'_N, y_o))\} \qquad (3)$$

This way, strong perturbations are encouraged in the hard-to-detect (textural) area, and weak perturbations are encouraged in the easy-to-detect area. **Algorithm 1** summarizes the proposed attack. Since the proposed scheme is implemented by integrating a weight term to a baseline attack, henceforward, we denote it with a prefix 'W-' to a baseline attack, e.g., W-IFGSM.

Algorithm 1. W-IFGSM

Input: Benign image I;
Parameter: iteration T, step size α, and maximum perturbation ϵ.
Output: Adversarial image I'.
1: Calculate the embedding suitability map ξ_I of I.
2: Get the normalized embedding suitability map $N\xi_I$ as (2).
3: for iteration $= 0, ..., T$ do
4:    Update I' according to (3).
5: end for
6: return I'.

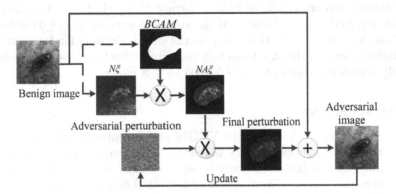

Fig. 4. The diagram of the 'WA-' attack. Our used adversarial perturbation is restricted to the most discriminative (from the perspective of classification) and undetectable (from the perspective of statistical analysis) region.

3.3 Combination with CAM

In the attentional attack [25], adversarial noise is only added to the most discriminative regions for a CNN model. In the proposed 'W-' attack, regions that are hard to predict are our focus. It is natural to combine them for an even better attack: both efficient and undetectable. We first use the CAM method [26] to generate the attention map CAM_I. Then CAM_I is binarized as [25]:

$$BCAM_I(i,j) = \begin{cases} 0. & CAM_I(i,j) < 1/3 \\ 1. & CAM_I(i,j) \geqslant 1/3 \end{cases} \qquad (4)$$

Lastly, replace the weight term $N\xi_I$ of (3) with its attentional version $NA\xi_I$

$$NA\xi_I = (BCAM_I \cdot \xi_I)/max(BCAM_I \cdot \xi_I) \qquad (5)$$

As such, the adversarial perturbations are concentrated in the most efficient yet hard-to-detect areas. The weighted attentional scheme is denoted with the prefix 'WA-' to a baseline attack henceforth. Figure 4 shows a general procedure of the 'WA-' attack. The attention mechanism avoids the 'WA-' attack to modify the non-critical areas, such as the ground at the bottom right, despite their complex textures.

4   Experimental Results

In this section, we evaluate our method from two aspects: attack ability and undetectability. We launch untargeted attacks against a Resnet50 model [33] on the ImageNet-compatible dataset. Two standardly trained models (Densenet121 [34], VGG16 [35]) and two adversarially trained models (Resnet50adv [36], InceptionV3adv [8]) are intentionally selected as the target models for calculating black-box attack success rate (ASR). Since attacking incorrectly classified images is less meaningful, we select 674 images that can be correctly classified by all five models to attack. IFGSM and its enhanced version TDI are used as the baselines. The perturbations are restricted by L_∞ norm with $\epsilon \in \{1, 2, 3, 4, 6, 8, 16\}$. The iteration number T is set to 50, and the step size α is set as $\epsilon/8$. Due to page limitation, we leave the ablation study on the distortion function for calculating the suitability map in the supplementary material: *Undetectable-attack/supp.pdf*, and only report the result of S-UNIWARD in this paper.

4.1   Attack Ability

Figure 5 visualizes the ASRs against VGG16 and InceptionV3adv (More results are provided in the supplementary material). The general pattern is that the adversarially trained InceptionV3adv is much resilient to attack than the standardly trained VGG16. Given the considerable difference in absolute values between Fig. 5(a, b) and (c, d), they share similar trends and relative relations between different methods. Hence, we focus on Fig. 5(a, b) to compare various attacks. As expected, the black-box ASRs of '-TDI' methods are significantly higher than their '-IFGSM' counterparts. Another conclusion is that the ASR of 'W-' is not far behind that of the baseline, especially for '-TDI' methods. Taking $\epsilon=8$, VGG16 for example, the ASR of W-TDI is 72.8%, which is only 16.6% lower than that of TDI. This finding means that the proposed method can deliver a similar attack ability to the baseline by increasing the attack strength ϵ slightly.

We further compare candidate attacks based on their robustness to common post-processings. In this experiment, AEs undergo JPEG compression (quality factor = 95) and bit-depth reduction (6 bits/pixel/channel) before evaluating their attack ability against two standardly trained models. Figure 6 compares the black-box ASRs with TDI as the baseline. Compared with the corresponding curves of Fig. 5(b), there is only a minor drop in attack ability after post-processing for both 'W-TDI' and 'WA-TDI.'

4.2   Undetectability

This subsection examines the undetectability of the generated AEs with a steganalysis-based detector [13]. We train an ensemble classifier [37] for each attack and strength, respectively. Undetectability is quantified using the corresponding ensemble's error probability (P_E). As shown in Table 2, the proposed scheme is much more undetectable than the baseline. Surprisingly, the P_E of

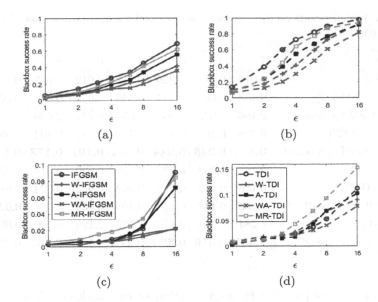

Fig. 5. ASR against different models. (a, b) VGG16, (c, d) InceptionV3adv. Figure 5(a, c) share a legend, and (b, d) share another legend for better visualization.

the proposed WA-TDI at ϵ=16 (15.0%) is even higher than that of TDI at ϵ=1 (11.2%). Since the AEs crafted with WA-TDI at ϵ=16 are much more transferable than the AEs crafted with TDI at ϵ=1 (See Fig. 5(b) and (d)), we believe that WA-TDI can be a suitable alternative to TDI if both attack ability and undetectability are taken into account.

4.3 Undetectability-Attack Ability Tradeoff

With the results provided in Sects. 4.1 and 4.2, we are ready to compare the undetectability of different methods under the same ASR. Recall that our ultimate goal is crafting undetectable AEs while preserving attack ability.

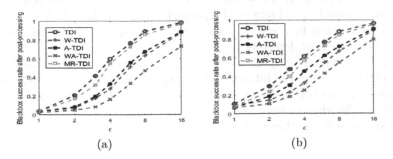

Fig. 6. Comparison of the attack robustness to common image processings. The baseline attack is TDI and the target models are Densenet121 (a) and VGG16 (b).

Table 2. Undetectability (P_E) against the steganalysis-based detection. The best results are in **bold**.

methods\ϵ	1	2	3	4	6	8	16
IFGSM	0.076	0.038	0.020	0.012	0.012	0.012	0.002
MR-IFGSM [27]	0.148	0.064	0.054	0.046	0.029	0.018	0.006
W-IFGSM (proposed)	0.166	0.072	0.060	0.054	0.042	0.036	0.014
A-IFGSM [25]	0.236	0.168	0.144	0.132	0.112	0.104	0.083
WA-IFGSM (proposed)	**0.310**	**0.248**	**0.214**	**0.202**	**0.192**	**0.178**	**0.149**
TDI	0.112	0.062	0.034	0.026	0.020	0.014	0.004
MR-TDI [27]	0.186	0.112	0.080	0.062	0.032	0.022	0.010
W-TDI (proposed)	0.180	0.144	0.126	0.100	0.059	0.044	0.020
A-TDI [25]	0.284	0.204	0.172	0.152	0.129	0.120	0.090
WA-TDI (proposed)	**0.356**	**0.316**	**0.300**	**0.272**	**0.219**	**0.192**	**0.150**

To this end, we plot the P_E as a function of the black-box ASR (against the standardly and adversarially trained models) for each method in Fig. 7. The proposed method's advantage over the baseline is quite evident. Take the standardly trained models for example; if we fix the black-box ASR as 0.4, the P_E of the baseline TDI is less than 6.2%, which means it is pretty detectable by the steganalysis-based detector. In contrast, the P_E of the proposed W-TDI is approximately 10.0%, more undetectable than the baseline TDI. Moreover, by combining with CAM, the proposed WA-TDI achieves $P_E \approx 22.5\%$, superior to A-TDI ($P_E \approx 16.2\%$). Such an advantage establishes the feasibility of integrating the proposed scheme with other enhanced schemes.

4.4 Visual Quality

We qualitatively and quantitatively compare the visual quality of AEs crafted by competing methods. Besides the example shown in Fig. 1, we further display several AEs generated by different methods with ϵ=16 in Fig. 8. The baseline attack is IFGSM for the first row and TDI for the second. The proposed 'W-' and 'WA-' schemes tend to add adversarial perturbations in rich texture regions that are hard for humans to perceive. This is also the key to crafting undetectable AEs against statistical detectors.

For a quantitative comparison, we use the Universal Image Quality Index (UIQI) [38] to measure the fidelity of AEs. An AE with a higher UIQI (with a maximum of 1) has better visual quality. Table 3 provides the averaged UIQI for different attacks when ϵ=16. The proposed 'W-' and 'WA-' schemes triumph the compared schemes, whether they are based on IFGSM or TDI.

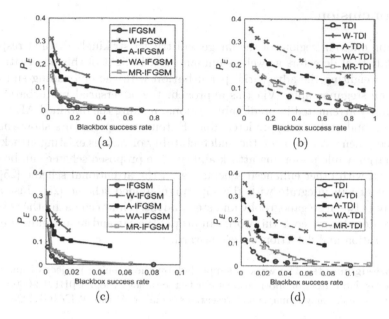

(a)   (b)

(c)   (d)

Fig. 7. Undetectability-ASR tradeoff. The ASR is averaged over two standardly-trained models (a, b) and two adversarially-trained models (c, d). Higher P_E under the same ASR is preferred.

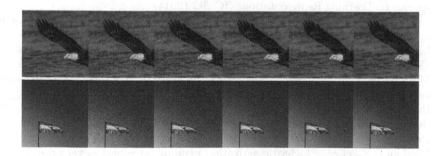

Fig. 8. Visual comparison of AEs. The baseline attack is IFGSM for the top and TDI for the bottom. From left to right: the original image, the AEs generated by the baseline attack, the 'MR-' attack, the 'W-' attack, the 'A-' attack, and the 'WA-' attack.

Table 3. Image quality comparison in terms of Universal Image Quality Index [38]. The best results are in **bold**.

method	IFGSM	MR-IFGSM	W-IFGSM	A-IFGSM	WA-IFGSM
UIQI	0.9637	0.9750	**0.9982**	0.9863	0.9948
method	TDI	MR-TDI	W-TDI	A-TDI	WA-TDI
UIQI	0.9625	0.9668	0.9871	0.9857	**0.9943**

5   Conclusion

Attacking different regions of the image will trigger strikingly distinct responses from statistical detectors as well as human eyes. In light of this observation, we propose weighting the adversarial perturbation with the embedding suitability map from steganography. Via this approach, the adversarial perturbations are significantly attenuated in predictable regions, making the obtained AE visually appealing and statistically undetectable. Extensive experiments show that the proposed scheme can improve the undetectability of various existing attacks by a large margin while preserving attack ability. The proposed scheme can be easily integrated with other enhanced schemes, e.g., the attentional scheme [25]. Our future work will integrate with the superpixel-guided scheme [25]. Just as the interplay between steganography and steganalysis reinforced each other over the past two decades, we hope this work can propel the cat-and-mouse game between AE generation and detection a step forward.

Acknowledgements. The work is supported by the opening project of guangdong province key laboratory of information security technology (no. 2020B1212-060078) and the network emergency management research special topic (no. WLYJGL2023ZD003).

References

1. Szegedy, C., et al.: Intriguing properties of neural networks. In: International Conference on Learning Representations (ICLR) (2014)
2. Liu, Y., Chen, X., Liu, C., Song, D.: Delving into transferable adversarial examples and black-box attacks. In: ICLR (2017)
3. Li, M., et al.: Towards transferable targeted attack. In: IEEE/CVF Conference on Computer Vision and Pattern Recognition, pp. 638–646 (2020)
4. Moosavi-Dezfooli, S., Fawzi, A., Fawzi, O., Frossard, P.: Universal adversarial perturbations. In: CVPR, pp. 86–94 (2017)
5. Papernot, N., McDaniel, P., Wu, X., et al.: Distillation as a defense to adversarial perturbations against deep neural networks. In: IEEE Symposium on Security and Privacy (SP), pp. 582–597 (2016)
6. Dhillon, G.S., et al.: Stochastic activation pruning for robust adversarial defense. In: ICLR (2018)
7. Athalye, A., Carlini, N., Wagner, D.: Obfuscated gradients give a false sense of security: circumventing defenses to adversarial examples. In: ICLR (2018)
8. Tramèr, F., et al.: Ensemble adversarial training: attacks and defenses. In: ICLR (2018)
9. Hendrycks, D.: Early methods for detecting adversarial images. In: ICLR (2017)
10. Li, X., Li, F.: Adversarial examples detection in deep networks with convolutional filter statistics. In: ICCV, pp. 5775–5783 (2017)
11. Liang, B., et al.: Detecting adversarial image examples in deep neural networks with adaptive noise reduction. IEEE Trans. Depend. Secure Comput. **18**(1), 72–85 (2018)
12. Deng, K., Peng, A., Dong, W., Zeng, H.: Detecting C&W adversarial images based on noise addition-then-denoising. In: ICIP, pp. 3607–3611 (2021)
13. Liu, J., et al.: Detection based defense against adversarial examples from the steganalysis point of view. In: CVPR, pp. 4820–4829 (2019)

14. Fridrich, J., Kodovsky, J.: Rich models for steganalysis of digital images. IEEE Trans. Inf. Forens. Secur. **7**(3), 868–882 (2012)
15. Peng, A., et al.: Gradient-based adversarial image forensics. In: the 27th International Conference on Neural Information Processing, pp. 417–428 (2020)
16. Zeng, H., et al.: How secure are the adversarial examples themselves? In: ICASSP, pp. 2879–2883 (2022)
17. Madry, A., et al.: Towards deep learning models resistant to adversarial attacks. In: ICLR (2017)
18. Kurakin, A., Goodfellow, I., Bengio, S.: Adversarial examples in the physical world. In: ICLR (2017)
19. Dong, Y., et al.: Boosting adversarial attacks with momentum. In: CVPR, pp. 9185–9193 (2018)
20. Carlini, N., Wagner, D.: Towards evaluating the robustness of neural networks. In: IEEE Symposium on Security and Privacy, pp. 39–57 (2017)
21. Xiao, C., et al.: Spatially transformed adversarial examples. In: ICLR (2018)
22. Xie, C., et al.: Improving transferability of adversarial examples with input diversity. In: CVPR, pp. 2730–2739 (2019)
23. Wang, X., He, X., Wang, J., He, K.: Admix: enhancing the transferability of adversarial attacks. In: IEEE International Conference on Computer Vision (2021)
24. Dong, Y., et al.: Evading defenses to transferable adversarial examples by translation-invariant attacks. In: CVPR, pp. 4312–4321 (2019)
25. Dong, X., et al.: Robust superpixel-guided attentional adversarial attack. In: CVPR, pp. 12892–12901 (2020)
26. Zhou, B., et al.: Learning deep features for discriminative localization. In: CVPR, pp. 2921–2929 (2016)
27. Zhong, N., Qian, Z., Zhang, X.: Undetectable adversarial examples based on microscopical regularization. In: IEEE International Conference on Multimedia and Expo, pp. 1–6 (2021)
28. Holub, V., Fridrich, J.: Designing steganographic distortion using directional filters. In: IEEE Workshop on Information Forensic and Security, pp. 234–239 (2012)
29. Holub, V., Fridrich, J., Denemark, T.: Universal distortion function for steganography in an arbitrary domain. EURASIP Journal on Information Security, (Section: SI: Revised Selected Papers of ACM III and MMS2013), no. 1 (2014)
30. Li, B., Wang, M., Huang, J., Li, X.: A new cost function for spatial image steganography. In: ICIP, pp. 4206–4210 (2014)
31. Sedighi, V., Cogranne, R., Fridrich, J.: Content-adaptive steganography by minimizing statistical detectability. IEEE Trans. Inf. Forens. Secur. **11**(2), 221–234 (2016)
32. https://github.com/cleverhans-lab/cleverhans/tree/master/cleverhans_v3.1.0/examples/nips17_adversarial_competition. Accessed 20 May 2023
33. He, K., et al.: Deep residual learning for image recognition. In: CVPR, pp. 770–778 (2016)
34. Huang, G., Liu, Z., Laurens, V., Weinberger, K.Q.: Densely connected convolutional networks. In: CVPR, pp. 2261–2269 (2017)
35. Simonyan, K., Zisserman, A.: Very deep convolutional networks for largescale image recognition. In: ICLR (2015)
36. https://github.com/MadryLab/robustness. Accessed 20 May 2023
37. Kodovsky, J., Fridrich, J., Holub, V.: Ensemble classifiers for steganalysis of digital media. IEEE Trans. Inf. Forens. Secur. **7**(2), 432–444 (2012)
38. Zhou, W., Bovic, A.C.: A universal image quality index. IEEE Signal Process. Lett. **9**(3), 81–84 (2002)

On Efficient Federated Learning for Aerial Remote Sensing Image Classification: A Filter Pruning Approach

Qipeng Song, Jingbo Cao[✉], Yue Li, Xueru Gao, Chengzhi Shangguan, and Linlin Liang

School of Cyber Engineering, Xidian University, Xi'an 710071, China
jbcao@stu.xidian.edu.cn

Abstract. To promote the application of federated learning in resource-constraint unmanned aerial vehicle swarm, we propose a novel efficient federated learning framework CALIM-FL, short for Cross-All-Layers Importance Measure pruning-based Federated Learning. In CALIM-FL, an efficient one-shot filter pruning mechanism is intertwined with the standard FL procedure. The model size is adapted during FL to reduce both communication and computation overhead at the cost of a slight accuracy loss. The novelties of this work come from the following two aspects: 1) a more accurate importance measure on filters from the perspective of the whole neural networks; and 2) a communication-efficient one-shot pruning mechanism without data transmission from the devices. Comprehensive experiment results show that CALIM-FL is effective in a variety of scenarios, with a resource overhead saving of 88.4% at the cost of 1% accuracy loss.

Keywords: Federated Learning · Filter Pruning · UAV · CNN

1 Introduction

Smart unmanned aerial vehicles (UAVs) have spawned many new intelligent applications and have become an indispensable part of ubiquitous remote sensing. They generally require massive data to train a high-quality task-specific model. However, the images captured by UAVs cannot be freely exchanged, due to the increasingly-strict privacy regulation [16] and constrained communication resources [7]. Thus, it is important to address how to ensure privacy-preserving aerial remote sensing image classification to completely unleash the power of UAV swarm.

Federated learning (FL) [10] is a distributed privacy-preserving machine learning framework, which allows participants to collaboratively train a shared global model without data sharing. Although the FL framework solves the privacy issue for UAV swarms, it still faces the following challenges. The first is **statistical heterogeneity**. There exists statistical heterogeneity among the

© The Author(s), under exclusive license to Springer Nature Singapore Pte Ltd. 2024
B. Luo et al. (Eds.): ICONIP 2023, LNCS 14450, pp. 184–199, 2024.
https://doi.org/10.1007/978-981-99-8070-3_15

UAVs with non-IID datasets. The non-IID degree is a metric to measure the difference between the distribution of a dataset and that of all the devices. Statistical heterogeneity hinders the efficiency of the FL training process and degrades the performance of the global model. The second is **limited computing and communication resources**. It is not a big deal to train neural networks (for example, VGG16 with more than 100 million parameters) within data centers, but it poses a great challenge for UAV swarms. Thus, it is of paramount importance to deal with the challenge of how to efficiently perform FL on the resource-constraint device at a reasonable cost.

Plenty of research efforts have been made to achieve efficient and low-cost FL so that FL can be deployed on resource-limited devices such as UAVs. The key idea is to approximate the original model with a compressed one, trading an acceptable accuracy loss for the reduction in computational and storage overhead. However, most of the existing solutions are proposed for centralized machine-learning scenarios, and cannot be directly integrated into UAV swarm-based FL. Conventional model compression methods are subject to challenges such as data limitations, computational costs, and communication efficiency.

Filter pruning (i.e., neuron pruning/model pruning) is a promising solution to mitigate the aforementioned challenges and achieve efficient FL, since it is not only able to reduce the communication costs between the devices and the server, but also to accelerate the training process [4]. To integrate filter pruning into UAV-based FL, there still exist two problems: 1) how to design a pruning criterion in the FL system; and 2) how to design a low-cost pruning mechanism for UAVs.

In this work, we propose a novel filter pruning-based FL paradigm CALIM-FL short for Cross-All-Layers Importance Measure pruning-based Federated Learning, which adapts the model size during the FL training process to reduce both communication and computation overhead at the cost of a slight accuracy loss. The contributions are summarized as follows:

- We propose a cross-all-layers importance measure (CALIM) on filters for pruning. The importance of a given filter simultaneously depends on the information capture capability for information flowed from the previous layer (i.e., information richness) and the impact on the predictive power of the model (i.e., error propagation). To the best knowledge of the authors, CALIM is the first importance measure to evaluate a filter from the whole network. Comprehensive experiments confirm the effectiveness of this novel importance measure.
- We design an efficient CALIM-based one-shot pruning mechanism intertwined within conventional FL, improving the efficiency of FL. One challenge is that no data is available on the central server. Activation in the model layer based on the data-driven acquisition can capture more information, thanks to the advantages of our approach, which is constructed to be robust enough to noise that the central node can be pruned without data. In addition, centralized pruning can save resource consumption on resource-constraint devices.

– Extensive experiments demonstrate the effectiveness of our proposal. Our proposal achieves good performance in the non-IID scenario, where the learned data in the model affects the model parameters, and the parameters affect the pruning, enabling CALIM-FL to be applicable in the non-IID scenario. Our proposed framework can guarantee the effective convergence of the model in multiple scale datasets and multiple non-IID dataset divisions, verifying its effectiveness.

The remaining of this article is organized as follows: Sect. 2 introduces some research efforts towards efficient FL. Section 3 explains the novel importance measure and the FL with integrated pruning mechanism in detail. Section 4 presents the experimental design and results analysis, and Sect. 5 concludes the article.

2 Related Work

2.1 Efficient Federated Learning

FL is first proposed to collaboratively train a global model without the need for data sharing among participated clients. There is no direct data transmission between multiple nodes to avoid privacy leakage as much as possible. As the de facto approach for FL, FedAvg [10] mitigates the non-IID issue by iterative model averaging with local stochastic gradient descent (SGD) on each local dataset. However, statistical heterogeneity is still a factor that degrades the training efficiency for FL and thus hinders the deployment of FL on resource-constrained devices.

Some research efforts focus on improving the FL training process by reducing non-IID degrees. For example, Li et al. [8] proposed FedProx, which can be viewed as a generalization and re-parametrization of FedAvg. This work provides convergence guarantees from a theoretical aspect when training over non-IID data. Karimireddy et al. [5] propose SCAFFOLD. This framework uses control variates (variance reduction) to correct for the "client drift" in its local updates to reduce the required communication rounds for convergence. SCAFFOLD can take advantage of similarities in the client's data to further improve convergence. Wang et al. [18] analyze the convergence of heterogeneous federated optimization algorithms, and propose a normalized averaging method that eliminates objective inconsistency while preserving fast error convergence. Because of incentive mechanisms [19], some insensitive data may reside on the server and can be exploited for non-IID degree mitigation. Zhang et al. [20] propose FedDU that accelerates the training process by leveraging the server data and taking into account the non-IID degree on each device.

2.2 Filter Pruning

Model pruning aims to remove the redundant parameters in models with acceptable accuracy degradation. It helps reduce the computation and communication cost in the training process of FL [4]. Model pruning results in a sparse model, which accelerates both training and inference time. Generally, the research related to model pruning can be categorized into the following aspects:

Depending on how to treat the neurons, pruning techniques can be classified into two categories: unstructured (weight) pruning [4] and structured (filter) pruning [9]. Unstructured pruning consists of setting weight parameters of less important neurons to 0 without changing the model structure, while structured pruning directly changes the structure of the model by removing a portion of less important neurons (i.e., filters in the convolution layers). Practically, unstructured pruning is mainly favored for model size compression, structured pruning is more favored for acceleration due to the hardware-friendly sparsity structure [17].

The pruning ratio indicates how many weights to remove from the network. Basically, there exist two schemes to determine the appropriate pruning ratio. The first one is a pre-defined pruning ratio with expertise. Namely, we know exactly how many parameters will be removed before the pruning operation. This scheme can be further specified into two sub-schemes. One is to set a global pruning ratio (i.e., how many weights will be pruned for the whole network); the other is to set layer-wise pruning ratios [9]. The second is adaptive pruning ratio [12,20]. This way mostly appears in defining an optimization problem towards some objectives. For example, PruneFL [4] searches the appropriate pruning ratio by minimizing the round time. Recent years also have seen some works that automatically search the optimal layer-wise pruning ratio [3]. It is worth indicating that there is no consensus about the pre-defined pruning ratio and adaptive pruning ratio which is better.

It is worth noting that the objective of filter pruning is to remove less important filters from the model, which is orthogonal to other acceleration methods, such as quantization [1] and low-rank decomposition [15]. Filter pruning can be applied together with these other methods.

3   Methodology

3.1   System Model

This section present the system model. The notations in this article are explained in Table 1. Bold symbols represent vectors and metrics.

We consider an FL system composed of a central node (e.g. a powerful server in a data center) and N devices[1]. Each device indexed by n has a local empirical risk $F_n(\mathbf{w}) := (1/|\mathcal{D}_n|) \sum_{i \in \mathcal{D}_n} f_i(\mathbf{w})$ defined on its local dataset $\mathcal{D}_n$ for model parameter vector $\mathbf{w}$, where $|\mathcal{D}_n|$ refers to the cardinality of dataset $\mathcal{D}_n$, $f_i(\mathbf{w})$ is the loss function that captures the difference between the model output and the desired output of data sample i. The objective of training an FL system is to find the parameter vector $\mathbf{w}$ that minimizes the global empirical risk:

$$\min_{\mathbf{w}} F(\mathbf{w}) := \sum_{n=1}^{N} p_n F_n(\mathbf{w}), \tag{1}$$

---

[1] Device and UAV are interchangeably used in this article.

Table 1. Main Notation

Notation	Definition
n, N	device index, total numbers of devices
i, K	convolution layer index, total numbers of convolution layers
k_i	kernel size
I	data sample set for filter pruning
g	size of data sample for filter pruning
$\mathcal{D}_n$	local dataset for device n
$\mathbf{w}_j^i$	j-th filter in i-th convolutional layer
$\mathbf{o}_j^i$	j-th feature map generated by $\mathbf{w}_j^i$
h_i, w_i	height and width of feature map
$\mathbf{s}_i$	filter prune indicator vector for i-th convolutional layer

where $p_n > 0$ are weights such that $\sum_{n=1}^{N} p_n = 1$. In this work, we let $p_n = |\mathcal{D}_n| / \sum_{n=1}^{N} |\mathcal{D}_n|$.

In FL, each device n has a local parameter $\mathbf{w}_n(k)$ in iteration k. The aggregation of these local parameters is defined as $\mathbf{w}(k) := \sum_{n \in [N]} p_n \mathbf{w}_n(k)$. The FL procedure usually involves multiple updates of $\mathbf{w}_n(k)$ using stochastic gradient descent (SGD) on the local empirical risk $F_n(\mathbf{w}_n(k))$, followed by a parameter aggregation step computing the aggregated parameter $\mathbf{w}(k)$ by server, on the basis of device' local parameters $\{\mathbf{w}_n(k) : \forall n \in [N]\}$. After parameter fusion, the local parameters $\{\mathbf{w}_n(k) : \forall n \in [N]\}$ are all set to be equal to the aggregated parameter $\mathbf{w}(k)$.

For the pruning process within the central node, we consider a global CNN model with K convolutional layers. Let $\mathcal{C}^i$ denote the i-th convolutional layer. The parameters in $\mathcal{C}^i$ can be represented as a set of filters $\mathbf{w}_{\mathcal{C}^i} = \{\mathbf{w}_1^i, \mathbf{w}_2^i, \ldots, \mathbf{w}_{n_i}^i\} \in \mathbb{R}^{n_i \times n_{i-1} \times k_i \times k_i}$, where the j-th filter is $\mathbf{w}_j^i \in \mathbb{R}^{n_{i-1} \times k_i \times k_i}$, n_i represents the number of filters in $\mathcal{C}^i$ and k_i denotes the kernel size. The outputs of filters, i.e., feature maps, are denoted as $\mathcal{O}^i = \{\mathbf{o}_1^i, \mathbf{o}_2^i, \ldots, \mathbf{o}_{n_i}^i\} \in \mathbb{R}^{n_i \times |I| \times h_i \times w_i}$, where the j-th feature map $\mathbf{o}_j^i \in \mathbb{R}^{|I| \times h_i \times w_i}$ is generated by $\mathbf{w}_j^i$. $|I|$ is the size of input data samples, h_i and w_i are the height and width of the feature map, respectively.

In filter pruning, the key challenge is to remove the less important filters. Let $\mathcal{L}(\mathbf{w}_{\mathcal{C}^i})$ be the importance measure (or information richness) for filters in the i-th convolutional layer, and let $\mathbf{s}_i$ refer to the prune indicator vector associated with i-th convolutional layer. Thus, the objective of filter pruning for the CNN model is defined as follows:

$$\min_{\mathbf{s}} \sum_{i=1}^{K} < \mathbf{s_i}, \mathcal{L}(\mathbf{w}_{\mathcal{C}^i}) >, \qquad (2)$$

where $\langle \cdot, \cdot \rangle$ is dot product.

Since the goal of efficient FL is to remove the redundant filters while minimizing accuracy loss. Therefore, the optimization objective of FL with filter pruning is to simultaneously find **w** and **s** so that Eq. 1 and Eq. 2 achieve the minimum value.

3.2 Cross-All-Layers Importance Measure for Pruning

The core of filter pruning integrated FL lies in the selection of less important filters, which should yield the highest compression ratio with the lowest degradation in accuracy. For a given filter (neuron), one hand it can capture the underlying information hidden in the input dataset. Thus, it is natural to keep filters capable of retrieving more information. However, as deep network training is a complex process, a slight change in shallower layers may have a significant impact on the prediction power. Thus, it is necessary to take into account the impact on the output layer when pruning a given filter. Existing methods that prune the less important filters layer-by-layer cannot accurately reflect the information richness. Therefore, a global measure of filter importance that takes into account the whole propagation from the input layer to the output layer is required to efficiently guide the filter selection procedure.

In this paper, we propose a cross-all-layers importance measure for the i-th filter $\mathcal{L}(\mathbf{w}_j^i)$ on the j-th convolutional layer. For a batch of data, let $\mathcal{I}_m^{(1)}(\mathbf{w}_j^i; I)$ be the importance measure from information richness that quantifies the information capture capability from all previous layers, while $\mathcal{I}_m^{(2)}(\mathbf{w}_j^i; I)$ is importance measure from error propagation that characterizes the impact to the predictive power of network. Thus, the importance measure for a given filter can be defined as follows:

$$\mathcal{L}(\mathbf{w}_j^i) = f(\mathcal{I}_m^{(1)}(\mathbf{w}_j^i; I), \mathcal{I}_m^{(2)}(\mathbf{w}_j^i; I)). \tag{3}$$

For the information richness, the rank of the feature map (e.g., the output of a neuron, refer to Fig. 1) has been proved to be able to efficiently evaluate the information richness of a filter [9]. A feature map captures the information from the shallow layer to the current convolutional layer. Its rank implies the redundancy level and is able to quantify the information richness. The filters generating feature maps with higher ranks are preferentially to be kept during pruning operation. The rank expectation is empirically proved to be robust to the distribution of the dataset used for pruning. Thus, inspired from Eq. (5) given in [9], we have:

$$\mathcal{I}_m^{(1)}(\mathbf{w}_j^i; I) = \mathbb{E}_{I \sim P(I)} \left[\text{Rank}(\mathcal{O}^i) \right]$$

$$\approx \sum_{t=1}^{|I|} \text{Rank} \left(\mathbf{o}_j^i(t, :, :) \right), \tag{4}$$

which $|I|$ refers to the size of data samples for pruning, Rank denotes the operation calculating the rank for feature map $\mathbf{o}_j^i$.

For the importance measure of a filter in terms of impact on the final response, i.e., classification result, it can be measured as a squared difference of prediction

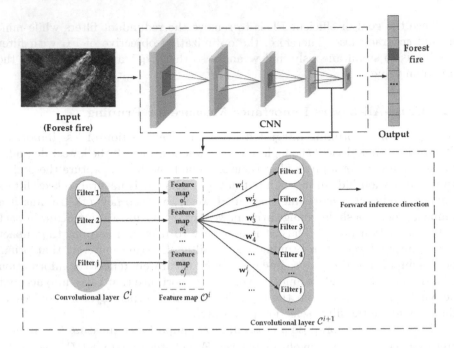

Fig. 1. Neuron and its corresponding feature map

errors with and without the parameter $\mathbf{w}_j^i$. A higher squared prediction error indicates a more importance level of the considered neuron:

$$\mathcal{I}_m^{(2)}(\mathbf{w}_j^i; I) := \left[\mathbb{E}_{I \sim P(I)}[I, \mathbf{w}] - \mathbb{E}_{I \sim P(I)}\left[I, \mathbf{w} \mid \mathbf{w}_j^i = 0\right]\right]^2. \tag{5}$$

Directly computing $\mathcal{I}_m^{(2)}(\mathbf{w}_j^i; I)$ for each parameter with Eq. (5) is computationally expensive. To reduce the computational complexity, $\mathcal{I}_m^{(2)}(\mathbf{w}_j^i; I)$ can be approximated in the vicinity of $\mathbf{w}$ by its first-order Taylor expansion, which is a frequently used mathematical practice in the literature [6]:

$$\mathcal{I}_m^{(2)}(\mathbf{w}_j^i; I) \approx \left[g(\mathbf{w}_j^i)\mathbf{w}_j^i\right]^2, \tag{6}$$

where $g(\cdot)$ denotes gradient operator. This approximation is rational since the gradient is obtained from backpropagation that implicitly incorporates the impact of all deeper layers to the current convolutional layer. In addition, the importance measure defined by Eq. (6) is significantly easier to be evaluated compared with Eq. (5).

How to precisely model f in Eq. (3) is a mathematical challenge. We observe that $\mathcal{I}_m^{(1)}(\mathbf{w}_j^i; I)$ in Eq. (4) mainly depends on all shallow layers before the current convolution layer, while $I_m^{(2)}(\mathbf{w}_j^i; I)$ in Eq. (6) is related to the deeper layers than the current one. In the literature, different convolution layers can be approximately mutually independent. Thus, we reasonably assume that Eq. (3) can be approximately by a linear combination of $\mathcal{I}_m^{(1)}(\mathbf{w}_j^i; I)$ and $\mathcal{I}_m^{(2)}(\mathbf{w}_j^i; I)$:

$$\mathcal{L}(\mathbf{w}_j^i; I) = \alpha \mathcal{I}_m^{(1)}(\mathbf{w}_j^i; I) + (1-\alpha)\mathcal{I}_m^{(2)}(\mathbf{w}_j^i; I)$$

$$= \sum_{t=1}^{|I|} \text{Rank}\left(\mathbf{o}_j^i(t,:,:)\right) + \alpha \left[g(\mathbf{w}_j^i)\mathbf{w}_j^i\right]^2, \qquad (7)$$

where α is a hyperparameter that tunes the weight between $\mathcal{I}_m^{(1)}(\mathbf{w}_j^i; I)$ and $\mathcal{I}_m^{(2)}(\mathbf{w}_j^i; I)$. It should be noted that both terms should be normalized into the same scale.

3.3   CALIM-FL Work Process

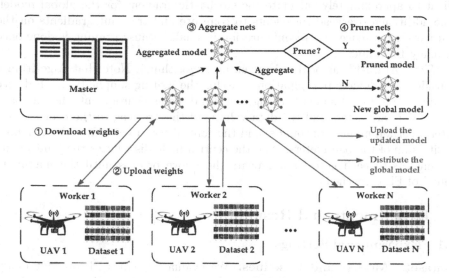

Fig. 2. Illustration of federated learning with model pruning

The proposed FL with model pruning guided by the importance measure defined in Eq. (7), which we dub as CALIM-FL, is illustrated in Fig. 2. The basic workflow is taken from the classic FedAvg framework [10]. The pruning operation is conducted within the central server (refer to step ④ in Fig. 2), since (1) the global model is maintained by the central server and pruning on the global model is more efficient; (2) pruning on edge UAVs requires extra scheduling, computing, and communication cost for resource-constraint devices. For the model pruning, a pruning stage for the global model is periodically scheduled within the whole training procedure, followed by a fine-tuning stage.

To evaluate the importance measure for each filter with CALIM quantified by Eq. (7) into FL, the key challenge is how to calculate $\mathcal{I}_m^{(1)}(\mathbf{w}; I)$ and $\mathcal{I}_m^{(2)}(\mathbf{w}; I)$ in a distributed way.

For term $\mathcal{I}_m^{(1)}(\mathbf{w}; I)$ representing the information capture capability, the problem to be solved is that data samples are required for pruning while no data collected by devices is available in the central server due to privacy-preserving constraint. Based on many experimental observations, we observe that the cosine distance between the importance measure obtained by feeding a batch of randomly generated synthetic data and that by real data during the training convergence process of the model is almost 0. This implies that the importance measure quantified by Eq. (7) is approximately independent of the input data sample. Therefore, CALIM-FL can calculate $\mathcal{I}_m^{(1)}(\mathbf{w}; I)$ by randomly generated synthetic data.

For term $\mathcal{I}_m^{(2)}(\mathbf{w}; I)$ within the central server, the problem faced is that the $\mathbf{g}(\mathbf{w})$ is not available on the server, which requires the intervention of edge devices. The solution is to use the weighted average for $\mathbf{g}(\mathbf{w})$ received from each client to approximately calculate the stochastic gradient for the global model. The additional overhead for clients to compute and transmit gradients on the full parameter space in a reconfiguration is small because pruning is done once in many FL rounds.

The CALIM-FL algorithm is given by Algorithm 1. Note that Algorithm 1 considers full devices participation case, for the writing simplicity. Each device conducts traditional model training in a stand-alone environment, then uploads the model parameters and the data volume of this training to the central node after a certain number of rounds, and the central server calculates the weighted parity sum of the parameters, then the central node distributes the joint model parameters to each node, and repeats the above process until the maximum round of FL is reached.

4   Experiments and Results

4.1   Experimental Settings

Datasets, Models and Baselines. We evaluate CALIM-FL on four image classification tasks: FashionMNIST on a Conv2 model; CIFAR10 on a VGG model [14]; CIFAR100 on a ResNet model [2]; UC Merced Land Use Dataset (UCM) on a MobileNetV2 model [13]. We compare the test accuracy versus the round of CALIM-FL with FedAvg [11] and HRank [9]. FedAvg is a conventional FL implementation without any pruning operation. Note that HRank is initially a pruning scheme for the centralized machine learning environment and cannot be directly applied within FL. As a baseline, we let one UAV from swarms compute the ranks for all feature maps, and transfer the results to the central server, so that the latter can conduct HRank-based pruning. Since HRank is a one-shot pruning method, we start the pruning operation at the end of the training stage.

Implementation and Non-IID Partition. The proposed CALIM-FL is implemented with PyTorch. For experiments, we set up an FL system composed of a central server and 100(or 10) devices. For each FL round, 10(or 2)

Algorithm 1. CALIM-FL algorithm

1: **function** SERVER EXECUTES:
2:     initialize $\mathbf{w}(0)$
3:     initialize global pruning indicator vector $\mathbf{s}(0)$
4:     **for** $r = 0, 1, 2...$ **do**
5:         **for** each device n, in parallel **do**
6:             $\mathbf{w}_n(r+1) \leftarrow \mathbf{ClientUpdate}(n, \mathbf{w}(r))$
7:             $\mathbf{w}(r+1) \leftarrow \sum_{n \in [N]} p_n \mathbf{w}_n(r+1)$
8:         **end for**
9:         **if** r is a pruning round **then**
10:             $I \leftarrow$ (randomly constructs a batch of synthetic samples)
11:             $\mathcal{I}_m^{(1)}(\mathbf{w}(r); I) \leftarrow \sum_{t=1}^{|I|} \text{Rank}(\mathbf{o}(t, :, :))$
12:             $\mathcal{I}_m^{(2)}(\mathbf{w}(r)) \leftarrow \sum_{n=1}^{N} p_n \overline{\mathcal{I}_m^{(2)}}(\mathbf{w}, n)$
13:             $\mathcal{L}(\mathbf{w}(r); I) \leftarrow \alpha \mathcal{I}_m^{(1)}(\mathbf{w}(r); I) + (1 - \alpha)\mathcal{I}_m^{(2)}(\mathbf{w}(r); I)$
14:             $\mathbf{s}(r+1) \leftarrow$ model pruning$(\mathcal{L}(\mathbf{w}(r)), \text{prune ratio})$
15:             $\mathbf{w}(r+1) \leftarrow \mathbf{w}(r+1) \odot \mathbf{s}(r+1)$
16:         **end if**
17:     **end for**
18: **end function**
19: **function** DEVICEUPDATE$(r, n, \mathbf{w}, \mathbf{s})$
20:     initialize $\mathcal{Z}_n$ as empty set
21:     $\mathbf{w} \leftarrow \mathbf{w} \odot \mathbf{s}$
22:     **for** each epoch **do**
23:         **for** each iteration (using a batch of local data I) **do**
24:             $\mathbf{w} \leftarrow \mathbf{w} - \eta \mathbf{g}_n(\mathbf{w})$
25:             calculate $\mathcal{I}_m^{(2)}(\mathbf{w}; I)$ according to Eq. (6)
26:             $\mathcal{Z}_n \leftarrow \mathcal{Z}_n \bigcup \left\{ \mathcal{I}_m^{(2)}(\mathbf{w}; I) \right\}$
27:         **end for**
28:         **if** r is a pruning round **then**
29:             $\overline{\mathcal{I}_m^{(2)}}(\mathbf{w}, n) \leftarrow$ average set $\mathcal{Z}_n$
30:             **return** $\mathbf{w}, \overline{\mathcal{I}_m^{(2)}}(\mathbf{w}, n)$
31:         **else**
32:             **return** $\mathbf{w}$
33:         **end if**
34:     **end for**
35: **end function**

---

clients are randomly selected to participate in the training process. The non-IID degree has an impact on the training efficiency and accuracy of FL. We consider two different non-IID dataset partitioning methods. One is to partition the target dataset according to the Dirichlet distribution (also known as the hetero division method). The other is to partition the dataset into shards, as used in FedAvg [10] (Table 2).

Table 2. Evaluation configurations

Dataset-Model	Conv2-FashionMNIST	CIFAR10-VGG16	CIFAR100-ResNet110	UC Merced-MobileNetV2
LR scheduler - Cos[1]	warm steps=10 min learning rate=0.0	warm steps=20 min learning rate=0.0	warm steps=50 min learning rate=0.0	warm steps=20 min learning rate=0.0
Optimizer - SGD	Learning rate=0.1	Learning rate=0.1	Learning rate=0.01	Learning rate=0.01
Workers (Each round)	100(10)			10(2)
Local epoch	5	10	10	3
Batch size	32	32	32	32
Federated rounds	50	150	500	150

[1] Learning rate scheduler - CosineAnnealing with warmup.

Hypermeters. The loss function is the cross-entropy loss function, the optimization function is SGD, and the learning rate scheduler is StepLR with a step of 1. We also use the CosineAnnealing scheduler to dynamically adjust the learning rate of the model, which brings the benefit of jitter prevention when the model is close to convergence by setting smaller learning rates. For the dataset used for pruning on the central server, we set the value of each pixel within the synthetic data with a discrete uniform distribution in the range of $[0, 255]$. CALIM-FL employs a pre-defined pruning ratio strategy. Thus, the pruning ratio is a carefully crafted hyperparameter. The related hyperparameters except pruning rate(inspired from [9]) are chosen empirically only with tuning by experience. A large number of repeated experiments were carried out and the optimal selection was selected empirically.

4.2 Result Discussion

Evaluation of CALIM Pruning. The proposed CALIM is able to increase the variance of importance measures on each filter. We conduct experiments on VGG-16 with CIFAR-10 to illustrate this effect. Figure 3 illustrates the performance comparison between HRank and our proposed CALIM, in terms of importance measure. The x-axis denotes the convolution layer index of the neural network model, while the y-axis reflects the importance measure level at the same layer. The box plot clearly visualizes the distribution of importance measure levels by indicating the maximum, minimum, and average for each layer. A better importance measure function is able to make the distribution of importance measure levels more scattered, corresponding to a longer box-in-box plot. We use a simple Conv2 model on the FashionMNIST dataset to observe the effect of hyperparameters on CALIMFL. As shown in Table 3, when the hyperparameter is set to 0.6, the CALIMFL method achieves a good accuracy(74.653), significantly better than that of HRank(72.856). This further suggests that there is some sort of trade-off between the importance of the forward and backward propagation metrics.

Taking the filter importance level of the first convolution layer as an example in Fig. 3, the values obtained by our proposal are in the range $[0, 80]$, while those of HRank are in the range $[10, 33]$. The remaining layers of the model have better maximum and minimum magnitude differences for the CALIM vector.

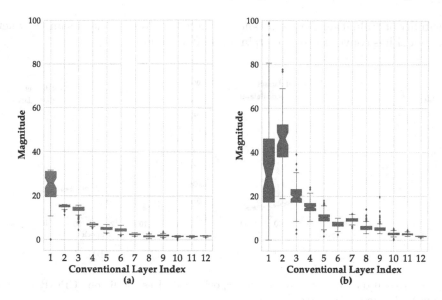

Fig. 3. Distribution of importance measure level for different layers with HRank and CALIM pruning criterion. (a) HRank; (b) CALIM

Table 3. Results on CALIMFL with different hyperparameters(α) vs. HRank

	0.1	0.2	0.3	0.4	0.5	0.6	0.7	0.8	0.9	1.0	HRank
Top-1 acc	64.757	75.000	71.528	69.445	68.230	**74.653**	69.444	71.007	72.694	71.465	**72.856**

Figure 3 demonstrates that the variance of the information measure of deeper layers in a neural model is small, which indicates that the redundancy of the shallower layers of the model is lower than that of the deeper layers, and the pruning rate of the deeper layers should be greater than that of the shallow layers when the model is pruned. This observation helps to design adaptive pruning algorithms. Since the focus of this work is to verify the importance of vertical channel-based pruning, a layer-aware pruning algorithm will be part of our future research work.

Evaluation on Synthetic Data Based Pruning. In a conventional FL setting, no dataset is available on a central server. We find that the cosine distance between the importance measure obtained by a batch of randomly generated synthetic data and that by real data is almost 0. Hence, we propose to conduct the pruning operation on a central server with a synthetic dataset. When the pruning operation is finished on the server. Figure 4 shows the loss function evolution versus epoch when fine-pruning with synthetic samples and real datasets. We conduct experiments for two cases: ResNet56 on CIFAR-10 and MobileNetV2 on CIFAR-100. The results are illustrated in Fig. 4. We observe that the loss evolution concerning epoch is almost similar, which confirms the feasibility of evaluating the importance measure for filter synthetic samples when the model has converged to a certain level after training.

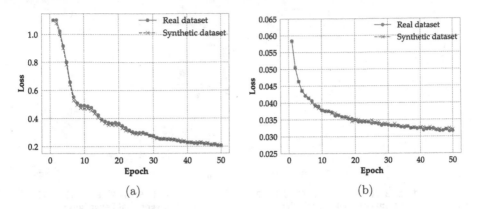

(a)   (b)

Fig. 4. Evolution of loss concerning epoch. (a) ResNet56 on CIFAR-10, (b) MobileNetV2 on CIFAR-100

The central server performs data-free pruning can avoid the information leakage problem during data transmission, and the pruning overhead is carried by the server. All UAVs can directly download the global parameters, which reduces the computational overhead of devices.

Evaluation on Accuracy and Model Size. To illustrate the generalizability of CALIM-FL on various data model scenarios, we conducted our image classification tasks, including Conv2 - FashionMNIST, VGG16 - CIFAR10, ResNet110 - CIFAR100 and MobileNetV2 - UC Merced. We use the shards Non-IId partitioning method in the first group and the Dirichlet partitioning method in other groups. Selected SOTA baselines, including traditional FL and Federated Pruning schemes, are FedAvg, FedProx, FedDUAP, PruneFL, and HRank-FL.

The Conv2 model consists of a single conv2d, batchnorm and linear layer. The FashionMNIST is a dataset containing single-channel data. So the task consumes only limited computational resources. On this task, we can quickly compare the performance of each solution. Notably, as shown in Table 4, CALIM-FL has an accuracy difference of ± 0.5% (+0.173 and -0.389) compared with the conventional federated learning algorithms (FedAvg and FedProx) after the same federated rounds. This confirm the effectiveness of CALIM-FL. Also, the model parameters were compressed by 49.6% under CALIM-FL. Because the communication process of federated learning involves the deserialization size of the model parameters, the communication overhead is greatly reduced. The running overhead FLOPs of the model are reduced from 44.37M to 22.18M, and CALIM-FL effectively reduces the computational overhead of end devices. PruneFL and FedDUAP are both federated pruning. PruneFL does not reduce the computational overhead metrics FLOPs of the model because it is unstructured pruning.

The FLOPs and the weights size of FedDUAP are bigger than ours (27.97M vs. 22.18M and 201.1KB vs. 162.0KB). Although CALIM-FL has the same computational overhead and communication overhead optimization as HRank-FL, the pruning process of HRank-FL requires full distribution data from the master node, while CALIM-FL is not limited by this restriction.

Table 4. Conv2 - FashionMNIST Performance Comparison between CALIM-FL, FedAvg, FedProx, FedDUAP, PruneFL, HRank-FL

	Accuracy (%)	FLOPs (M)	Model size (KB)	Compress ratio (%)
FedAvg [10]	75.521	44.3	321.5	0
FedProx [8]	76.083	44	321.5	0
FedDUAP [20]	75.761	27	201.1	37.5
PruneFL [4]	76.320	44	49.5	84.6
HRank-FL [9]	74.917	22	162.0	49.6
CALIM-FL (ours)	75.694	22.18	162.0	49.6

Table 5. VGG16 - CIFAR10 Performance Comparison between CALIM-FL, FedAvg, FedProx, PruneFL, HRank-FL

	Accuracy (%)	FLOPs (M)	Model size (MB)	Compress ratio (%)
FedAvg [10]	84.375	10066	57.293	0
FedProx [8]	84.417	10066	57.293	0
PruneFL [4]	84.285	10066	11.874	79.3
HRank-FL [9]	81.125	2123	6.611	88.4
CALIM-FL (ours)	83.854	2123	6.611	88.4

Table 6. ResNet110 - CIFAR100 Performance Comparison between CALIM-FL, FedAvg, FedProx, FedDUAP, PruneFL, HRank-FL

	Accuracy (%)	FLOPs (M)	Model size (MB)	Compress ratio (%)
FedAvg [10]	51.042	8227	6.908	0
FedProx [8]	51.812	8227	6.908	0
FedDUAP [20]	51.475	3702	3.109	55.0
HRank-FL [9]	50.042	3494	3.813	44.8
CALIM-FL (ours)	50.521	3494	3.183	44.8

Further, as shown in Table 5, for the CIFAR10 dataset, using the VGG16 model, the accuracy difference of CALIM-FL relative to the no-pruning learning is −0.6%. The parameter size and computational overhead are better than

Table 7. MobileNetV2 - UC Merced Performance Comparison between CALIM-FL, FedAvg, FedProx, FedDUAP, PruneFL, HRank-FL

	Accuracy (%)	FLOPs (M)	Model size (MB)	Compress ratio (%)
FedAvg [10]	53.125	10439	9.081	0
FedProx [8]	55.104	10439	9.081	0
HRank-FL [9]	50.521	5723	5.033	44.6
CALIM-FL (ours)	54.979	5723	5.033	44.6

the PruneFL. As shown in Table 6, we tested the ResNet110 model on the dataset CIFAR100, which is more difficult to converge. The accuracy difference of CALIM-FL is -1.5%, and the optimization is also better than that of FedDUAP. In the remote sensing scenario, we used the mobile efficient model MobileNetV2 with UC Merced remote sensing dataset. Table 7, CALIM-FL can reduce the computational overhead and communication overhead by about 50%.

5 Conclusion

This paper proposes an efficient federated learning framework CALIM-FL enabled by a filter pruning approach. CALIM-FL first proposes a novel importance measure CALIM (short for cross-all-layers importance measure) that evaluates the importance of a neuron simultaneously considering its information richness and impact on the final output. To support pruning on a central server and reduce the resource consumption on the UAV side, CALIM-FL proposes a synthetic-dataset-based pruning method, capable of efficiently compressing models. The experimental results demonstrate that CALIM-FL can greatly improve the model computing efficiency with limited accuracy loss of the model under both VGG and ResNet network structures.

Acknowledgments. This work was supported in part by the National Key Research and Development Program of China (2021YFB3101304), in part by the Natural Science Basic Research Program of Shaanxi (2022JQ-621, 2022JQ-658), in part by the National Natural Science Foundation of China (62001359, 62002278), in part by the Fundamental Research Funds for the Central Universities (ZYTS23163).

References

1. Gong, Y., Liu, L., Yang, M., Bourdev, L.: Compressing deep convolutional networks using vector quantization. arXiv preprint arXiv:1412.6115 (2014)
2. He, K., Zhang, X., Ren, S., Sun, J.: Deep residual learning for image recognition. In: Proceedings of the IEEE Conference on Computer Vision and Pattern Recognition, pp. 770–778 (2016)
3. He, Y., Lin, J., Liu, Z., Wang, H., Li, L.J., Han, S.: AMC: automl for model compression and acceleration on mobile devices. In: Proceedings of the European Conference on Computer Vision (ECCV), pp. 784–800 (2018)

4. Jiang, Y., et al.: Model pruning enables efficient federated learning on edge devices. IEEE Trans. Neural Netw. Learn. Syst. (2022)

5. Karimireddy, S.P., Kale, S., Mohri, M., Reddi, S., Stich, S., Suresh, A.T.: Scaffold: stochastic controlled averaging for federated learning. In: International Conference on Machine Learning, pp. 5132–5143 (2020)

6. Lee, N.T.A., Torr, P.: SNIP: single-shot network pruning based on connection sensitivity. In: ICLR (2019)

7. Lee, W.: Federated reinforcement learning-based UAV swarm system for aerial remote sensing. Wirel. Commun. Mob. Comput. **2022** (2022)

8. Li, T., Sahu, A.K., Zaheer, M., Sanjabi, M., Talwalkar, A., Smith, V.: Federated optimization in heterogeneous networks. Proc. Mach. Learn. Syst. **2**, 429–450 (2020)

9. Lin, M., et al.: Hrank: filter pruning using high-rank feature map. In: Proceedings of the IEEE/CVF Conference on Computer Vision and Pattern Recognition, pp. 1529–1538 (2020)

10. McMahan, B., Moore, E., Ramage, D., Hampson, S., Arcas, B.A.: Communication-efficient learning of deep networks from decentralized data. In: Artificial Intelligence and Statistics, pp. 1273–1282. PMLR (2017)

11. McMahan, B., Moore, E., Ramage, D., Hampson, S., Arcas, B.A.: Communication-efficient learning of deep networks from decentralized data. In: Artificial Intelligence and Statistics, pp. 1273–1282. PMLR (2017)

12. Ruan, X., Liu, Y., Li, B., Yuan, C., Hu, W.: DPFPS: dynamic and progressive filter pruning for compressing convolutional neural networks from scratch. In: Proceedings of the AAAI Conference on Artificial Intelligence, vol. 35, pp. 2495–2503 (2021)

13. Sandler, M., Howard, A., Zhu, M., Zhmoginov, A., Chen, L.C.: Mobilenetv 2: inverted residuals and linear bottlenecks. In: Proceedings of the IEEE Conference on Computer Vision and Pattern Recognition, pp. 4510–4520 (2018)

14. Simonyan, K., Zisserman, A.: Very deep convolutional networks for large-scale image recognition. arXiv preprint arXiv:1409.1556 (2014)

15. Vogels, T., Karimireddy, S.P., Jaggi, M.: PowerSGD: practical low-rank gradient compression for distributed optimization. Adv. Neural Inf. Process. Syst. **32** (2019)

16. Voigt, P., Von dem Bussche, A.: The EU general data protection regulation (GDPR). In: A Practical Guide, 1st edn, vol. 10, no. 3152676, p. 105555 . Springer, Cham (2017)

17. Wang, H., Qin, C., Bai, Y., Zhang, Y., Fu, Y.: Recent advances on neural network pruning at initialization. In: Proceedings of the International Joint Conference on Artificial Intelligence, IJCAI, Vienna, pp. 23–29 (2022)

18. Wang, J., Liu, Q., Liang, H., Joshi, G., Poor, H.V.: Tackling the objective inconsistency problem in heterogeneous federated optimization. Adv. Neural. Inf. Process. Syst. **33**, 7611–7623 (2020)

19. Yoshida, N., Nishio, T., Morikura, M., Yamamoto, K., Yonetani, R.: Hybrid-FL for wireless networks: cooperative learning mechanism using non-IID data. In: ICC 2020–2020 IEEE International Conference On Communications (ICC), pp. 1–7. IEEE (2020)

20. Zhang, H., Liu, J., Jia, J., Zhou, Y., Dai, H., Dou, D.: FedDUAP: Federated Learning with Dynamic Update and Adaptive Pruning Using Shared Data on the Server. arXiv preprint arXiv:2204.11536 (2022)

ASGNet: Adaptive Semantic Gate Networks for Log-Based Anomaly Diagnosis

Haitian Yang[1,2](✉), Degang Sun[2], Wen Liu[1,2], Yanshu Li[1,2], Yan Wang[1,2], and Weiqing Huang[1,2]

[1] Institute of Information Engineering, Chinese Academy of Sciences, Beijing, China
[2] School of Cyber Security, University of Chinese Academy of Sciences, Beijing, China
{yanghaitian,sundegang}@iie.ac.cn

Abstract. Logs are widely used in the development and maintenance of software systems. Logs can help engineers understand the runtime behavior of systems and diagnose system failures. For anomaly diagnosis, existing methods generally use log event data extracted from historical logs to build diagnostic models. However, we find that existing methods do not make full use of two types of features, (1) statistical features: some inherent statistical features in log data, such as word frequency and abnormal label distribution, are not well exploited. Compared with log raw data, statistical features are deterministic and naturally compatible with corresponding tasks. (2) semantic features: Logs contain the execution logic behind software systems, thus log statements share deep semantic relationships. How to effectively combine statistical features and semantic features in log data to improve the performance of log anomaly diagnosis is the key point of this paper. In this paper, we propose an adaptive semantic gate networks (ASGNet) that combines statistical features and semantic features to selectively use statistical features to consolidate log text semantic representation. Specifically, ASGNet encodes statistical features via a variational encoding module and fuses useful information through a well-designed adaptive semantic threshold mechanism. The threshold mechanism introduces the information flow into the classifier based on the confidence of the semantic features in the decision, which is conducive to training a robust classifier and can solve the overfitting problem caused by the use of statistical features. The experimental results on the real data set show that our method proposed is superior to all baseline methods in terms of various performance indicators.

Keywords: Anomaly Diagnosis · Semantic features · Statistical Features · Diagnose System Failures

1 Introduction

With the rapid development and evolution of information technology, during the past few years, we witness that large-scale distributed systems and cloud

B. Luo et al. (Eds.): ICONIP 2023, LNCS 14450, pp. 200–212, 2024.
https://doi.org/10.1007/978-981-99-8070-3_16

computing systems gradually become critical technical support of the IT industry [1–3]. Anomaly detection and diagnosis play an important role in the event management of large-scale systems [4,5], which aim to detect abnormal behavior of the system in time. Timely anomaly detection enables system developers (or engineers) to pinpoint problems the first time and resolve them immediately, thereby reducing system downtime [6–8]. However, as the scale of modern software become larger and more complex, the traditional log anomaly detection and diagnosis approaches based on specialized domain knowledge or manually constructed and maintained rules become less and less inefficient [9–12]. Benefiting from the development of deep learning technology, a number of effective log anomaly detection and diagnosis methods emerge in recent years, but these methods ignore two issues [13,14], (1) Statistical features: some inherent statistical features in log data, such as word frequency and abnormality label distribution, is not well utilized by deep learning based methods. Statistical features consist of statistical characteristics deterministic compared with log raw data, which is naturally compatible with the corresponding task. (2) Semantic features: Logs contain the execution logic behind the software system, thus log statements share deep semantic relationships. Hence, this paper focuses on how to effectively combine statistical features and semantic features in log data to improve the performance of log anomaly diagnosis. In order to demonstrate the problems that we raised in log data, we list different examples in Table 1 and Fig. 1.

Table 1. Statistics on the number of occurrences of words in the seven log datasets

Dataset	BGL	BGP	Tbird	Spirit	HDFS	Liberty	Zokeeper
Dataset size	1.207GB	1.04GB	27.367GB	30.289GB	1.58GB	29.5GB	10.4M
total number of distinct words	5632912	4491076	23330854	12793353	3585666	14682493	53094
appear only once	5173492	4330501	7789598	3861462	2576220	2020068	28954
	(91.8%)	(96.4%)	(33.4%)	(30.2%)	(71.8%)	(13.8%)	(54.5%)
appear less than 5 times	5298053	4424922	16863767	9914926	2853030	6656253	51443
	(94.05%)	(98.5%)	(72.3%)	(77.5%)	(79.6%)	(45.3%)	(96.9%)
appear less than 10 times	5403556	4463325	19948834	10815055	2885414	9882010	52686
	(95.9%)	(99.4%)	(85.5%)	(84.5%)	(80.5%)	(67.3%)	(99.23%)
appear less than 20 times	5465876	4472566	21112093	11311509	3353119	11425515	52797
	(97.03%)	(99.6%)	(90.5%)	(88.4%)	(93.5%)	(77.8%)	(99.4%)
appear at least once per 10000 lines	3825	1299	79063	137224	1998	10052	564
	(0.068%)	(0.029%)	(0.34%)	(1.07%)	(0.056%)	(0.068%)	(1.06%)
appear at least once per 1000 lines	573	316	6243	2663	258	5365	172
	(0.01%)	(0.007%)	(0.027%)	(0.021%)	(0.007%)	(0.037%)	(0.32%)

From Table 1, we can see that most of the words are infrequent, and most of these infrequent words appear only once. For example, in the BGP dataset, 96.4% of words appear only one time. In the Liberty dataset, 77.8% of the words have a frequency lower than 20. In the BGP and Zookeeper datasets, the number of words with a frequency lower than 20 accounts for more than 99%. At the same time, only a small fraction of words are frequent, i.e. they occur at least once in every 10000 or 1000 lines of logs. We can observe that most of the logs generated during the system operation are normal, and the abnormal logs are limited. Based on this, we infer that the abnormal words are not frequent, further we conclude that statistical features in logs are necessary for anomaly diagnosis.

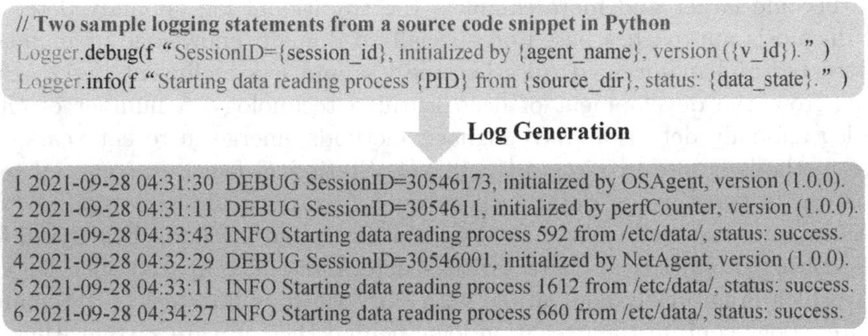

```
// Two sample logging statements from a source code snippet in Python
Logger.debug(f "SessionID={session_id}, initialized by {agent_name}, version ({v_id})." )
Logger.info(f "Starting data reading process {PID} from {source_dir}, status: {data_state}." )
```

Log Generation

```
1 2021-09-28 04:31:30  DEBUG SessionID=30546173, initialized by OSAgent, version (1.0.0).
2 2021-09-28 04:31:11  DEBUG SessionID=3054611, initialized by perfCounter, version (1.0.0).
3 2021-09-28 04:33:43  INFO Starting data reading process 592 from /etc/data/, status: success.
4 2021-09-28 04:32:29  DEBUG SessionID=30546001, initialized by NetAgent, version (1.0.0).
5 2021-09-28 04:33:11  INFO Starting data reading process 1612 from /etc/data/, status: success.
6 2021-09-28 04:34:27  INFO Starting data reading process 660 from /etc/data/, status: success.
```

Fig. 1. The process from initialization to sending data.

From Fig. 1, we can see that the first log to the sixth log shows the complete process from initialization to sending data, demonstrating the complete program execution logic. Therefore, log sequence contains the execution logic behind the program and has rich semantic information. We can see from Table 1 and Fig. 1 that the statistical features and text semantic features used in this paper are widely present in log data.

To deal with the two above-mentioned issues, this paper designs an Adaptive Semantic Gate Networks (ASGNet) which consists of three parts, log statistics information representation (V-Net), log deep demantic representation(S-Net) and adaptive semantic threshold mechanism(G-Net). Specifically, V-Net leverages an unsupervised autoencoder [15] to learn a global representation of each statistical feature vector, where we note that employing variational inference can further improve model performance compared to vanilla autoencoders. S-Net extracts latent semantic representations from text input by pre-trained RoBERTa [16]. G-Net aligns information from two sources, then adjusts the information flow.

The main contributions of this paper are summarized as follows:

(1) To the best of our knowledge, we are the first to attempt to explicitly leverage statistical features of system log data for anomaly diagnosis in a deep learning architecture. We also demonstrate that our proposed approach is very effective based on various experiments.
(2) To fuse statistical features into low-confidence semantic features, we propose a novel adaptive semantic threshold mechanism to retrieve necessary and useful global information. The experimental results prove that the threshold mechanism is very effective for the log-based anomaly diagnosis task.

(3) We conduct extensive experiments on 7 datasets of different scales and subjects. Results show that our proposed model yields a significant improvement over the baseline models.

2 Related Work

The main purpose of anomaly diagnosis is to help O&M engineers analyse the cause of anomalies and understand the operational status of the system or network. A number of excellent methods have emerged in recent years.

Yu et al. [17] utilized log templates to build workflows offline. Such models can provide contextual log messages of problems and diagnose problems buried deep within log sequences, which can provide the correct log sequence and tell engineers what is happening. Jia et al. [18] proposed a black-box diagnostic method TCFG (Timed-weight Control Flow Graph) for control flow graphs with temporal weights, which does not require prior domain knowledge and any assumptions and achieves better performance on relevant datasets.

Fu et al. [19] proposed the use of a Finite Automata Machine (FSA) to simulate the execution behaviour of a log-based system model. The model first clusters logs to generate log templates and removes all parameters based on regular expressions. Each log is then labelled by its log template type to construct sequences from which the FSA is learned to capture the system's normal work flow, achieving better performance on relevant datasets. Beschastnikh et al. [20] generated finite state machines from the execution trace of a concurrent system to infer a concise and accurate model of that system's behaviour. Engineers can use the inferred finite state machine model of communication to understand complex behaviour and detect anomalies for developers. Lou et al. [21] proposed an automaton model and a corresponding mining algorithm for reconstructing concurrent workflows. The algorithm can automatically discover program workflows and can construct concurrent workflows based on traces of interleaved events.

Although these methods have achieved better performance, none of these explored statistical feature and semantic feature of log sequence. Our study breaks the conventional thinking of treating logs as general objects of time series and introduces novel ideas in natural language processing to anomaly detection and diagnosis, investigating log sequences as text sequences with semantic information.

3 The Proposed Model

In this section, we present our proposed log-based anomaly diagnosis ASGNet model. Model structure is depicted in Fig. 2.

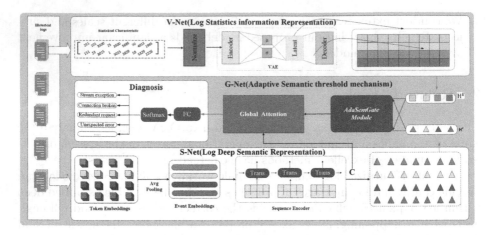

Fig. 2. Overview of our proposed ASGNet model.

3.1 Task Description

In this research, the log-based anomaly diagnosis task can be described as a tuple of three elements (S, I, y), where $S = [s^1, s^2, \ldots, s^g]$ represents the log sequence whose length is g. I denotes the current log message task ID and $y \in Y$ conveys the anomaly diagnosis specific labels for logging exceptions, currently diagnosable exceptions are Stream exception, Connection broken, Redundant request, Unexpected error, etc.

3.2 Definition of Terms

In this section, we first give the formal definition of the log statistics vector, which is defined as follows. Given a word w in a log message and a set of log exception labels $C = \{c_1, c_2, \cdots, c_n, \}$, the statistical information vector of w is:

$$E^w = [e_1, e_2, \cdots, e_n] \qquad (1)$$

where e_i represents the count of word w on label c_i. Given a log message $L = \{w_i\}_{i=1}^m$, the word statistics matrix in the log message is as follows:

$$E^L = [E^{w_1}, E^{w_2}, ..., E^{w_m}] \qquad (2)$$

The statistical information vector captures the global distribution of log anomaly labels as a feature of the words in the log.

These statistical features are primitive but can be used for feature selection by determining word-relatedness. Intuitively, if a word w has very high or very low frequency across all labels, then we can assume that w has a limited contribution to the anomaly diagnosis task. In contrast, if a word appears more frequently in specific label class, we assume this word is discriminative. Here, the term statistical label vector dictionary E is only obtained from the training set.

3.3 Log Statistics Information Representation

Log Statistics Information Representation (V-Net) aims to convert statistical features into an efficient representation. Log statistics vectors contain integer counts of words, thus initially incompatible with semantic features in both dimension and scale. Therefore, V-Net employs an autoencoder to map discrete vectors of log statistics into a latent continuous space to obtain a global representation of the statistics. In this work, we employ a variational autoencoder(VAE) [15] to encode log statistics vectors.

We generate log statistics vectors for all log messages in the data set to obtain $S = \{E_{(i)}^L\}_{i=1}^N$, which consists of N discrete log statistics vector composition. We assume that all log statistics vectors are generated by a stochastic process $p\theta(E|S)$ involving a latent variable S sampled from a prior distribution $p\theta(S)$. Since the posterior $p\theta(S|E)$ is intractable, we cannot directly learn the generative model parameters θ. Next, we employ a variational approximation $q\phi(S|E)$ to jointly learn the variational parameters ϕ and θ. Therefore, we can optimize the model by maximizing the marginal likelihood that is composed of a sum over the marginal likelihoods of individual E:

$$logp_\theta(E) = D_{KL}(q_\phi(S|E)||p_\theta(S|E)) + \text{L}(\theta, \phi; E) \qquad (3)$$

Because the KL divergence term is non-negative, we can derive the likelihood term $\text{L}(\theta, \phi; E)$ to obtain a variational lower bound on the marginal likelihood, i.e.:

$$\text{L}(\theta, \phi; E) = -D_{KL}(q_\phi(S|E)||p_\theta(S)) + E_{(q_\phi(S|E))}[logp_\theta(E|S)]) \qquad (4)$$

where the KL term is the closed solution, and the expected term is the reconstruction error. We employ a reparameterization approach to adapt the variational framework to an autoencoder. We use two encoders to generate two sets of μ and σ as the mean and standard deviation of the prior distributions, respectively. Since our approximate prior is a multivariate Gaussian distribution, we represent the variational posterior with a diagonal covariance structure:

$$logq_\phi(S|E) = logN(S; \mu, \sigma^2 I) \qquad (5)$$

By training an unsupervised VAE model, we can obtain latent variables E^s through a probabilistic encoder, which would be a global representation of statistical features. The training of V-Net is independent of the main classifier, and the representation E^s is generated in the preprocessing stage and is fed to the classifier through the adaptive semantic threshold mechanism.

3.4 Log Deep Semantic Representation

Log Deep Semantic Representation (S-Net) extracts semantic features from log message input and projects the semantic features into the information space for confidence evaluation. The input of S-Net is a log message $W = [w_1, w_2, \cdots, w_m]$ with fixed length m.

In this paper, we use the pre-training model to obtain the semantic representation of the log. Specifically, we use the pre-trained RoBERTa [16] to extract the feature map of the input log text:

$$C = RoBERTa(W) \tag{6}$$

Then, we map the semantic feature map C into the information space through dense layers. We use the sigmoid to activate values in the representation.

$$H^C = W^C \cdot C + b^C \tag{7}$$

where $H'^C = \sigma(H^C)$, where $\sigma(\cdot)$ is the sigmoid function, are used to evaluate the confidence of the corresponding semantic features in the decision-making process.

3.5 Adaptive Semantic Threshold Mechanism

As described in Sect. 2.3, the semantic representation of log statistics features E^s is obtained offline. To flexibly utilize statistical features, we apply dense layers to project E^s into the information space shared with semantic features:

$$H^E = W^E \odot (E^s) + b^E \tag{8}$$

The valve component is fused with H^C and H^E, and the semantic feature map H^O enhanced by statistical information is output through the AdaSemGate function,

$$H^O = AdaSemGate(H^C, H'^C, H^E, \epsilon) = ReLU(H^C) + Gate(H'^C, \epsilon) \odot H^E \tag{9}$$

where $ReLU(.)$ is the activation function, and $\odot$ represents element-wise point multiplication. The values in H'^C are in probabilistic form, and the Gate function is designed to recover less confident entries (probability close to 0.5) to match elements in $H\zeta$. Specifically, for each unit $a \in H'^C$,

$$Gate(\alpha, \epsilon) = \begin{cases} \alpha, & "if" 0.5 - \epsilon \geq \alpha \leq 0.5 + \epsilon \\ 0, & "otherwise" \end{cases} \tag{10}$$

where ϵ is a vulnerable hyperparameter for tuning the confidence threshold. Specifically, if $\epsilon = 0$, all statistics are rejected, and if $\epsilon = 0.5$, statistics are accepted. Therefore, the $Gate(\alpha, \epsilon)$ function uses element-wise multiplication as a filter to extract only the necessary information.

We employ global attention to combine the consolidated semantic representation H^O with the original feature map C:

$$GlobalAttention(\mathbf{H}^O, \mathbf{C}) = \text{softmax}(\mathbf{H}^O \mathbf{C}^\mathsf{T}) \mathbf{C} \tag{11}$$

Note that if we reject all statistical information (i.e., $\epsilon = 0$), Eqn.(11) will become self-attention [22] as $H^O = C$.

After passing through fully-connected layers and a softmax layer, feature vectors are mapped to the label space for label prediction and loss calculation. To maximize the probability of the correct label Y_{True}, we deploy an optimizer tominimize cross-entropy loss L.

$$L = \mathrm{CrossEntropy}(Y_{\mathrm{True}}, Y_{\mathrm{Pred.}}) \qquad (12)$$

4 Experimental Setup

4.1 Dataset and Hyper-parameters

We evaluate our model ASGNet on seven public datasets. The statistics of these seven datasets are listed in Table 2.

Table 2. The statistics of the seven datasets

Datasets	Size	#of logs	#of anomalies	#of anomalies types
BGL	1.207GB	4,747,963	949,024	5
BGP	1.04GB	11,428,282	1,276,742	11
Tbird	27.367GB	211,212,192	43,087,287	7
Spirit	30.289GB	272,298,969	78,360,273	8
HDFS	1.58GB	11,175,629	362,793	6
Liberty	29.5GB	266,991,013	191,839,098	17
Zookeeper	10.4M	74,380	49,124	10

4.2 Training and Hyperparameters

We fix all the hyper-parameters applied to our model. Specifically, We use the basic version of RoBERTa as the pre-trained embeddings in our experiments. The ϵ set to 0.2, which empirically shows the best performance. The algorithm we choose for optimization is Adam Optimizer with the first momentum coefficient $\beta_1 = 0.9$ and the second momentum coefficient $\beta_2 = 0.999$. We use the best parameters on development sets and evaluate the performance of our model on test sets.

5 Experimental Results

In this section, we elaborate the experimental setup and analyze the experimental results, aiming to answer:

RQ1: Can ASGNet achieve better log-based anomaly diagnosis performance than the state-of-the-art methods for log-based anomaly diagnosis task?

RQ2: How do the key model components and information types used in ASGNet contribute to the overall performance?

RQ3: How does the size of these parameters, specifically, hidden state dimension of global attention and the gate function ϵ, affect the performance of the entire model?

5.1 Model Comparisons (RQ1)

To analyze the effectiveness of our model, we take some of the most advanced methods as baselines on the above-mentioned seven datasets, to validate the performance of our ASGNet model. The results are demonstrated as follows.

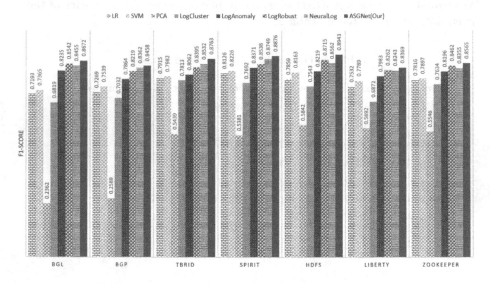

Fig. 3. Overall perform of our proposed ASGNet model.

As shown in Fig. 3, **NeuralLog** [23], **LogRobust** [11] and **LogAnomaly** [24] are strongest baseline models of the Log-Based Anomaly Detection task. The experiment results show that our model ASGNet achieves best performance on seven datasets. For instance, on the BGL dataset, our model outperforms baseline model **NeuralLog** [23] by 2.17% in F1-Score (p < 0.05 on student t-test). **LogRobust** [11] by 1.3% in F1-Score (p < 0.05 on student t-test). **LogAnomaly** [24] by 4.37% in F1-Score (p < 0.05 on student t-test). It also showed good performance compared to the comparison method on the other six datasets.

The reason is our proposed model takes into account not only the semantic features behind the log execution logic but also the statistical features of the log text in the log exception diagnosis task. In order to better fit the two features, we propose well-designed threshold mechanism which effectively selects the statistical features to consolidate the semantic features, instead of using all statistical features. The threshold mechanism introduces the information flow into the classifier based on the confidence of the semantic feature in the decision, which is conducive to training a robust classifier and can solve the overfitting problem caused by the use of statistical features. Therefore the final experimental results demonstrate that our method achieves remarkable performance in the log-based anomaly diagnosis task.

5.2 Ablation Study (RQ2)

To thoroughly figure out the effect of each key model component, we carry out a series of ablation study to decompose the whole model into three derived models.

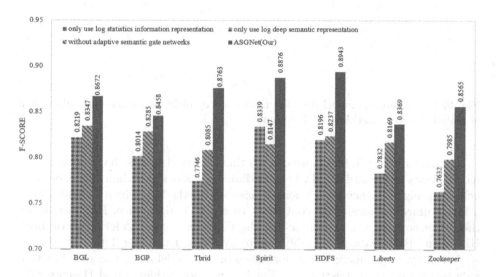

Fig. 4. Ablation study on seven datasets

Model (1): only use log statistics information representation – The entire model only uses the statistics information representation of the log sequence.

Model (2): only use log deep semantic representation – The entire model only uses the log deep semantic representation of the log sequence.

Model (3): without adaptive semantic gate networks – The entire model excludes the adaptive semantic gate networks.

As shown in Fig. 4, we can see that in our model, the log deep semantic representation module contributes more to task log-based anomaly diagnosis than the log statistics information representation module, mainly because the semantic information of the text of the logs can better express the deep logical semantic information of the logs, while the statistical features can only express the high and low frequency distribution of words, so log deep semantic representation module shows more advantages. In addition, it can be seen that the adaptive semantic gate module is indispensable in the whole component of our model, and removing the adaptive semantic gate module will affect the overall performance of the text proposed model.

5.3   Parameter Sensitivity (RQ3)

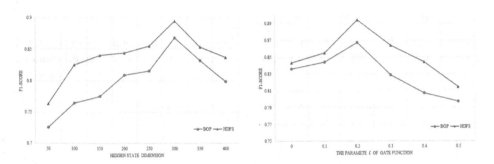

Fig. 5. Performance of model ASGNet influenced by different hidden state dimension of global attention and the gate function ϵ

As shown in Fig. 5, firstly, we can see that on two datasets the f1-score of our model shows an upward trend when the dimension size is less than 300, especially achieving highest when the dimension size is exactly 300, which indicates that a large dimension size could contribute to model performance. However, when the dimension size is larger than 300, the F1-Score of the model drops on both HDFS and BGL Dataset, possibly due to insufficient training data.

Secondly, we compare the performance of our model using the gate function ϵ on two datasets. As illustrated in Fig. 5, our model achieves best f1-score with the $\epsilon = 0.2$ thus demonstrating that not all statistical features are useful for the model.

6   Conclusion

Anomaly detection and diagnosis play an important role in the event management of large-scale systems, which aims to detect abnormal behavior of the system in time. Timely anomaly detection enables system developers (or engineers) to pinpoint problems the first time and resolve them immediately, thereby reducing system downtime. In this paper we design an Adaptive Semantic Gate Networks (ASGNet) which consists of three parts, log statistics information representation (V-Net), log deep semantic representation (S-Net) and adaptive semantic threshold mechanism (G-Net). Specifically, V-Net leverages an unsupervised autoencoder to learn a global representation of each statistical feature vector. S-Net extracts latent semantic representations from text input by pretrained RoBERTa. G-Net aligns information from two sources, then adjusts the information flow. The final experimental results demonstrate that our method achieves remarkable performance in the log-based anomaly diagnosis task.

Acknowledgement. This work is partially supported by E0HG043104.

References

1. Chen, J., et al.: An empirical investigation of incident triage for online service systems. In: ICSE-SEIP. IEEE (2019)
2. Yan, M., et al.: Exposing numerical bugs in deep learning via gradient back-propagation. In: ESEC/FSE (2021)
3. Yang, L., et al.: Plelog: semi-supervised log-based anomaly detection via probabilistic label estimation. In: ICSE. IEEE (2021)
4. Chen, R., et al.: Logtransfer: cross-system log anomaly detection for software systems with transfer learning. In: ISSRE. IEEE (2020)
5. Zhang, S., et al.: Cat: beyond efficient transformer for content-aware anomaly detection in event sequences. In: SIGKDD (2022)
6. Chen, J., et al.: How incidental are the incidents? characterizing and prioritizing incidents for large-scale online service systems. In: ASE (2020)
7. Zhao, N.,et al.: Real-time incident prediction for online service systems. In: ESEC/FSE (2020)
8. Zhao, N., et al.: Understanding and handling alert storm for online service systems. In: ICSE-SEIP. IEEE (2020)
9. Zhi, C., et al.: An exploratory study of logging configuration practice in java. In: ICSME. IEEE (2019)
10. Wang, L., et al.: Root-cause metric location for microservice systems via log anomaly detection. In: ICWS. IEEE (2020)
11. Zhang, X., et al.: Robust log-based anomaly detection on unstable log data. In: ESEC/FSE (2019)
12. Decker, L., et al.: Real-time anomaly detection in data centers for log-based predictive maintenance using an evolving fuzzy-rule-based approach. In: FUZZ-IEEE. IEEE (2020)
13. Huang, S., et al.: Hitanomaly: hierarchical transformers for anomaly detection in system log. IEEE Trans. Netw. Service Manag. (2020)
14. Xie, Y., Zhang, H., Zhang, B., Babar, M.A., Lu, S.: LogDP: combining dependency and proximity for log-based anomaly detection. In: Hacid, H., Kao, O., Mecella, M., Moha, N., Paik, H.-Y. (eds.) ICSOC 2021. LNCS, vol. 13121, pp. 708–716. Springer, Cham (2021). https://doi.org/10.1007/978-3-030-91431-8_47
15. Kingma, D.P., Welling, M.: Auto-encoding variational bayes. stat (2014)
16. Liu, Y., et al.: Roberta: a robustly optimized bert pretraining approach. arXiv preprint arXiv:1907.11692 (2019)
17. Yu, X., et al.: Cloudseer: workflow monitoring of cloud infrastructures via interleaved logs. In: SIGARCH (2016)
18. Jia, T., et al.: Logsed: anomaly diagnosis through mining time-weighted control flow graph in logs. In: CLOUD. IEEE (2017)
19. Qiang, F., et al.: Execution anomaly detection in distributed systems through unstructured log analysis. In: ICDM. IEEE (2009)
20. Beschastnikh, I., et al.: Inferring models of concurrent systems from logs of their behavior with csight. In: ICSE, pp. 468–479 (2014)

21. Lou, J.-G., et al.: Mining program workflow from interleaved traces. In: SIGKDD (2010)
22. Vaswani, A., et al.: Attention is all you need. In: NIPS (2017)
23. Le, V.-H., Zhang, H.: Log-based anomaly detection without log parsing. In: ASE, pp. 492–504. IEEE (2021)
24. Meng, W., et al.: Loganomaly: unsupervised detection of sequential and quantitative anomalies in unstructured logs. In: IJCAI (2019)

Propheter: Prophetic Teacher Guided Long-Tailed Distribution Learning

Wenxiang Xu[1,5], Yongcheng Jing[2], Linyun Zhou[1], Wenqi Huang[3],
Lechao Cheng[4(✉)], Zunlei Feng[1,6], and Mingli Song[1,6]

[1] Zhejiang University, Hangzhou 310027, China
{xuwx1996,zhoulyaxx,zunleifeng,brooksong}@zju.edu.cn
[2] The University of Sydney, Darlington, Camperdown, NSW 2008, Australia
yjin9495@uni.sydney.edu.au
[3] Digital Grid Research Institute, China Southern Power Grid, Guangzhou 510663,
China
huangwq@csg.cn
[4] Zhejiang Lab, Hangzhou 311121, China
chenglc@zhejianglab.com
[5] Zhejiang University - China Southern Power Grid Joint Research Centre on AI,
Hangzhou 310058, China
[6] ZJU-Bangsun Joint Research Center, Hangzhou, China

Abstract. The problem of deep long-tailed learning, a prevalent challenge in the realm of generic visual recognition, persists in a multitude of real-world applications. To tackle the heavily-skewed dataset issue in long-tailed classification, prior efforts have sought to augment existing deep models with the elaborate class-balancing strategies, such as class rebalancing, data augmentation, and module improvement. Despite the encouraging performance, the limited class knowledge of the tailed classes in the training dataset still bottlenecks the performance of the existing deep models. In this paper, we propose an innovative long-tailed learning paradigm that breaks the bottleneck by guiding the learning of deep networks with external prior knowledge. This is specifically achieved by devising an elaborated "prophetic" teacher, termed as "Propheter", that aims to learn the potential class distributions. The target long-tailed prediction model is then optimized under the instruction of the well-trained "Propheter", such that the distributions of different classes are as distinguishable as possible from each other. Experiments on eight long-tailed benchmarks across three architectures demonstrate that the proposed prophetic paradigm acts as a promising solution to the challenge of limited class knowledge in long-tailed datasets. The developed code is publicly available at https://github.com/tcmyxc/propheter.

Keywords: Unbalanced data · Long-tailed learning · Knowledge distillation

1   Introduction

Deep long-tailed learning is a formidable challenge in practical visual recognition tasks. The goal of long-tailed learning is to train effective models from a vast number of images, but most involving categories contain only a minimal number of samples. Such a long-tailed data distribution is prevalent in various real-world applications, including image classification [20], object detection [15], and segmentation [21]. As such, for the minority classes, the lack of sufficient instances to describe the intra-class diversity leads to the challenge of heavily skewed models towards the head classes. To tackle this challenge, various class balancing techniques have been proposed in the community, which can be broadly divided into three categories, *i.e.*, *Data Resampling*, *Loss Re-weighting*, and *Transfer Learning* based methods. Specifically, *Data Resampling* techniques involve either downsampling the samples in the head class or upsampling those in the tailed class [2,5], whereas the *Loss Re-weighting* methods design the elaborated loss functions to boost the accuracy of unbalanced data, especially those in the tailed classes [4,20]. In contrast, the transfer-learning-based approaches enhance the performance of the tailed classes by transferring the knowledge either from the head classes [11] to the tailed ones or from a teacher to a student through knowledge distillation [10].

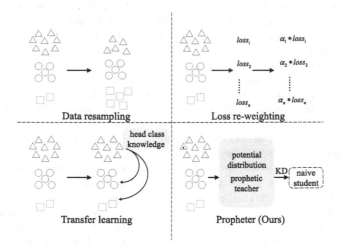

Fig. 1. Illustrations of the proposed *Propheter* paradigm as compared with other prevalent long-tailed learning schemes, including data resampling, loss re-weighting, and transfer learning.

Despite the plausible performance, most of the existing approaches heavily rely on well-designed sophisticated data samplers to alleviate the dilemma of the scarce tail classes, with some of them even retaining a large number of auxiliary parameters. As such, even at the expense of the heavy computational burden, the challenge of limited tailed classes is not fundamentally addressed. Besides, there is a lack of explicit considerations in the differences of the activation distributions among various long and tail categories in the high-dimensional feature space.

In this paper, we strive to make one further step towards breaking the performance bottleneck of long-tailed representation learning, by devising a novel *"Propheter"* paradigm that explores the long-tailed problem from the perspective of the *latent space*. Unlike the existing three prevalent schemes illustrated in Fig. 1, including data resampling, loss re-weighting, and transfer learning, the proposed *"Propheter"* aims to explicitly learn the potential feature distributions of various long and tail categories, based on which the challenge of deficient tailed data is solved by looking into the learned latent space.

To achieve this goal, we start by proposing a reasonable assumption that the activation distributions of various categories should be distinguishable in the latent space. Based on this hypothesis, we develop a residual distribution learning scheme to bridge the gap between the learned and the real class distributions. Specifically, we develop a parameter-efficient plug-and-play propheter module that adaptively learns the residual between the real and the present head/tailed class distributions in the dataset, by incorporating the captured class-specific distribution knowledge in the devised propheter module. We then transfer the learned class-specific distribution knowledge from the elaborated propheter module to the vanilla student long-tailed model, through the well-studied knowledge distillation scheme, leading to an innovative prophetic teacher guided long-tailed distribution learning paradigm.

In aggregate, the key contributions of this work can be summarized as follows:

- We propose the conjecture that the performance bottleneck of the existing deep long-tailed representation learning lies in the lack of the explicit considerations in the potential category-specific activation distributions, and accordingly devise a novel *"Propheter"* paradigm that aims to learn the potential real feature distribution of each head/tail class in the dataset;
- We propose a dedicated two-phase prophetic training strategy for the proposed *"Propheter"* paradigm, comprising the prophetic teacher learning and the propheter-guided long-tailed prediction stages. The first phase is backed by residual distribution learning, whereas the following propheter-instructed learning stage is built upon knowledge distillation;
- Experiments on eight prevalent long-tailed datasets demonstrate that our propheter-guided paradigm delivers encouraging results, with a performance gain of over 4% on CIFAR-10-LT as compared with the existing methods.

2   Related Work

Long-Tailed Image Categorization. In recent years, the field of computer vision has made remarkable strides across various domains, such as depth estimation [36,37], intelligent transportation [26–28], image recognition [22–25], image detection [29,34,35], 3D object analysis [6,12,38], model reusing [30,32], dataset condensation [17,18,33], facial expression recognition [19,39], and generative models [7,31]. Among these tasks, the challenge of addressing long-tailed visual recognition has garnered significant attention as a trending topic. Several approaches [2,3,13,15] have been proposed to tackle the class imbalance issue, including decoupling representation and classifier learning. For example, Kang *et al.* [13] employ multiple normalization techniques on the linear classifier layer. Chu *et al.* [3] enhance the feature representation of the tail classes. Another direction focuses on promoting the learning of the tail classes, such as re-sampling techniques [2], loss re-weighting [4], loss balancing [15], and knowledge transfer from the head classes to the tail classes [11].

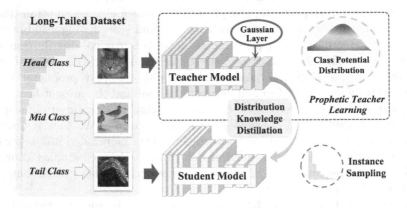

Fig. 2. The proposed *Propheter-Guided Long-Tailed Learning.*

Knowledge Distillation. The concept of knowledge distillation, as first introduced in [8], pertains to the transfer of information from a teacher model to a student model. This technique has been widely employed across various deep learning and computer vision tasks [8–10]. Iscen *et al.* [10] leveraged *Class-Balanced Distillation* (CBD) to enhance feature representations. This method trains multiple teacher models through different seed or data augmentation techniques. Feature distillation has also been demonstrated to be effective in tasks such as asymmetric metric learning and reducing catastrophic forgetting in incremental learning [9].

3   Proposed Method

3.1   Prophetic Teacher Learning

Learning Potential Class Distributions. In the real world, data often exhibit a specific distribution, such as Gaussian, uniform, or chi-square distributions. When such data are fed into a deep learning model, the features generated by the embedding layer should similarly reflect this distribution. Conventional deep learning models only learn the categorical distribution of the training dataset. However, while commonly used datasets are often balanced, real-world data often exhibit imbalanced distributions. Consequently, deep learning models trained on balanced datasets may not perform well on skewed data. Hence, it has been observed that relying solely on the original distribution of the training data results in limited generalization ability for deep learning models. By incorporating prior knowledge of the potential distribution of the dataset into the training process of the deep learning model, its performance can be improved as demonstrated in recent works such as [20,21].

Balanced Propheter Learning. While datasets may vary in their distributions across categories, features within a single category are generally assumed to follow a Gaussian distribution [20].

Suppose the features of different class samples conform to a Gaussian distribution. We can derive the feature representation $\mathbf{f}_i^{pred}$ for class i from the embedding layer (convolution layer). However, there exists a disparity between the learned features distribution and the ground-truth distribution $\mathbf{f}_i^{gt}$. Given that Gaussian distribution is additive, we can augment the learned Gaussian distribution $\mathbf{E}_i^{learned}$ after the convolution layer with the aim of:

$$\mathbf{f}_i^{pred} + \mathbf{E}_i^{learned} \triangleq \mathbf{f}_i^{gt}, \tag{1}$$

where $\mathbf{E} \sim \mathcal{N}(\mathbf{u}, \boldsymbol{\sigma}^2)$. From the formula, it can be seen that our goal is to learn the residual between two distributions $\mathbf{E}_i^{learned}$. Specifically, we will learn an independent Gaussian distribution $\mathbf{E}_i^{learned}$ for each class. For simplicity, the following article we will replace $\mathbf{E}_i^{learned}$ with $\mathbf{E}_i$.

About the Gaussian distribution, we will sample from the standard normal distribution $\mathbf{E} \sim \mathcal{N}(0,1)$. Then we scale and translate the distribution through two learnable parameters a_i and b_i. So the $\mathbf{E}_i$ will be:

$$\mathbf{E}_i := a_i \cdot \mathbf{E} + b_i, \tag{2}$$

and then $\mathbf{E}_i \sim \mathcal{N}(b_i, (a_i)^2)$. It can be seen that we can change the distribution through these two learnable parameters. That is to say, these two parameters control the shape of the class distribution in the latent space.

However, we discovered that a straightforward implementation of this approach did not yield the desired results. In response, we adopt an alternating training scheme, inspired by Liang *et al.* [14], to incrementally incorporate the target Gaussian noise into the model. Specifically, as outlined in Algorithm 1, during epochs where epoch % period = 0, we inject Gaussian noise into the features generated by the convolution layer, producing a modified feature vector which is then supplied to the classifier. This pathway is referred to as the Gaussian Noise (GN) path. Conversely, during non-zero values of epoch % period, the standard (STD) path, which does not alter the feature representation, is employed.

During the training phase, the known class labels of the images are utilized to incorporate the Gaussian distribution $\mathbf{E}_i$ into the feature representation of

218   W. Xu et al.

Algorithm 1. Training a prophetic teacher with balanced *Propheter* learning.

1: **for** e in epochs **do**
2:　　**if** e % period $= 0$ **then**
3:　　　$\tilde{y}_\theta \leftarrow$ prediction through the GN path
4:　　　$\mathcal{L} \leftarrow \ell(y||\tilde{y}_\theta)$
5:　　**else**
6:　　　$\tilde{y}_\theta \leftarrow$ prediction through the STD path
7:　　　$\mathcal{L} \leftarrow \ell(y||\tilde{y}_\theta)$
8:　　$\theta \leftarrow \theta - \epsilon\frac{\partial \mathcal{L}}{\partial \theta}$

---

each class. The addition of the Gaussian distribution $\mathbf{E}_i$ is performed subsequent to the final convolution layer and prior to the fully connected layer. The mean and variance of each category are utilized as learnable parameters and are incorporated into the model's parameters. This enables the optimization of these parameters through backpropagation, allowing for the model to adjust and improve its predictions over the training set X by minimizing the loss function with respect to the model's parameters θ and W:

$$L(X,Y;\theta,W) := \sum_{i=1}^{n} \ell\left(\sigma(\mathbf{z}_i), y_i\right),\tag{3}$$

where $\mathbf{z}_i = \phi_{\theta,W}(x_i)$ is the output of the model, $\sigma(.)$ is the softmax activation function, and $\ell(.)$ is the loss function, like the cross-entropy loss and others.

However, another challenge arises as the labels of samples are unknown during testing, precluding the addition of distinct Gaussian distributions to individual samples, thereby resulting in an inconsistent training-testing scenario. To overcome this, we employ the technique of knowledge distillation. Typically, knowledge distillation aims to align the logits of the student model with those of the teacher model. In our implementation, we adopt a variation that facilitates information transfer at the feature level.

To clarify, we adopt the model that incorporates Gaussian noise during training as our teacher model. Subsequently, through the utilization of knowledge distillation, we empower the student model to acquire the Gaussian distribution learned by the teacher model in the first stage. The objective loss function described in Eq. 3 transforms as follows:

$$L(X,Y;\theta,W) := \sum_{i=1}^{n} \ell\left(\sigma(\mathbf{z}_i), y_i\right) + \alpha \cdot \ell_F\left(\mathbf{v}_i, \widehat{\mathbf{v}}_i\right).\tag{4}$$

Specifically, $\widehat{\mathbf{v}}_i$ represents the feature representation produced by the teacher model, $\mathbf{v}_i$ for the student model, while $\ell_F(\mathbf{v}_i, \widehat{\mathbf{v}}_i)$ represents the mean squared error loss that seeks to minimize the dissimilarity between two feature representations. The hyperparameter α serves as a scaling factor, and unless specified otherwise, is set to 1.0. Our second stage algorithm is outlined in Algorithm 2.

Algorithm 2. Training a long-tailed student classification model, guided by the learned *Propheter* in Algorithm 1.

1: **for** e in epochs **do**
2:     **if** teacher model **then**
3:         $\hat{\tilde{\mathbf{v}}}_i \leftarrow$ feature descriptor produced by teacher model through the GN path
4:     **else if** student model **then**
5:         $\tilde{y}_{\theta_s} \leftarrow$ prediction through the STD path
6:         $\tilde{\mathbf{v}}_i \leftarrow$ feature descriptor produced by student model
7:         $\mathcal{L} \leftarrow \ell(y||\tilde{y}_\theta) + \alpha \cdot \ell_F(\mathbf{v}_i||\hat{\mathbf{v}}_i)$
8:     $\theta_s \leftarrow \theta_s - \epsilon \frac{\partial \mathcal{L}}{\partial \theta_s}$

3.2 Propheter-Guided Long-Tailed Classification

With the well-trained Prophetic Teacher (Propheter), the target long tail classification will be trained on the benchmark dataset under Propheter's guidance.

In general, the two-stage approaches involve the utilization of instance sampling in the first stage and class-balanced sampling in the second stage [3,13]. However, it was observed that class-balanced sampling can lead to a reduction in accuracy during the second stage when employing knowledge distillation. As a result, this study only employs instance sampling, which not only enhances the efficiency of training but also mitigates the overfitting of the tail classes.

For the second stage, three methods were employed to directly transfer information at the feature level.

▶ **1. Classical Decoupling.** The first method resembles the classical decoupling approach, with the learning rate being restarted and trained for a specified number of epochs, followed by the loading of the weights of the teacher model produced in the first stage by the student model.

▶ **2. Distillation with High-confidence Kernels.** The second method is similar to the first, but it involves selecting a pre-determined number of high-confidence pictures from the training set, counting their activation values on the last convolution layer, and then selecting the top k (where $k = 10$) convolution kernels for each category. During fine-tuning, only the Gaussian distribution is added to the convolution kernels corresponding to these categories on the teacher model. This approach was found to slightly improve the performance of the student model and is considered an extension of the first method.

▶ **3. Distilling from Scratch.** The third method involves training the student model from scratch while incorporating a feature distillation loss. Unlike the previous methods, this approach does not employ the use of highconfidence images to select the convolution kernels of each category, as it is believed that adding this prior knowledge on the teacher model would result in undue prejudice to the student model when training it from scratch. Experiments show that this scheme can maximize the performance of the model. In the following experimental results, we will only present the results of this method.

4   Experiments

4.1   Datasets and Implementation Details

Datasets. In our experiments, we utilize eight long-tailed benchmark datasets.

- *CIFAR-10-LT/CIFAR-100-LT*, are modified versions of CIFAR-10/CIFAR-100, as introduced in [1]. They have been generated by downsampling per-class training examples with exponential decay functions, resulting in long-tailed datasets. We perform experiments with imbalance factors of 100, 50, and 10, thereby obtaining a total of six datasets.
- *ImageNet-LT*, was introduced by Liu *et al.* [16] and is created by artificially truncating the balanced version of ImageNet. It comprises 115.8K images from 1000 categories, with a maximum of 1280 images per class and a minimum of 5 images per class.
- *Places-LT*, was introduced by Liu *et al.* [16], is a long-tailed variant of Places365. It comprises 62,500 training images distributed over 365 categories with a severe imbalance factor of 996.

Network Architectures. To maintain parity with prior literature, we adopt the network architectures utilized in prior studies [13]. Specifically, we utilize the ResNet32 architecture on the CIFAR-10-LT/CIFAR-100-LT datasets, the ResNeXt50 architecture on the ImageNet-LT dataset, and the ResNet152 architecture on the Places-LT datasets.

Evaluation Protocol. Our evaluation protocol entails training on the class-imbalanced, long-tailed training set of each dataset and evaluating performance on its corresponding balanced validation or test set. The performance metrics are reported on the validation sets.

Parameter Settings. The most crucial parameters in the first stage were the parameters for scaling a_i and translating b_i. We adopt a randomized approach to initialize these two parameters and impose constraints on a_i to ensure that its value does not fall below 0. In regards to b_i, we clamp it to the range $[0, 1]$ in order to prevent the occurrence of exploding gradients. The cycle period is set to 7. More implementation details can be found in the source code.

4.2   Experimental Results

Results on CIFAR-10-LT/CIFAR-100-LT. In this study, we conduct extensive evaluations on long-tailed CIFAR datasets with varying degrees of class imbalance. The results, depicted in Table 1, showcase the classification accuracy of ResNet-32 on the test set. Our evaluations encompass various loss functions, including Cross-Entropy Loss with softmax activation, Focal Loss, Class-Balanced Loss, and Balanced Softmax Loss, among others. The two entries in

Table 1. Classification accuracy of ResNet32 trained with different loss functions on long-tailed CIFAR-10 and CIFAR-100.

Dataset Name	Long-Tailed CIFAR-10			Long-Tailed CIFAR-100		
Imbalance	100	50	10	100	50	10
CE Loss	75.24	79.73	88.90	40.29	45.06	58.69
+Ours	79.27 (+4.03)	83.34 (+3.61)	89.52 (+0.62)	42.07 (+1.78)	47.12 (+2.06)	60.14 (+1.45)
Focal Loss ($\gamma = 1.0$) [15]	75.61	79.17	87.90	40.69	45.57	58.06
+Ours	77.12 (+1.51)	81.73 (+2.56)	89.05 (+1.15)	41.94 (+1.25)	46.83 (+1.26)	60.02 (+1.96)
Focal Loss ($\gamma = 2.0$) [15]	74.26	78.25	87.31	40.22	45.90	57.08
+Ours	76.57 (+2.31)	81.12 (+2.87)	88.54 (+1.23)	41.50 (+1.28)	46.72 (+0.82)	59.75 (+2.67)
Class-Balanced Loss [4]	73.48	80.60	87.66	39.79	45.89	58.97
+Ours	76.71 (+3.23)	81.43 (+0.83)	88.45 (+0.79)	41.77 (+1.98)	48.91 (+3.02)	59.94 (+0.97)
Balanced Softmax Loss [21]	81.91	84.27	89.92	46.93	51.24	60.66
+Ours	83.99 (+2.08)	86.05 (+1.78)	90.23 (+0.31)	48.33 (+1.40)	52.76 (+1.52)	62.23 (+1.57)

the table correspond to two different scenarios. The first shows the baseline performance without any additional modifications, the second corresponds to the result of distillation for 200 epochs (Distillation from scratch). All results are averages of three runs.

The results indicate that the addition of a Gaussian distribution layer and the utilization of knowledge distillation enhances the overall performance of the model. Specifically, for the case of Cross-Entropy Loss, the performance improvement surpasses 4% relative to the baseline. The results of our experiments show that our approach is suitable for long-tailed datasets with various imbalance factors. At the same time, our strategy can increase the model precision of the majority of known loss functions. That demonstrates our generalizability.

Table 2. Classification accuracy on the ImageNet-LT and Places-LT dataset. Here, [†] indicates that the hyperparameter α is 2.0 for Eq. 4.

	ImageNet-LT			Places-LT		
	Baseline	Ours	Gains	Baseline	Ours	Gains
CE Loss	43.97	44.57	**0.60** ↑	29.32	30.11	**0.79** ↑
Focal Loss ($\gamma = 1.0$) [15]	43.47	43.76	**0.29** ↑	29.28	30.12	**0.84** ↑
Focal Loss ($\gamma = 2.0$) [15]	43.29	44.26[†]	**0.97** ↑	29.18	29.78	**0.60** ↑
Balanced Softmax Loss [21]	47.95	48.55	**0.60** ↑	35.60	36.55	**0.95** ↑

Results on Large-Scale Datasets. We give the results for the baseline and the results following certain distillation epochs (90 epochs for ImageNet-LT and 30 epochs for Places-LT) for the large-scale datasets. For ImageNet-LT and Places-LT, the student model does not load the pre-training weight generated in the first stage (Distillation from scratch). Unfortunately, several methods do not provide hyperparameters for ImageNet-LT and Places-LT, so we only trained the model from scratch using softmax Cross-Entropy Loss, Focal Loss, and Balanced Softmax Loss. The results are shown in Table 2.

The experimental results show that our method still improves the performance of the model on large-scale datasets. This demonstrates that our strategy is still useful for large datasets and that it can enhance the performance of the model based on existing methods.

4.3 Ablation Study

We investigate the effect of (1) the proposed three methods by us, (2) the number and placement of Gaussian distribution layers. Our study focuses on the CIFAR-10-LT dataset with an imbalance factor of 100 and the use of the Balanced Softmax Loss.

Table 3. Overall classification accuracies of various knowledge transfer methods elaborated in Sect. 3.2.

#	Methods	Accuracy	Gains
1	Baseline	81.91	–
2	Classical Decouple	82.63	**0.72** ↑
3	Distillation with High-confidence Kernels	82.81	**0.90** ↑
4	Distilling from Scratch	83.99	**2.08** ↑

Various Knowledge Transfer Methods in Sect. 3.2. We contrast the outcomes of the three suggested approaches with the baseline, as indicated in Table 3. We can find that the three methods we proposed have improved the performance of the model, but the performance of method 3 (Distillation from scratch) has improved the most. At the same time, by comparing the second and third rows of the table, if we add Gaussian distribution on the class-related convolution kernel in the classical distillation scheme, we can slightly improve the performance of the model. This also proves the effectiveness of method 2 (Distillation with high-confidence kernels) from the side.

Position and Number of Gaussian Distribution Layers. In the ResNet32 architecture, there are three blocks. To assess the impact of the number and position of Gaussian distribution layers, we conduct experiments by altering their placement. The results are shown in Table 4. The first row serves as the baseline for comparison.

Our findings indicate that adding a single Gaussian distribution layer after the third block resulted in the most favorable outcome. This is evident when comparing the table's first, second, and third rows. Additionally, the comparison between the first, fourth, and fifth rows highlights that placing a single Gaussian distribution layer after the third block appears to be the optimal choice. We believe that this is because the features generated by the last block are more strongly associated with the classes, thereby leading to optimal performance.

Table 4. The influence of the position and number of Gaussian distribution layers on performance. Here, ✓ means that a Gaussian distribution layer is added after the corresponding block.

#	Block1	Block2	Block3	Accuracy (%)
1	–	–	✓	83.99
2	–	✓	✓	81.11
3	✓	✓	✓	80.40
4	✓	–	–	80.48
5	–	✓	–	80.87

5 Conclusions

In this work, we present a novel two-stage long-tailed recognition strategy that incorporates instance balance sampling in both stages and utilizes a singular teacher model in the process of knowledge distillation. Our method essentially learns the residual between the distribution learned by the model for each class and the true distribution in the latent space, from the perspective of the latent space. Our proposed method has been demonstrated to enhance the performance of existing long-tailed classification techniques and achieve substantial gains, as validated through a comprehensive series of experiments. In our future work, we will strive to explore the extension of our method to transformer-based models, and even the fields of object detection and image segmentation.

Acknowledgements. This work is funded by National Key Research and Development Project (Grant No: 2022YFB2703100), National Natural Science Foundation of China (61976186, U20B2066), Zhejiang Province High-Level Talents Special Support Program "Leading Talent of Technological Innovation of Ten-Thousands Talents Program" (No. 2022R52046), Ningbo Natural Science Foundation (2022J182), Basic Public Welfare Research Project of Zhejiang Province (LGF21F020020), and the Fundamental Research Funds for the Central Universities (2021FZZX001-23, 226-2023-00048). This work is partially supported by the National Natural Science Foundation of China (Grant No. 62106235), the Exploratory Research Project of Zhejiang Lab (2022PG0AN01), and the Zhejiang Provincial Natural Science Foundation of China (LQ21F020003).

References

1. Cao, Y., Long, M., Wang, J., Zhu, H., Wen, Q.: Deep quantization network for efficient image retrieval. In: AAAI (2016)
2. Chawla, N.V., Bowyer, K.W., Hall, L.O., Kegelmeyer, W.P.: SMOTE: synthetic minority over-sampling technique.J. Artif. Intell. Res. **16**, 321–357 (2002)
3. Chu, P., Bian, X., Liu, S., Ling, H.: Feature space augmentation for long-tailed data. In: Vedaldi, A., Bischof, H., Brox, T., Frahm, J.-M. (eds.) ECCV 2020. LNCS, vol. 12374, pp. 694–710. Springer, Cham (2020). https://doi.org/10.1007/978-3-030-58526-6_41

4. Cui, Y., Jia, M., Lin, T.Y., Song, Y., Belongie, S.: Class-balanced loss based on effective number of samples. In: CVPR (2019)
5. Feng, C., Zhong, Y., Huang, W.: Exploring classification equilibrium in long-tailed object detection. In: ICCV (2021)
6. Feng, M., et al.: Exploring hierarchical spatial layout cues for 3D point cloud based scene graph prediction. IEEE Trans. Multimedia **99**, 1–13 (2023)
7. Feng, Z., Jing, Y., Zhang, C., Xu, R., Lei, J., Song, M.: Graph-based color gamut mapping using neighbor metric. In: ICME (2017)
8. Hinton, G., Vinyals, O., Dean, J.: Distilling the knowledge in a neural network. In: NIPS Deep Learning and Representation Learning Workshop (2015)
9. Hou, S., Pan, X., Loy, C.C., Wang, Z., Lin, D.: Learning a unified classifier incrementally via rebalancing. In: CVPR (2019)
10. Iscen, A., Araujo, A., Gong, B., Schmid, C.: Class-balanced distillation for long-tailed visual recognition. In: BMVC (2021)
11. Jamal, M.A., Brown, M., Yang, M.H., Wang, L., Gong, B.: Rethinking class-balanced methods for long-tailed visual recognition from a domain adaptation perspective. In: CVPR (2020)
12. Jing, Y., Yuan, C., Ju, L., Yang, Y., Wang, X., Tao, D.: Deep graph reprogramming. In: CVPR (2023)
13. Kang, B., et al.: Decoupling representation and classifier for long-tailed recognition. In: ICLR (2019)
14. Liang, H., et al.: Training interpretable convolutional neural networks by differentiating class-specific filters. In: Vedaldi, A., Bischof, H., Brox, T., Frahm, J.-M. (eds.) ECCV 2020. LNCS, vol. 12347, pp. 622–638. Springer, Cham (2020). https://doi.org/10.1007/978-3-030-58536-5_37
15. Lin, T.Y., Goyal, P., Girshick, R., He, K., Dollár, P.: Focal loss for dense object detection. In: ICCV (2017)
16. Liu, S., Garrepalli, R., Dietterich, T.G., Fern, A., Hendrycks, D.: Open category detection with PAC guarantees. In: ICML (2018)
17. Liu, S., Wang, K., Yang, X., Ye, J., Wang, X.: Dataset distillation via factorization. NeurIPS (2022)
18. Liu, S., Ye, J., Yu, R., Wang, X.: Slimmable dataset condensation. In: CVPR (2023)
19. Luo, B., et al.: Learning deep hierarchical features with spatial regularization for one-class facial expression recognition. In: AAAI (2023)
20. Mengke Li, Yiu-ming Cheung, Y.L.: Long-tailed visual recognition via gaussian clouded logit adjustment. In: CVPR, pp. 6929–6938 (2022)
21. Ren, J., et al.: Balanced meta-softmax for long-tailed visual recognition. In: NeurIPS (2020)
22. Su, X., et al.: Prioritized architecture sampling with monto-carlo tree search. In: CVPR (2021)
23. Su, X., et al.: Locally free weight sharing for network width search. arXiv preprint arXiv:2102.05258 (2021)
24. Su, X., You, S., Wang, F., Qian, C., Zhang, C., Xu, C.: BCNet: searching for network width with bilaterally coupled network. In: CVPR (2021)
25. Su, X., et al.: ViTAS: vision transformer architecture search. In: Avidan, S., Brostow, G., Cissé, M., Farinella, G.M., Hassner, T. (eds.) Computer Vision – ECCV 2022. ECCV 2022. Lecture Notes in Computer Science, vol. 13681. Springer, Cham. https://doi.org/10.1007/978-3-031-19803-8_9
26. Xi, H.: Data-driven optimization technologies for MaaS. In: Big Data and Mobility as a Service (2022)

27. Xi, H., Liu, W., Waller, S.T., Hensher, D.A., Kilby, P., Rey, D.: Incentive-compatible mechanisms for online resource allocation in mobility-as-a-service systems. Trans. Res. Part B Methodol. **170**, 119-147 (2023)
28. Xi, H., Tang, Y., Waller, S.T., Shalaby, A.: Modeling, equilibrium, and demand management for mobility and delivery services in mobility-as-a-service ecosystems. Comput-Aided Civ. Infrastruct. Eng. **38**(11), 1403–1423 (2023)
29. Xi, H., Zhang, Y., Zhang, Y.: Detection of safety features of drivers based on image processing. In: 18th COTA International Conference of Transportation Professionals (2018)
30. Yang, X., Ye, J., Wang, X.: Factorizing knowledge in neural networks. In: Avidan, S., Brostow, G., Cissé, M., Farinella, G.M., Hassner, T. (eds.) Computer Vision – ECCV 2022. ECCV 2022. Lecture Notes in Computer Science, vol. 13694. Springer, Cham (2022). https://doi.org/10.1007/978-3-031-19830-4_5
31. Yang, X., Zhou, D., Feng, J., Wang, X.: Diffusion probabilistic model made slim. In: CVPR (2023)
32. Yang, X., Zhou, D., Liu, S., Ye, J., Wang, X.: Deep model reassembly. NeurIPS (2022)
33. Yu, R., Liu, S., Wang, X.: Dataset distillation: a comprehensive review. arXiv preprint arXiv:2301.07014 (2023)
34. Zhai, W., Cao, Y., Zhang, J., Zha, Z.J.: Exploring figure-ground assignment mechanism in perceptual organization. NeurIPS (2022)
35. Zhai, W., Luo, H., Zhang, J., Cao, Y., Tao, D.: One-shot object affordance detection in the wild. Int. J. Comput. Vis. **130**, 2472–2500 (2022). https://doi.org/10.1007/s11263-022-01642-4
36. Zhao, H., Bian, W., Yuan, B., Tao, D.: Collaborative learning of depth estimation, visual odometry and camera relocalization from monocular videos. In: IJCAI (2020)
37. Zhao, H., Zhang, J., Zhang, S., Tao, D.: JPerceiver: joint perception network for depth, pose and layout estimation in driving scenes. In: Avidan, S., Brostow, G., Cissé, M., Farinella, G.M., Hassner, T. (eds.) Computer Vision – ECCV 2022. ECCV 2022. Lecture Notes in Computer Science, vol. 13698. Springer, Cham (2022). https://doi.org/10.1007/978-3-031-19839-7_41
38. Zhao, H., Zhang, Q., Zhao, S., Zhang, J., Tao, D.: BEVSimDet: simulated multimodal distillation in bird's-eye view for multi-view 3D object detection. arXiv preprint arXiv:2303.16818 (2023)
39. Zhu, J., Luo, B., Yang, T., Wang, Z., Zhao, X., Gao, Y.: Knowledge conditioned variational learning for one-class facial expression recognition. IEEE Trans. Image Process. **32**, 4010–4023 (2023)

Sequential Transformer for End-to-End Person Search

Long Chen and Jinhua Xu$^{(\boxtimes)}$

East China Normal University, Shanghai, China
longchen@stu.ecnu.edu.cn, jhxu@cs.ecnu.edu.cn

Abstract. Person Search aims to simultaneously localize and recognize a target person from realistic and uncropped gallery images. One major challenge of person search comes from the contradictory goals of the two sub-tasks, *i.e.*, person detection focuses on finding the commonness of all persons so as to distinguish persons from the background, while person re-identification (re-ID) focuses on the differences among different persons. In this paper, we propose a novel Sequential Transformer (SeqTR) for end-to-end person search to deal with this challenge. Our SeqTR contains a detection transformer and a novel re-ID transformer that sequentially addresses detection and re-ID tasks. The re-ID transformer comprises the self-attention layer that utilizes contextual information and the cross-attention layer that learns local fine-grained discriminative features of the human body. Moreover, the re-ID transformer is shared and supervised by multi-scale features to improve the robustness of learned person representations. Extensive experiments on two widely-used person search benchmarks, CUHK-SYSU and PRW, show that our proposed SeqTR not only outperforms all existing person search methods with a 59.3% mAP on PRW but also achieves comparable performance to the state-of-the-art results with an mAP of 94.8% on CUHK-SYSU.

Keywords: Person Search · Person Re-identification · Pedestrian Detection · Transformer

1 Introduction

Practical applications of person search, such as searching for suspects and missing people in intelligent surveillance, require separating people from complex backgrounds and discriminating target identities (IDs) from other IDs. It involves two fundamental tasks in computer vision, *i.e.*, pedestrian detection and person re-identification (re-ID). Pedestrian detection aims at detecting the bounding boxes (Bboxes) of all candidates in the image. Person re-ID aims at retrieving a person of interest across multiple non-overlapping cameras. Person search has recently attracted the tremendous interest of researchers in the computer vision community for its importance in building smart cities. However, it remains a difficult task that suffers from many challenges, such as jointly optimizing contradictory objectives of two sub-tasks in a unified framework, scale/pose variations, background clutter and occlusions and so on.

B. Luo et al. (Eds.): ICONIP 2023, LNCS 14450, pp. 226–238, 2024.
https://doi.org/10.1007/978-981-99-8070-3_18

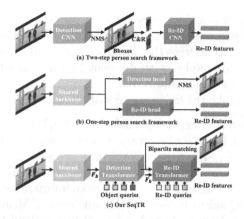

Fig. 1. Comparison of person search frameworks. (a) The two-step framework. (b) The one-step framework. (c) Our proposed SeqTR adopts the sequential framework to perform detection and re-ID in order.

According to training manners, existing person search methods can be generally grouped into two categories: two-step frameworks [4,7,8,18] and one-step frameworks [5,6,15,21–23]. Two-step methods typically perform detection and re-ID with two separate independent models. As shown in Fig. 1(a), pedestrians are first detected by an off-the-shelf detection model. After non-maximum suppression (NMS), the person patches are cropped and resized (C&R) into a fixed size. Then the person re-ID model is applied to produce ID feature embeddings, which will be used to calculate the similarity between the query persons and the candidates. The two-step frameworks can achieve satisfactory performance since each step focuses on one task and no contradictory is involved. However, this pipeline is time-consuming and resource-consuming. In contrast, one-step methods simultaneously optimize two sub-tasks in a joint framework (Fig. 1(b)). The two sub-tasks first share a common backbone for features extraction and then the detection head and re-ID head are applied in parallel.

In terms of architecture, the sequential architecture combines the merits of two-step and one-step frameworks. It not only inherits the better performance of two-stage frameworks via providing accurate bounding boxes (Bboxes) for the re-ID stage but also preserves the efficiency of the end-to-end training manner of one-step frameworks. However, as Li *et al.* [13] has pointed out, the performance bottleneck of the sequential CNNs framework lies in the design of the re-ID subnetwork. In addition, we find that NMS, commonly used in CNNs-based detection models, primarily hinders the inference speed of this architecture, especially in crowded scenes.

As transformers [17] become popular in many vision tasks, transformers-based person search methods [1,24] also show advantages over CNNs-based models, such as no NMS needed and powerful capability of learning fine-grained features.

Motivated by the above observations, we propose a novel Sequential transformer (SeqTR) for end-to-end Person Search (Fig. 1(c)). Two transformers are integrated seamlessly to address the detection and re-ID task. Meanwhile, the two transformers are decoupled with different features to alleviate the contradictory objectives of detection and re-ID.

In summary, we make the following contributions:

- We propose a novel Sequential Transformer (SeqTR) model for end-to-end person search, which utilizes two transformers to sequentially perform pedestrian detection and re-ID without NMS post-processing.
- We propose a novel re-ID transformer to generate discriminative re-ID feature embeddings. To make full use of context information, we introduce the self-attention mechanism in our re-ID transformer. Meanwhile, we employ multiple cross-attention layers to learn local fine-grained features. To obtain scale-invariant person representations, our re-ID transformer is shared by multi-scale features.
- We achieve a remarkable result on two datasets. Comprehensive experiments show the effectiveness of our proposed sequential framework and the re-ID transformer. With PVTv2-B2 [19] backbone, SeqTR achieves 59.3% mAP that outperforms all existing person search models on the PRW [25] dataset.

2 Method

2.1 SeqTR Architecture

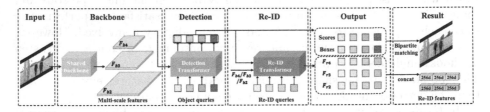

Fig. 2. Architecture of our proposed SeqTR, which comprises a backbone, a detection transformer and a re-ID transformer.

The overall architecture of our SeqTR is depicted in Fig. 2. It contains three main components: a backbone to extract multi-scale feature maps of the input image, a detection transformer to predict Bboxes, and a novel re-ID transformer to learn robust person feature embeddings.

Backbone. Starting from the original image $x_{img} \in \mathbb{R}^{3 \times H_0 \times W_0}$ (with 3 color channels). The backbone extracts original multi-scale feature maps $\{x^l\}_{l=1}^3$ from stages P_2 through P_4 in PVTv2-B2 [19] (or from stages C_3 through C_5 in RestNet [10]). The resolution of x^l is 2^{l+2} lower than the input image.

Detection Transformer. We introduce the transformer-based detector, deformable DETR [27], into our framework to predict the pedestrian bounding boxes. However, The difference with the original deformable DETR [27] is the input features. First, the channel dimensions of all feature maps $\{x^l\}_{l=1}^3$ from the backbone are mapped to a smaller dimension $d = 256$ by 1×1 convolution. Then, a 3×3 deformable convolution is used to generate more accurate feature maps. Finally, $\{F_{bi} \in \mathbb{R}^{d \times H \times W}\}_{i=2}^4$ are transformed from original feature maps $\{x^l\}_{l=1}^3$ by the above two steps and fed into a standard deformable DETR [27].

re-ID Transformer. Our re-ID transformer aims to adaptively learn discriminative re-ID features around the human body center. Motivated by object queries in DETR [2], we set a fixed number of learnable re-ID queries Q_r to reconcile the relationship between detection and re-ID and obtain re-ID feature embeddings.

2.2   re-ID Transformer

The architecture of the re-ID transformer is shown in Fig. 3. Suppose that the detection transformer decodes N objects in each image. The re-ID query number is also set as N. Taking the enhanced backbone features $\{F_{bi}\}_{i=2}^4$, N reference points P_q from the detection transformer and N re-ID queries Q_r as input, the re-ID transformer outputs N instance-level re-ID embeddings F_{ri} that have the same dimension as the pixel features. These instance-level re-ID feature embeddings are highly associated with pedestrian locations. Furthermore, to aggregate multi-scale features, multi-scale feature maps $\{F_{bi}\}_{i=2}^4$ are used to generate multi-scale re-ID embeddings $\{F_{ri} \in \mathbb{R}^{d \times H \times W}\}_{i=2}^4$ ($d{=}256$ in experiments) by the re-ID transformer. During inference, all multi-scale re-ID embeddings $\{F_{ri}\}_{i=2}^4$ are concatenated into a final 768-dimensional vector for calculating similarity.

Re-ID Queries. To mitigate the objective contradictory problem, we set re-ID queries Q_r, like object queries, to obtain re-ID features. Specifically, re-ID queries guarantees that the final re-ID embeddings $\{F_{ri}\}_{i=2}^4$ are instance-level fine-grained features learned from the augmented multi-scale backbone features $\{F_{bi}\}_{i=2}^4$. Through this design, the final learned re-ID feature embeddings are highly correlated with the detected pedestrian locations, but not affected by the detection features. This is different from the re-ID decoder in PSTR [1], in which the re-ID queries come from the output features of the detection decoder.

Self-attention Layer. To produce discriminative re-ID feature embeddings, we introduce the self-attention layer into the re-ID transformer to learn contextual information. This is different from the re-ID decoder in PSTR [1], in which no self-attention layer is used. From the ablation study in experiments, the performance is improved with the self-attention layer. Specifically, we adopt a standard multi-head self-attention (with H heads) in the Transformer [17]. We

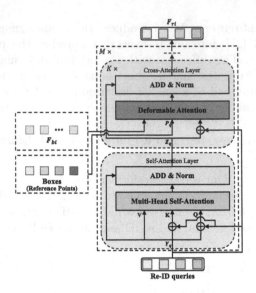

Fig. 3. Architecture of our proposed re-ID transformer.

denote the input of the self-attention layer as Y_q. The initial input $Y_q = Q_r$. Y_q are transformed into query vectors $Q \in \mathbb{R}^{N \times d}$, key vectors $K \in \mathbb{R}^{N \times d}$ and value vectors $V \in \mathbb{R}^{N \times d}$ by three different linear projections. The output embeddings then are generated by performing the multi-head self-attention module.

$$\text{head}_i = \text{Attention}(QW_i^Q, KW_i^K, VW_i^V), \tag{1}$$

where $W_i^Q \in \mathbb{R}^{d \times d_k}$, $W_i^K \in \mathbb{R}^{d \times d_k}$, $W_i^V \in \mathbb{R}^{d \times d_v}$, $d_k = d_v = d/H$. The self-attention module use Scaled Dot-Product Attention in each head:

$$\text{Attention}(Q, K, V) = \text{softmax}\left(\frac{QK^\top}{\sqrt{d_k}}\right) V. \tag{2}$$

The embeddings from all heads are concatenated and projected to yield d-demensional embeddings:

$$\text{MultiHead}(Q, K, V) = \text{Concat}(\text{head}_1, ..., \text{head}_H) W^O, \tag{3}$$

where $W^O \in \mathbb{R}^{Hd_k \times d}$. At last, we use a layer normalization to get the final embeddings $\hat{Y}_q$.

$$\hat{Y}_q = \text{layernorm}(Y_q + \text{dropout}(\text{MultiHead}(Q, K, V))). \tag{4}$$

The self-attention layer in the first re-ID transformer layer can be skipped. After passing through the first re-ID transformer layer, the output features are correlated with reference points. N feature embeddings correspond to N locations respectively. These embeddings interact with each other for learning spatial relationship by the self-attention layer in the mth ($m \in [2, M]$) re-ID transformer layer, resulting to enhance feature embeddings by instances in the same scene.

Cross-Attention Layer. Different from the previous works that use the RoI-Align layer on detection features, we employ and stack several cross-attention layers to address the region misalignment. In the cross-attention layer, there is a deformable attention module and a layer normalization. The deformable attention module proposed by deformable DETR [27], only attends to a small set of key sampling points around a reference point. It is useful for learning fine-grained features. Given an input feature map $F_{bi} \in \mathbb{R}^{C \times H \times W}$, a set of detected bounding boxes, $i.e.$, reference points (denoted P_q), and query features (denoted Z_q), the output feature embeddings $\hat{Z}_q$ can be calculated:

$$\hat{Z}_q = \text{layernorm}(Z_q + \text{dropout}(\text{DeformAttn}(Z_q, P_q, F_{bi}))), \tag{5}$$

$$\text{DeformAttn}(Z_q, P_q, F_{bi}) = \sum_{h=1}^{H} W_h \left[\sum_{s=1}^{S} A_{hs} \cdot W'_h F_{bi}(P_q + \Delta P_{hs}) \right] \tag{6}$$

where H is the total attention heads, S is the total sampled key number. A_{hs} and ΔP_{hs} denote attention weight of the sth sampling point in the hth attention head and the sampling offset, respectively. The sampling offset, $\Delta P_{hs} \in \mathbb{R}^2$ is learnable. The attention weight $A_{hs} \in \mathbb{R}$ is a scalar in the range [0, 1], normalized by $\sum_{s=1}^{S} A_{hs} = 1$. Both are obtained via linear projection over $(Z_q + Y_q)$, respectively. In this way, each query feature corresponds to one detected bounding boxes and integrates the features of the surround sampling points. In PSTR [1], features at sampling points are averaged rather than using the attention weight A_{hs} as in Eq. 6 because it was observed that the attention weights from the query struggle to effectively capture the features of a person instance. We think it may be caused by the coupling of the two decoders since the re-ID queries in PSTR [1] are from the detection decoder.

2.3 Training and Inference

For each image, our SeqTR predicts N classification scores, bounding boxes and re-ID feature embeddings $\{F_{ri}\}_{i=2}^4$. In the training phase, $\{F_{ri}\}_{i=2}^4$ are supervised separately. They are concatenated into a 768-dim vector during inference.

During training, our SeqTR is trained end-to-end for detection and re-ID. Specifically, detection transformer is supervised with loss functions of deformable DETR [27] for classification (L_{cls}), bounding-box IoU loss (L_{iou}), bounding-box Smooth-L1 loss (L_{l1}). While the re-ID transformer is supervised by the Focal OIM loss (L_{oim}) [22].

The overall loss is given by:

$$L = \lambda_1 L_{cls} + \lambda_2 L_{iou} + \lambda_3 L_{l1} + \lambda_4 L_{oim} \tag{7}$$

where $\lambda_1, \lambda_2, \lambda_3, \lambda_4$, responsible for the relative loss importance, are set as 2.0, 5.0, 2.0, 0.5, respectively.

During inference, our SeqTR predicts Bboxes and corresponding re-ID feature embeddings for gallery images. For the query person, we get predictions of the query image in the same way and then choose the one that has maximum overlap with its annotated bounding box.

3   Experiments

3.1   Datasets and Settings

CUHK-SYSU. There are a total of 18,184 realistic and uncropped images, 96,143 annotated bounding boxes and 8,432 different identities. The training set includes 11,206 images, 55,272 pedestrians, and 5,532 identities. The test set contains 6,978 images, 40,871 pedestrians, and 2,900 identities.

PRW. There are 11,816 video frames and 43,110 annotated bounding boxes. 34,304 of these boxes are annotated with 932 labelled identities and the rest are marked as unknown identities. The training set contains 5,704 images, 18,048 pedestrians, and 482 identities. The test set has 6,112 images and 2,057 query persons with 450 identities.

Evaluation Metrics. Following the previous works [21], we employ Mean Average Precision (mAP) and Cumulative Matching Characteristics (CMC top-K) to evaluate the performance of the person search.

3.2   Implementation Details

We adopt ResNet50 [10] and transformer-based PVTv2-B2 [19] that are pre-trained on ImageNet [16] as backbone. To train our model, we adopt the AdamW optimizer with a weight decay rate of 0.0001. The initial learning rate is set to 0.0001 that is warmed up during the first epoch and decreased by a factor of 10 at 19th and 23th epoch, with a total of 24 epochs. For CUHK-SYSU/PRW, the circular queue size of OIM is set to 5000/500. During training, we employ a multi-scale training strategy, where the longer side of the image is randomly resized from 400 to 1666. For inference, we rescale the test images to a fixed size of 1500 × 900 pixels. For our SeqTR with ResNet50 [10] backbone, we use one NVIDIA GeForce RTX 3090 to run all experiments and batch size set to 2. Our SeqTR with PVTv2-B2 [19] backbone is trained on two RTX 3090 GPUs with batch size set to 1 because of the limitation of GPU memory.

3.3   Comparison to the State-of-the-arts

Results on CUHK-SYSU. As shown in Table 1, our SeqTR outperforms most one-step methods and achieves comparable performance to two-step methods on the CUHK-SYSU [21]. Our SeqTR with the ResNet50 backbone, which achieves comparable 93.4% mAP and 94.1% top-1 accuracy, outperforms AlignPS [22] by 0.3% and 0.7% in mAP and top-1 accuracy, respectively. Our results are slightly worse than the transformer-based COAT [24] and PSTR [1].

Then, based on PVTv2-B2 [19] backbone, the performance of our SeqTR is significantly improved to 94.8% mAP and 95.5% top-1 accuracy. Our method

Table 1. Comparison with the state-of-the-art methods on CUHK-SYSU and PRW test sets. * denotes our reproduced result. The highest scores in each group are highlighted in bold.

Method	Backbone	CUHK-SYSU		PRW	
		mAP(%)	Top-1(%)	mAP(%)	Top-1(%)
Two-step methods					
MGTS [4]	VGG16	83.0	83.7	32.6	72.1
CLSA [12]	ResNet50	87.2	88.5	38.7	65.0
RDLR [8]	ResNet50	93.0	94.2	42.9	70.2
IGPN [7]	ResNet50	90.3	91.4	**47.2**	87.0
TCTS [18]	ResNet50	**93.9**	**95.1**	46.8	**87.5**
One-step methods with CNNs					
OIM [21]	ResNet50	75.5	78.7	21.3	49.4
NPSM [14]	ResNet50	77.9	81.2	24.2	53.1
RCAA [3]	ResNet50	79.3	81.3	–	–
IAN [20]	ResNet50	76.3	80.1	23.0	61.9
CTXGraph [23]	ResNet50	84.1	86.5	33.4	73.6
QEEPS [15]	ResNet50	88.9	89.1	37.1	76.7
BI-Net [6]	ResNet50	90.0	90.7	45.3	81.7
APNet [26]	ResNet50	88.9	89.3	41.9	81.4
NAE [5]	ResNet50	91.5	92.4	43.3	80.9
NAE+ [5]	ResNet50	92.1	92.9	44.0	81.1
PGSFL [11]	ResNet50	90.2	91.8	42.5	83.5
SeqNet [13]	ResNet50	93.8	94.6	46.7	83.4
DMRN [9]	ResNet50	93.2	94.2	46.9	83.3
AlignPS [22]	ResNet50	93.1	93.4	45.9	81.9
One-step methods with transformers					
COAT [24]	ResNet50	94.2	94.7	53.3	87.4
PSTR [1]	ResNet50	93.5	95.0	49.5	87.8
SeqTR(Ours)	ResNet50	93.4	94.1	52.0	86.5
PSTR [1]	PVTv2-B2	**95.2**	**96.2**	56.5	89.7
PSTR* [1]	PVTv2-B2	94.6	95.6	57.6	**90.1**
SeqTR(Ours)	PVTv2-B2	94.8	95.5	**59.3**	89.4
COAT [24]+CBGM	ResNet50	94.8	95.2	54.0	89.1
PSTR [1]+CBGM	PVTv2-B2	**95.8**	**96.8**	58.1	**92.0**
PSTR* [1]+CBGM	PVTv2-B2	95.2	96.1	58.2	91.5
SeqTR(Ours)+CBGM	PVTv2-B2	95.4	96.3	**59.8**	90.6

outperforms the reproduced results of PSTR by 0.2% in mAP. Moreover, the post-processing strategy Context Bipartite Graph Matching (CBGM) [13] is widely used to improve mAP and top-1 accuracy. By employing CBGM, our SeqTR achieves 95.4% mAP and 96.3% top-1, which outperforms the reproduced results of PSTR* with CBGM.

Results on PRW. The PRW dataset [25] is more challenging than the CUHK-SYSU dataset [21] for less training data and larger gallery size. Furthermore, there is a large number of people wearing similar uniforms and there are more scale variations, pose/viewpoint changes and occlusions. Nevertheless, our method achieves strong performance in Table 1. With ResNet50 [10] backbone, our SeqTR achieves 52.0% mAP and 86.5% top-1 accuracy, outperforming all two-step methods and with a significant gain of 2.5% mAP than PSTR [1] with the same backbone. The performance of our method is slightly lower than COAT [24] by 1.3% mAP and 0.9% top-1.

With PVTv2-B2 backbone [19], our SeqTR achieves 59.3% mAP and 89.4% top-1 accuracy, outperforming all existing methods with a clear margin on mAP. Finally, our SeqTR is improved to the best 59.8% mAP and comparable 90.6% top-1 accuracy with CBGM.

3.4 Ablation Study

We perform a series of ablation studies on the PRW [25] dataset. The ResNet50 [10] backbone is used in the ablation study.

Table 2. Ablation study for different shared re-ID transformer structures on the PRW dataset.

Transformer layers M	Cross-attention layers K	mAP(%)	Top-1(%)
2	2	50.8	86.5
	3	50.4	86.0
	4	49.9	85.5
3	2	50.4	86.7
	3	52.0	86.5
	4	50.7	86.8
4	2	50.5	86.7
	3	50.3	86.1
	4	50.6	86.7

Re-ID Transformer Structure. Setting the number of the self-attention layer in each transformer layer to 1, we evaluate the impact of the number of transformer layers M and the number of cross-attention layers K. As shown in Table 2, when the number of transformer layers is greater than 2, different combinations of M and K have a slight impact on performance. Among these configurations, when $M = 3$ and $K = 3$, our SeqTR achieves the best performance of 52.0% mAP and 86.5% top-1 accuracy.

Importance of Self-attention Layer. We also evaluate the importance of the self-attention layer. We find that adding self-attention layers yields improvements of 2.4% on mAP and 1% on top-1 accuracy respectively.

Table 3. Ablations on sequential architecture on the PRW dataset.

Method	Re-identifier	mAP(%)	Top-1(%)
Sequential Transformer	re-ID transformer	52.0	86.5
Sequential Transformer	original transformer	45.2	82.6
Non-sequential	res5	35.2	76.8

Sequential Transformer Architecture. To verify the advantages of our proposed main innovation point "sequence transformer", we design two other methods to compare. The detector of these three methods is the same detection transformer. First, the "original transformer" is the vanilla transformer with 6 layers [17], in which each layer consists of a multi-head self-attention layer, a cross-attention layer and a feed-forward network. As shown in Table 3, our proposed re-ID transformer is 6.8% higher on mAP and 3.9% higher on Top-1. In the "Non-sequential" method, there is only a detection transformer. It exploits RoI-Align on $\{F_{bi}\}_{i=2}^{4}$ to get $1024 \times 14 \times 14$ regions on three-scale features and feeds them into res5 to extract 2048-dim features, as in OIM [21]. A 256-dim re-ID vector, which consists of a 128-dim vector from 1024-dim features and a 128-dim vector from 2048-dim features is obtained from each scale and then concatenated to get the final 768-dim re-ID vector. Table 3 shows that the sequential transformer frameworks outperform the non-sequential one significantly.

Choices of Input Features to re-ID transformer. We conduct experiments on employing different input features to the shared re-ID transformer, including single-scale and multi-scale features. The results are reported in Table 4. Specifically, we first evaluate the single-level feature respectively. Among these single-level features, the output feature of the encoder in the detection transformer provides less information for the re-ID task and is discarded in later experiments. Relatively, F_{b3} yields the best performance. Furthermore, we also show the performance of utilizing multi-scale features. As can be observed, the best performance is achieved by using three-scale features.

Table 4. Comparative results by employing different input features on the PRW dataset. "✓" means using the corresponding feature. "E" denotes the output feature of the encoder in the detection transformer.

Input feature	E	F_{b4}	F_{b3}	F_{b2}	mAP(%)	Top-1(%)
Single-scale feature	✓				26.5	66.4
		✓			41.6	79.5
			✓		45.6	82.9
				✓	41.7	82.6
Multi-scale feature	✓				41.6	79.5
	✓	✓			47.8	82.7
	✓	✓	✓		52.0	86.5

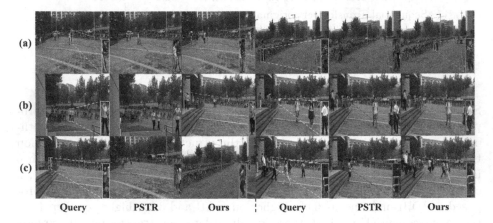

Fig. 4. Qualitative comparison with PSTR [1]. The yellow bounding boxes denote the queries, while the green and red bounding boxes denote correct and incorrect top-1 matches, respectively. We scale up the bounding box to the bottom right. Row (a) are two cases to illustrate the strength of the sequential transformer framework. Row (b) are two cases to show the importance of self-attention layers in our re-ID transformer. Row (c) are two cases to show the advantages of cross-attention layers in our re-ID transformer.

Qualitative Results. To demonstrate the advantages of our SeqTR, we show some qualitative comparisons between our SeqTR with PSTR [1] on the PRW [25] dataset. As shown in Fig. 4(a), our SeqTR achieves more accurate pedestrian localizations in both examples, because the sequential framework can better relieve the contradictory problem. In both cases of Fig. 4(b), compared to PSTR [1], our SeqTR accurately identifies the query persons, whose co-travellers wear similar uniforms. It is attributed to the self-attention layer that employs contextual information. In addition, the cross-attention layers in our re-ID transformer contribute to focusing on meaningful regions, although occlusions occur in the given query person in Fig. 4(c). The above examples illustrate that our SeqTR further alleviates some challenges, such as occlusions and distinguishing similar appearances.

4   Conclusion

In this paper, we propose a novel Sequential Transformer (SeqTR) for end-to-end person search. Within our SeqTR, a detection transformer and a re-ID transformer are integrated to solve the two contradictory tasks sequentially. We design a re-ID transformer that contains self-attention layers and cross-attention layers to generate discriminative re-ID feature embeddings. Furthermore, our re-ID transformer adopts a share strategy for employing multi-scale features. Extensive experiments demonstrate the performance of our proposed transformer framework, which achieves state-of-the-art results on PRW [25] dataset and comparable performance on the CUHK-SYSU [21] dataset.

References

1. Cao, J., et al.: PSTR: end-to-end one-step person search with transformers. In: Proceedings of the IEEE/CVF Conference on Computer Vision and Pattern Recognition, pp. 9458–9467 (2022)
2. Carion, N., Massa, F., Synnaeve, G., Usunier, N., Kirillov, A., Zagoruyko, S.: End-to-end object detection with transformers. In: Vedaldi, A., Bischof, H., Brox, T., Frahm, J.-M. (eds.) ECCV 2020. LNCS, vol. 12346, pp. 213–229. Springer, Cham (2020). https://doi.org/10.1007/978-3-030-58452-8_13
3. Chang, X., Huang, P.-Y., Shen, Y.-D., Liang, X., Yang, Y., Hauptmann, A.G.: RCAA: relational context-aware agents for person search. In: Ferrari, V., Hebert, M., Sminchisescu, C., Weiss, Y. (eds.) ECCV 2018. LNCS, vol. 11213, pp. 86–102. Springer, Cham (2018). https://doi.org/10.1007/978-3-030-01240-3_6
4. Chen, D., Zhang, S., Ouyang, W., Yang, J., Tai, Y.: Person search via a mask-guided two-stream CNN model. In: Ferrari, V., Hebert, M., Sminchisescu, C., Weiss, Y. (eds.) ECCV 2018. LNCS, vol. 11211, pp. 764–781. Springer, Cham (2018). https://doi.org/10.1007/978-3-030-01234-2_45
5. Chen, D., Zhang, S., Yang, J., Schiele, B.: Norm-aware embedding for efficient person search. In: Proceedings of the IEEE/CVF Conference on Computer Vision and Pattern Recognition, pp. 12615–12624 (2020)
6. Dong, W., Zhang, Z., Song, C., Tan, T.: Bi-directional interaction network for person search. In: Proceedings of the IEEE/CVF Conference on Computer Vision and Pattern Recognition, pp. 2839–2848 (2020)
7. Dong, W., Zhang, Z., Song, C., Tan, T.: Instance guided proposal network for person search. In: Proceedings of the IEEE/CVF Conference on Computer Vision and Pattern Recognition, pp. 2585–2594 (2020)
8. Han, C., et al.: Re-id driven localization refinement for person search. In: Proceedings of the IEEE/CVF International Conference on Computer Vision, pp. 9814–9823 (2019)
9. Han, C., Zheng, Z., Gao, C., Sang, N., Yang, Y.: Decoupled and memory-reinforced networks: towards effective feature learning for one-step person search. In: Proceedings of the AAAI Conference on Artificial Intelligence, vol. 35, pp. 1505–1512 (2021)

10. He, K., Zhang, X., Ren, S., Sun, J.: Deep residual learning for image recognition. In: Proceedings of the IEEE conference on computer vision and pattern recognition, pp. 770–778 (2016)
11. Kim, H., Joung, S., Kim, I.J., Sohn, K.: Prototype-guided saliency feature learning for person search. In: Proceedings of the IEEE/CVF Conference on Computer Vision and Pattern Recognition, pp. 4865–4874 (2021)
12. Lan, X., Zhu, X., Gong, S.: Person search by multi-scale matching. In: Ferrari, V., Hebert, M., Sminchisescu, C., Weiss, Y. (eds.) ECCV 2018. LNCS, vol. 11205, pp. 553–569. Springer, Cham (2018). https://doi.org/10.1007/978-3-030-01246-5_33
13. Li, Z., Miao, D.: Sequential end-to-end network for efficient person search. In: Proceedings of the AAAI Conference on Artificial Intelligence, vol. 35, pp. 2011–2019 (2021)
14. Liu, H., et al.: Neural person search machines. In: Proceedings of the IEEE International Conference on Computer Vision, pp. 493–501 (2017)
15. Munjal, B., Amin, S., Tombari, F., Galasso, F.: Query-guided end-to-end person search. In: Proceedings of the IEEE/CVF Conference on Computer Vision and Pattern Recognition, pp. 811–820 (2019)
16. Russakovsky, O., et al.: ImageNet large scale visual recognition challenge. Int. J. Comput. Vision 115(3), 211–252 (2015)
17. Vaswani, A., et al.: Attention is all you need. arXiv preprint arXiv:1706.03762 (2017)
18. Wang, C., Ma, B., Chang, H., Shan, S., Chen, X.: TCTS: a task-consistent two-stage framework for person search. In: Proceedings of the IEEE/CVF Conference on Computer Vision and Pattern Recognition, pp. 11952–11961 (2020)
19. Wang, W., et al.: PVT v2: improved baselines with pyramid vision transformer. Comput. Vis. Media 8(3), 415–424 (2022)
20. Xiao, J., Xie, Y., Tillo, T., Huang, K., Wei, Y., Feng, J.: IAN: the individual aggregation network for person search. Pattern Recogn. 87, 332–340 (2019)
21. Xiao, T., Li, S., Wang, B., Lin, L., Wang, X.: Joint detection and identification feature learning for person search. In: Proceedings of the IEEE Conference on Computer Vision and Pattern Recognition, pp. 3415–3424 (2017)
22. Yan, Y., et al.: Anchor-free person search. arXiv preprint arXiv:2103.11617 (2021)
23. Yan, Y., Zhang, Q., Ni, B., Zhang, W., Xu, M., Yang, X.: Learning context graph for person search. In: Proceedings of the IEEE/CVF Conference on Computer Vision and Pattern Recognition, pp. 2158–2167 (2019)
24. Yu, R., et al.: Cascade transformers for end-to-end person search. In: Proceedings of the IEEE/CVF Conference on Computer Vision and Pattern Recognition, pp. 7267–7276 (2022)
25. Zheng, L., Zhang, H., Sun, S., Chandraker, M., Yang, Y., Tian, Q.: Person re-identification in the wild. In: Proceedings of the IEEE Conference on Computer Vision and Pattern Recognition, pp. 1367–1376 (2017)
26. Zhong, Y., Wang, X., Zhang, S.: Robust partial matching for person search in the wild. In: Proceedings of the IEEE/CVF conference on computer vision and pattern recognition, pp. 6827–6835 (2020)
27. Zhu, X., Su, W., Lu, L., Li, B., Wang, X., Dai, J.: Deformable DETR: deformable transformers for end-to-end object detection. arXiv preprint arXiv:2010.04159 (2020)

Multi-scale Structural Asymmetric Convolution for Wireframe Parsing

Jiahui Zhang[1], Jinfu Yang[1,2](✉), Fuji Fu[1], and Jiaqi Ma[1]

[1] Faculty of Information, Beijing University of Technology, Beijing 100124, China
jfyang@bjut.edu.cn
[2] The Beijing Key Laboratory of Computational Intelligence and Intelligent System,
Beijing University of Technology, Beijing 100124, China

Abstract. Extracting salient line segments with their corresponding junctions is a promising method for structural environment recognition. However, conventional methods extract these structural features using square convolution, which greatly restricts the model performance and leads to unthoughtful wireframes due to the incompatible geometric properties with these primitives. In this paper, we propose a Multi-scale Structural Asymmetric Convolution for Wireframe Parsing (MSACWP) to simultaneously infer prominent junctions and line segments from images. Benefiting from the similar geometric properties of asymmetric convolution and line segment, the proposed Multi-Scale Asymmetric Convolution (MSAC) effectively captures long-range context feature and prevents the irrelevant information from adjacent pixels. Besides, feature maps obtained from different stages in decoder layers are combined using Multi-Scale Feature Combination module (MSFC) to promote the multi-scale feature representation capacity of the backbone network. Sufficient experiments on two public datasets (Wireframe and YorkUrban) are conducted to demonstrate the advantages of our proposed MSACWP compared with previous state-of-the-art methods.

Keywords: Wireframe Parsing · Asymmetric Convolution ·
Long-Range Context · Multi-Scale Feature

1 Introduction

Wireframe parsing is a fundamental but critical computer vision task, which aims to extract salient and prominent line segments with their corresponding junctions [1–3]. As a high level geometric primitive, line segment contains much structural information of surrounding environment than feature points in a compact representation manner. Therefore, line segment becomes a representative geometry feature in many low-level tasks, such as structure-from-motion (SfM)

This work was partly supported by the National Natural Science Foundation of China
under Grant No.61973009.

[4,5], Visual Simultaneous Localization and Mapping (VSLAM) [6,7], 3D reconstruction [8,9] and vanishing point detection [10].

Traditional line segment detection methods focus on exploring local feature in the given images using edge detector (e.g. Canny or Sobel) and generate long straight lines by Hough Transform (HT). Although HT, with its various methods, has become one category of the representative line segment detection method, it still suffers from the interference from noise and many false negative results. As a counterpart of HT-based methods, gradient-based methods constitute another branch of line segment detection task. [11] proposed a line segment detection method LSD, which greatly enhances the line segment extraction accuracy. Specifically, LSD extended the straight line detection method in [12] with line validation method based on Helmholtz principle [13], which enables LSD to infer concise results in a parameter-less way. However, the robustness to image noise and too much time consumption lead LSD an improper method in many real-time tasks. EDlines [14] presented a parameter-less line segment detector with linear time complexity. Concretely, edge pixels detected by edge drawing (ED) [15] algorithm are connected to generate clean and continuous line segments. Nevertheless, both LSD and EDlines concentrate on local pixel gradient but ignore the high level semantic information of line segments, which limits their feature extraction performance.

To mitigate the afore-mentioned problems, data-driven methods are designed for better extraction effect. [1] first introduced a large-scale well annotated wireframe parsing benchmark and presented the first learning-based wireframe parsing method DWP, which separately predicts junctions with their corresponding line segments in a parallel way. Although has achieved promising improvement, DWP is not a fully end-to-end network, and cannot predict vectorized parsing results, which are the shortcomings of this method. [2] proposed a fully differentiable wireframe parser L-CNN that can generate vectorized representation of line segments without heuristic methods. Further, L-CNN presented static and dynamic line sampler to meet the need of cold-start problem during the network training. Moreover, it proposed LOI Pooling inspired by the well-known ROIPooling and ROIAlign operation in object detection [16,17]. Different from the previous methods, HAWP [3] developed an efficient parsing method using 4-dimensional attraction field map. By doing so, the line segment detection problem is eventually transformed to model the relation between line segments with their corresponding pixels in the attraction map. To construct a unified line segment detection method across different kinds of camera, ULSD [18] is presented to achieve better accuracy with preferable real-time performance. Specifically, Bezier curve [19] is introduced to fit the distorted lines existing in images. Recently, [20] designed a fully convolutional line parsing network F-Clip. A customized line feature extraction residual block is proposed for better feature representation effect. In spite of achieving better trade-off between accuracy and real-time performance, the model convergence still an inevitable shortcoming for F-Clip. On the other hand, the multi-scale property of line segments in images creates opportunities for further improvement of existing methods.

In this paper, we present a Multi-scale Structural Asymmetric Convolution Wireframe Parser (MSACWP). First, we propose a Multi-Scale Asymmetric Convolution (MSAC) for better concentrating on the long-range dependencies of line segments and avoiding the interferences from neighboring pixels. Second, we concatenate the multi-stage features in decoder layers of the improved Hourglass network to enhance the representation learning performance. Finally, we perform extensive experiments on two public datasets (Wireframe [1] and YorkUrban [21]) to verify the superiority of our proposed MSACWP that produces impressive feature extraction accuracy with an acceptable real-time performance.

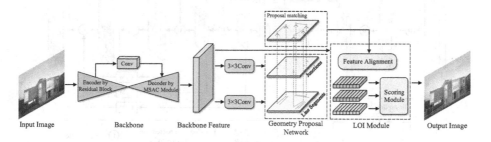

Fig. 1. The overall architecture of our proposed MSACWP. The whole model contains three parts including Backbone, Geometry Proposal Network and LOI Module.

2 Methodology

2.1 Overall Network Architecture

The overall architecture of our proposed MSACWP is shown in Fig. 1. MSACWP is a fully end-to-end network which can simultaneously extract remarkable junctions with line segments in parallel. Specifically, MSAC is leveraged as the decoder layers of the backbone network for better extracting the structural features. Meanwhile, stage-wise decoder features are combined using MSFC module for better representation of multi-scale information. Subsequently, junction and line segment proposals are inferred by two convolution layers. After the proposal matching procedure, the Line of Interest (i.e. LOI) module is used to evaluate the correctness of the acquired line segments.

2.2 Customized Backbone

The backbone network serves as a fundamental feature extractor, which affects the inference performance to a great extent. Taking consideration of the long-span and narrow geometric properties of line segment, the conventional square convolution cannot precisely capture this characteristic due to its inherent structure. Therefore, different from the other state-of-the-art wireframe parsing methods that uses the stacked Hourglass Network [22] as the backbone, we propose an effective and concise backbone network to acquire the implicitly existing structural information, and the detailed structure is illustrated in Fig. 2.

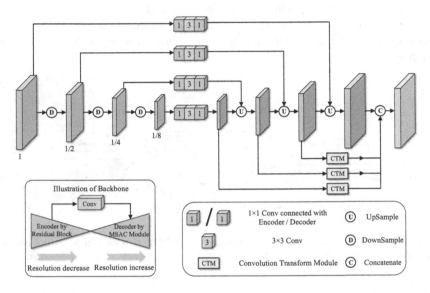

Fig. 2. The detailed illustration of our customized backbone network in MSACWP. Residual blocks are used as the encoder and the MSAC blocks are leveraged as the decoder. Meanwhile, stage-wise features in decoder layers are combined to generate the final output.

We first propose the MSAC module as the stage-wise decoder to capture long-range dependencies of line segment and prevent the irrelevant information, as shown in Fig. 3. The MSAC module consists of four types of structural convolution kernels belong to horizontal and vertical directions. We denote the input tensor of MSAC module as $\mathbf{X}^{in}_{MSAC} \in \mathbb{R}^{C \times H \times W}$, $\mathbf{X}^{in}_{MSAC}$ is then transmitted to the 1×1 convolution layer for channel reduction. Subsequently, the structural convolution layers are leveraged to capture the multi-scale structural feature and purify the irrelevant features with less computation costs. After that, the acquired feature maps are concatenated to get the diversity information before BatchNorm and ReLU activations. Finally, a 1×1 convolution layer follows the intermediate result is designed to align the channel dimension.

Let $\mathbf{c}_s \in \mathbb{R}^{2s+1}$ denotes the multi-scale asymmetric convolution kernel at size $2s+1$, $\mathbf{D} = (D_H, D_V)$ represents the horizontal and vertical direction of kernel respectively, and the output is represented as $\mathbf{X}_{MSAC}^{out} \in \mathbb{R}^{C \times H \times W}$. We formulate the structural feature extraction procedure in MSAC as:

$$\mathbf{X}_{D_{H_1}}^{out}(i,j) = (\mathbf{X}_{MSAC}^{in} * \mathbf{c}_{s_1})_{D_{H_1}}(i,j) = \sum_{k=-s_1}^{s_1} x(i+k,j+k) \cdot c(i+k,j+k) D_{H_1} \tag{1}$$

$$\mathbf{X}_{D_{H_2}}^{out}(i,j) = (\mathbf{X}_{MSAC}^{in} * \mathbf{c}_{s_2})_{D_{H_2}}(i,j) = \sum_{k=-s_2}^{s_2} x(i+k,j+k) \cdot c(i+k,j+k) D_{H_2} \tag{2}$$

$$\mathbf{X}_{D_{V_1}}^{out}(i,j) = (\mathbf{X}_{MSAC}^{in} * \mathbf{c}_{s_1})_{D_{V_1}}(i,j) = \sum_{k=-s_1}^{s_1} x(i+k,j+k) \cdot c(i+k,j+k) D_{V_1} \tag{3}$$

$$\mathbf{X}_{D_{V_2}}^{out}(i,j) = (\mathbf{X}_{MSAC}^{in} * \mathbf{c}_{s_2})_{D_{V_2}}(i,j) = \sum_{k=-s_2}^{s_2} x(i+k,j+k) \cdot c(i+k,j+k) D_{V_2} \tag{4}$$

$$\mathbf{X}_{MSAC}^{out} = concat(\mathbf{X}_{D_{H_1}}^{out}, \mathbf{X}_{D_{H_2}}^{out}, \mathbf{X}_{D_{V_1}}^{out}, \mathbf{X}_{D_{V_1}}^{out}) \tag{5}$$

where $*$ refers to the convolution operation, s_1 and s_2 are set to 3 and 5 which means the asymmetric convolution kernels are set to 1×7, 7×1, 1×11 and 11×1 respectively. To sum up, the proposed MSAC module is a goal-oriented feature extractor for line feature extraction with the consideration of feature scale diversity.

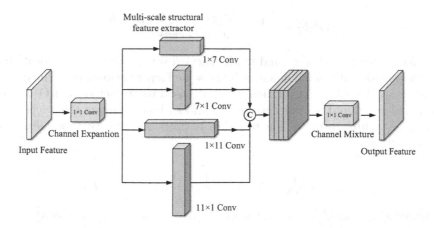

Fig. 3. The Multi-Scale Asymmetric Convolution module (MSAC). MSAC contains four kinds of structural convolution kernel.

Theoretically, the integration of multi-scale feature benefits for extracting the salient objects from the images, including the line segments in wireframe parsing task. Therefore, to strengthen the multi-scale representative capacity of backbone, we combine the stage-wise feature from different decoder layers using the Convolution Transformation Module (CTM), as shown in the bottom right part of backbone in Fig. 2. The multi-scale feature obtained by decoder layers are up-sampled using bi-linear interpolation to same size, and are concatenated to generate the compound intermediate feature. Finally, a 1×1 convolution kernel is leveraged for channel compression to get the final output.

2.3   Geometry Proposal Network

The geometry proposal network is proposed to infer junctions and line segments in a parallel way. Specifically, line segments are fitted by Bezier Curve [19] using several equipartition points (4-order Bezier Curve is used in this paper). We conduct classification and regression successively for proposal prediction. Specifically, the input image at resolution of W×H is divided into $W_b \times H_b$ bins, which has the same resolution of the obtained feature map. Then, the network predicts whether a junction $\hat{\mathbf{J}}$ existing in each bin with its offset to the center point of bin b. Therefore, a likelihood map $\hat{\mathbf{L}}^j$ and a offset map $\hat{\mathbf{O}}^j$ are needed to be predicted. Mathematically, we can formulate the relation between predicted maps and ground-truth as:

$$L^j(b) = \begin{cases} 1 & \exists \mathbf{p}_i \in \mathcal{V} : \mathbf{p}_i \in b \\ 0 & otherwise \end{cases} \tag{6}$$

and

$$\mathbf{O}^j(b) = \begin{cases} (\mathbf{p}_i - \mathbf{b})/w & \exists \mathbf{p}_i \in \mathcal{V} : \mathbf{p}_i \in b \\ 0 & otherwise \end{cases} \tag{7}$$

where $\mathcal{V}$ refers to the ground-truth junction set, $\mathbf{p}_i$ denotes the location of the ith ground-truth junction and w is the scale factor between input image and feature map. Two 3×3 convolution layers are used to predict $\hat{\mathbf{L}}^j$ and $\hat{\mathbf{O}}^j$ respectively, and we adopt the cross-entropy loss and L_1 loss to predict them separately during the training process. The total loss of junction prediction module can be formulated as:

$$\mathbb{L}_{junc} = \lambda_{like}^j \mathbb{L}(L^j, \hat{L}^j) + \lambda_{off}^j \cdot \mathbb{L}(\mathbf{O}^j, L^j \odot \hat{\mathbf{O}}^j) \tag{8}$$

where λ_{like}^j and λ_{off}^j refer to the weight parameters (λ_{like}^j=8 and λ_{off}^j=0.25 in this paper), and $\odot$ denotes the element-wise product. After getting the predicted junctions, a standard non-maximum suppression (NMS) [23,24] is applied to avoid the dense candidates, and we choose the top-k_{junc} with the highest probabilities as the final junctions.

For line segment prediction, we perform a similar method to junction prediction. Besides of predicting the existing probability and offset of the mid-point of line segment, we further predict the offset vectors of each equipartition point. Hence, we can formulate the line segment prediction process as:

$$L^s(b) = \begin{cases} 1 & \exists l_i \in E : \mathbf{c}_i \in b \\ 0 & otherwise \end{cases} \tag{9}$$

$$\mathbf{O}^s(b) = \left\{ \begin{matrix} (\mathbf{c}_i - \mathbf{b})/w & \exists l_i \in E : \mathbf{c}_i \in b \\ 0 & otherwise \end{matrix} \right\} \tag{10}$$

$$\mathbf{O}_k^e(b) = \left\{ \begin{matrix} (\mathbf{p}_k - \mathbf{c}_i) & \exists l_i \in E : \mathbf{c}_i \in b \\ 0 & otherwise \end{matrix} \right\} \tag{11}$$

where E refers to the ground-truth line segment map and $\mathbf{p}_k$ indicates the kth equipartition point of line segment l_i. We also perform NMS and choose the top-k_{line} with the highest probabilities as the final line segments. Meanwhile, the binary cross-entropy loss, L_1 loss and smooth L_1 loss are adopted separately to predict $\hat{\mathbf{L}}^s$, $\hat{\mathbf{O}}^s$ and $\hat{\mathbf{O}}^e$. The total loss of line segment prediction can be formulated as:

$$\mathbb{L}_{line} = \lambda_{like}^s \mathbb{L}(L^s, \hat{L}^s) + \lambda_{off}^s \mathbb{L}(\mathbf{O}^s, L^s \odot \hat{\mathbf{O}}^s) + \lambda_{off}^e \sum_{k=0, k \neq \frac{K}{2}}^{K} \mathbb{L}(\mathbf{O}_k^e, L^s \odot \hat{\mathbf{O}}_k^e) \tag{12}$$

where λ_{like}^s, λ_{off}^s and λ_{off}^e are the weight parameters (λ_{like}^s=8, λ_{off}^s=0.25 and λ_{off}^e=1 in this paper). In the proposal matching process, similar to [18], we leverage the weighted matching strategy of each equipartition points for precise matching results. For LOI module, we uniformly calculate the sample points in feature map using bi-linear interpolation and concatenate them to generate the feature vector of each line segment. Subsequently, a max-pooling layer follows the integrated feature for dimensional reduction. The binary cross-entropy is used during the training process, and we calculate the positive and negative samples separately as the total loss of LOI head as:

$$\mathbb{L}_{LOI} = \alpha_P \mathbb{L}_P + \alpha_N \mathbb{L}_N \tag{13}$$

where α_P and α_N are the weight parameters of positive and negative samples ($\alpha_P = \alpha_N = 1$ in this paper). As a fully end-to-end neural network, MSACWP is trained leveraging the total loss of junctions, line segments and LOI head, which can be formulated as:

$$\mathbb{L}_{total} = \mathbb{L}_{junc} + \mathbb{L}_{line} + \mathbb{L}_{LOI} \tag{14}$$

3   Experiments

3.1   Datasets and Metrics

We conduct all the experiments on two public datasets (Wireframe [1] and YorkUrban [21]) to demonstrate the outstanding performance of our proposed MSACWP. Wireframe dataset contains 5000 images as training set and 462 images as testing set. For YorkUrban dataset, it only contains 102 images for testing. To fairly evaluate all the methods, we refer to the evaluation metrics in [18] including structural average precision (sAP), mean structural average precision (msAP), mean average precision (mAP^J) and frames per second (FPS). Specifically, sAP is used to evaluate the precision between all the equipartition points and ground-truth under different threshold (e.g. sAP^5, sAP^{10}, sAP^{15}). The inferred junctions are evaluated by mAP^J metric, and FPS is used to measure the real-time performance of each method.

3.2   Implementation Details

For the input images, regardless of training or testing phase, MSCAWP first resizes them into 512×512 and feeds them to the backbone network for feature extraction. To get fair comparison results, hyper-parameters like stacked number, depth and convolution blocks are set according to L-CNN [2], HAWP [3] and ULSD [18]. In geometry proposal network, we set k_{junc} and k_{line} to 300 and 5000 for the final junction and line segment proposals. The learning rate, weight decay and training batch size are set to 4×10^{-4}, 1×10^{-4} and 4 respectively. Meanwhile, the network is trained for 30 epochs, and the learning rate is decayed by 10 at the 20th and 25th epoch. All the experiments are conducted on a single NVIDIA GTX 2080Ti GPU with PyTorch deep learning framework.

3.3   Ablation Study

In this section, we discuss the effectiveness of each proposed module and explore the impacts by adjusting the network structure. As shown in Table 1, the structural feature extraction performance has different levels of decrement when using a single direction of asymmetric convolution (i.e. "H" and "V" refers to 7×1 and 1×7 convolution kernel respectively). Besides, we compared MSACWP with the single scale version SACWP* which only contains 1×7 and 7×1 asymmetric convolution kernel as decoder layers. Consistently, MSACWP achieves better performance due to its delicate designation on multi-scale structural feature extraction. Meanwhile, the considerable performance enhancement on YorkUrban dataset demonstrates MSACWP equips with preferable generalization capability. Finally, we remove the MSFC module in MSACWP, and the parsing results are shown in the fourth row in Table 1. By getting rid of the MSFC module, the junction and line segment extraction performance have decreased to a certain extent. Specifically, there are 1.30% and 4.22% decline on mAP^J evaluation metric on Wireframe and YorkUrban datasets respectively.

Table 1. Ablation study (%) of each proposed module in MSACWP.

Method	Wireframe Dataset					YorkUrban Dataset				
	sAP^5	sAP^{10}	sAP^{15}	msAP	mAP^J	sAP^5	sAP^{10}	sAP^{15}	msAP	mAP^J
H	63.6	67.5	69.1	66.8	60.3	24.6	27.1	28.7	26.8	29.9
V	64.2	68.1	69.7	67.4	60.7	24.8	27.4	29.1	27.1	30.3
SACWP*	66.0	69.7	71.3	69.0	61.0	27.0	29.7	29.7	29.4	32.0
MSACWP w/o MSFC	65.8	69.6	71.1	68.8	60.9	26.8	29.6	31.3	29.2	31.8
MSACWP	**66.4**	**70.1**	**71.7**	**69.4**	**61.7**	**27.9**	**30.8**	**32.7**	**30.4**	**33.2**

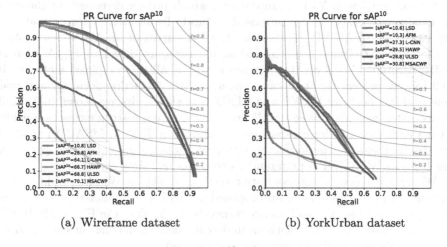

(a) Wireframe dataset   (b) YorkUrban dataset

Fig. 4. Precision-Recall (PR) curve of sAP^{10} on Wireframe and YorkUrban datasets.

Table 2. Quantitative comparison results (%) on Wireframe and YorkUrban datasets with other state-of-the-art wireframe parsing methods.

Method	Wireframe Dataset					YorkUrban Dataset				
	sAP^5	sAP^{10}	sAP^{15}	msAP	mAP^J	sAP^5	sAP^{10}	sAP^{15}	msAP	mAP^J
LSD [11]	8.3	10.8	12.7	10.6	17.2	8.5	10.6	12.2	10.4	15.4
AFM [25]	21.2	26.8	30.2	26.1	24.3	8.0	10.3	12.1	10.1	12.5
L-CNN [2]	60.7	64.1	65.6	63.5	59.3	25.3	27.3	28.5	27.1	30.3
HAWP [3]	64.5	66.7	69.2	67.1	60.2	27.3	29.5	30.8	29.2	31.7
ULSD [18]	65.0	68.8	70.4	68.1	61.3	26.0	28.8	30.6	28.5	31.8
MSACWP(Ours)	**66.4**	**70.1**	**71.7**	**69.4**	**61.7**	**27.9**	**30.8**	**32.7**	**30.4**	**33.2**

3.4 Comparison with Other Methods

In this section, we conduct quantitative and qualitative comparisons experiment with several representative wireframe parsing methods, including LSD[1] [11], AFM [25], L-CNN [2], HAWP [3] and ULSD [18]. Noticed that all the learning-based methods are trained on Wireframe dataset for fair comparison, and the experiment results are shown in Table 2. When comparing with other

---

[1] The built-in LSD in OpenCV v3.4.2 is used for evaluation.

methods, MSACWP achieves the best results in all the evaluation metrics. The better results mainly attribute to the efficient structural feature extraction paradigm that can capture meaningful structural primitives. Meanwhile, the proposed MSFC module aggregates the stage-wise multi-scale feature from decoder layers which benefits for structural feature representation. Specifically, when comparing with L-CNN and HAWP, MSACWP achieves 9.36% and 5.10% improvement on sAP^{10} evaluation metric on Wireframe dataset. On the other hand, there are 6.67% and 4.40% promotion than ULSD on msAP and mAP^J evaluation metrics on YorkUrban dataset, which further demonstrate the comprehensive advantages of our proposed method. To intuitively demonstrate the superiority of MSACWP, we visualize the PR curves of MSACWP with other methods on Wireframe and YorkUrban datasets, as shown in Fig. 4. Consistently, MSACWP achieves the best result among all the compared methods. In addition, we present the trade-off between feature extraction accuracy and real-time performance with HAWP and ULSD, and the results are shown in Table 3. Noticed that we only choose msAP and mAP^J as the feature extraction evaluation metrics for concise comparison. As shown in Table 3, MSACWP realizes the absolute real-time performance with preferable feature extraction precision, which makes MSACWP a preferential choice on real-time visual tasks.

We visualize several representative testing results on Wireframe dataset to further verify the reliability of our proposed method. The predicted junctions and line segments are in cyan and orange respectively. As shown in Fig. 5, MSACWP gets more analogical results with ground-truth data than other methods, which illustrates the superiority of our proposed method.

Table 3. Trade-off analysis between feature extraction accuracy and real-time performance among different methods.

Method	Wireframe Dataset		YorkUrban Dataset		FPS
	msAP	mAP^J	msAP	mAP^J	
HAWP [3]	67.1	60.2	29.2	31.7	32.0
ULSD [18]	68.1	61.3	28.5	31.8	36.8
MSACWP (Ours)	**69.4**	**61.7**	**30.4**	**33.2**	**34.1**

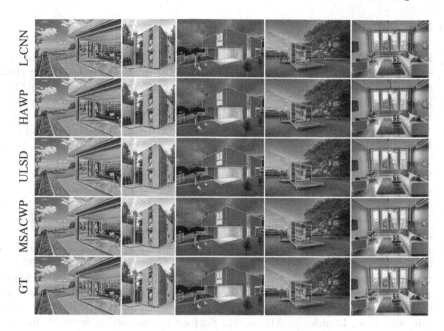

Fig. 5. Qualitative comparison of MSACWP with other methods on Wireframe dataset.

4 Conclusions

In this paper, we have presented a Multi-scale Structural Asymmetric Convolution for Wireframe Parsing (MSACWP). Benefiting from the similar geometric properties of asymmetric convolution and line segment, our method effectively captures long-range dependencies while preventing the inclusion of irrelevant information. On the other hand, the multi-scale designations in both the MSAC and MSFC modules make MSACWP a scale-aware algorithm for wireframe parsing tasks. Sufficient experiments on two public datasets are conducted to demonstrate the preferable feature extraction performance of MSACWP compared with other state-of-the-art methods. Although MSACWP has achieved promising performance, it still encounters challenges in motion blur environments and produces unreliable line segments in low-contrast images. In the future, we will continuously focus on the structural feature extraction methods with their real-world application such as VSLAM, 3D reconstruction and robot navigation.

Acknowledgements. This work was partly supported by the National Natural Science Foundation of China under Grant No.61973009.

References

1. Huang, K., Wang, Y., Zhou, Z., Ding, T., Gao, S., Ma, Y.: Learning to parse wireframes in images of man-made environments. In: Proceedings of the IEEE Conference on Computer Vision and Pattern Recognition, pp. 626–635 (2018)
2. Zhou, Y., Qi, H., Ma, Y.: End-to-end wireframe parsing. In: Proceedings of the IEEE/CVF International Conference on Computer Vision, pp. 962–971 (2019)
3. Xue, N., et al.: Holistically-attracted wireframe parsing. In: Proceedings of the IEEE/CVF Conference on Computer Vision and Pattern Recognition, pp. 2788–2797 (2020)
4. Schonberger, J.L., Frahm, J.M.: Structure-from-motion revisited. In: Proceedings of the IEEE Conference on Computer Vision and Pattern Recognition, pp. 4104–4113 (2016)
5. Chen, W., Kumar, S., Yu, F.: Uncertainty-driven dense two-view structure from motion. IEEE Rob. Autom. Lett. **8**(3), 1763–1770 (2023)
6. Liu, X., Wen, S., Zhang, H.: A real-time stereo visual-inertial slam system based on point-and-line features. IEEE Trans. Veh. Technol. **72**, 5747–5758 (2023)
7. Pumarola, A., Vakhitov, A., Agudo, A., Sanfeliu, A., Moreno-Noguer, F.: Pl-slam: real-time monocular visual slam with points and lines. In: 2017 IEEE International Conference on Robotics and Automation (ICRA), pp. 4503–4508. IEEE (2017)
8. Langlois, P.A., Boulch, A., Marlet, R.: Surface reconstruction from 3D line segments. In: 2019 International Conference on 3D Vision (3DV), pp. 553–563. IEEE (2019)
9. Geiger, A., Ziegler, J., Stiller, C.: Stereoscan: dense 3D reconstruction in real-time. In: 2011 IEEE Intelligent Vehicles Symposium (IV), pp. 963–968. IEEE (2011)
10. Lin, Y., Wiersma, R., Pintea, S.L., Hildebrandt, K., Eisemann, E., van Gemert, J.C.: Deep vanishing point detection: geometric priors make dataset variations vanish. In: Proceedings of the IEEE/CVF Conference on Computer Vision and Pattern Recognition, pp. 6103–6113 (2022)
11. Von Gioi, R.G., Jakubowicz, J., Morel, J.M., Randall, G.: LSD: a fast line segment detector with a false detection control. IEEE Trans. Pattern Anal. Mach. Intell. **32**(4), 722–732 (2008)
12. Burns, J.B., Hanson, A.R., Riseman, E.M.: Extracting straight lines. IEEE Trans. Pattern Anal. Mach. Intell. **4**, 425–455 (1986)
13. Desolneux, A., Moisan, L., More, J.M.: A grouping principle and four applications. IEEE Trans. Pattern Anal. Mach. Intell. **25**(4), 508–513 (2003)
14. Akinlar, C., Topal, C.: Edlines: a real-time line segment detector with a false detection control. Pattern Recogn. Lett. **32**(13), 1633–1642 (2011)
15. Topal, C., Akinlar, C.: Edge drawing: a combined real-time edge and segment detector. J. Vis. Commun. Image Represent. **23**(6), 862–872 (2012)
16. Ren, S., He, K., Girshick, R., Sun, J.: Faster r-cnn: towards real-time object detection with region proposal networks. Adv. Neural Inf. Process. Syst. **28** (2015)
17. Girshick, R.: Fast r-cnn. In: Proceedings of the IEEE International Conference on Computer Vision, pp. 1440–1448 (2015)
18. Li, H., Yu, H., Wang, J., Yang, W., Yu, L., Scherer, S.: ULSD: unified line segment detection across pinhole, fisheye, and spherical cameras. ISPRS J. Photogramm. Remote. Sens. **178**, 187–202 (2021)
19. Bezier, P.E.: The first years of cad/cam and the unisurf cad system. In: Fundamental Developments of Computer-Aided Geometric Modeling, pp. 13–26 (1993)

20. Dai, X., Gong, H., Wu, S., Yuan, X., Yi, M.: Fully convolutional line parsing. Neurocomputing **506**, 1–11 (2022)
21. Denis, P., Elder, J.H., Estrada, F.J.: Efficient edge-based methods for estimating manhattan frames in urban imagery. In: Forsyth, D., Torr, P., Zisserman, A. (eds.) ECCV 2008. LNCS, vol. 5303, pp. 197–210. Springer, Heidelberg (2008). https://doi.org/10.1007/978-3-540-88688-4_15
22. Newell, A., Yang, K., Deng, J.: Stacked hourglass networks for human pose estimation. In: Leibe, B., Matas, J., Sebe, N., Welling, M. (eds.) ECCV 2016. LNCS, vol. 9912, pp. 483–499. Springer, Cham (2016). https://doi.org/10.1007/978-3-319-46484-8_29
23. Felzenszwalb, P., McAllester, D., Ramanan, D.: A discriminatively trained, multi-scale, deformable part model. In: 2008 IEEE Conference on Computer Vision and Pattern Recognition, pp. 1–8. IEEE (2008)
24. Liu, W., et al.: SSD: single shot multibox detector. In: Leibe, B., Matas, J., Sebe, N., Welling, M. (eds.) ECCV 2016. LNCS, vol. 9905, pp. 21–37. Springer, Cham (2016). https://doi.org/10.1007/978-3-319-46448-0_2
25. Xue, N., Bai, S., Wang, F., Xia, G.S., Wu, T., Zhang, L.: Learning attraction field representation for robust line segment detection. In: Proceedings of the IEEE/CVF Conference on Computer Vision and Pattern Recognition, pp. 1595–1603 (2019)

S3ACH: Semi-Supervised Semantic Adaptive Cross-Modal Hashing

Liu Yang[1], Kaiting Zhang[1], Yinan Li[2], Yunfei Chen[2], Jun Long[2], and Zhan Yang[2](✉)

[1] School of Computer Science and Engineering, Central South University, Changsha, Hunan, China
{yangliu,software-zkt}@csu.edu.cn
[2] Big data institute, Central South University, Changsha, Hunan, China
{liyinan,yunfeichen,junlong,zyang22}@csu.edu.cn

Abstract. Hash learning has been a great success in large-scale data retrieval field because of its superior retrieval efficiency and storage consumption. However, labels for large-scale data are difficult to obtain, thus supervised learning-based hashing methods are no longer applicable. In this paper, we introduce a method called **S**emi-**S**upervised **S**emantic **A**daptive **C**ross-modal **H**ashing (S3ACH), which improves performance of unsupervised hash retrieval by exploiting a small amount of available label information. Specifically, we first propose a higher-order dynamic weight public space collaborative computing method, which balances the contribution of different modalities in the common potential space by invoking adaptive higher-order dynamic variable. Then, less available label information is utilized to enhance the semantics of hash codes. Finally, we propose a discrete optimization strategy to solve the quantization error brought by the relaxation strategy and improve the accuracy of hash code production. The results show that S3ACH achieves better effects than current advanced unsupervised methods and provides more applicable while balancing performance compared with the existing cross-modal hashing.

Keywords: Hashing · Cross-modal retrieval · Semi-Supervised

1 Introduction

With the rapid development of Big Data, plenty of multimodal data have been produced, such as text, images, audio, etc. Many practiced scenarios require multimodal data processing, and the cross-modal retrieval techniques that have turned a hot research theme, which rely on semantic similarity calculation between data. Hash-based methods become a practical solution to deal with massive heterogeneous data, aiming to reduce data dimensions to a binary code while retaining the original semantic information. This can reduce time and memory overhead. The basic principle of hash methods is that multimodal data be projected into a uniform low-dimensional Hamming space, so that achieve efficient similarity search with Hamming distance.

Traditional hash methods led by Local Sensitive Hash series [21] are data-independent. They generate hash codes by random projection without considering data distribution, so it is difficult to keep high accuracy with short coding at the same time. As machine learning techniques develop rapidly, data-dependent methods come into mainstream, while multimodal hashing gradually become the most promising research direction, and many cross-modal hash methods have been offered [1,31]. The key to cross-modal hashing is retaining similarity of hash code in multimodal. These methods [19,24] exploit semantic information for supervised learning and show very good performance, but ignore the high time and labor cost of acquiring labels. In contrast, unsupervised cross-modal hashing [7,13,16] is able to find modalities relationships without label information. Despite the significant progress of these methods, they still have the limitation of lacking label supervision. The reality is that only a very tiny part of data is labeled, so semi-supervised methods using few labels is the most realistic solution. However, semi-supervised hash retrieval methods currently compensate for the lack of supervised capacity by deep learning [30], which is not only expensive but also difficult to reuse and interpret.

To address the above limitations, we propose Semi-supervised Semantic Adaptive Cross-modal Hashing (S3ACH). Firstly, we design a public potential space learning framework with higher-order dynamic weights to collaborative computing the contributions of different modalities. To fully utilize semantic information from sparse labels, we design an adaptive label enhancement module to enhance representation learning. Meanwhile, S3ACH obtains a potentially consistent representation of different modalities based on a matrix decomposition strategy with dynamic weights, which ensures the semantic completeness among modalities. Moreover, an efficient iterative optimization algorithm is offered for discrete constraints with direct one-step hash coding. Summarily, the core contributions of this work include:

1. A higher-order dynamic variable collaborative computing method is proposed to adaptively balance the different modalities' contribution, so as to improve the learning stability of common potential representation.
2. A label enhancement framework is proposed to directly exploit the labeled data to mine semantic information and maximize similarity differences between modalities.
3. A fast iterative optimization method for solving discrete constrained problems is proposed, where the time consumption of the method scales linearly related to data size.
4. Extensive experiments are conducted on the MIRFlickr and NUS-WIDE datasets, and our proposed S3ACH shows better retrieval performance and higher applicability in real-world scenarios with large-scale data compared to state-of-the-art methods.

The rest of the paper includes the following. Reviews the work related to cross-mode hash retrieval in Sect. 2, and Sect. 3 details our proposed S3ACH. In

Sect. 4, we conduct comparative and analytical experiments. Finally, the conclusion drawn in Sect. 5.

2 Related Works

Cross-mode hashing mainly consists of (un)supervised methods. Unsupervised cross-modal hashing methods generate hash codes from data distribution, rather than semantic labeling information, and focus on inter-modal and intra-modal correlations. They can be classified into shallow and deep methods. In some early studies, CMFH [6] handles different modalities by collective matrix decomposition method, and LSSH [32] preserves specific properties by constructing inter-modal sparse representations. In recent years, the similarity in the common subspace is optimized from different perspectives. For example, CUH [18] uses a novel optimization strategy for multi-modal clustering and hash learning. RUCMH [2] preserves both the deterministic continuous shared space and discrete hamming space. JIMFH [17] retains both the shared properties and the properties specific to each modality. FUCMSH [25] ensures inter-modal consistent representation and intra-modal specific potential representation by shared matrix decomposition and individual self-coding, respectively.

In addition, the latest research in deep cross-modal hashing [14,20] has showed superior performance thanks to the strong nonlinear representation of deep learning. But the massive resource consumption also becomes an obvious drawback. Therefore, one of the focuses of this paper is to design an interpretable objective function and an effective discrete optimization method based on shallow method.

Supervised cross-modal hashing, in contrast, enhances the correlation between modalities by supervising the learning process using semantic labeling information. For example, SRLCH [12] transforms class labels in the Hamming subspace into relational information. LEADH [23] designs a label-binary mutual mapping architecture to fully exploit and utilize multi-label semantic information, and SPECH [22] uses a likelihood loss technique to measure the semantic similarity of paired data. However, artificial semantic annotation is costly with massive data, and the tiny percentage of labeled data lead to the unavailability of supervised methods in real-world situations. We consider this problem and enhance the process by a label learning framework that utilizes as few labels as possible, so as to improve the semi-supervised learning.

Research in semi-supervised cross-modal hashing, for example, SCH-GAN [30] fits the correlation distribution of unlabeled data by adversarial networks, and UMCSH [3] uses uncertainty estimation methods to select label information which gives discriminative features to unlabeled data, but they are both deep methods. Recent research in shallow methods such as FlexCmh [27] allows learning hash codes from weakly paired data, and WASH [28] uses weakly supervised enhanced learning by regularizing the noise label matrix. In this paper, we propose shallow semi-supervised cross-modal hashing to satisfies real-world scenarios concisely and effectively.

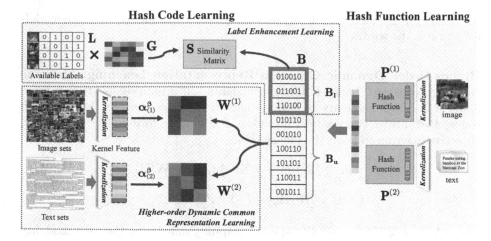

Fig. 1. The framework of S3ACH.

3 The Proposed Method

3.1 Notation and Problem Formulation

Support $\mathbb{O} = \{\mathbf{x}_1^{(v)}, \mathbf{x}_2^{(v)}, ..., \mathbf{x}_n^{(v)}\} \in \mathbb{R}^{d^{(v)} \times n}$ denote the cross-modal dataset, n is the total instances amount, $d^{(v)}$ refers to the v-th modality. Specifically, $\mathbf{X}^{(v)}$ is decomposed into two parts: the labeled instances $\mathbf{X}_l^{(v)} \in \mathbb{R}^{d^{(v)} \times n_l}$ and the unlabeled instances $\mathbf{X}_u^{(v)} \in \mathbb{R}^{d^{(v)} \times n_u}$, where n_l and n_u are the numbers of labeled and unlabeled datasets, respectively. $\mathbf{B} \in \mathbb{R}^{k \times n}$ is the binary codes, k is the hash codes length. The aim of this paper is to learn a hash function $\mathcal{H}$: $\mathbf{x} \to \mathbf{b} \in \{-1, 1\}^k$, where $\mathbf{x}$ is the input instance. The framework of this paper is shown in Fig. 1, which consists of two main processes: hash code learning process include Higher-order Dynamic Common Representation and Label Enhancement Learning, and hash function learning process rely on the generated hash code. All two processes will be kernelized initially.

In this paper, bold lowercase and uppercase letters denote vectors and matrices, respectively, and ordinary lowercase denote Scalars. $tr(\mathbf{U})$ represents the trace, and $\mathbf{U}^\top$ represents the transposition of matrix $\mathbf{U}$.

3.2 S3ACHMethod

Kernelization. Kernelization is one of the frequently used nonlinear relationship modeling techniques in machine learning domain. Thus, for more accurate represent the potentially complex relationships between different modalities, we use the RBF-based kernel technique to process raw features of different modalities. For a instance $x_i^{(v)}$, the kernelized features $\phi(x_i^{(v)})$ can be expressed as,

$$\phi(x_i^{(v)}) = \left[\exp(\frac{-||x_i^{(v)} - a_1^{(v)}||_2^2}{2\sigma^2}), ..., \exp(\frac{-||x_i^{(v)} - a_q^{(v)}||_2^2}{2\sigma^2}) \right]^\top \quad (1)$$

where $a_q^{(v)}$ represents the randomly selected q anchor instances of the v-th modality and σ is the width.

Higher-Order Dynamic Common Representation Learning. Since the "heterogeneous gap" between different modalities, it is not possible to fuse them directly, and it is necessary to find a common subspace (i.e., the learned hash codes) $\mathbf{B}$ to bridge two modals. As shown in Fig. 1, we formulate the following objective function to learn the binary codes,

$$\min_{\mathbf{W}^{(v)},\mathbf{B}} \| \mathbf{W}^{(v)}\mathbf{B} - \phi(\mathbf{X}^{(v)})\|_F^2 + \mathcal{R}(\mathbf{W}^{(v)}),$$
$$s.t.\ \mathbf{B} = [\mathbf{B}_l; \mathbf{B}_u] \in \{-1,1\}^{k \times n}, \tag{2}$$

where $\mathbf{W}^{(v)} \in \mathbb{R}^{q \times k}$ is modality-specific mapping matrix for the v-th modality and $\mathcal{R}(\cdot) = \delta||\cdot||_F^2$ is the regularization term. It is noteworthy that there exist an opposite learning manner [11], i.e., $\ell(\phi(\mathbf{X}^{(v)})\mathbf{W}^{(v)}, \mathbf{B})$, which encoding different modalities information into a common latent representation. But this means that each modality can be individually encode common latent representation, which weaken the completeness of the information substantially.

However, the above approach treats the contributions of all modalities to the common latent representation as the same. In fact, the contributions of different modalities to the common representation should be different, thus we introduce a self-learning dynamic weight parameter α_v to represent the contribution of v-th modality. Therefore, Eq. (2) can be rewritten as,

$$\min_{\alpha_v,\mathbf{W}^{(v)},\mathbf{B}} \alpha_v||\mathbf{W}^{(v)}\mathbf{B} - \phi(\mathbf{X}^{(v)})||_F^2 + \mathcal{R}(\mathbf{W}^{(v)}),$$
$$s.t.\ \sum_v^V \alpha_v = 1, \alpha_v \geq 0,\ \mathbf{B} = [\mathbf{B}_l; \mathbf{B}_u] \in \{-1,1\}^{k \times n} \tag{3}$$

Although the solution of Eq. (3) can fix the problem of unbalance contribution, there are still an issue need to fix, i.e., if the feature values of a modality $\mathbf{X}^{(v)}$ are sparse, then it will make the corresponding weight parameter take the maximum value, i.e., $\alpha_v = 1$. In other word, other modalities will be ignored directly. Therefore, we introduce a smooth exponential factor $\beta > 0$ to avoid the problem, that is,

$$\min_{\alpha_v,\mathbf{W}^{(v)},\mathbf{B}} \alpha_v^{\beta}||\mathbf{W}^{(v)}\mathbf{B} - \phi(\mathbf{X}^{(v)})||_F^2 + \mathcal{R}(\mathbf{W}^{(v)}),$$
$$s.t.\ \sum_v^V \alpha_v = 1, \alpha_v \geq 0, \beta > 0,\ \mathbf{B} = [\mathbf{B}_l; \mathbf{B}_u] \in \{-1,1\}^{k \times n} \tag{4}$$

where $\mathbf{X}^{(v)}$ is composed of labeled instances $\mathbf{X}_l^{(v)}$ and unlabeled instances $\mathbf{X}_u^{(v)}$.

Label Enhancement Learning. Since some labeled data exists in real scenarios, it is important to make full use of labeled data to improve a recognition of the learned hash codes. As shown in Fig. 1, we construct a pairwise similar matrix $\mathbf{S} \in \mathbb{R}^{n_l \times n_l}$, i.e., $\mathbf{S} = 2\mathbf{L}^\top \mathbf{L} - \mathbf{1}_{n_l} \mathbf{1}_{n_l}^\top$, where $\mathbf{L} \in \mathbb{R}^{c \times n_l}$ represents label matrix and $\mathbf{1}_{n_l}$ denotes an all-one column vector with length n_l. Note that, the implementation of this approach can solve the problem of high time consumption caused by the direct use of pairwise similarity matrices. Then we build a bridge to link the semantic information and the corresponding hash codes $\mathbf{B}_l$. Inspired by a popular symmetric framework KSH [10] method, which the definition is $\min_{\mathbf{B}} ||\mathbf{B}^\top \mathbf{B} - k\mathbf{S}||^2$, $s.t.\ \mathbf{B} \in \{-1,1\}^{k \times n}$. This strategy, however, is a difficult quadratic optimization to solve. Luckily, a few researchers [5,26] propose asymmetric learning frameworks to address this problem in terms of accuracy and speed. Therefore, we design an asymmetric learning framework as follows:

$$\min_{\mathbf{B}_l, \mathbf{G}} \gamma ||k\mathbf{S} - \mathbf{B}_l{}^\top (\mathbf{GL})||_F^2 + \rho ||\mathbf{B}_l - \mathbf{GL}||_F^2,$$
$$s.t.\ \mathbf{B}_l \in \{-1,1\}^{k \times n_l}, \tag{5}$$

where γ, ρ are the parameters to balance the asymmetric learning framework, and $\mathbf{G} \in \mathbb{R}^{k \times c}$ is a transformation matrix.

Joint Hash Learning Framework. Combing Eqs. (4) Eq(5), we obtain the overall objective function of S3ACH as follows.

$$\min_{\alpha_v, \mathbf{W}^{(v)}, \mathbf{G}, \mathbf{B}_u, \mathbf{B}_l} \sum_v^V \alpha_v^\beta ||\mathbf{W}^{(v)}\mathbf{B} - \phi(\mathbf{X}^{(v)})||_F^2 + \mathcal{R}(\mathbf{W}^{(v)}, \mathbf{GL})$$
$$+ \gamma ||k\mathbf{S} - \mathbf{B}_l{}^\top (\mathbf{GL})||_F^2 + \rho ||\mathbf{B}_l - \mathbf{GL}||_F^2, \tag{6}$$
$$s.t.\ \sum_v^V \alpha_v = 1, \alpha_v \geq 0, \beta > 0, \mathbf{B} = [\mathbf{B}_l; \mathbf{B}_u] \in \{-1,1\}^{k \times n}.$$

3.3   Optimization

Equation (6) is an NP-hard issue due to the multivariate and discrete constraints. Therefore, some strategies involve initially approximating the discrete variables using the *sgn* function. Such an approach can cause huge quantization errors and affect the quality of hash code generation. Some strategies use the DCC (Discrete Cyclic Coordinate Descent) strategy to optimize each hash bit in hash code by circular iteration. Although this approach does not cause the problem of quantization loss, the time consumption of optimization is proportional to the hash code length and the solution is inefficient. This paper uses discrete optimization methods to learn complete hash codes in one step, addressing issues with the mentioned strategies. Specifically, we solve for the other variables by fixing one of them. The overall optimization process of Eq. (6) is as follows.

α_v-step We fix $\mathbf{G}, \mathbf{B}_l, \mathbf{B}_u, \mathbf{W}^{(v)}$ variables, the updating for variable α_v can be reformulated as,

$$\min_{\alpha_v} \alpha_v^\beta ||\mathbf{W}^{(v)}\mathbf{B} - \phi(\mathbf{X}^{(v)})||_F^2,$$
$$s.t. \sum_v^V \alpha_v = 1, \alpha_v \geq 0. \tag{7}$$

The Lagrange multiple scheme is used to construct the Lagrange arithmetic formulation, that is,

$$\min_{\alpha_v} \alpha_v^\beta ||\mathbf{W}^{(v)}\mathbf{B} - \phi(\mathbf{X}^{(v)})||_F^2 - \nu(1^\top\alpha - 1), \tag{8}$$

where $\alpha = [\alpha_1, \alpha_2, ..., \alpha_V]^\top \in \mathbb{R}^V$ is the vector of weights for the related modalities, and ν is the Lagrange arithmetic.

Setting the derivative with respect to α_v and ν to 0, we get,

$$\alpha_v = \frac{||\mathbf{W}^{(v)}\mathbf{B} - \phi(\mathbf{X}^{(v)})||_F^{2^{1/1-\beta}}}{\sum_v^V (||\mathbf{W}^{(v)}\mathbf{B} - \phi(\mathbf{X}^{(v)})||_F^{2^{1/1-\beta}})}. \tag{9}$$

$\mathbf{W}^{(v)}$-step We fix $\mathbf{G}, \mathbf{B}_l, \mathbf{B}_u, \alpha_v$ variables, the updating for variable $\mathbf{W}^{(v)}$ can be reformulated as,

$$\min_{\mathbf{W}^{(v)}} \alpha_v^\beta ||\mathbf{W}^{(v)}\mathbf{B} - \phi(\mathbf{X}^{(v)})||_F^2 + \delta||\mathbf{W}^{(v)}||_F^2. \tag{10}$$

Then Eq. (10) can be simplified as,

$$\min_{\mathbf{W}^{(v)}} \alpha_v^\beta tr(\mathbf{W}^{(v)}\mathbf{B}\mathbf{B}^\top\mathbf{W}^{(v)^\top} - 2\phi(\mathbf{X}^{(v)})\mathbf{B}^\top\mathbf{W}^{(v)^\top})$$
$$+ \delta tr(\mathbf{W}^{(v)}\mathbf{W}^{(v)^\top}). \tag{11}$$

Setting the derivative with respect to $\mathbf{W}^{(v)}$ to 0, we get,

$$\mathbf{W}^{(v)} = \alpha_v^\beta \phi(\mathbf{X}^{(v)})\mathbf{B}^\top(\alpha_v^\beta\mathbf{B}\mathbf{B}^\top + \delta\mathbf{I})^{-1}. \tag{12}$$

G-step We fix $\mathbf{W}^{(v)}, \mathbf{B}_l, \mathbf{B}_u, \alpha_v$ variables, the updating for variable $\mathbf{G}$ can be reformulated as,

$$\min_{\mathbf{G}} \delta||\mathbf{G}\mathbf{L}||_F^2 + \gamma||k\mathbf{S} - \mathbf{B}_l^\top(\mathbf{G}\mathbf{L})||_F^2 + \rho||\mathbf{B}_l - \mathbf{G}\mathbf{L}||_F^2, \tag{13}$$

Then Eq. (13) can be simplified as,

$$\min_{\mathbf{G}} \delta \, tr(\mathbf{G}\mathbf{L}\mathbf{L}^\top\mathbf{G}^\top) + \gamma \, tr(-2k\mathbf{S}\mathbf{L}^\top\mathbf{G}^\top\mathbf{B}_l + \mathbf{B}_l^\top\mathbf{G}\mathbf{L}\mathbf{L}^\top\mathbf{G}^\top\mathbf{B}_l)$$
$$+ \rho \, tr(-2\mathbf{B}_l\mathbf{L}^\top\mathbf{G}^\top + \mathbf{G}\mathbf{L}\mathbf{L}^\top\mathbf{G}^\top). \tag{14}$$

Setting the derivative with respect to $\mathbf{G}$ to 0, we get,

$$\mathbf{G} = ((\delta + \rho)\mathbf{I} + \gamma\mathbf{B}_l\mathbf{B}_l^\top)^{-1}(\gamma k\mathbf{B}_l\mathbf{S}\mathbf{L}^\top + \rho\mathbf{B}_l\mathbf{L}^\top)(\mathbf{L}\mathbf{L}^\top)^{-1}. \tag{15}$$

$\mathbf{B}_l, \mathbf{B}_u$-step We fix $\mathbf{W}^{(v)}, \mathbf{G}, \mathbf{B}_u, \alpha_v$ variables, the updating for variable $\mathbf{B}_l$ can be reformulated as,

$$\min_{\mathbf{B}_l} \sum_v^V \alpha_v^\beta \|\mathbf{W}^{(v)}\mathbf{B}_l - \phi(\mathbf{X}_l^{(v)})\|_F^2 + \gamma\|k\mathbf{S} - \mathbf{B}_l^\top(\mathbf{GL})\|_F^2 + \rho\|\mathbf{B}_l - \mathbf{GL}\|_F^2,$$

$$s.t. \ \mathbf{B}_l \in \{-1, 1\}^{k \times n_l}.$$

(16)

Eq. (16) can be reformulated as

$$\min_{\mathbf{B}_l} \sum_v^V \alpha_v^\beta tr \ (\mathbf{W}^{(v)}\mathbf{B}_l\mathbf{B}_l^\top\mathbf{W}^{(v)\top} - 2\phi(\mathbf{X}_l^{(v)})\mathbf{B}_l^\top\mathbf{W}^{(v)\top})$$

$$+ \gamma tr \ (-2k\mathbf{B}_l^\top\mathbf{GLS}^\top + \mathbf{B}_l^\top\mathbf{GLL}^\top\mathbf{G}^\top\mathbf{B}_l) + \rho tr \ (-2\mathbf{GLB}_l^\top),$$

$$s.t. \ \mathbf{B}_l \in \{-1, 1\}^{k \times n_l}.$$

(17)

The discrete constraints of the discrete variables to be solved make the above problem difficult to solve. Therefore, we use the ALM (Augmented Lagrange Multiplier method) to separate the discrete variables to be solved, i.e., we introduce an auxiliary discrete variable $\mathbf{K}_l$ to substitute the first $\mathbf{B}_l$ in $\mathbf{W}^{(v)}\mathbf{B}_l\mathbf{B}_l^\top\mathbf{W}^{(v)\top}$ and $\mathbf{B}_l^\top\mathbf{GLL}^\top\mathbf{G}^\top\mathbf{B}_l$. We obtain,

$$\min_{\mathbf{B}_l} \sum_v^V \alpha_v^\beta tr \ (\mathbf{W}^{(v)}\mathbf{K}_l\mathbf{B}_l^\top\mathbf{W}^{(v)\top} - 2\phi(\mathbf{X}_l^{(v)})\mathbf{B}_l^\top\mathbf{W}^{(v)\top})$$

$$+ \gamma tr \ (-2k\mathbf{B}_l^\top\mathbf{GLS}^\top + \mathbf{B}_l^\top\mathbf{GLL}^\top\mathbf{G}^\top\mathbf{K}_l) + \rho tr \ (-2\mathbf{GLB}_l^\top)$$

$$+ \frac{\xi}{2}\|\mathbf{B}_l - \mathbf{K}_l + \frac{\mathbf{H}_l}{\xi}\|_F^2,$$

$$s.t. \ \mathbf{B}_l \in \{-1, 1\}^{k \times n_l},$$

(18)

where $\mathbf{H}$ denotes the differences between $\mathbf{B}_l$ and $\mathbf{K}_l$, and the last term $\frac{\xi}{2}\|\mathbf{B}_l - \mathbf{K}_l + \frac{\mathbf{H}_l}{\xi}\|_F^2$ can be rewritten as

$$\min_{\mathbf{B}_l} tr \ (-\xi\mathbf{K}_l\mathbf{B}_l^\top + \mathbf{H}_l\mathbf{B}_l^\top).$$

(19)

Then we optimize the function for $\mathbf{B}_l$ rewritten as

$$\max_{\mathbf{B}_l} tr \ (\sum_v^V (2\alpha_v^\beta\mathbf{W}^{(v)\top}\phi(\mathbf{X}_l^{(v)}))\mathbf{B}_l^\top + 2k\gamma\mathbf{GLS}^\top\mathbf{B}_l^\top + 2\rho\mathbf{GLB}_l^\top$$

$$+ \xi\mathbf{K}_l\mathbf{B}_l^\top - \alpha_v^\beta\mathbf{W}^{(v)\top}\mathbf{W}^{(v)}\mathbf{K}_l\mathbf{B}_l^\top - \gamma\mathbf{GLL}^\top\mathbf{G}^\top\mathbf{K}_l\mathbf{B}_l^\top - \mathbf{H}_l\mathbf{B}_l^\top).$$

(20)

The closed-solution of $\mathbf{B}_l$ as

$$\mathbf{B}_l = sgn(2\sum_v^V \alpha_v^\beta \mathbf{W}^{(v)^\top} \phi(\mathbf{X}_l^{(v)}) + 2k\gamma \mathbf{GLS}^\top + 2\rho \mathbf{GL}$$

$$+ \xi \mathbf{K}_l - \sum_v^V (\alpha_v^\beta \mathbf{W}^{(v)^\top} \mathbf{W}^{(v)}) \mathbf{K}_l - \gamma \mathbf{GLL}^\top \mathbf{G}^\top \mathbf{K}_l - \mathbf{H}_l). \tag{21}$$

Similarly, the variable $\mathbf{B}_u$ can be computed as

$$\mathbf{B}_u = sgn(2\sum_v^V \alpha_v^\beta \mathbf{W}^{(v)^\top} \phi(\mathbf{X}_u^{(v)}) + \xi \mathbf{K}_u - \sum_v^V (\alpha_v^\beta \mathbf{W}^{(v)^\top} \mathbf{W}^{(v)}) \mathbf{K}_u - \mathbf{H}_u). \tag{22}$$

Finally, the learned hash codes $\mathbf{B}$ can be obtained by $\mathbf{B} = [\mathbf{B}_l; \mathbf{B}_u]$.
$\mathbf{K}_l, \mathbf{K}_u$-step We fix other variables, and the updating for variables $\mathbf{K}_l, \mathbf{K}_u$ can be reformulated as,

$$\min_{\mathbf{K}_l} tr \left(\sum_v^V \alpha_v^\beta \mathbf{W}^{(v)} \mathbf{K}_l \mathbf{B}_l^\top \mathbf{W}^{(v)^\top} + \gamma \mathbf{B}_l^\top \mathbf{GLL}^\top \mathbf{G}^\top \mathbf{K}_l\right)$$

$$+ \frac{\xi}{2} \|\mathbf{B}_l - \mathbf{K}_l + \frac{\mathbf{H}_l}{\xi}\|_F^2, \tag{23}$$

$$s.t. \ \mathbf{K}_l \in \{-1, 1\}^{k \times n_l},$$

Then we optimize the function for $\mathbf{K}_l$ rewritten as

$$\min_{\mathbf{K}_l} tr \left((\sum_v^V (\alpha_v^\beta \mathbf{W}^{(v)^\top} \mathbf{W}^{(v)}) \mathbf{B}_l + \gamma \mathbf{GLL}^\top \mathbf{G}^\top \mathbf{B}_l - \xi \mathbf{B}_l - \mathbf{H}_l) \mathbf{K}_l^\top\right)$$

$$s.t. \ \mathbf{K}_l \in \{-1, 1\}^{k \times n_l}, \tag{24}$$

Finally, the optimal solution of $\mathbf{K}_l$ can be obtained as

$$\mathbf{K}_l = sgn(-\sum_v^V (\alpha_v^\beta \mathbf{W}^{(v)^\top} \mathbf{W}^{(v)}) \mathbf{B}_l - \gamma \mathbf{GLL}^\top \mathbf{G}^\top \mathbf{B}_l + \xi \mathbf{B}_l + \mathbf{H}_l). \tag{25}$$

Similarly, the variable $\mathbf{K}_u$ can be obtained as

$$\mathbf{K}_u = sgn(-\sum_v^V (\alpha_v^\beta \mathbf{W}^{(v)^\top} \mathbf{W}^{(v)}) \mathbf{B}_u + \xi \mathbf{B}_u + \mathbf{H}_u). \tag{26}$$

$\mathbf{H}_l, \mathbf{H}_u$-step According to ALM scheme, the variables $\mathbf{H}_l, \mathbf{H}_u$ can be updated by,

$$\mathbf{H}_l = \mathbf{H}_l + \xi(\mathbf{B}_l - \mathbf{K}_l), \tag{27}$$

$$\mathbf{H}_u = \mathbf{H}_u + \xi(\mathbf{B}_u - \mathbf{K}_u). \tag{28}$$

3.4 Hash Function Learning

In hash function learning process, we need to use the hash codes learned in previous sections for hash function generation. The hash function can be a linear function, deep neural network, support vector machine and other models. Due to the consideration of training time, we use a linear model as base model of the hash function in this paper, and in fact other deep nonlinear models are trained in a similar way. Specifically, the hash function can be learned by the following solution,

$$\min_{\mathbf{P}^{(v)}} ||\mathbf{B} - \mathbf{P}^{(v)}\phi(\mathbf{X}^{(v)})||_F^2 + \omega||\mathbf{P}^{(v)}||_F^2, \tag{29}$$

where ω is a regularization parameter. The optimal solution of the variable $\mathbf{P}^{(v)} \in \mathbb{R}^{k \times q}$ is,

$$\mathbf{P}^{(v)} = \mathbf{B}\phi(\mathbf{X}^{(v)})^\top (\phi(\mathbf{X}^{(v)})\phi(\mathbf{X}^{(v)})^\top + \omega\mathbf{I})^{-1}. \tag{30}$$

The overall training procedures of S3ACH are described in Algorithm 1.

Algorithm 1 S3ACH

Input: Training labeled instances $\mathbf{X}_l^{(v)}$, label matrix $\mathbf{L}$, Training unlabeled instances $\mathbf{X}_u^{(v)}$, parameter $k, \beta, \delta, \rho, \gamma, \xi, \omega$, maximum iteration number $\mathbf{I}$.
Output: Binary codes $\mathbf{B}$.
Procedure:
1.Construct $\phi(X^{(v)})$ with randomly selected q anchors;
2.Initialize $\mathbf{B}, \mathbf{W}^{(v)}, \mathbf{G}, \mathbf{K}^{(v)}$ randomly with a standard normal distribution;
3.Initialize $\mathbf{H}^{(v)} = \mathbf{B} - \mathbf{K}^{(v)}$;
4.Initialize $\mathbf{S} = 2\mathbf{L}^\top\mathbf{L} - \mathbf{1}\mathbf{1}^\top$;
% *step 1: Hash code learning*
5.Repeat
    α_v-step: Update α_v via Eq. (9).
    $\mathbf{W}^{(v)}$-step: Update $\mathbf{W}^{(v)}$ via Eq. (12).
    G-step: Update $\mathbf{G}$ via Eq. (15).
    B-step: Update $\mathbf{B}_l, \mathbf{B}_u$ via Eq. (21) and Eq. (22).
    Obtain $\mathbf{B} = [\mathbf{B}_l; \mathbf{B}_u]$.
    $\mathbf{K}_l, \mathbf{K}_u$-step: Update $\mathbf{K}_l, \mathbf{K}_u$ via Eq. (25) and Eq. (22).
    $\mathbf{H}_l, \mathbf{H}_u$-step: Update $\mathbf{H}_l, \mathbf{H}_u$ via Eq. (27) and Eq. (28).
  Until up to $\mathbf{I}$
6.End
% *step 2: Hash function learning*
7.Learn the hash mapping matrix $\mathbf{P}^{(v)}$ via Eq. (30).
Return Hash function

3.5 Time Cost Analysis

In this subsection, we analyze the time consumption required to close the solution with different parameters of S3ACH, i.e., $\mathbf{W}^{(v)}$, $\mathbf{G}$ and $\mathbf{B}_l, \mathbf{B}_u$. Specifically, the training time of $\mathbf{W}^{(v)}$ is $\mathcal{O}(qnk + k^2n + k^3 + k^2q)$, the training time of $\mathbf{G}$ is $\mathcal{O}(k^2n_l + k^3 + kcn_l + c^3 + k^2c)$, the training time of $\mathbf{B}_l$ and $\mathbf{B}_u$ are $\mathcal{O}(kqn_l + kcn_l + k^2q + k^2n_l + kcn_l)$ and $\mathcal{O}(kqn_u + kcn_l + k^2q + k^2n_u)$ respectively. It can be seen that the overall time complexity of S3ACH in training parameters is linearly proportional to the number of samples, i.e., n, indicating that our proposed optimization algorithm can satisfy efficient learning under large-scale data environment.

4   Experiments

4.1   Datasets

We validate our proposed S3ACH with two publicly available datasets of MIR-Flickr [8] and NUS-WIDE [4].

1. **MIRFlickr dataset** consists of 25,000 instances tagged in 24 categories. Each instance contains a pair of image modality and text modality. The image modality uses 512-dim GIST features and the text modality uses 1,386-dim Bag-of Words (BoW) features. We select 20,015 instances and randomly choose 18,015 of these pairs as the train set and the rest 2,000 as the test set.
2. **NUS-WIDE dataset** contains 269,648 instances with 81 different tags. For each pair, the image modality uses a 500-dim SIFT vector and the text modality uses 1,000-dim BoW features. We select 186,577 instances of 10 of these common concept labels and randomly choose 1867 of them as test set and remaining ones as training set. In addition, we cannot load total training data at once due to the limitations in our experimental conditions, so we split them equally into four subsets for parallel experiments and take the average as the result.

4.2   Compared Baselines and Evaluation Metrics

Compared with unsupervised learning methods, our method makes full use of a small number of labels for semi-supervised hash learning, which is more applicable in real world. To evaluate the effectiveness, therefore, we select some classical and advanced unsupervised cross-modal hashing methods for comparison, including: CVH [9], IMH [15], CMFH [6], LSSH [32], UGACH [29], RUCMH [2], JIMFH [17], CUH [18], FUCMSH [25]. We focus on two cross-modal retrieval tasks, such as text retrieval by image (I T) and image retrieval by text (T I). We use the mean average precision (mAP) and top-k precision (P@k), and these evaluation metrics can evaluate retrieval performance. It should be noted that retrieving instances returned number in mAP is set to 100 in our experiment, while the top k is set from 0 to 1000 with 50 per step.

4.3   Implementation Details

There are parameters to be set to implement our proposed S3ACH, where β is the smoothing parameter, γ and ρ are used to balance the different terms, δ and ξ are used to optimize the solution process, and ω is used for the regularization of the hash function. We perform a grid search for all parameters (β from 2 to 9 and the rest from 10^{-5} to 10^5, with 10 times per step) and set anchor $q = 2500$ for kernelization, if not specified. In our experiments, the optimal parameter combinations is obtained, when $\{\beta = 9, \gamma = 10^4, \rho = 10^{-1}, \delta = 10^5, \xi = 10^4, \omega = 10^5\}$ and $\{\beta = 9, \gamma = 10^{-2}, \rho = 10^5, \delta = 10^5, \xi = 10^3, \omega = 10^{-5}\}$, respectively, corresponding to MIRFlickr and NUS-WIDE datasets. Note that our method is selected to use 20% labeled data for the experiments.

In addition, we implement FUCMSH, JIMFH and RUCMH ourselves with the parameters they provided. For the other baseline methods, we implement them directly with open source codes. The environment for experiments is a server with an Intel Xeon Gold 5220R @2.20 GHz with 24 cores and 64G RAM.

4.4   Results

We get the results that the mAP scores of our method and the comparison methods on two datasets with hash code lengths from 16 bits to 128 bits, as shown in Table 1, while Fig. 2 shows P@k curves of these methods with a hash code length of 64 bit. It can be seen as follows:

1. Our method outperform all comparative baselines on both datasets. Compared to the best baseline FUCMSH, our method improves on average 19.9% and 7.2% for I2T task, 5.8% and 2.7% for T2I task, respectively on MIRFlickr and NUS-WIDE datasets. These improvements in mAP scores indicate that our method enhances the learning process using a small amount of labeled data. In general, semi-supervised methods compared to unsupervised ones can achieve better retrieval performance.
2. The mAP scores of these methods are related to the hash code length. This may be because that longer hash code lengths are able to distinguish more semantic information. Nevertheless, the retrieval performance does not improve significantly when the encoding length is increased to a certain level, and the reason might be the longer hash codes enlarge the accumulation of quantization errors or other factors. In addition, our method result have no advantage at 16 and 32 bits length on T2I task of NUS-WIDE. This is probably due to the less quantity of label categories, which leads to insufficient use of semantic information at short hash codes.
3. Compared with graph-based methods (CVH, IMH), the matrix factorization based methods (including the remaining baseline and S3ACH) can better extract potential representations of multimodal data.
4. The trend of the P@k curve is similar to that of the mAP score, and our method outperforms the comparative baseline in retrieval performance. From Fig. 2, it can be seen that the effect of our method gradually approaches the

optimal baseline FUCMSH with increasing the number of retrieval instances. This might be caused by the limited performance improvement of label-enhanced learning, where too many retrieval instances weaken the semi-supervised effect.

Table 1. The mAP results for all methods on MIRFlickr and NUS-WIDE datasets.

task	method	MIRFlikr				NUS-WIDE			
		16 bits	32 bits	64 bits	128 bits	16 bits	32 bits	64 bits	128 bits
I T	CVH	0.6317	0.6356	0.6229	0.6003	0.5201	0.5162	0.4899	0.4610
	IMH	0.6018	0.6089	0.6012	0.5981	0.4952	0.4871	0.4813	0.4309
	CMFH	0.6295	0.6232	0.6231	0.6045	0.4249	0.4601	0.4778	0.4592
	LSSH	0.5993	0.6123	0.6407	0.6471	0.4592	0.4251	0.4738	0.4606
	UGACH	0.5906	0.6206	0.6218	0.6178	0.5367	0.5545	0.5562	0.5580
	RUCMH	0.6447	0.6588	0.6567	0.6513	0.5433	0.5675	0.5691	0.5593
	JIMFH	0.6371	0.6572	0.6504	0.6489	0.5469	0.5710	0.5593	0.5672
	CUH	0.6312	0.6513	0.6433	0.6542	0.5375	0.5361	0.5449	0.5326
	FUCMSH	0.6679	0.6702	0.6809	0.6845	0.5532	0.5799	0.6108	0.6320
	S3ACH	**0.7931**	**0.7954**	**0.8186**	**0.8350**	**0.5635**	**0.6081**	**0.6481**	**0.7110**
T I	CVH	0.6322	0.6317	0.6206	0.6111	0.5423	0.5271	0.4909	0.4647
	IMH	0.6227	0.6231	0.6119	0.6105	0.4979	0.4991	0.4821	0.4395
	CMFH	0.6893	0.7110	0.7331	0.7428	0.4048	0.5552	0.5805	0.5872
	LSSH	0.6779	0.7151	0.7412	0.7486	0.6631	0.6668	0.6805	0.6983
	UGACH	0.6390	0.6420	0.6472	0.6524	0.6259	0.6406	0.6673	0.6679
	RUCMH	0.6907	0.7364	0.7536	0.7501	0.6727	0.6859	0.6922	0.6846
	JIMFH	0.6885	0.7302	0.7476	0.7600	0.6614	0.6937	0.6998	0.7177
	CUH	0.6818	0.6773	0.6541	0.6570	0.6584	0.6527	0.6489	0.6417
	FUCMSH	0.7288	0.7460	0.7681	0.7745	**0.6822**	**0.7153**	0.7228	0.7289
	S3ACH	**0.7666**	**0.7792**	**0.8187**	**0.8292**	0.6482	0.7056	**0.7664**	**0.8065**

4.5   Ablation Experiments

In order to further proof the effectiveness of our method, ablation studies are conducted as follows.

Effects of Kernelization. We design S3ACH-K, which uses the original data features instead of kernel-based features, i.e., replacing $\phi(\mathbf{X}^{(v)})$ in Eq. (6) with $\mathbf{X}^{(v)}$. As shown in Table 2, the comparison shows that the performance after kernelization is better than the one without kernelization.

Effects of Label Enhancement Learning. We conduct experiments on different numbers of labels, and introduce a factor τ to measure the percentage of labeled data among all data. Note that when $\tau = 0$ means no labeled data, the label-enhanced learning framework is invalidated, i.e., the $\gamma||k\mathbf{S} - \mathbf{B}_l^{\top}(\mathbf{GL})||_F^2 + \rho||\mathbf{B}_l - \mathbf{GL}||_F^2$ in Eq. (6), when our method is degenerated to simple unsupervised learning. As shown in Table 2, comparing the unsupervised case, our label-enhanced framework plays a significant role in the effect,

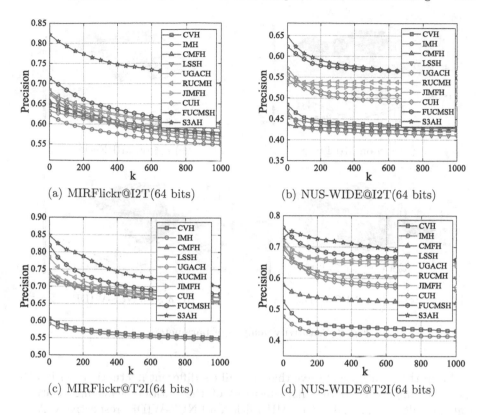

Fig. 2. Top-k precision curves on MIRFlickr and NUS-WIDE datasets.

which prove its effectiveness. Moreover, the effect improves as the percentage of labeled data increases, and it indicates that the framework can adapt to varying amounts of labeled data.

4.6   Parameter Sensitivity Analysis

In this subsection, we conduct a sensitivity analysis of the parameters in the method, and we divide them into three groups: (1) β is the smoothing parameter, which has a relatively small impact on other parameters and is therefore experimented independently; (2) γ and ρ are used to balance different parts of the asymmetric learning framework, and their values interact with each other, so a grid search is performed; (3) δ and ξ are parameters introduced to solve discrete constraint, and a grid search is conducted to observe their comprehensive impact on the results. By varying the parameter values in each group while keeping all other parameters constant, we perform these three sets of experiments on MIRFlickr and NUS-WIDE datasets with 64 bits hash codes. The search ranges of β is $\{2,3,4...,8,9\}$, that of γ, ρ, δ and ξ is $\{10^{-5},10^{-4},...,10^{3},10^{4},10^{5}\}$. For convenience of observation, we take the average mAP values of the two tasks (I2T and T2I) obtained from the experiments and plot them in Fig. 3 and Fig. 4.

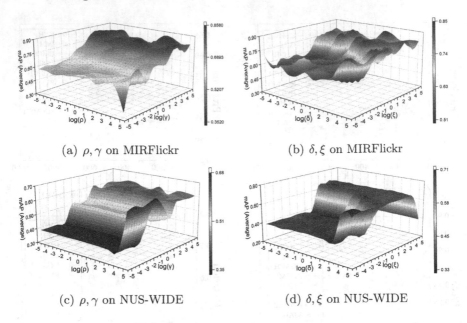

(a) ρ, γ on MIRFlickr

(b) δ, ξ on MIRFlickr

(c) ρ, γ on NUS-WIDE

(d) δ, ξ on NUS-WIDE

Fig. 3. Sensitivity analysis of parameters sets.

From the figure, we observe that β exhibits different fluctuations within the specified range due to the characteristics of the dataset, achieving relatively stable results at 2, 9 and 6, 9 on MIRFlickr and NUS-WIDE, respectively. Additionally, we can observe that ρ has a relatively small impact on the overall performance within a large range, while γ shows a generally increasing trend in a stepwise manner as it increases. Furthermore, it is evident that δ and ξ have an interaction, and they yield better results when their values differ significantly. Through comprehensive observation, it can be concluded that the parameter values exhibit certain regularities within the global range and are not sensitive to the retrieval performance within the given range. This also indicates the robustness and practical applicability of our proposed method in real-world deployments.

4.7 Convergence Analysis

In this subsection, we use the NUS-WIDE dataset as an example to analyze the convergence of our proposed S3ACH . The experiments are performed with different hash code lengths, and the objective values obtained from each iteration of Eq. (6) are calculated. The first 50 values are selected and plotted in Fig. 5. It should be noted that to better demonstrate the convergence of different hash code lengths, we normalize these objective values. From the figure, we observe that S3ACH achieves rapid convergence within 10 iterations, confirming the effectiveness of the optimization algorithm.

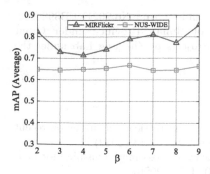

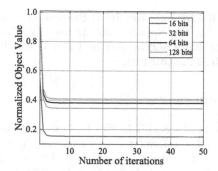

Fig. 4. Sensitivity analysis of parameter β on MIRFlickr and NUS-WIDE datasets.

Fig. 5. Convergence curves on NUS-WIDE dataset.

Table 2. The mAP results of S3ACH for different labeling ratio.

task	method	MIRFlikr				NUS-WIDE			
		16 bits	32 bits	64 bits	128 bits	16 bits	32 bits	64 bits	128 bits
I T	S3ACH-K	0.7187	0.7549	0.7739	0.7826	0.4920	0.5317	0.5899	0.6625
	S3ACH$_{(\tau=0)}$	0.5744	0.5588	0.5682	0.5984	0.3752	0.3747	0.3751	0.3788
	S3ACH$_{(\tau=0.2)}$	0.7931	0.7954	0.8186	0.8350	0.5635	0.6081	0.6781	0.7110
	S3ACH$_{(\tau=0.5)}$	0.8153	0.8266	0.8406	0.8554	0.6309	0.7133	0.7574	0.7496
T I	S3ACH-K	0.7905	0.7853	0.8111	0.8215	0.5184	0.5851	0.6631	0.7427
	S3ACH$_{(\tau=0)}$	0.5671	0.5516	0.5543	0.5871	0.3759	0.3762	0.3805	0.3947
	S3ACH$_{(\tau=0.2)}$	0.7666	0.7792	0.8187	0.8292	0.6482	0.7056	0.7764	0.8065
	S3ACH$_{(\tau=0.5)}$	0.8069	0.8397	0.8724	0.8833	0.7604	0.8244	0.8578	0.8587

5   Conclusion

In this paper, we propose an semi-supervised adaptive cross-modal hashing method called S3ACH. we add a self-learning dynamic weight parameter to the unsupervised representation of the potential public semantic space for balancing the contributions of each modality. Besides, an asymmetric learning framework is designed for semi-supervised hash learning process, so that it can make full use of limited labels to enhance the accuracy of hash codes. This framework can adapt to different amounts of labeled data. Afterwards, we propose a discrete optimization method to improve hash code learning process. We introduce an augmented Lagrange multiplier to separate the solved discrete variables and converge in fewer times. We perform experimental evaluations on two datasets and the results show that S3ACH is better than the existing advanced baseline methods with stability and high practicality. In the future, we will further focus on the effect of few labels for potential public semantic learning and try to apply the theory to the deep transformation models.

Acknowledgements. This work is supported in part by the National Natural Science Foundation of China under the Grant No.62202501, No.62172451 and No.U2003208, in part by the National Key R&D Program of China under Grant No.2021YFB3900902, in part by the Science and Technology Plan of Hunan Province under Grant No.2022JJ40638 and in part by Open Research Projects of Zhejiang Lab under the Grant No.2022KG0AB01.

References

1. Cao, M., Li, S., Li, J., Nie, L., Zhang, M.: Image-text retrieval: a survey on recent research and development. arXiv preprint arXiv:2203.14713 (2022)
2. Cheng, M., Jing, L., Ng, M.K.: Robust unsupervised cross-modal hashing for multimedia retrieval. ACM Trans. Inf. Syst. **38**(3), 1–25 (2020)
3. Cheng, S., et al.: Uncertainty-aware and multigranularity consistent constrained model for semi-supervised hashing. IEEE Trans. Circuits Syst. Video Technol. **32**(10), 6914–6926 (2022)
4. Chua, T.S., Tang, J., Hong, R., Li, H., Luo, Z.: Nus-wide: a real-world web image database from national university of Singapore. In: ACM International Conference on Image & Video Retrieval (2009)
5. Da, C., Xu, S., Ding, K., Meng, G., Xiang, S., Pan, C.: AMVH: asymmetric multivalued hashing. In: 2017 IEEE, CVPR, pp. 898–906 (2017)
6. Ding, G., Guo, Y., Zhou, J.: Collective matrix factorization hashing for multimodal data. In: IEEE on CVPR (2014)
7. Hu, P., Zhu, H., Lin, J., Peng, D., Zhao, Y.P., Peng, X.: Unsupervised contrastive cross-modal hashing. IEEE Trans. Pattern Anal. Mach. Intell. **45**, 3877–3889 (2022)
8. Huiskes, M.J., Lew, M.S.: The MIR flickr retrieval evaluation. In: ACM International Conference on Multimedia Information Retrieval, p. 39 (2008)
9. Kumar, S., Udupa, R.: Learning hash functions for cross-view similarity search. In: International Joint Conference on Artificial Intelligence (2011)
10. Liu, W., Wang, J., Ji, R., Jiang, Y., Chang, S.: Supervised hashing with kernels. In: 2012 IEEE Conference on Computer Vision and Pattern Recognition, Providence, RI, USA, June 16–21, 2012, pp. 2074–2081 (2012)
11. Meng, M., Wang, H., Yu, J., Chen, H., Wu, J.: Asymmetric supervised consistent and specific hashing for cross-modal retrieval. IEEE Trans. Image Process. **30**, 986–1000 (2021)
12. Shen, H.T., et al.: Exploiting subspace relation in semantic labels for cross-modal hashing. IEEE Trans. Knowl. Data Eng. **33**(10), 3351–3365 (2020)
13. Shi, D., Zhu, L., Li, J., Zhang, Z., Chang, X.: Unsupervised adaptive feature selection with binary hashing. IEEE Trans. Image Process. **32**, 838–853 (2023)
14. Shi, Y., et al.: Deep adaptively-enhanced hashing with discriminative similarity guidance for unsupervised cross-modal retrieval. IEEE Trans. Circuits Syst. Video Technol. **32**(10), 7255–7268 (2022)
15. Song, J., Yang, Y., Yang, Y., Huang, Z., Shen, H.T.: Inter-media hashing for large-scale retrieval from heterogeneous data sources. IN: Proceedings of the 2013 ACM SIGMOD International Conference on Management of Data (2013)
16. Tu, R.C., Jiang, J., Lin, Q., Cai, C., Tian, S., Wang, H., Liu, W.: Unsupervised cross-modal hashing with modality-interaction. IEEE Trans. Circ. Syst. Video Technol. (2023)
17. Wang, D., Wang, Q., He, L., Gao, X., Tian, Y.: Joint and individual matrix factorization hashing for large-scale cross-modal retrieval. Pattern Recogn. **107**, 107479 (2020)

18. Wang, L., Yang, J., Zareapoor, M., Zheng, Z.: Cluster-wise unsupervised hashing for cross-modal similarity search. Pattern Recogn. **111**(5), 107732 (2021)
19. Wang, Y., Chen, Z.D., Luo, X., Li, R., Xu, X.S.: Fast cross-modal hashing with global and local similarity embedding. IEEE Trans. Cybern. **52**(10), 10064–10077 (2021)
20. Wu, F., Li, S., Gao, G., Ji, Y., Jing, X.Y., Wan, Z.: Semi-supervised cross-modal hashing via modality-specific and cross-modal graph convolutional networks. Pattern Recogn. **136**, 109211 (2023)
21. Wu, W., Li, B.: Locality sensitive hashing for structured data: a survey. arXiv preprint arXiv:2204.11209 (2022)
22. Yang, F., Ding, X., Liu, Y., Ma, F., Cao, J.: Scalable semantic-enhanced supervised hashing for cross-modal retrieval. Knowl.-Based Syst. **251**, 109176 (2022)
23. Yang, F., Han, M., Ma, F., Ding, X., Zhang, Q.: Label embedding asymmetric discrete hashing for efficient cross-modal retrieval. Eng. Appl. Artif. Intell. **123**, 106473 (2023)
24. Yang, Z., Deng, X., Guo, L., Long, J.: Asymmetric supervised fusion-oriented hashing for cross-modal retrieval. IEEE Transactions on Cybernetics (2023)
25. Yang, Z., Deng, X., Long, J.: Fast unsupervised consistent and modality-specific hashing for multimedia retrieval. Neural Comput. Appl. **35**(8), 6207–6223 (2023). https://doi.org/10.1007/s00521-022-08008-4
26. Yang, Z., Raymond, O.I., Huang, W., Liao, Z., Zhu, L., Long, J.: Scalable deep asymmetric hashing via unequal-dimensional embeddings for image similarity search. Neurocomputing **412**, 262–275 (2020)
27. Yu, G., Liu, X., Wang, J., Domeniconi, C., Zhang, X.: Flexible cross-modal hashing. IEEE TNNLS **33**(1), 304–314 (2022)
28. Zhang, C., Li, H., Gao, Y., Chen, C.: Weakly-supervised enhanced semantic-aware hashing for cross-modal retrieval. IEEE Trans. Knowl. Data Eng. **35**, 6475–6488 (2022)
29. Zhang, J., Peng, Y., Yuan, M.: Unsupervised generative adversarial cross-modal hashing. In: National Conference on Artificial Intelligence (2018)
30. Zhang, J., Peng, Y., Yuan, M.: SCH-GAN: semi-supervised cross-modal hashing by generative adversarial network. IEEE Trans. Cybern. **50**(2), 489–502 (2020)
31. Zhang, P.F., Li, Y., Huang, Z., Yin, H.: Privacy protection in deep multi-modal retrieval. In: ACM SIGIR, pp. 634–643 (2021)
32. Zhou, J., Ding, G., Guo, Y.: Latent semantic sparse hashing for cross-modal similarity search. In: ACM SIGIR, pp. 415–424 (2014)

Intelligent UAV Swarm Planning Based on Undirected Graph Model

Tianyi Lv, Qingyuan Xia[✉], and Qiwen Zheng

Nanjing University of Science and Technolog, Nanjing, China
{lvtianyi,xqy,mr_zhqw}@njust.edu.cn

Abstract. The coordination of multiple drones for formation flight and collaborative task execution in the air, known as drone cluster control, has become a key research focus in recent years. Collision avoidance during formation maintenance remains a challenging aspect of cluster control, as traditional cluster control and path planning algorithms struggle to enable individual drones to independently avoid obstacles and maintain formation within the cluster. To address this issue, this paper proposes a cluster modeling method based on an undirected graph, which optimizes the entire model by adding constraints. A distributed system is also utilized to plan the entire cluster, improving the efficiency and robustness of the system and allowing individual drones to execute their flight tasks independently. Experimental verification was conducted in a ROS-based simulation environment, and the results demonstrate that our proposed algorithm effectively maintains the formation of drone clusters with high performance and stability.

Keywords: Drone cluster control · Formation flight · Collaborative task execution

1 Introduction

The maneuverability and flexibility of quadrotor UAVs endow them with the ability to navigate independently in unknown environments [1–6], and with the development of UAV technology, it is also relatively easy to achieve accurate formation control for UAV swarms in known or open environments. However, achieving autonomous navigation and task execution for multiple UAVs in shared unknown spaces, especially for distributed systems, is still a challenging task. The main difficulties include but are not limited to the difficulty in quantifying obstacles, limited sensor range, communication robustness and bandwidth issues, and common localization drift. In recent years, researchers have made further efforts [7,8] to promote obstacle avoidance for UAV swarms in real environments.

Gradient-based motion planning is the mainstream approach for local planning of quadrotor UAVs. Based on the pioneering work that formulates the local planning problem as unconstrained nonlinear optimization [9,10], however, these

© The Author(s), under exclusive license to Springer Nature Singapore Pte Ltd. 2024
B. Luo et al. (Eds.): ICONIP 2023, LNCS 14450, pp. 270–283, 2024.
https://doi.org/10.1007/978-981-99-8070-3_21

methods require high computational power to perform operations such as dense matrix inversion, which can lead to numerical issues.

To address these issues, trajectory representation methods can be further improved. Among them, Bezier curves [11] and B-spline curves [12,13] are two commonly used trajectory representation methods. Both methods have the property of sharing convex hulls, which allows for convenient addition of constraints and constrains the constraint points within the feasible convex hull region. The advantage of this approach is that it can reduce computational complexity while also avoiding numerical issues. These two methods can effectively prevent trajectories from becoming too aggressive near the physical limits of the UAV [14–16]. For Bezier curves, the time distribution can be adjusted through basis adjustment, while for B-splines, the time-adjusted distribution evaluation is highly nonlinear and may lead to changes in the shape of the curve trajectory.

MINCO [17] is a trajectory representation method specifically designed for cluster systems, which is characterized by its ability to effectively handle various constraints while maintaining spatial and temporal optimality. In this paper, we propose improvements to this method by adding constraints through discretization of the trajectory and further checking the obtained solutions. The main advantage of this approach is that it can handle multiple types of constraints, such as obstacles, velocity and acceleration limits, and obstacle avoidance, while also achieving various spatial and temporal constraints, such as maintaining relative distance and trajectory synchronization. In addition, this method can achieve good scalability and efficiency when dealing with large-scale cluster systems.

2 Methods

In this chapter, we will introduce our improved MINCO algorithm and UAV cluster modeling. As mentioned in the previous chapter, polynomial curves are commonly used as decision variables for path planning of unmanned aerial vehicles. However, to create more complex trajectories, these polynomial curves are often segmented and controlled through the addition of intermediate waypoints to shape the trajectory.

Our improved MINCO algorithm builds upon this approach by discretizing the trajectory and adding constraints to handle various types of constraints such as obstacles, velocity and acceleration limits, and obstacle avoidance. This allows for the creation of more intricate trajectories while maintaining spatial and temporal optimality. The algorithm is highly scalable and efficient, making it well-suited for large-scale cluster systems.

In addition to our improved algorithm, we will also discuss UAV cluster modeling. Modeling UAV clusters is essential for effective path planning and coordination. We will introduce a comprehensive model that considers the dynamics and kinematics of each UAV in the cluster, as well as the interactions between UAVs. This model allows for the prediction of UAV behavior and the optimization of cluster trajectories. Overall, this chapter will provide a comprehensive overview of our improved MINCO algorithm and UAV cluster modeling, high-lighting their advantages and demonstrating their effectiveness in path planning and coordination of UAV clusters.

2.1   Improved MINCO Algorithm

The definition of improved MINCO trajectory is as follows:

$$f_{MINCO} = p(t) : [0, T] \mapsto R^m$$
$$\{c = c(q, T), \forall q \in R^{m \times (M-1)}, T \in R_{>0}^M\} \tag{1}$$

The midpoint $q = (q_1, ..., q_{M-1})^T$ representing the trajectory is denoted as the waypoint vector q_i at point $t = t_i$; T_i in $T = (T_1, ..., T_M)^T$ represents the time assigned to the i-th segment of the trajectory. Therefore, a trajectory can be represented as follows:

$$p(t) = p_i(t - t_{i-1}), \forall t \in [t_{i-1}, t_i] \tag{2}$$

The trajectory of the aforementioned segment is illustrated as follows:

$$p_i(t) = c_i^T \beta(t), \forall t \in [0, T_i] \tag{3}$$

The natural basis for the polynomial is denoted by $\beta(t) = [1, t, ..., t^N]^T$, where the coefficients of the polynomial c_i^T in the trajectory $p(t)$ are represented by $c = (c_1^T, ..., c_M^T)^T$, $T_i = t_i - t_{i-1}$, $T = \sum_{i=1}^M T_i$. In essence, the key to constructing the function f_{MINCO} lies in mapping based on the construction parameters $c = \varphi(q, T)$, which directly controls the minimum control trajectory of any specified initial and final conditions of the m-dimensional integrator chain.

By constructing functions $\kappa(c, T)$ with solvable gradients, we represent any defined objective or constraint along the trajectory. Consequently, f_{MINCO} can be further expressed as:

$$\varphi(q, T) = \kappa(c(q, T), T) \tag{4}$$

In order to achieve the desired transformation, it is necessary for the function φ to compute its gradient to optimize the objective. This can be accomplished by reformulating the system of linear equations as follows:

$$M(T)c(q, T) = b(q) \tag{5}$$

Transforming the problem of computing $\frac{\partial \varphi}{\partial q}$ and $\frac{\partial \varphi}{\partial T}$ inverse to the problem of computing $\frac{\varphi \kappa}{\partial c}$ and $\frac{\partial \kappa}{\partial T}$ transpose. Here, $M(T) \in R^{2Ms \times 2Ms}$ is any $T > 0$ non-singular matrix. The transformation from (c, T) to (q, T) transpose can be achieved through the use of a linear time and space complexity PLU decomposition.

2.2   UAV Cluster Modeling

To better describe the formation relationship between drone swarms, an undirected graph $G = (V, E)$ is used to represent the relative positions and relationships between the drones. As shown in Fig. 1. In this graph, each drone can be seen as a node V_i in the graph, and the relative positions and relationships between drones can be represented by edges E. The weight $\omega_{ij} = \|p_i - p_j\|^2$ is used to indicate the distance or cost between the drones.

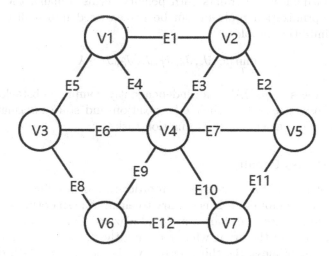

Fig. 1. Undirected graph of drone cluster

To better quantify the structure of drone formations, we utilize the Laplacian matrix to represent an undirected graph. The Laplacian matrix is a crucial tool in graph theory used to describe the topological structure of a graph. In order to preserve the topological structure of the graph while better reflecting the relative positional relationships among nodes, we further regularize the Laplacian matrix to enhance the efficiency and accuracy of path planning algorithms.

In the context of UAV formation flight, a formation similarity metric is proposed to quantify the similarity between different formations. Specifically, the metric is defined as follows:

$$f = \left\| \hat{L} - \hat{L}_{des} \right\|_F^2 = tr\left\{ \left(\hat{L} - \hat{L}_{des} \right)^T \left(\hat{L} - \hat{L}_{des} \right) \right\} \tag{6}$$

The regularized Laplacian matrix represents $\hat{L}_{des}$ the desired formation corresponding to the formation of the fleet. As shown in the above equation, this metric has rotational invariance for drones within the formation, and translation and scaling invariance for the formation itself. Moreover, this metric is differentiable with respect to the position of each drone, allowing for gradient calculations to satisfy the constraint conditions.

3   Constraints in Cost Functions

In autonomous drone swarms, drones need to collaborate to complete tasks. However, in complex environments, they also need to possess independent obstacle avoidance and path planning capabilities, which is a key challenge for drone swarm technology.

To solve the continuous constrained optimization problem of cluster control, we use the optimization variables of MINCO to eliminate all equality constraints and handle inequality constraints with penalty terms. Finally, the continuous constrained optimization problem can be transformed into a discrete unconstrained optimization problem:

$$\min_{c,T} [J_e, J_t, J_o, J_f, J_r, J_d, J_u] \cdot \lambda \tag{7}$$

During this process, the UAVs can independently complete obstacle avoidance and route planning based on their own position and state information, while satisfying the requirements of the formation task.

3.1   Smoothness Penalty

In the path planning of a single drone, in order to improve flight efficiency and reduce energy consumption, it is necessary to ensure the smoothness of the path planning, avoiding excessive sharp turns and zigzags. This can also improve the stability and safety of the drone while reducing the pressure on flight control.

To ensure smoothness, the third derivative, i.e. jerk, of the i-th drone's trajectory $p_i(t)$ is used as the control input for the segment, and the cost function is as follows:

$$J_e = \int_0^{T_i} \left\| p_i^{(3)}(t) \right\|^2 dt \tag{8}$$

The gradient calculation corresponding to it is as follows:

$$\frac{\partial J_e}{\partial c_i} = 2 \left(\int_0^{T_i} \beta^{(3)}(t) \beta^{(3)}(t)^T dt \right) \cdot c_i \tag{9}$$

$$\frac{\partial J_e}{\partial T_i} = c_i^T \beta^{(3)}(T_i) \beta^{(3)}(T_i)^T c_i \tag{10}$$

3.2   Total Time Penalty

In order to enhance the efficiency and speed of the formation flight of drone swarms, it is necessary to minimize the weighted total execution time of the entire formation to reach the target location by shortening the time for each drone to reach its expected position as much as possible. The cost function is as follows:

$$J_t = \sum_{i=1}^{M} T_i \tag{11}$$

The gradient calculation corresponding to it is as follows:

$$\frac{\partial J_t}{\partial \mathbf{c}} = 0 \tag{12}$$

$$\frac{\partial J_t}{\partial \mathbf{T}} = 1 \tag{13}$$

3.3   Collision Penalty

In drone trajectory planning, the ESDF mapping method can be used to detect whether the drone's motion path collides with obstacles. This is achieved by observing the constrained points of the drone's trajectory:

$$\psi_o \left(p_{i,j} \right) = \begin{cases} d_{thr} - d \left(p_{i,j} \right), & d \left(p_{i,j} \right) < d_{thr} \\ 0, & d \left(p_{i,j} \right) \geq d_{thr} \end{cases} \tag{14}$$

In this paper, we introduce the safety threshold d_{thr}, which is the minimum safe distance between the constraint point and any obstacle $d \left(p_{i,j} \right)$. When the distance between the constraint point and the closest obstacle is less than the safety threshold, the generated trajectory of the unmanned aerial vehicle will be adjusted to avoid the obstacle, ensuring the safety of the UAV. The cost function is defined as follows:

$$J_o = \frac{T_i}{k_i} \sum_{k=0}^{k_i} \bar{w}_j \max \left\{ \psi \left(p_{i,j} \right), 0 \right\}^3 \tag{15}$$

Vector $\left(\bar{\omega}_0, \bar{\omega}_1, ..., \bar{\omega}_{k-1}, \bar{\omega}_k \right) = \left(\frac{1}{2}, 1, ..., 1, \frac{1}{2} \right)$ constitutes an orthogonal set of coefficients that adhere to the principles of gradient rules.

The gradient calculation corresponding to it is as follows:

$$\frac{\partial J_o}{\partial c_i} = \frac{\partial J_o}{\partial \psi_o} \frac{\partial \psi_o}{\partial c_i} \tag{16}$$

$$\frac{\partial J_o}{\partial T_i} = \frac{J_o}{T_i} + \frac{\partial J_o}{\partial \psi_o} \frac{\partial \psi_o}{\partial t} \frac{\partial t}{\partial T_i} \tag{17}$$

The variable t represents the relative time for each path segment:

$$t = \frac{j}{k_i} T_i \tag{18}$$

$$\frac{\partial t}{\partial T_i} = \frac{j}{k_i} \tag{19}$$

For each constraint point that satisfies $d\left(p_{i,j}\right) < d_{thr}$, the gradient can be further computed as:

$$\frac{\partial \psi_o}{\partial c_i} = -\beta\left(t\right) \bigtriangledown d^T \tag{20}$$

$$\frac{\partial \psi_o}{\partial t} = -\bigtriangledown d^T \dot{p}\left(t\right) \tag{21}$$

$\bigtriangledown d$ represents the gradient of point $p_{i,j}$ on the ESDF distance field, which serves as a constraint.

When $d\left(p_{i,j}\right) \geq d_{thr}$:

$$\frac{\partial \psi_o}{\partial c_i} = 0 \tag{22}$$

$$\frac{\partial \psi_o}{\partial t} = 0 \tag{23}$$

3.4   Cluster Formation Penalty

As previously stated, the Laplacian matrix can be utilized to compute the minimum cost required to transform the drone formation to the desired configuration. The function that describes the current state in comparison to the desired target state is as follows:

$$\psi_f = f\left(p\left(t\right), \bigcup_{\phi} p_\phi\left(\tau\right)\right) \tag{24}$$

The variable "f" denotes the metric of formation similarity, whereas "ϕ" represents the trajectories of other unmanned aerial vehicles collected through broadcast.

The cost function J_f involves both the trajectory of oneself and that of other drones, hence it is imperative to consider the local time $t = \dfrac{j}{k_i}T_i$ of one's own trajectory and the global time-stamps of other trajectories:

$$\tau = T_1 + T_2 + \ldots + T_{i-1} + \frac{jT_i}{\kappa_i} \tag{25}$$

Thus, the cost function J_f can be expressed for any moment $1 \leq l \leq i$ of T_i as:

$$J_f = \frac{T_i}{\kappa_i}\sum_{j=0}^{\kappa_i}\bar{\omega}_j \max\left\{\psi_f\left(p\left(t\right), \bigcup_{\phi} p_\phi\left(\tau\right)\right), 0\right\}^3 \tag{26}$$

The gradient calculation corresponding to it is as follows:

$$\frac{\partial J_f}{\partial c_i} = \frac{\partial J_f}{\partial \psi_f}\frac{\partial \psi_f}{\partial c_i} \tag{27}$$

$$\frac{\partial J_f}{\partial T_l} = \frac{J_f}{T_l} + \frac{\partial J_f}{\partial \psi_f}\frac{\partial \psi_f}{\partial T_l} \tag{28}$$

Further computation is possible and can be expressed as:

$$\frac{\partial \psi_f}{\partial T_i} = \frac{\partial \psi_f}{\partial t}\frac{\partial t}{\partial T_l} + \frac{\partial \psi_f}{\partial \tau}\frac{\partial \tau}{\partial T_l} \tag{29}$$

$$\frac{\partial \psi_f}{\partial c_i} = \frac{\partial \psi_f}{\partial p(t)}\frac{\partial p(t)}{\partial c_i} \tag{30}$$

where:

$$\frac{\partial t}{\partial T_l} = \begin{cases} \dfrac{j}{\kappa_i}, & l = i \\ 0, & l < i \end{cases} \tag{31}$$

$$\frac{\partial \tau}{\partial T_l} = \begin{cases} \dfrac{j}{\kappa_i}, & l = i \\ 1, & l < i \end{cases} \tag{32}$$

$$\frac{\partial \psi_f}{\partial t} = \frac{\partial \psi_f}{\partial p(t)}\dot{p}(t) \tag{33}$$

$$\frac{\partial \psi_f}{\partial \tau} = \sum_\phi \frac{\partial \psi_f}{\partial p_\phi(\tau)}\dot{p}_\phi(\tau) \tag{34}$$

3.5   Penalty for Collisions Between Unmanned Aerial Vehicles

In order to generate a safe formation flight trajectory, a drone can accept trajectory broadcasts from other drones as constraint conditions to avoid collisions with them. Specifically, at a global timestamp, each drone will accept trajectory broadcasts from other drones and push the trajectory constraint points away from the trajectories of other drones at that timestamp.

Within the global timestamp range of τ, the set of constraint points that are in close proximity to other unmanned aerial vehicle trajectory constraint points is selected as a penalty term. The cost function J_f is defined as follows:

$$J_f = \sum_\phi \frac{T_i}{\kappa_i}\sum_{j=0}^{\kappa_i} \bar{\omega}_j \max\left\{\phi_{\tau_\psi}(p(t),\tau),0\right\}^3 \tag{35}$$

where:

$$\psi_{\tau_\phi}(p(t),\tau) = D_r^2 - d(p(t),p_\phi(\tau))^2 \tag{36}$$

$$d(p(t),p_\phi(\tau)) = \|p(t) - p_\phi(\tau)\| \tag{37}$$

The distance between each drone is represented by "D_r".

The gradient calculation corresponding to it is as follows:

$$\frac{\partial J_r}{\partial c_i} = \frac{\partial J_r}{\partial \psi_{\tau_\phi}}\frac{\partial \psi_{\tau_\phi}}{\partial c_i} \tag{38}$$

$$\frac{\partial J_r}{\partial T_l} = \frac{J_r}{T_l} + \frac{\partial J_r}{\partial \psi_{\tau_\phi}} \frac{\partial \psi_{\tau_\phi}}{\partial T_l} \tag{39}$$

When $D_r^2 \geq d\left(p\left(t\right), p_\phi\left(\tau\right)\right)^2$, the gradient is further computed as:

$$\frac{\partial \psi_{\tau_\phi}}{\partial c_i} = -2\beta\left(t\right)\left(p\left(t\right) - p_\phi\left(\tau\right)\right)^T \tag{40}$$

$$\frac{\psi_{\tau_\phi}}{\partial t} = -2\left(p\left(t\right) - p_\phi\left(\tau\right)\right)^T \dot{p}\left(t\right) \tag{41}$$

$$\frac{\partial \psi_{\tau_\phi}}{\partial \tau} = 2\left(p\left(t\right) - p_\phi\left(\tau\right)\right)^T \dot{p}_\phi\left(t\right) \tag{42}$$

3.6 Dynamic Feasibility Penalty

To ensure that the trajectories generated by the algorithm can be smoothly executed by the drone, a dynamic feasibility penalty term has also been introduced. This penalty term restricts the amplitude of the drone trajectory and trajectory derivatives, i.e., the amplitude of the drone's velocity, acceleration, and jerk.

3.7 Penalty for Uniform Distribution of Constraint Points

The generated trajectory of the drone comprises multiple constraint points, which should be uniformly distributed in space. If the distribution of constraint points is uneven, it may cause the algorithm to skip some smaller obstacles, thereby compromising the safety of the trajectory. Therefore, a penalty term is introduced to constrain the variance of the square distance between adjacent points, thereby achieving a uniform distribution of constraint points.

4   Results and Discussion

4.1 Experimental Environment

Our unmanned aerial vehicle cluster control experiments were conducted in a simulation environment based on ROS. We used quadrotor UAVs as the experimental platform and designed the experiments according to different cluster scales and varying levels of environmental complexity. The purpose of the experiments was to study the performance and stability of UAV cluster control and optimize control algorithms in different environments. We used multiple indicators to evaluate the experimental results, including the overall motion trajectory of the UAV cluster, the relative position, speed, and acceleration between UAVs.

The UAV flight parameters are shown in the following Table 1:

The flowchart of the unmanned aerial vehicle cluster control system is shown in Fig. 2:

Table 1. UAV flight parameters

UAV flight parameter	Value
Weight	0.98 kg
Size	264 mm*264 mm*496 mm
Maximum tilt angle	48
Rotor radius	62 mm
Blade Thrust coefficient	8.98*10e-9
Arm length	260 mm
Motor time constant	1/30 s
Minimum speed	1200 rpm
Maximum speed	35000 rpm

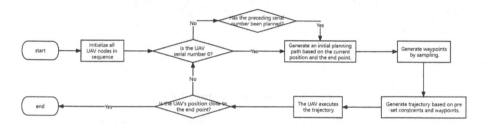

Fig. 2. UAV cluster system flowchart

4.2 Experimental Analysis

This article presents experiments designed for clusters of varying sizes. At the onset of the simulation, random obstacles are generated on the map. As depicted in Fig. 3, one group of simulation cluster trajectories demonstrates the drone cluster's ability to maintain its formation and move forward, while individual drones navigate complex environments by avoiding obstacles through the cluster planning algorithm.

In Fig. 4, the relative distance between any drone and other bodies is shown. The relative position remains relatively stable in the figure, with moments of fluctuations representing the drone cluster's avoidance actions against obstacles in the trajectory, resulting in deceleration or detours. The cluster algorithm imposes constraints on the dynamic feasibility of drones, limiting their maximum speed and acceleration. During the obstacle avoidance process of a single drone, other drones continue to perform flight tasks, resulting in corresponding changes in the relative distance. However, in the same complex environment, the drones can still maintain their basic formation to achieve the current task goal, pass through the obstacle area, and eventually reach convergence at the formation level.

The average planning time for UAV swarms of different scales is shown in the TABLE 2. It can be observed that the larger the swarm, the longer the planning

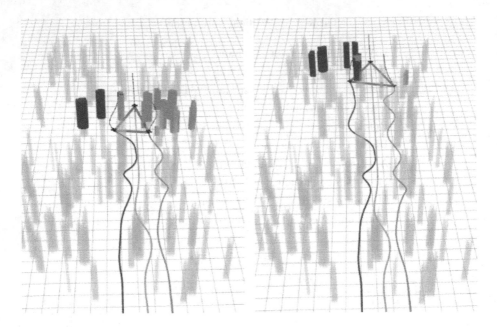

Fig. 3. Drone cluster trajectory

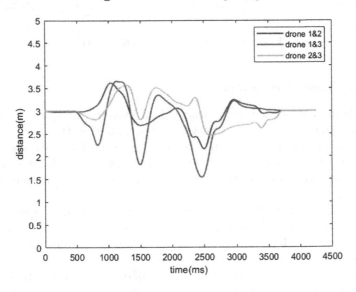

Fig. 4. Relative distance between UAVs

time required.

As shown in Fig. 5 and Fig. 6, a comparison between the velocities and accelerations of each UAV reveals that the UAVs are subject to feasibility constraints

Table 2. UAV flight parameters

Cluster size	Average flight time(s)
1	36.18
3	37.12
5	37.14
7	61.04

and will not exceed the set maximum speed and acceleration. They fly independently without affecting each other, and are able to maintain a basic synchronized formation.

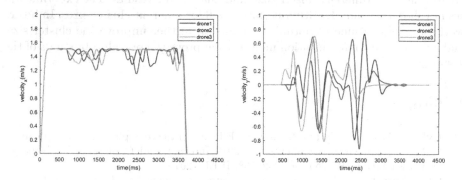

Fig. 5. Comparison of UAV velocity

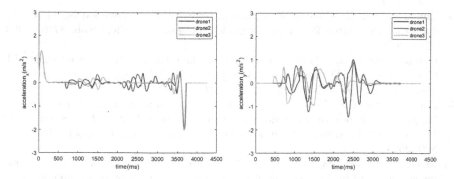

Fig. 6. Comparison of UAV acceleration

5   Conclusion and Future

In this paper, we propose an improved cluster model and control algorithm for the MINCO algorithm to achieve autonomous planning of UAV clusters. Traditional path planning and cluster control algorithms are separated from each other, making it difficult for autonomous UAV cluster formation to be achieved due to real-world factors and integration difficulties. To solve this problem, we propose a cluster model based on an undirected graph. This model can quantify the UAV cluster using the Laplacian matrix and extract effective information from it. Based on this information, corresponding constraints are given for differential equations. By calculating the overall differential of the cost function, we can obtain the corresponding gradient to make the cost function converge to the desired cluster state. We validate our approach in a ROS-based simulation environment for different cluster sizes and analyze the results. The experimental results show that our algorithm performs well for different cluster sizes. In future work, we will continue to optimize the model, further improve the cluster size, and enhance the communication module to improve the overall robustness of the system.

References

1. Ryll, M., Ware, J., Carter, J., Roy, N.: Efficient trajectory planning for high speed flight in unknown environments. In: 2019 International Conference on Robotics and Automation (ICRA), pp. 732–738. IEEE (2019)
2. Oleynikova, H., et al.: An open-source system for vision-based micro-aerial vehicle mapping, planning, and flight in cluttered environments. J. Field Rob. 37(4), 642–666 (2020)
3. Zhang, J., Hu, C., Chadha, R.G., Singh, S.: Falco: fast likelihood-based collision avoidance with extension to human-guided navigation. J. Field Rob. 37(8), 1300–1313 (2020)
4. Campos-Macías, L., Aldana-López, R., de la Guardia, R., Parra-Vilchis, J.I., Gómez-Gutiérrez, D.: Autonomous navigation of MAVs in unknown cluttered environments. J. Field Rob. 38(2), 307–326 (2021)
5. Zhou, X., Wang, Z., Ye, H., Xu, C., Gao, F.: Ego-planner: an ESDF-free gradient-based local planner for quadrotors. IEEE Rob. Autom. Lett. 6(2), 478–485 (2020)
6. Foehn, P., et al.: Alphapilot: autonomous drone racing. Auton. Robot. 46(1), 307–320 (2022)
7. McGuire, K., De Wagter, C., Tuyls, K., Kappen, H., de Croon, G.C.: Minimal navigation solution for a swarm of tiny flying robots to explore an unknown environment. Sci. Rob. 4(35), eaaw9710 (2019)
8. Zhou, D., Wang, Z., Schwager, M.: Agile coordination and assistive collision avoidance for quadrotor swarms using virtual structures. IEEE Trans. Rob. 34(4), 916–923 (2018)
9. Ratliff, N., Zucker, M., Bagnell, J.A., Srinivasa, S.: Chomp: gradient optimization techniques for efficient motion planning. In: 2009 IEEE International Conference on Robotics and Automation, pp. 489–494. IEEE (2009)

10. Kalakrishnan, M., Chitta, S., Theodorou, E., Pastor, P., Schaal, S.: Stomp: stochastic trajectory optimization for motion planning. In: 2011 IEEE International Conference on Robotics and Automation, pp. 4569–4574. IEEE (2011)
11. Gao, F., Wu, W., Lin, Y., Shen, S.: Online safe trajectory generation for quadrotors using fast marching method and bernstein basis polynomial. In: 2018 IEEE International Conference on Robotics and Automation (ICRA), pp. 344–351. IEEE (2018)
12. Rueckert, D., Sonoda, L.I., Hayes, C., Hill, D.L., Leach, M.O., Hawkes, D.J.: Nonrigid registration using free-form deformations: application to breast MR images. IEEE Trans. Med. Imaging **18**(8), 712–721 (1999)
13. Zhou, B., Gao, F., Wang, L., Liu, C., Shen, S.: Robust and efficient quadrotor trajectory generation for fast autonomous flight. IEEE Rob. Autom. Lett. **4**(4), 3529–3536 (2019)
14. Van Den Berg, J., Guy, S.J., Lin, M., Manocha, D.: Reciprocal n-body collision avoidance. In: Robotics Research: The 14th International Symposium ISRR, pp. 3–19. Springer, Heidelberg (2011). https://doi.org/10.1007/978-3-642-19457-3_1
15. Van Den Berg, J., Snape, J., Guy, S.J., Manocha, D.: Reciprocal collision avoidance with acceleration-velocity obstacles. In: 2011 IEEE International Conference on Robotics and Automation, pp. 3475–3482. IEEE (2011)
16. Arul, S.H., Manocha, D.: DCAD: decentralized collision avoidance with dynamics constraints for agile quadrotor swarms. IEEE Rob. Autom. Lett. **5**(2), 1191–1198 (2020)
17. Wang, Z., Zhou, X., Xu, C., Gao, F.: Geometrically constrained trajectory optimization for multicopters. IEEE Trans. Rob. **38**(5), 3259–3278 (2022)

Learning Item Attributes and User Interests for Knowledge Graph Enhanced Recommendation

Zepeng Huai[1]([✉]), Guohua Yang[2], Jianhua Tao[3], and Dawei Zhang[2]

[1] School of Artificial Intelligence, University of Chinese Academy of Sciences, Beijing, China
zepenghuai6@gmail.com
[2] Institute of Automation, Chinese Academy of Sciences, Beijing, China
{guohua.yang,dawei.zhang}@nlpr.ia.ac.cn
[3] Department of Automation, Tsinghua University, Beijing, China
jhtao@tsinghua.edu.cn

Abstract. Knowledge Graphs (KGs) manifest great potential in recommendation. This is ascribable to the rich attribute information contained in KG, such as the price attribute of goods, which is further integrated into item and user representations and improves recommendation performance as side information. However, existing knowledge-aware methods leverage attribute information at a coarse-grained level in two aspects: (1) item representations don't accurately learn the distributional characteristics of different attributes, and (2) user representations don't sufficiently recognize the pattern of user preferences towards attributes. In this paper, we propose a novel *attentive knowledge graph attribute network*(AKGAN) to learn item attributes and user interests via attribute information in KG. Technically, AKGAN adopts a novel graph neural network framework, which has a different design between the first layer and the latter layer. The first layer merges one-hop neighbors' attribute information by concatenation operation to avoid breaking down the independence of different attributes, and the latter layer recursively propagates attribute information without weight decrease of high-order significant neighbors. With one attribute placed in the corresponding range of element-wise positions, AKGAN employs a novel interest-aware attention unit, which releases the limitation that the sum of attention weight is 1, to model the complexity and personality of user interests. Experimental results on three benchmark datasets show that AKGAN achieves significant improvements over the state-of-the-art methods. Further analyses show that AKGAN offers interpretable explanations for user preferences towards attributes.

Keywords: Recommendation · Knowledge Graph · Graph Neural Network

1 Introduction

In recent years, introducing knowledge graphs (KGs) into recommender systems as side information has been effective for improving recommendation perfor-

B. Luo et al. (Eds.): ICONIP 2023, LNCS 14450, pp. 284–297, 2024.
https://doi.org/10.1007/978-981-99-8070-3_22

mance. With the attribute information iteratively propagating in KGs, GNN-based methods aggregate multi-hop neighbors into representations and have achieved the state-of-the-art recommendation results.

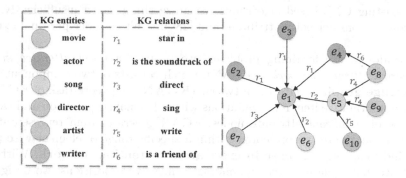

Fig. 1. An example of how KG contains multiple attributes information. Best viewed in color.

Despite the success of GNN in exploiting multi-hop attribute information, we argue that there are still three shortcomings: (1) **The pollution caused by weighted sum operation in merging different attribute information.** Different attributes are independent in terms of semantics and user preference. Take the actor attribute e_2 and the singer attribute e_8 of movie e_1 in Fig. 1 as an example, semantic independence means e_2 doesn't co-occur with e_8 in each movie, since an actor and a singer are invited to work for a movie respectively. Preference independence means whether a user prefers actor e_2 is independent of whether he likes singer e_8. However, existing GNN-based methods pool embeddings from different attribute nodes with weighted sum (i.e., sum, mean, attention) operation, which brings about semantic pollution and the difficulty to distill user-interested attribute information. (2) **The nonlinearity between the distance and importance of attribute node relative to item node.** Usually, high-order neighbors are less relative to the center node in graph data, but this is not absolute. For example, singer attribute e_8 is more valued than director attribute e_7 by some movie viewers who like music, while e_8, a 2-hop neighbor, is further than e_7, a 1-hop neighbor. This problem is caused by inherent graph topology, which means some significant attribute nodes don't connect to item nodes directly, such that it requires multiple passes to aggregate these high-order neighbors into the center node. However, existing GNN-based algorithms neglect this issue and decrease the weight of significant high-order neighbors coupled with the increase of propagation times. (3) **The complexity and personality of user interests towards attributes.** User interests show the following pattern: a user just gets interested in a part of rather than all attributes of an item, and different users prefer different attributes even towards the same item. For example, both users u_1 and u_2 watch movie e_1 because u_1 likes e_1's actor e_2 rather than director e_7 while u_2 prefers the theme song e_5 rather than actor e_2.

Therefore, we should integrate e_2's rather than e_7's embedding into u_1's representation and integrate e_5's rather than e_2's embedding into u_2's representation. In other words, the item representation learned after GNN propagation contains noisy signals and we should distill user-interested attributes personally. However, existing GNN-based algorithms recognize this pattern insufficiently, like [30] doesn't consider noisy attributes and [28] doesn't consider personal extraction.

To address the foregoing problems, we propose a novel *attentive knowledge graph attribute network* (AKGAN), which consists of two components: (1) **knowledge graph attribute network (KGAN)**. The core task of KGAN is to learn informative item representations without semantic pollution and weight decrease of significant attribute nodes. KGAN generates item representations where one attribute is represented within a specific range of element-wise positions independently. (2) **user interest-aware attention network**. With different attributes placed in corresponding element-wise positions, we design an interest-aware attention layer to distill user-interested attributes. A novel activation unit is introduced to release the limitation that the sum of attention weight is 1, which aims to reserve the intensity of user interests [38]. Finally all interest scores and item representations are further combined to infer user representations.

2   Problem Formalization

We begin by introducing some related notations and then defining the KG enhanced recommendation problem. Let $\mathcal{U} = \{u\}$ and $\mathcal{I} = \{i\}$ separately denote the user and item sets. A typical recommender system usually has historical user-item interactions, which is defined as $\mathcal{O}^+ = \{(u,i)|u \in \mathcal{U}, i \in \mathcal{I}\}$. Each (u,i) pair indicates user u has interacted with item i before, such as clicking, review, or purchasing. We also have a knowledge graph that stores the structured semantic information of real-world facts. We denote the entity and relation sets in KG as $\mathcal{R} = \{r\}$ and $\mathcal{E} = \{e\}$. KG is presented as $\mathcal{G} = \{(h,r,t)|h \in \mathcal{E}, r \in \mathcal{R}, t \in \mathcal{E}\}\}$, where each triple means there is a relation r from head entity h to tail entity t. For example, $(James\ Cameron, Direct, Avatar)$ describes the fact that James Cameron is the director of the movie Avatar. Note that $\mathcal{R}$ contains relations in both canonical direction (e.g., Direct) and inverse direction (e.g., DirectedBy). The bridge between KG and the recommender system is that items are contained in entities. Specifically, $\mathcal{E}$ consists of item node $\mathcal{I}(\mathcal{I} \subseteq \mathcal{E})$ as well as attribute node $\mathcal{E} \setminus \mathcal{I}$.

3   Methodology

3.1   Model Overview

In this subsection, we introduce the framework of AKGAN. With a widely used two-tower structure in recommender model [5,18,36], AKGAN consists of two

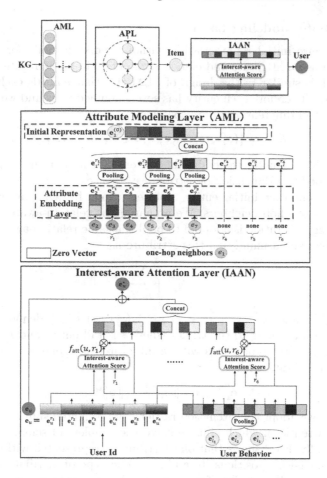

Fig. 2. The overall architecture of AKGAN. Best viewed in color.

main modules: (1) knowledge graph attribute network, and (2) user interest-aware attention module. Figure 2 gives its framework.

The core task of KGAN is to learn KG enhanced item representations that contain multiple attribute information. KGAN adopts GNN framework and has a heterogeneous design in the first layer and the latter layer, which are named attribute modeling layer and attribute propagation layer respectively. **Attribute modeling layer** (AML) is to construct initial entity representations using one-hop neighbors' embeddings $e_{j \in \mathcal{N}_i}^{atr}$. **Attribute propagation layer** (APL) recursively propagates attribute information to acquire more informative item representations. When item representations have been learned, the **interest-aware attention** (IAA) module is to learn user representations via interaction data. Finally, a scoring function is used to predict their matching score.

We illustrate the architecture of AML (cf. Sect. 3.2), APL (cf. Sect. 3.3) and IAA (cf. Sect. 3.4) in detail in the following sections.

3.2   Attribute Modeling Layer

To model a pure initial representation without semantic pollution, we regard each relation as an attribute and embed each entity in all attribute embedding spaces, which is similar to the design of FFM [11] that embeds each feature in different fields. Therefore each entity has several embeddings and we denote the embedding set in different attribute spaces as

$$\mathcal{E}^{atr} = \{(\mathbf{e}_i^{r_1}, \mathbf{e}_i^{r_2}, ..., \mathbf{e}_i^{r_M})|i \in \mathcal{E}, M = |\mathcal{R}|\} \tag{1}$$

where $\mathbf{e}_i^{r_m} \in \mathbb{R}^{d^m}$ denotes the embedding of entity i in attribute r_m space and d^m is the embedding dimension of attribute r_m space.

Then we construct initial entity representations by aggregating attribute embeddings of one-hop neighbors. We prepared the relation-aware embeddings and the average operation is adopted to pool the same relation-aware neighbors to acquire the main semantics of this attribute as

$$\mathbf{e}_{i'}^{r_j} = \frac{1}{|\mathcal{N}_i^{r_j}|} \sum_{j \in \mathcal{N}_i^{r_j}} \mathbf{e}_j^{r_j} \tag{2}$$

where $j \in \mathcal{N}_i^{r_j}$ and $\mathcal{N}_i^{r_j} = \{j|j \in (j, r_j, i), (j, r_j, i) \in \mathcal{G}\}$ denotes the set of head entities that belong to the triplet where i is tail entity and r_j is relation. Finally, all attribute embeddings will be integrated into one representation by concatenation operation as

$$\mathbf{e}_i^{(0)} = \mathbf{e}_{i'}^{r_1} \| \mathbf{e}_{i'}^{r_2} \| \cdots \| \mathbf{e}_{i'}^{r_M} \tag{3}$$

Note that we adopt concatenation rather than weighted sum operation (i.e., sum, mean, attention), which aims to solve the problem of semantic pollution. More specifically, Eq. 3 shows that one attribute is represented within a specific range of element-wise positions. In addition, one type of attribute is placed in the same element-wise positions for all entities. For example, two representations $\mathbf{e}_1^{(0)} = \mathbf{e}_{1'}^{r_1} \| \cdots \| \mathbf{e}_7^{r_3} \| \cdots$ and $\mathbf{e}_{11}^{(0)} = \mathbf{e}_{11'}^{r_1} \| \cdots \| \mathbf{e}_7^{r_3} \| \cdots$ mean movie e_1 and movie e_{11} have the same director and different actors. In a word, a concatenated representation avoids the interaction and preserves the semantic independence of different attributes, which will be further advantageous to maintain the weight of significant high-order neighbors in Sect. 3.3 and distill user-interested attributes in Sect. 3.4.

3.3   Attribute Propagation Layer

Here we regard a KG as a heterogeneous graph and adopt a widely used two-step scheme [2] to aggregate the representations of neighbors: (1) same relation-aware neighbors aggregation; (2) relations combination. The pooling methods in the above two steps are average and sum operation, respectively. More formally, in the $l-$th layer, we recursively formulate the representation of an entity as:

$$\mathbf{e}_i^{(l)} = \sum_{r_j \in \mathcal{R}} \frac{1}{|\mathcal{N}_i^{r_j}|} \sum_{j \in \mathcal{N}_i^{r_j}} \mathbf{e}_j^{(l-1)} \tag{4}$$

We re-examine $\mathbf{e}_i^*$ from the perspective of element-wise position. Without loss of generality, $\mathbf{e}_i$ is contructed as

$$\mathbf{e}_i^* = \mathbf{e}_{i''}^{r_1} \| \mathbf{e}_{i''}^{r_2} \| \cdots \| \mathbf{e}_{i''}^{r_M} \tag{5}$$

where $\mathbf{e}_{i''}^{r_m} \in \mathbb{R}^{d^m}$ is a truncated vector in $\mathbf{e}_i^* \in \mathbb{R}^D$ and $D = \sum_{r_m \in \mathcal{R}} d^m$. The start and ending index of $\mathbf{e}_{i''}^{r_m}$ in $\mathbf{e}_i^*$ are $\sum_{p=1}^{m-1} d^p$ and $\sum_{p=1}^{m} d^p$, respectively. Reviewing the generation process of $\mathbf{e}_{i''}^{r_m}$, we can see that $\mathbf{e}_{i''}^{r_m}$ is learned by neighbors' embeddings in attribute r_m space and doesn't include any embeddings from other attribute spaces. For example, in Fig. 2, $\mathbf{e}_{1''}^{r_1}$ is learned by $\mathbf{e}_2^{r_1}$, $\mathbf{e}_3^{r_1}$, and $\mathbf{e}_4^{r_1}$. This means $\mathbf{e}_{i''}^{r_m}$ maintains the independence of attribute r_m and KGAN remains semantics unpolluted after multi-layer propagations. Furthermore, we check specific neighbors which are aggregated into $\mathbf{e}_{i''}^{r_m}$. Take $\mathbf{e}_{1''}^{r_5}$ as an example, after 1-order propagation (one AML), $\mathbf{e}_{1''}^{r_5}$ is a zero vector since attribute r_5 doesn't occur in one-hop neighbors of movie e_1. Then after 2-order propagation (one AML and one APL), $\mathbf{e}_{1''}^{r_5} = \mathbf{e}_{10}^{r_5}$, which seems as if there were a link (e_{10}, r_5, e_1) and e_{10} connected to e_1 directly. In general, when one relation, a.k.a attribute (i.e., r_5), firstly appears in the receptive field of center node (i.e., e_1) at $l-$hop (i.e., $2-$hop) position, the weight of its corresponding node (i.e., e_{10}) will not decrease, which proves that KGAN maintains the weight of significant high-order neighbors.

3.4 Interest-Aware Attention Layer

After item representations have been obtained by KGAN, a typical idea in recommender system is to enhance user representations by clicked items, like [13, 25, 32]. We take average pooling as an example, as follows:

$$\mathbf{e}_u^* = \mathbf{e}_u + \frac{1}{|\mathcal{N}_u|} \sum_{i \in \mathcal{N}_u} \mathbf{e}_i^* \tag{6}$$

where $\mathcal{N}_u = \{i | (u, i) \in \mathcal{O}^+\}$ and $\mathbf{e}_u$ represents user embedding for collaborative filtering. However, average pooling doesn't consider user preferences personally, thereby recent works aims to model user interests via attention mechanism, such as [14, 19, 28, 30, 37].

Here we propose a novel attention module to model user interests towards different attributes. We assign an interest score $f_{\text{att}}(u, r_m)$ to each pair of attribute r_m and user u, and personally generate user representation by combining interest scores and interacted items. We firstly introduce how to calculate $f_{\text{att}}(u, r_m)$ and then illustrate how to combine $f_{\text{att}}(u, r_m)$ with interacted items.

With different attributes placed in corresponding element-wise positions as shown in Eq. 5, we truncate user vector $\mathbf{e}_u$ with the same strategy as

$$\mathbf{e}_u = \mathbf{e}_u^{r_1} \| \mathbf{e}_u^{r_2} \| \cdots \| \mathbf{e}_u^{r_M} \tag{7}$$

where $\mathbf{e}_u^{r_m} \in \mathbb{R}^{d^m}$ is a truncated vector in $\mathbf{e}_u \in \mathbb{R}^D$. The start and ending index of $\mathbf{e}_u^{r_m}$ in $\mathbf{e}_u$ are $\sum_{p=1}^{m-1} d^p$ and $\sum_{p-1}^{m} d^p$, respectively. Then the interest score of user u towards attribute r_m is calculated by

$$f_{\text{att}}(u, r_m) = \sigma \left(\tau \frac{\frac{1}{|\mathcal{N}_u|} \sum_{i \in \mathcal{N}_u} \mathbf{e}_u^{r_m \top} \mathbf{e}_{i''}^{r_m}}{\frac{1}{|\mathcal{I}|} \sum_{i \in \mathcal{I}} \mathbf{e}_u^{r_m \top} \mathbf{e}_{i''}^{r_m}} \right) \tag{8}$$

where τ is the temperature coefficient as a hyperparameter.

After $f_{\text{att}}(u, r_m)$ has been prepared, we need to combine it with interacted items to learn user representations. We firstly re-express Eq. 6 as

$$\mathbf{e}_u^* = \mathbf{e}_u + \underset{r_m \in \mathcal{R}}{\|} \left(\frac{1}{|\mathcal{N}_u|} \sum_{i \in \mathcal{N}_u} \mathbf{e}_{i'''}^{r_m} \right) \tag{9}$$

where $\|$ is the concatenation operation. Therefore, an intuitive idea is to use $f_{\text{att}}(u, r_m)$ to control how much attribute information, a.k.a. $\frac{1}{|\mathcal{N}_u|} \sum_{i \in \mathcal{N}_u} \mathbf{e}_{i'''}^{r_m}$, will be passed to user. Consider the limit case, when $f_{\text{att}}(u, r_m) = 0$ which means user u doesn't pay attention to attribute r_m at all, any information of attribute r_m should not be contained in user representation. Formally, user representation is obtained by

$$\mathbf{e}_u^* = \mathbf{e}_u + \underset{r_m \in \mathcal{R}}{\|} \left(\frac{f_{\text{att}}(u, r_m)}{|\mathcal{N}_u|} \sum_{i \in \mathcal{N}_u} \mathbf{e}_{i'''}^{r_m} \right) \tag{10}$$

Note that here we relax the constraint that the sum of attention weights towards all attributes is 1, a.k.a. $\sum_{r_m \in \mathcal{R}} f_{\text{att}}(u, r_m) \neq 1$. The reason is as follows: when the number of members participating in the attention calculation is large and the limitation that the sum of attention weights is 1 is still reserved at this time, it will cause the attention weight of each member to be dispersed, making it difficult to learn the coefficients of important nodes. This problem has also appeared in Graphair [8], which shows estimating $\mathcal{O}(|\mathcal{N}|^2)$ coefficients exposes the risk of overfitting. Therefore, we introduce the above novel activation unit to release the limitation that the sum of attention weights is 1, and the same idea is adopted by DIN [38].

3.5 Model Optimization

The objective function is formulated as

$$\mathcal{L} = \sum_{(u, i^+, i^-) \in \mathcal{O}} -\ln \sigma \left(\hat{y}(u, i^+) - \hat{y}(u, i^-) \right) + \lambda \|\Theta\|_2^2 \tag{11}$$

where $\mathcal{O} = \{(u, i^+, i^-) | (u, i^+) \in \mathcal{O}^+, (u, i^-) \in \mathcal{O}^-\}$ denotes the training set, which contains the observed interactions $\mathcal{O}^+$ and the unobserved interactions $\mathcal{O}^-$; $\sigma(\cdot)$ is the sigmoid function. $\Theta = \{\mathbf{e}_i^{r_m}, \mathbf{e}_u | i \in \mathcal{I}, u \in \mathcal{U}, r_m \in \mathcal{R}\}$ is the model parameter set.

Table 1. Statistics of the datasets.

Datasets	Amazon-book	Last-FM	Alibaba-iFashion
# users	70,679	23,566	114,737
# items	24,915	48,123	30,040
# interactions	847,733	3,034,796	1,781,093
# ave./int.	11.99	128.78	15.52
# entities	88,572	58,266	59,156
# relations	39	9	51
# triples	2,557,746	464,567	279,155

4 Experiments

- **RQ1**: Does our proposed AKGAN outperform the state-of-the-art recommendation methods?
- **RQ2**: How do different components (i.e., attribute modeling layer, attribute propagation layer, and interest-aware attention layer) affect AKGAN?
- **RQ3**: Can AKGAN provide potential explanations about user preferences towards attributes?

4.1 Experimental Settings

Dataset Description. We choose three benchmark datasets to evaluate our method: Amazon-Book, Last-FM, and Alibaba-iFashion. The former two datasets are released in [28] and the last one is released in [30]. All of them are publicly available. Each dataset consists of two parts: user-item interactions and a corresponding knowledge graph. The basic statistics of the three datasets are presented in Table 1. We follow the same data partition used in [28] to split the datasets into training and testing sets. For each observed user-item interaction, we randomly sample one negative item that the user has not interacted with before, and pair it with the user as a negative instance.

Evaluation Metrics. We evaluate our method in the task of top-K recommendation: Recall@K and NDCG@K, where K is set as 20 by default.

Baselines. To demonstrate the effectiveness, we compare AKGAN with KG-free (MF [20]), attribute-based (FM [21] and FFM [11]), knowledge-aware embedding-based (CKE [34]), knowledge-aware path-based (RippleNet [24]), and knowledge-aware GNN-based (KGAT [28], KGIN [30], PRKG [39]) methods.

Parameter Settings. We implement our AGKAN model in Pytorch and Deep Graph Library (DGL), which is a Python package for deep learning on graphs. We released all implementations (code, datasets, parameter settings, and training logs) to facilitate reproducibility[1].

---

[1] https://github.com/huaizepeng2020/AKGAN.

Table 2. Overall Performance Comparison. The best performance is boldfaced; the runner up is labeled with '*'. '%Imp.' indicates the relative improvements.

Dataset	Metrics	KG-free	Attribute-based		Embedding-based	Path-based
		MF	FM	FFM	CKE	RippleNet
Amazon-Book	Recall	0.1241	0.1366	0.1601	0.1287	0.1355
	NDCG	0.0650	0.0755	0.0912	0.0674	0.0763
Last-FM	Recall	0.0774	0.0771	0.0935	0.0780	0.0842
	NDCG	0.0669	0.0648	0.0810	0.0659	0.0766
Alibaba-iFashion	Recall	0.0921	0.1003	0.1089	0.1068	0.1121
	NDCG	0.0562	0.0577	0.0681	0.0633	0.0695

Dataset	Metrics	GNN-based				Imp.
		KGAT	KGIN	PRKG	AKGAN	
Amazon-Book	Recall	0.1473	0.1687*	0.1562	**0.1783**	5.69%
	NDCG	0.0782	0.0915*	0.0894	**0.0994**	8.63%
Last-FM	Recall	0.0876	0.0978	0.1022*	**0.1209**	18.30%
	NDCG	0.0745	0.0848	0.0890*	**0.1072**	20.45%
Alibaba-iFashion	Recall	0.1015	0.1147*	0.1125	**0.1253**	9.24%
	NDCG	0.0616	0.0716*	0.0705	**0.0801**	11.87%

4.2 Performance Comparison (RQ1)

We report the empirical results in Table 2 and use %Imp. to denote the percentage of relative improvement on each metric. The observations are as follows:

- AKGAN consistently achieves the best performance on three datasets in terms of all measures. Specifically, it achieves significant improvements over the strongest baselines w.r.t. NDCG@20 by 8.63%, 20.45%, and 11.87% in Amazon-Book, Last-FM, and Alibaba-iFashion, respectively.
- Jointly analyzing AKGAN across the three datasets, we find that the improvement on Last-FM is more significant than that on Alibaba-iFashion and Amazon-Book. The main reason is that the average number of interactions per user of Last-FM (128.78) is much larger than that of the other two datasets (11.99, 15.52).
- In four GNN-based methods, AKGAN performs best, PRKG and KGIN are the second-best, while KGAT achieves the worst results. The decreasing performance is because that the level of how a recommender learns item attributes and user interests is in descending order, from fine-grained to coarse-grained.

4.3 Study of AKGAN (RQ2)

Impact of Concatenation Operation & Interest Score. To demonstrate the necessity of concatenation operation and interest score, we compare the performance of AKGAN with the following three variants: (1) combing different attributes with the average operation and discarding interest score, termed AKGAN-mean, (2) combing different attributes with the sum operation and discarding interest score, termed AKGAN-sum, (3) only discarding interest score, termed AKGAN-w/o att. The results are shown in Table 3 and the major findings

Table 3. Impact of concatenation operation and interest score.

	Amazon-Book		Last-FM		Alibaba-iFashion	
	Recall	NDCG	Recall	NDCG	Recall	NDCG
AKGAN-mean	0.1504	0.0799	0.0849	0.0713	0.1064	0.0660
AKGAN-sum	0.1489	0.0784	0.0886	0.0736	0.1045	0.0643
AKGAN-w/o att	0.1758	0.0969	0.1141	0.1012	0.1213	0.0774

Table 4. Impact of the number of layers L.

	Amazon-Book		Last-FM		Alibaba-iFashion	
	Recall	NDCG	Recall	NDCG	Recall	NDCG
AKGAN-1	0.1737	0.0974	0.1192	0.1060	0.1245	0.0791
AKGAN-2	0.1768	0.0987	0.1205	0.1069	0.1253	0.0801
AKGAN-3	0.1783	0.0994	0.1209	0.1066	0.1221	0.0776

are as below: (1) Comparing AKGAN-mean/AKGAN-sum with AKGAN-w/o att, we can clearly see that replacing concatenation operation with weighted sum operation dramatically degrades performance of recommendation, which indicates the superiority of concatenation operation. (2) The improvement from AKGAN-w/o att to AKGAN verifies the necessity of interest score.

Impact of Model Depth. We investigate the influence of depth of receptive field in AKGAN by searching L in the range of $\{1, 2, 3\}$. Particularly, $L = 1$ means AKGAN has only one attribute modeling layer. The results are reported in Table 4. In Amazon-Book and Last-FM, increasing propagation times of attributes can boost the performance, because more attribute information is aggregated into the center node to learn more informative user and item representations. While in Alibaba-iFashion, AKGAN-2 is the best and AKGAN-3 is the worst.

Impact of Temperature Coefficient. We vary the temperature coefficient τ in the range of $\{0.01, 0.1, 0.2, ..., 0.9, 1\}$ to study its influence on the performance of AKGAN and summarize the results on partial datasets in Fig. 3. We can see that the best performance occurs when τ is set to the optimal value. The reason is that τ describes the sensitivity of the interest score relative to a given divisor in Eq. 8. A large τ will make the output of Eq. 8 saturate too early, while a small τ will cause underfitting.

4.4 Case Study (RQ3)

In this section, we visualize the interest score to show how AKGAN learns user preferences towards attributes and use Alibaba-iFashion as the example dataset since it helps to provide an intuitive explanation.

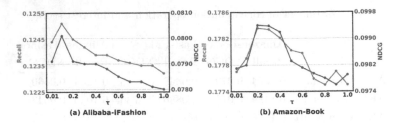

Fig. 3. Impact of temperature coefficient τ. Best viewed in color.

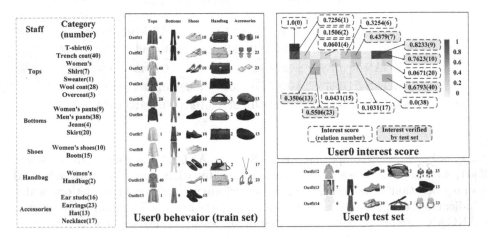

Fig. 4. An example of the interest score of user0 to verify the interpretability of AKGAN with its training set and testing set.

We randomly select a user called *User0* as the example user. Figure 4 introduces his historical behaviors, including both training set and testing set and displays the interest score learned from the training set and we use the testing set to validate the effectiveness of these scores. We have the following observations:

- The interest score accurately models user0's preferences towards attributes. Among 11 outfits clicked by user0, the tops contain four main kinds: T-shirt, shirt, trench coat, and sweater, which means user0 prefers the above four types of tops. The corresponding interest score are 0.3254, 0.4379, 0.6793, 0.7256, respectively.
- Testing set proves the effectiveness of interest score. For example, $f_{att}(u_0, r_{40}) = 0.6793$ is proved to be reasonable since outfit12 which contains a trench coat appears in the testing set. And $f_{att}(u_0, r_9) = 0.8233$ is validated by outfit13 and outfit14 with same reason.
- In one staff, the interest degree varies according to category. Take accessories as an example, user0 prefers earrings than hat and $f_{att}(u_0, r_{23}) = 0.5506 > f_{att}(u_0, r_{13}) = 0.3506$. This result is also proved by the testing set, where two outfits contain earrings while only one outfit has a hat.

5   Related Work

Our work is highly related with the knowledge-aware recommendation, which can be grouped into three categories.

Embedding-Based Methods [9,25,34,35] hire KG embedding algorithms (i.e., TransE [1] and TransH [31]) to model prior representations of item, which are used to guide the recommender model. For example, DKN [25] learns knowledge-level embedding of entities in news content via TransD [10] for news recommendation. KSR [9] utilizes knowledge base information learned with TransE as attribute-level preference to enhance the sequential recommendation. However, they ignore higher-order connectivity in a knowledge graph.

Path-Based Methods [16,22] usually predefine a path scheme (i.e., meta-path) and leverages path-level semantic similarities of entities to refine the representations of users and items. For example, MCRec [7] learns the explicit representations of meta-paths to depict the interaction context of user-item pairs. RKGE [23] mines the path relation between user and item automatically and encodes the entire path using a recurrent network to predict user preference towards this item. However, they rely on expert knowledge and are limited by path quality.

GNN-Based Methods [26,28,30] utilize the message-passing mechanism in graph to aggregate high-order attribute information into item representation for enhanced and explainable recommendation. KGAT [28] regards user-item interaction as a new relation added to KG and then employs attentive mechanism to propagate attribute information. KGIN [30] learn user interest via an attentive combination of attributes and integrates relational information from multi-hop paths to refine the representations. However, they all ignore the semantic pollution problem and the weight decrease of significant high-order nodes.

6   Conclusion

In this paper, we study the attribute information in knowledge graphs intending to improve the recommendation performance. On the item side, the proposed AKGAN can learn more high-quality item representation by remaining the independency of attributes and maintaining the weight of high-order significant attribute nodes. On the user side, AKGAN mines user interests towards attributes and provides a personal recommendation. Extensive experiments demonstrate the effectiveness and explainability of AKGAN.

References

1. Bordes, A., Usunier, N., Garcia-Duran, A., Weston, J., Yakhnenko, O.: Translating embeddings for modeling multi-relational data. In: NIPS, vol. 26 (2013)
2. Fu, X., Zhang, J., Meng, Z., King, I.: MAGNN: metapath aggregated graph neural network for heterogeneous graph embedding. In: WWW, pp. 2331–2341 (2020)

3. Glorot, X., Bengio, Y.: Understanding the difficulty of training deep feedforward neural networks. In: JMLR. JMLR Workshop and Conference Proceedings, pp. 249–256 (2010)

4. Guo, Q., et al.: A survey on knowledge graph-based recommender systems. TKDE **34**(8), 3549–3568 (2020)

5. He, X., Deng, K., Wang, X., Li, Y., Zhang, Y., Wang, M.: Lightgcn: simplifying and powering graph convolution network for recommendation. In: SIGIR, pp. 639–648 (2020)

6. He, X., Liao, L., Zhang, H., Nie, L., Hu, X., Chua, T.-S.: Neural collaborative filtering. In: WWW, pp. 173–182 (2017)

7. Hu, B., Shi, C., Zhao, W.X., Yu, P.S.: Leveraging meta-path based context for top-n recommendation with a neural co-attention model. In: KDD, pp. 1531–1540 (2018)

8. Fenyu, H., Zhu, Y., Shu, W., Huang, W., Wang, L., Tan, T.: Graphair: graph representation learning with neighborhood aggregation and interaction. Pattern Recogn. **112**(2021), 107745 (2021)

9. Huang, J., Zhao, W. X., Dou, H., Wen, J.-R., Chang, E.Y.: Improving sequential recommendation with knowledge-enhanced memory networks. In: SIGIR, pp. 505–514 (2018)

10. Ji, G., He, S., Xu, L., Liu, K., Zhao, J.: Knowledge graph embedding via dynamic mapping matrix. In: ACL, pp. 687–696 (2015)

11. Juan, Y., Zhuang, Y., Chin, W.-S., Lin, C.-J.: Field-aware factorization machines for CTR prediction. In: RecSys, pp. 43–50 (2016)

12. Kingma, D.P., Ba, J.: Adam: a method for stochastic optimization. arXiv preprint arXiv:1412.6980 (2014)

13. Koren, Y.: Factorization meets the neighborhood: a multifaceted collaborative filtering model. In: KDD, pp. 426–434 (2008)

14. Li, C., et al.: Multi-interest network with dynamic routing for recommendation at Tmall. In: ICKM, pp. 2615–2623 (2019)

15. Lin, Y., Liu, Z., Sun, M., Liu, Y., Zhu, X.: Learning entity and relation embeddings for knowledge graph completion. In: AAAI (2015)

16. Luo, C., Pang, W., Wang, Z., Lin, C.: HETE-CF: social-based collaborative filtering recommendation using heterogeneous relations. In: ICDM, pp. 917–922. IEEE (2014)

17. Ma, H., Yang, H., Lyu, M.R., King, I.: Sorec: social recommendation using probabilistic matrix factorization. In: CIKM, pp. 931–940 (2008)

18. Niu, X., et al.: A dual heterogeneous graph attention network to improve long-tail performance for shop search in e-commerce. In: KDD, pp. 3405–3415 (2020)

19. Qin, J., Zhang, W., Wu, X., Jin, J., Fang, Y., Yu, Y.: User behavior retrieval for click-through rate prediction. In: SIGIR, pp. 2347–2356 (2020)

20. Rendle, S., Freudenthaler, C., Gantner, Z., Schmidt-Thieme, L.: BPR: Bayesian personalized ranking from implicit feedback. arXiv preprint arXiv:1205.2618 (2012)

21. Rendle, S., Gantner, Z., Freudenthaler, C., Schmidt-Thieme, L.: Fast context-aware recommendations with factorization machines. In: Proceedings of the 34th International ACM SIGIR Conference on Research and Development in Information Retrieval, pp. 635–644 (2011)

22. Shi, C., Zhang, Z., Luo, P., Yu, P.S., Yue, Y., Wu, B.: Semantic path based personalized recommendation on weighted heterogeneous information networks. In: CIKM, pp. 453–462 (2015)

23. Sun, Z., Yang, J., Zhang, J., Bozzon, A., Huang, L.-K., Xu, C.: Recurrent knowledge graph embedding for effective recommendation. In: RecSys, pp. 297–305 (2018)
24. Wang, H., et al.: Ripplenet: propagating user preferences on the knowledge graph for recommender systems. In: ICKM, pp. 417–426 (2018)
25. Wang, H., Zhang, F., Xie, X., Guo, M.: DKN: deep knowledge-aware network for news recommendation. In: WWW, pp. 1835–1844 (2018)
26. Wang, H., et al.: Knowledge-aware graph neural networks with label smoothness regularization for recommender systems. In: KDD, pp. 968–977 (2019)
27. Wang, J., Huang, P., Zhao, H., Zhang, Z., Zhao, B., Lee, D.L.: Billion-scale commodity embedding for e-commerce recommendation in Alibaba. In: KDD, pp. 839–848 (2018)
28. Wang, X., He, X., Cao, Y., Liu, M., Chua, T.-S.: KGAT: knowledge graph attention network for recommendation. In: KDD, pp. 950–958 (2019)
29. Wang, X., He, X., Wang, M., Feng, F., Chua, T.-S.: Neural graph collaborative filtering. In: SIGIR, pp. 165–174 (2019)
30. Wang, X., et al.: Learning intents behind interactions with knowledge graph for recommendation. In: WWW, pp. 878–887 (2021)
31. Wang, Z., Li, J., Liu, Z., Tang, J.: Text-enhanced representation learning for knowledge graph. In: IJCAI, pp. 4–17 (2016)
32. Wu, L., Sun, P., Fu, Y., Hong, R., Wang, X., Wang, M.: A neural influence diffusion model for social recommendation. In: SIGIR, pp. 235–244 (2019)
33. Ying, R., He, R., Chen, K., Eksombatchai, P., Hamilton, W.L., Leskovec, J.: Graph convolutional neural networks for web-scale recommender systems. In: KDD, pp. 974–983 (2018)
34. Zhang, F., Yuan, N.J., Lian, D., Xie, X., Ma, W.-Y.: Collaborative knowledge base embedding for recommender systems. In: KDD, pp. 353–362 (2016)
35. Zhang, Y., Ai, Q., Chen, X., Wang, P.: Learning over knowledge-base embeddings for recommendation. arXiv preprint arXiv:1803.06540 (2018)
36. Zhang, Y., Cheng, D.Z., Yao, T., Yi, X., Hong, L., Chi, E.H.: A model of two tales: dual transfer learning framework for improved long-tail item recommendation. In: WWW, pp. 2220–2231 (2021)
37. Zhou, G., et al.: Deep interest evolution network for click-through rate prediction. In: AAAI, vol. 33, pp. 5941–5948 (2019)
38. Zhou, G., et al.: Deep interest network for click-through rate prediction. In: KDD, pp. 1059–1068 (2018)
39. Yang, Y., Zhu, Y., Li, Y.: Personalized recommendation with knowledge graph via dual-autoencoder. Appl. Intell. 1–12 (2022)

Multi-view Stereo by Fusing Monocular and a Combination of Depth Representation Methods

Fanqi Yu[1]([✉]) and Xinyang Sun[2]

[1] School of Electronic and Computer Engineering, Peking University, Beijing, China
fqyu@stu.pku.edu.cn
[2] School of Software and Microelectronics, Peking University, Beijing, China
xinyangsun@stu.pku.edu.cn

Abstract. The design of plane-sweep deep MVS primarily relies on patch-similarity based matching. However, this approach becomes impractical when dealing with low-textured, similar-textured and reflective regions in the scene, resulting in inaccurate matching results. One of the methods to avoid this kind of error is incorporating semantic information in matching process. In this paper, we propose an end-to-end method that uses monocular depth estimation to add semantic information to deep MVS. Additionally, we analyze the advantages and disadvantages of two main depth representations and propose a collaborative method to alleviate their drawbacks. Finally, we introduce a novel filtering criterion named Distribution Consistency, which can effectively filter out outliers with poor probability distribution, such as uniform distribution, to further enhance the reconstruction quality.

Keywords: Multi-view Stereo · monocular depth estimation · depth representations · Distribution Consistency

1 Introduction

Although learning-based MVS has been continuously developed in recent years, most plane-sweep deep MVS design adopts the principle of matching based on patch similarity. However, in areas with weak texture, the similarity between the features of different patches is very close, resulting in difficulties to obtain the correct depth. One of the ideal solutions is to add semantic information to learning-based MVS. But in current MVS datasets, there is no ground truth about semantic information. Even with pre-trained semantic segmentation networks, producing precise semantic labels is almost unattainable, and this inevitably affects the answers that were already accurate enough.

In this paper, we first suggest a fresh scheme that substitutes semantic information with monocular depth estimation for learning-based MVS. There are several main reasons accounting for our idea. To begin with, different from geometric cues, semantic information is regarded as pattern cues, which implies that it produces outcomes based on the mode of single-view features. As Fig. 1 shows, features from EfficientNet-based monocular depth estimation are smoother and more comprehensive, which obviously differentiate between different objects. Besides, our proposed pipeline is end-to-end, which is different from most two-stage methods that integrate monocular and multi-view depth estimation.

In addition, in current learning-based MVS, the mainstream depth supervision methods are often based on probability or integrated representations like UniMVSNet [1]. It is because the common practice is to find a way to make the probability distribution of each depth hypothesis more reasonable. However, simple classification and regression have their own advantages. Classification is more robust by directly constraining the cost volume, but it cannot estimate the continuous depth due to its discrete predictions. Regression can achieve sub-pixel estimation but tends to overfit due to its indirect learning cost volume. After experimentation, we propose a combined regression and classification method to compensate for their respective shortcomings, and it can achieve performance similar to that of depth representation based on UniMVSNet form.

Finally, to fuse the depth maps of all views into a consistent point cloud, we propose a novel filtering criterion called Distribution Consistency. Although existing photometric and geometric consistencies can filter out enough outliers, the distribution of the estimated probability among the hypothesis planes is also an important indicator of pixel reliability. Due to the matching essence of MVS, the closer the hypothesis plane to the ground-truth depth, the smaller the matching cost and the greater the probability. Therefore, those points, whose probability distribution tends to be unimodal, have higher confidence. And those points that are biased towards a uniform distribution should be discarded.

2   Related Work

2.1   Multi-view Depth Estimation Based on Monocular Assistance

The output of monocular depth estimation is smoother and more complete, but the accuracy of the prediction result is relatively lower than multi-view depth estimation. Many methods combined monocular and multi-view depth estimation to get a better result. Bae et al. [2] employed monocular depth estimation to initialize a MVS, and utilized an iterative approach for its optimization without a fixed number of iterations. Although effective, this is not a one-stage training approach. Wang [3] and Zhang [4] both employ monocular features to supervise multi-view features. However, in an end-to-end fashion, they exclusively rely on monocular features for supervising multi-view features, resulting in limited utilization of monocular information in the context of MVS.

2.2   Regression and Classification MVS Methods

As aforementioned, regression methods [5–7] is difficult to achieve perfect result due to the indirect learning cost volume and its one-to-many relationship between the depth and the weight combination of the hypothesis planes. To solve this, Zhang et al. [8] apply an extra cross-entropy constraint to the cost volume in the stereo matching field. In this paper, we only apply the regression in the finest stage with a quite narrow depth range to minimize the risk of overfitting.

Classification methods [9,10] can not accurately predict the continuous depth due to its discrete predictions. To maximize the performance of the model, Yao et al. [10] sample denser hypothesis planes and refine the depth through post-processing, and Yan et al. [9] and Wei et al. [11] take benefits from the more complex network. Nevertheless, none of these methods can essentially avoid the shortcomings of classification methods.

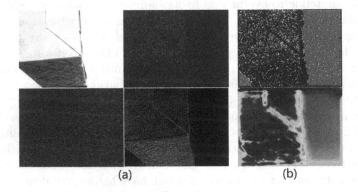

(a)   (b)

Fig. 1. Feature maps from the multi-view depth estimation network CasMVSNet [7] in part a and monocular depth estimation network EfficientNet [12] in part b. In part a, the images in the top-right, bottom-left, and bottom-right of the figure respectively come from the first-stage feature, second-stage feature, and third-stage feature extracted from the top-left image. In part b, these two features are from the encoder.

3   Methods

3.1   Inherited MVS Pipeline

The design of our backbone network mainly refers to deep MVS based on pyramid structure [5,7]. Supposing N input images with size $H \times W$ are denoted as $\{I_i | i = 1 \cdots N\}$, while I_1 refers to the reference image and the others represent the source image. The extracted multi-scale features through our weight-sharing feature extractor are denoted as $\{f_i^j \in R^{\frac{C}{2^{j-1}} \times \frac{H}{2^{L-j}} \times \frac{W}{2^{L-j}}} | i = 1 \cdots N, j = 1 \cdots L\}$, where L refers to the total number of stages and C is the maximum number of feature channels. We extract the multi-scale feature through a lightweight Feature Pyramid Network (FPN) as [7]. After having the multiple features at a certain stage, we warp all features into the fronto-parallel plane of reference view to construct the feature volumes utilizing the differentiable homography. Concretely, the homography between the i_{th} source view and the reference view at depth d is:

$$H_i(d) = dK_iT_iT_1^{-1}K_1^{-1} \tag{1}$$

where K and T denote as camera intrinsics and extrinsics respectively.

We finally apply the variance-based method and a 3D U-Net to generate and regularize the cost volume and don't share weights among different stages. The regularized volume $Y \in \mathbb{R}^{M \times H_1 \times W_1}$ through the 3D U-Net is then delivered to a *softmax* operator to predict depth probability volume $P \in \mathbb{R}^{M \times H_1 \times W_1}$.

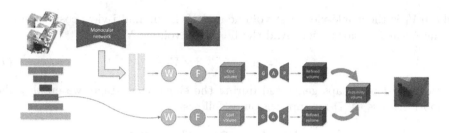

Fig. 2. The design of single-view and multi-view parts in the first stage of our proposed network, where W, F and GAP represent warping, fusing and global average pooling respectively.

The process of obtaining the final depth map from the probability volume will be implemented using our proposed combined regression and classification method.

3.2 MVS Assisted by Monocular Depth Estimation

The monocular-assisted MVS consists of two parts, the single-view part and the multi-view part. Single-view and multi-view networks are fused in the first stage as Fig. 2 shows.

Considering the limited GPU memory, we designe a encoder-decoder structure as the primary framework for the single-view component. The encoder is the same as FPN and decoder is implemented using deconvolution, mirroring the encoder's structure. This could enable the single-view part to produce features with the same receptive field as multi-view part, thereby facilitating feature correspondence between the single-view and multi-view parts. Specifically, in the first stage, we feed the images from the reference view that are one-quarter the size of the original into the monocular depth network to obtain all the monocular features and depth maps of the single-view stage. These features are then upsampled to the same size and concatenated into the output of the single-view stage, which enables us to obtain all different information. Subsequently, the output features undergoes the process of cost volume construction similar to the multi-view stage, resulting in a single-view cost volume. Once we have obtained the cost volumes from both the single-view and multi-view stages, we adaptively aggregate them into a multi-view cost volume using global average pooling [13]. The specific formula is as follows:

$$\alpha = Sigmoid(f_2(Relu(f_1(\frac{1}{H \times W} \cdot (\sum_{j=1}^{W}\sum_{i=1}^{H} V_{1_{i,j}}))))) \tag{2}$$

$$\beta = Sigmoid(f_2(Relu(f_1(\frac{1}{H \times W} \cdot (\sum_{j=1}^{W}\sum_{i=1}^{H} V_{2_{i,j}}))))) \tag{3}$$

where V_1 is the single-view cost volume and V_2 is the multi-view cost volume. f is the linear transformation. And the final cost volume $\mathbf{V}$ is:

$$V = \alpha \times V_1 + \beta \times V_2 \tag{4}$$

As for the depth maps generated during the single-view stage, we employ the Smooth L1 loss as [7] for supervision as follows:

$$ML(D_{gt}, D_{mono}) = \begin{cases} 0.5(D_{gt} - D_{mono})^2, & if |D_{gt} - D_{mono}| < 1 \\ |D_{gt} - D_{mono}| - 0.5, & else \end{cases} \tag{5}$$

where D_{mono} is the depth map from the single-view stage. And take the average value of all pixels as the final loss in single-view pipeline:

$$MML = \frac{1}{H_1 \times W_1} \sum_{(x,y)=(1,1)}^{(H_1,W_1)} ML(D_{gt}, D_{mono})^{x,y} \tag{6}$$

By supervising the depth maps generated in the monocular stage, we can maintain the smoothness of the features in the single-view stage while also adjusting them in the direction that leads to more accurate results based on the supervision from the multi-view stage.

3.3 Collaboration of Regression and Classification

We have introduced the pipeline from images to probability volumes in Sec.A. Here, we introduce the stage from probability volumes to depth maps with a collaborative method.

We first apply classification to estimate the sub-interval in the first several coarse stages and we don't care about its accuracy but only take to narrow the depth range. In the last fine stage, we take regression to estimate the continuous depth from the classified interval. In this way, we can reduce the risk of overfitting through the direct learning robust cost volume in coarse stages and can still achieve sub-pixel estimation through the last regression stage (Fig. 3).

The classification pipeline here is used to infer the depth interval that contains the ground-truth depth D_{gt}. Given the depth range $(d_{min} \sim d_{max})$ of a

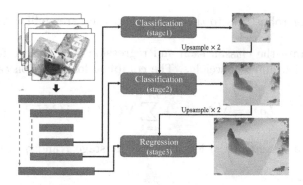

Fig. 3. Framework of EfficientMVS. We implement EfficientMVS in 3 stages and apply classification in the first two coarse stages to narrow the depth range of the final regression stage.

certain stage, we divide it evenly into M intervals of length $r = \frac{(d_{max} - d_{min})}{M}$. The construction of i_{th} interval at pixel (x, y) can be modeled as:

$$Q_i^{x,y} = \begin{cases} 1, & \text{if } d_i^{x,y} - \frac{r^{x,y}}{2} \le D_{gt}^{x,y} \text{ and } d_i^{x,y} + \frac{r^{x,y}}{2} > D_{gt}^{x,y} \\ 0, & \text{else} \end{cases} \tag{7}$$

where P is the estimated probability volume and $Q \in \{0,1\}^{M \times H_1 \times W_1}$ is the one-hot ground-truth volume constructed from the ground-truth depth, which assigns the interval containing ground-truth depth to 1 and the rest to 0.

To avoid the model become too confident to generalize to other scenarios, we adopt label smoothing to mitigate this drawback, which is a cost-effective solution through introducing noise into one-hot distribution just as:

$$\widehat{Q} = (1 - \epsilon) \times Q + \epsilon / M \tag{8}$$

where ϵ is a tunable factor and is set to 0.1 in this paper. Finally, the direct supervisory signal on cost volume can be generated through the *cross-entropy* loss:

$$CE(\widehat{Q}, P) = \frac{1}{H_1 \times W_1} \sum_{(x,y)=(1,1)}^{(H_1, W_1)} \sum_{i=1}^{M} Q_i^{x,y} \log(P_i^{x,y}) \tag{9}$$

Here, we suppose all pixels are valid for simplicity.

Following the traditional regression methods [5,7], the final depth map $\mathbf{D}$ is obtained through a *soft-argmax* operator as:

$$D = \sum_{d=d_1}^{d_M} dP(d) \tag{10}$$

Different from the classification method, the optimization here is executed on the final regressed depth map through the *Smooth L1* loss (SL) and its average value (MSL) for robustness to the outlier, and they are similar with Eq. 5 and Eq. 6.

To collaborate the classification and regression strategy, we fuse the stage loss function with different weights. The complete loss function can be defined as:

$$Loss = \lambda^L MSL^L + \sum_{i=1}^{L-1} \lambda^i CE^i + \gamma MML \qquad (11)$$

where λ^i is the weight of the loss function at stage i and γ is the weight of monocular loss.

3.4   Distribution Consistency

Filtering out unreliable matching points is a necessary step before depth map fusion. All previous methods [5,7,10] rely on two existing criteria: photometric consistency and geometric consistency.

In this paper, we propose a novel filtering criterion, that we call **Distribution Consistency** (DC). The motivation comes from our aforementioned matching essence of MVS. The solution to MVS is essentially matching on different hypothesis planes. Theoretically, the closer to the ground-truth depth, the greater the probability of the hypothesis plane, which ultimately generates a probability volume with a unimodal distribution, especially for our classification stages. Therefore, we can discard those pixels whose distribution is terrible, e.g., uniform distribution. Concretely, we measure our distribution consistency through the *entropy*. Supposing the probability distribution of all M hypothesis planes at a certain pixel (x, y) is $p = \mathbf{P}^{x,y}$, the distribution consistency at this pixel is defined as:

$$C_{dis} = 1 + \frac{1}{\log(M)} \sum_{i=1}^{M} p_i \log p_i \qquad (12)$$

The distribution consistency is $C_{dis} = 0$ for those pixels whose probability abides by uniform probability, and $C_{dis} = 1$ for one-hot distribution, which is the ideal case for a unimodal distribution. Finally, we filter out those pixels with $C_{dis} < \theta$.

4   Implementation and Result

4.1   Implementation

As for training on DTU, the number of even interval or hypothesis plane from the coarsest to the finest stage is set to $M^1 = 48$, $M^2 = 32$ and $M^3 = 8$, and the depth range of each stage is set to 1, 1/3 and 1/24 times as that of [5]. We adopt Adam optimizer [14] with the initial learning rate of 0.001 and decayed by 2 at 10_{th}, 12_{th}, 14_{th} and 16_{th} epoch to optimize our model for 17 epochs. As for testing on Tanks and Temples [15], we first finetune the model on BlendedMVS [16]. The number of input images is set to 7. Other settings are similar as [1].

As for filtering and fusion, we don't consider confidence as [17] for DTU depth map but we consider it for Tanks and Temples'. We apply our proposed distribution consistency to filter the Tanks and Temples depth map and set θ to 0.005, 0.05 and 0.6 from the coarsest to the finest stage. For depth map fusion, visibility-based depth map fusion [18] and mean average fusion [5] are adopted.

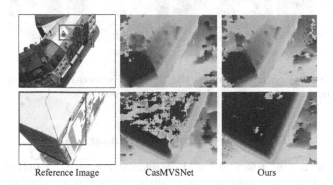

Reference Image   CasMVSNet   Ours

Fig. 4. Depth comparison with [7] **of several scenes on DTU.** Our model produces more accurate and sharper results.

4.2   Results on DTU Benchmark

We validate the effectiveness of our proposed methods on the DTU dataset here. We first compare the depth map with our baseline [7]. As shown in Fig. 4 and 5, the boundaries of different objects in our depth map are sharper and our model is more robust on the challenging areas, e.g., low-texture regions and reflections. The qualitative results of the 3D point cloud reconstruction shown in Table 1 further verify the advantages of our model. Besides, the comparison between our collaboration methods and the Unified Focal Loss [1] shows that the combination of classification and regression can generate similar results compared with the modern depth representation methods.

Table 1. Quantitative results on DTU evaluation set. BC represents the collaboration of classification and regression. Mono and UFL mean fusing monocular and Unified Focal Loss respectively.

Method	ACC.(mm)	Comp.(mm)	Overall(mm)
Baseline	0.300	0.360	0.330
Baseline + BC	0.359	0.276	0.318
Baseline + BC + Mono	0.353	**0.271**	**0.312**
Baseline + UFL	0.351	0.285	0.318

R-MVSNet   CasMVSNet   Ours   GT

Fig. 5. Qualitative comparisons with [7,10] **of several scenes on DTU.** EfficientMVS generates the most complete 3D point cloud.

4.3   Results on Tanks and Temples Benchmark

As our baseline mainly refers to CasMVSNet, we compare our performance with MVS models that were contemporary with CasMVSNet. Similar to existing methods [9,19], we evaluate our generalization ability on Tanks and Temples benchmark using the model finetuned on the BlendedMVS dataset. As reported in Table 2, our method yields a mean F-score 13% higher than UCS-Net [20] and 9.7% higher than CasMVSNet. Some qualitative results are shown in Fig. 6.

Fig. 6. Qualitative results on Tanks and Temples dataset.

4.4   Ablation Studies

1. Fusing Methods. We attempted various schemes for fusing single-view stage and multi-view stage features, including element-wise addition(EA) or multiplication(EM) of single-view and multi-view features, element-wise adaptive addition of the single-view stage and multi-view stage cost volumes using mean(EAM) or variance(EAV), and global adaptive addition of the single-view stage and multi-view stage cost volumes using mean(OAM), variance(OAV), or both(OAB) as coefficients. The results are shown in Table 3, and it can be observed that the best performance is achieved when the mean coefficient is used for global adaptive addition of the single-view and multi-view stage loss terms.

Table 2. Quantitative results of F-score on Tanks and Temples benchmark. Our model outperforms most published MVS methods and is a 9.8% improvement compared to our baseline [7]. DC refers to distribution consistency.

	Method	Mean	Family	Francis	Horse	Lighthouse	M60	Panther	Playground	Train
Others	ACMP [21]	58.41	70.30	54.06	**54.11**	61.65	54.16	57.60	58.12	**57.25**
	AttMVSNet [22]	60.05	73.90	**62.58**	44.08	64.88	56.08	59.39	**63.42**	56.06
	AA-RMVSNet [9]	**61.51**	**77.77**	59.53	51.53	64.02	**64.05**	**59.47**	60.85	55.50
Coarse-to-fine	PatchmatchNet [23]	53.15	66.99	52.64	43.24	54.87	52.87	49.54	54.21	50.81
	UCS-Net [20]	54.83	76.09	53.16	43.03	54.00	55.60	51.49	57.38	47.89
	CVP-MVSNet [6]	54.03	76.50	47.74	36.34	55.12	57.28	54.28	57.43	47.54
	CasMVSNet [7]	56.84	76.37	58.45	46.26	55.81	56.11	54.06	58.18	49.51
	VisMVSNet [19]	60.03	77.40	60.23	47.07	**63.44**	62.21	57.28	**60.54**	52.07
	Ours w/o DC	60.93	78.74	64.39	53.19	59.78	60.18	59.45	57.98	53.73
	Ours	**62.39**	**79.70**	**64.63**	**54.28**	60.79	**64.01**	**62.01**	59.53	**54.20**

Table 3. Ablation studies on methods of combinations of single-view stage and multi-view stage.

Method	ACC.(mm)	Comp.(mm)	Overall(mm)
EM	0.362	0.288	0.325
EA	0.355	0.285	0.320
OAV	0.357	0.279	0.320
EAV	0.352	0.282	0.317
EAM	0.356	0.276	0.316
OAB	0.359	0.273	0.316
OAM	0.355	0.269	0.312

Table 4. Ablation studies on stage pipeline. "Cla" refers to classification and "Reg" refers to regression.

Stage1	Stage2	Stage3	ACC.(mm)	Comp.(mm)	Overall(mm)
Cla	Cla	Cla	0.424	**0.284**	0.354
Reg	Cla	Cla	0.417	0.297	0.357
Reg	Reg	Cla	0.671	0.461	0.566
Reg	Reg	Reg	0.329	0.371	0.350
Cla	Reg	Reg	**0.325**	0.353	0.339
Cla	Cla	Reg	**0.325**	0.348	**0.336**

2. Collaboration is the Best. To fairly prove collaboration is the best, we adopt the same configuration as [7] for those methods whose final stage is regression, and as [10] for those methods whose final stage is classification. The ablation results in Table 4 verify that appropriate classification and regression collaboration can yield huge performance benefits.

3. Distribution Consistency. As shown in the last two experiments in Table 2, our proposed distribution consistency criterion can promote the reconstruction performance of all scenes significantly and it greatly improves the mean F-score from 60.93 to 62.39.

5   Conclusion

In this paper, we present an efficient collaboration model, which not only fuses monocular networks but also encodes the advantages of classification and regression strategies into a single model. Moreover, we apply label smoothing to the MVS field to improve the generalization of classification pipelines. Meanwhile, we propose a novel filtering criterion called distribution consistency to discard those pixels with a bad probability distribution and improve the reconstruction quality of all scenes. In the future, we will investigate more uniform and powerful strategies to MVS.

References

1. Peng, R., Wang, R., Wang, Z., Lai, Y., Wang, R.: Rethinking depth estimation for multi-view stereo: a unified representation. In: Proceedings of the IEEE/CVF Conference on Computer Vision and Pattern Recognition, pp. 8645–8654 (2022)
2. Bae, G., Budvytis, I., Cipolla, R.: Multi-view depth estimation by fusing single-view depth probability with multi-view geometry. In: Proceedings of the IEEE/CVF Conference on Computer Vision and Pattern Recognition, pp. 2842–2851 (2022)
3. Wang, X., et al.: MVSTER: epipolar transformer for efficient multi-view stereo. In: Avidan, S., Brostow, G., Cissé, M., Farinella, G.M., Hassner, T. (eds.) ECCV 2022. LNCS, pp. 573–591. Springer, Cham (2022). https://doi.org/10.1007/978-3-031-19821-2_33
4. Zhang, C., Meng, G., Bing, S., Xiang, S., Pan, C.: Monocular contextual constraint for stereo matching with adaptive weights assignment. Image Vis. Comput. **121**, 104424 (2022)
5. Yao, Y., Luo, Z., Li, S., Fang, T., Quan, L.: Mvsnet: depth inference for unstructured multi-view stereo. In: Proceedings of the European Conference on Computer Vision (ECCV), pp. 767–783 (2018)
6. Yang, J., Mao, W., Alvarez, J.M., Liu, M.: Cost volume pyramid based depth inference for multi-view stereo. In: Proceedings of the IEEE/CVF Conference on Computer Vision and Pattern Recognition, pp. 4877–4886 (2020)
7. Gu, X., Fan, Z., Zhu, S., Dai, Z., Tan, F., Tan, P.: Cascade cost volume for high-resolution multi-view stereo and stereo matching. In: Proceedings of the IEEE/CVF Conference on Computer Vision and Pattern Recognition, pp. 2495–2504 (2020)
8. Zhang, Y., et al.: Adaptive unimodal cost volume filtering for deep stereo matching. In: Proceedings of the AAAI Conference on Artificial Intelligence, vol. 34, pp. 12926–12934 (2020)
9. Wei, Z., Zhu, Q., Min, C., Chen, Y., Wang, G.: Aa-rmvsnet: adaptive aggregation recurrent multi-view stereo network. In: Proceedings of the IEEE/CVF International Conference on Computer Vision, pp. 6187–6196 (2021)

10. Yao, Y., Luo, Z., Li, S., Shen, T., Fang, T., Quan, L.: Recurrent mvsnet for high-resolution multi-view stereo depth inference. In: Proceedings of the IEEE/CVF Conference on Computer Vision and Pattern Recognition, pp. 5525–5534 (2019)
11. Yan, J., et al.: Dense hybrid recurrent multi-view stereo net with dynamic consistency checking. In: Vedaldi, A., Bischof, H., Brox, T., Frahm, J.-M. (eds.) ECCV 2020. LNCS, vol. 12349, pp. 674–689. Springer, Cham (2020). https://doi.org/10.1007/978-3-030-58548-8_39
12. Tan, M., Le, Q.: Efficientnet: rethinking model scaling for convolutional neural networks. In: International Conference on Machine Learning, pp. 6105–6114. PMLR (2019)
13. Lin, M., Chen, Q., Yan, S.: Network in network. arXiv preprint arXiv:1312.4400 (2013)
14. Kingma, D.P., Ba, J.: Adam: a method for stochastic optimization. arXiv preprint arXiv:1412.6980 (2014)
15. Knapitsch, A., Park, J., Zhou, Q.-Y., Koltun, V.: Tanks and temples: benchmarking large-scale scene reconstruction. ACM Trans. Graph. (ToG) 36(4), 1–13 (2017)
16. Yao, Y., et al.: Blendedmvs: a large-scale dataset for generalized multi-view stereo networks. In: Proceedings of the IEEE/CVF Conference on Computer Vision and Pattern Recognition, pp. 1790–1799 (2020)
17. Yu, F., Pang, J., Wang, R.: Sub-pixel convolution and edge detection for multi-view stereo. In: 2022 IEEE 8th International Conference on Computer and Communications (ICCC), pp. 1864–1868. IEEE (2022)
18. Merrell, P., et al.: Real-time visibility-based fusion of depth maps. In: 2007 IEEE 11th International Conference on Computer Vision, pp. 1–8. IEEE (2007)
19. Zhang, J., Yao, Y., Li, S., Luo, Z., Fang, T.: Visibility-aware multi-view stereo network. arXiv preprint arXiv:2008.07928 (2020)
20. Cheng, S., et al.: Deep stereo using adaptive thin volume representation with uncertainty awareness. In: Proceedings of the IEEE/CVF Conference on Computer Vision and Pattern Recognition, pp. 2524–2534 (2020)
21. Xu, Q., Tao, W.: Planar prior assisted patchmatch multi-view stereo. In: Proceedings of the AAAI Conference on Artificial Intelligence, vol. 34, pp. 12516–12523 (2020)
22. Luo, K., Guan, T., Ju, L., Wang, Y., Chen, Z., Luo, Y.: Attention-aware multi-view stereo. In: Proceedings of the IEEE/CVF Conference on Computer Vision and Pattern Recognition, pp. 1590–1599 (2020)
23. Wang, F., Galliani, S., Vogel, C., Speciale, P., Pollefeys, M.: Patchmatchnet: learned multi-view patchmatch stereo. In: Proceedings of the IEEE/CVF Conference on Computer Vision and Pattern Recognition, pp. 14194–14203 (2021)

A Fast and Scalable Frame-Recurrent Video Super-Resolution Framework

Kaixuan Hou(ID) and Jianping Luo(✉)(ID)

Guangdong Key Laboratory of Intelligent Information Processing, Shenzhen Key Laboratory of Media Security and Guangdong Laboratory of Artificial Intelligence and Digital Economy (SZ), Shenzhen University, Shenzhen, China
houkaixuan2021@email.szu.edu.cn, ljp@szu.edu.cn

Abstract. The video super-resolution(VSR) methods based on deep learning have become the mainstream VSR methods and have been widely used in various fields. Although many deep learning-based VSR methods have been proposed, they cannot be applied to real-time VSR tasks due to the vast computation and memory occupation. The lightweight VSR networks have faster inference speeds, but their super-resolution performance could be better. In this paper, we analyze the explicit and implicit motion compensation methods commonly used in VSR networks and design a fast and scalable frame-recurrent VSR network(FFRVSR). FFRVSR incorporates the Frame-Recurrent Network and Recurrent-Residual Network. This network structure can extract information from low-resolution video frames more efficiently and alleviate error accumulation during inference. We also design a super-resolution flow estimation network(SRFnet) that can more accurately estimate optical flow between video frames while reducing error information ingress. Extensive experiments demonstrate that the proposed FFRVSR surpasses state-of-the-art methods in terms of inference speed. FFRVSR also has strong scalability and can be adapted for both real-time video super-resolution tasks and high-quality video super-resolution tasks.

Keywords: video super-resolution · deep learning · recurrent network · scalability

1 Introduction

Super-resolution is a classic topic in computer vision, which aims to recover single or multiple high-resolution(HR) images from low-resolution(LR) images. In recent years, with the widespread popularity of high-resolution display devices, the demand for super-resolution technology has been increasing, and there are more and more application scenarios. So far, super-resolution technology has been widely used in video monitoring [36], high-definition television [9], medical image super-resolution [22], face Identify [10]. In 2014, Dong et al. proposed the first image super-resolution network based on deep learning SRCNN [5]. SRCNN reveals the great potential of deep learning technology in super-resolution. Since then, super-resolution technology has developed rapidly, and now the super-resolution methods based on deep learning have become mainstream.

Video super-resolution(VSR) originated from single-image super-resolution. Compared with single-image super-resolution, video super-resolution must extract the spatiotemporal information between video frames. According to the method of extracting inter-frame information, VSR methods based on deep learning can be divided into two categories methods with explicit motion compensation and methods with implicit motion compensation [16,21]. VSR networks using explicit motion compensation methods [1,3,4,25,27,28] mostly use the optical flow estimation module to extract the optical flow between adjacent video frames, and then use the optical flow to align to the reference video frames. VSR networks using implicit motion compensation methods mostly use 2D CNN [11,29], 3D CNN [15,17] or Recurrent Neural Network(RNN) [8,16]. They directly feed multiple video frames into the VSR networks without alignment. Although VSR networks using explicit motion compensation methods are easier to converge, they require complex network structures for excellent VSR performance and cannot be used for real-time VSR tasks due to the vast computation and memory occupation. The VSR networks using the implicit motion compensation methods are challenging to converge during training because of their abstract network frameworks. They are not easy to achieve better VSR performance.

After comprehensively analyzing the advantages and disadvantages of VSR networks using explicit and implicit motion compensation methods, we try to fuse the two network structures. In this paper, we propose a fast and scalable frame-recurrent VSR network (FFRVSR), which outperforms existing VSR methods under the condition of the same inference speed or the same VSR performance, and can be adapted for real-time video super-resolution tasks and high-quality video super-resolution tasks (see Fig. 1). Our contributions are summarized as follows:

1. We propose a fast and scalable VSR network that improves VSR inference speed and VSR performance. Experiments show that the proposed network is highly competitive.

2. We design a new network structure that enables the model to utilize the spatiotemporal information between video frames more efficiently and can effectively alleviate error accumulation during inference.

3. We design a more efficient super-resolution optical flow estimation module, eliminating the error information generated by traditional interpolation methods during optical flow estimation.

2   Related Work

Super-resolution is considered a classic ill-posed inverse problem [25,35]. Super-resolution methods can be divided into traditional methods and deep learning-based super-resolution methods based on the development history of super-resolution. Traditional methods include interpolation methods, example-based super-resolution [6,7], self-similarity methods [33], etc. In 2014, Dong et al. proposed the first image super-resolution network based on deep learning SRCNN [5], and the performance of super-resolution surpassed the traditional methods. SRCNN proved that deep learning technology has great potential in super-resolution, and since then, deep learning-based super-resolution methods have developed rapidly.

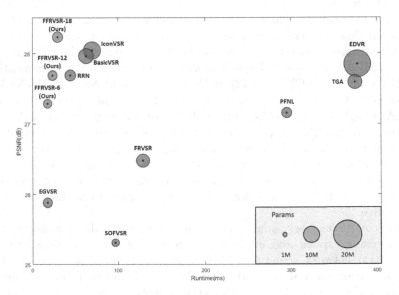

Fig. 1. Inference speed and performance comparison. 6, 12, and 18 are the number of residual blocks in the proposed framework. The runtime is calculated on an LR frame of size 320×180, and all models run on the same 2080Ti GPU. Comparisons of PSNR are performed on Vid4 dataset [20] for ×4 VSR.

Explicit Motion Compensation Methods. Since the successful application of SRCNN, the VSR methods based on deep learning have also developed rapidly. In 2016, Kappelerd et al. proposed the first deep learning-based VSR network [18], which uses the Druleas algorithm to calculate the optical flow information of inter-frames, and uses CNN for super-resolution. VESPCN [1] uses a spatial motion compensation transformer module (MCT) for motion estimation and compensation, followed by CNN and sub-pixel convolution for VSR. DRVSR [27] proposed a sub-pixel motion compensation module (SPMC) to align adjacent video frames and used ConvLSTM [26] for feature fusion and VSR. FRVSR [25] uses an optical flow estimation module (FNet) to perform optical flow estimation on adjacent LR video frames and adopts a frame-recurrent network framework, taking estimated previous SR video frame as input for subsequent iterations. SOFVSR [28] uses an optical flow reconstruction net(OFRnet) to extract the optical flow information between adjacent video frames from coarse to fine. EDVR [29] proposes the pyramid, cascading and deformable(PCD) alignment module and the temporal-spatial attention(TSA) fusion module to align video frames and fuse multiple frames, respectively. After that, more and more VSR networks use motion estimation and motion compensation methods, and the extraction of motion information between adjacent video frames becomes more and more sufficient. BasicVSR and IconVSR [3] uses the bidirectional recurrent network structure to extract optical flow information of forward and backward video frames, significantly improving the model's VSR performance. EGVSR [2] proposes an efficient and generic VSR system and uses the matrix multiplication (MatMul) algorithm to improve the inference speed of the model. ETDM [14] proposes a new information extraction scheme, which divides pixels into two subsets according to the level of temporal difference between frames, and extracts complementary information from them through two different branches. Explicit alignment methods require optical flow information of adjacent video frames. Still, even the most advanced optical flow estimation network cannot easily obtain sufficiently accurate optical flow information. There will be artifacts in the aligned frames, and these artifacts may also adversely affect the final super-resolution results.

Implicit Motion Compensation Methods. VSR networks using implicit motion compensation methods can overcome these problems. FFCVSR [32] directly inputs video frames into 2D CNN and achieves video super-resolution by reasonably combining the features information of the previous frame and the current frame. DUF [17] uses 3D CNN to generate dynamic upsampling filters and a residual image, uses upsampling filters to super-resolution the target frame, and adds the reconstruction result and residual image. RRN [16] inputs the reference video frame, the estimated previous SR video frame and the feature information of the estimated previous HR video frame into the recurrent residual network, and passes the estimated SR video frame and feature information to the following super-resolution process. PFNL [34] proposed a progressive fusion network and used a non-local residual block (NLRB) to capture the spatiotemporal correlation information between long-distance video frames. TGA [15] proposes temporal grouping and intra-group fusion methods to achieve implicit alignment.

The VSR models using explicit motion compensation methods [1,3,25,27,28] can quickly extract the optical flow between adjacent video frames. During the training, compared with the models using the implicit motion compensation methods [16,17,32,34], the models using the explicit motion compensation methods [1,3,25,27,28] converge faster. However, the VSR performance of the models using the explicit motion compensation methods is utterly dependent on the optical flow estimation modules. The better the performance of the optical flow estimation modules, the better the VSR performance of the models, but even the most advanced optical flow estimation modules cannot accurately estimate optical flow between adjacent video frames. The VSR models using the implicit motion compensation methods do not need to rely on the optical flow estimation modules, which can autonomously learn the relevant information in adjacent video frames and perform the implicit alignment. However, due to the abstract nature of the models, the models' training results using the implicit motion compensation methods are often unpredictable. On the other hand, the recurrent network [3,13,16,25] has a longer effective information transmission distance than the networks that conduct VSR within a local window [27,28,31]. Compared with bidirectional recurrent networks [3,13], unidirectional recurrent networks [16,25] have faster inference speeds and can be applied to real-time VSR tasks.

Based on the above research, our method combines the advantages of explicit motion compensation methods and implicit motion compensation methods. Related to [25], our method also uses the frame-recurrent network structure, but our method can effectively alleviate the inaccurate estimation of optical flow between adjacent video frames and the error accumulation during inference. In [25], they use the traditional interpolation method in the optical flow estimation network, and only use multiple 2D convolution layers for feature fusion and VSR, which will lead to the performance degradation of the optical flow estimation network and the error accumulation. In our work, we redesigned the optical flow estimation network, and used the residual recurrent network for error information filtering and VSR, which can improve the VSR performance and inference speed of our method.

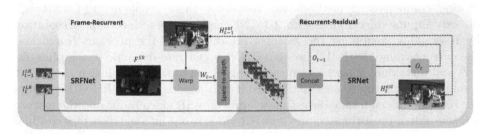

Fig. 2. The FFRVSR framework. It can be trained in an end-to-end manner. I_{t-1}^{LR} represents the previous LR(low-resolution) video frame, I_t^{LR} represents the current LR video frame, and F^{SR} represents the SR(super-resolution) optical flow map. H_{t-1}^{est} is the estimated previous SR video frame, W_{t-1} is the aligned SR video frame. O_t is the feature information of the current SRNet's output, O_{t-1} is the feature information of the previous SRNet's output.

3   Method

In video super-resolution, video can be considered an image sequence, the recurrent neural network can predict the current video frame more accurately based on the relevant information extracted from the previous video frames. Compared with the networks that conduct VSR within a local window, recurrent neural networks can more efficiently extract the spatiotemporal information between adjacent video frames and transmit it backward for greater distances. Moreover, recurrent neural networks often only need to input two adjacent video frames, which has a lower computational cost and faster inference speed. Therefore, recurrent neural networks have been widely used in video super-resolution, and various recurrent VSR networks have been proposed.

Inspired by [25], we designed a more efficient and scalable frame-recurrent network, called FFRVSR (see Fig. 2). In order to alleviate the error accumulation during inference, we used the Recurrent-Residual network in FFRVSR. FFRVSR consists of two parts: the Frame-Recurrent part and the Recurrent-Residual part.

3.1   Frame-Recurrent Part

In Frame-Recurrent part, the previous low-resolution(LR) frame and current LR frame are fed to our designed Super-Resolution Flow Net (SRFNet), SRFnet can extract the optical flow map between adjacent LR video frames and subsequently generate a more detailed super-resolution(SR) optical flow map of the same size as the high-resolution(HR) video frame. Then we use the SR optical flow map to warp the SR video frame estimated by the previous frame and obtain the aligned SR video frame. Finally, we split it into an LR draft cube using the space-to-depth transformation [25] (see Fig. 3).The process can be described as follows:

$$F^{SR} = SRFNet(I_{t-1}^{LR}, I_t^{LR}) \tag{1}$$

$$W_{t-1} = Warp(H_{t-1}^{est}, F^{SR}) \tag{2}$$

$$[W_{t-1}^{sH \times sW \times C}] \rightarrow [W_{t-1}^{H \times W \times s^2 C}] \tag{3}$$

where I_{t-1}^{LR} represents the previous LR video frame, I_t^{LR} represents the current LR video frame, and F^{SR} represents the super-resolution optical flow map. H_{t-1}^{est} is the estimated previous SR video frame, $Warp(\cdot)$ represents the warping operation. W_{t-1} is the aligned SR video frame, $W_{t-1}^{H \times W \times s^2 C}$ is the LR draft cube. $\rightarrow$ represents the space-to-depth transformation. s, H, W, C are the magnification factor, video frame's height, width and channel number.

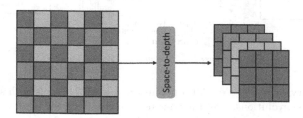

Fig. 3. Illustration of the space-to-depth transformation. The space-to-depth transformation split the aligned SR video frame into an LR draft cube.

3.2   Super-Resolution Flow Net (SRFNet)

We found that most researchers use interpolation methods in VSR networks [3,4,14,16,25,28,30,31], which are used for upsampling LR frames, LR optical flow, or feature maps. The interpolation methods use different formulas to calculate the value of the inserted point according to the values of surrounding points. However, these formulas cannot accurately describe the distribution of image pixels in real situations. Using interpolation methods will generate error information, which is why the image becomes blurry after upsampling using interpolation methods. Due to the cumulative effect in the recurrent networks, the interpolation methods will severely impact the performance of the VSR networks.

In order to solve the above problem, we avoid using interpolation methods in our VSR network to reduce the generation of error information. We design a more efficient super-resolution flow estimation network(SRFNet) inspired by the encoding-decoding structure of the recurrent network (see Fig. 4). SRFNet uses three encoder units and three decoder units. The encoder unit is composed of $\{Conv2d \rightarrow ReLU \rightarrow Conv2d \rightarrow ReLU \rightarrow MaxPool2X\}$, decoder unit is formed by $\{ConvTransposeX2 \rightarrow ReLU \rightarrow Conv2d \rightarrow ReLU\}$. Finally, the feature maps are transformed into an SR optical flow map the same size as the HR video frame through a 2D convolution layer and a sub-pixel convolution layer. SRFNet can autonomously learn the spatiotemporal information between adjacent video frames and output a more detailed super-resolution optical flow map, which will be verified in the experimental part.

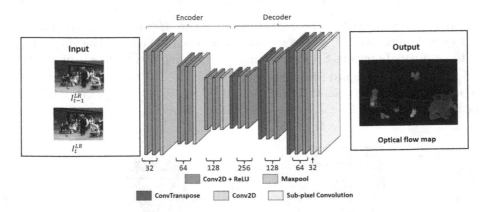

Fig. 4. The overview of SRFNet. All convolutions use 3×3 kernels with stride 1, except the transposed convolutions. The Encoder uses maxpooling for x2 downsampling, the Decoder uses the transposed convolutions with stride 2 for x2 upsampling.

3.3   Recurrent-Residual Part

In Recurrent-Residual part, Super-Resolution Net(SRNet) consists of residual blocks [12] (see Fig. 5) and each residual block contains {Conv2d → ReLU → Conv2d}. The current LR video frame, LR draft cube, and previous output O_{t-1} are fed to SRNet. In the final output part, one branch uses sub-pixel convolution to generate the SR video frame. The other branch directly outputs O_t after convolution and ReLU, which is fed to SRNet in the next cycle. Since the optical flow estimation network cannot accurately estimate the optical flow between adjacent video frames, the LR draft cube contains error information caused by inaccurate optical flow estimation. However, the optical flow estimation network does not affect the LR video frame and the previous output O_{t-1}. Therefore, we input the LR draft cube, LR video frame and O_{t-1} into SRNet, let SRNet learn independently, and filter out the error information contained in the LR draft cube so that the error information can be alleviated. The process can be described as follows:

$$\left(H_t^{est}, O_t\right) = SRNet\left(I_t^{LR} \copyright W_{t-1}^{H \times W \times s^2 C} \copyright O_{t-1}\right) \tag{4}$$

where $W_{t-1}^{H \times W \times s^2 C}$ represents the LR draft cube. $\copyright$ represents concatenate. H_t^{est} is the current SR video frame. O_t is the feature information of the current SRNet's output, O_{t-1} is the feature information of the previous SRNet's output.

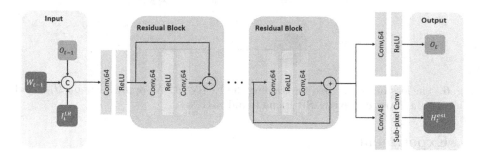

Fig. 5. An overview of SRNet, all convolutions use 3×3 kernels with stride 1, and the number of kernels of each convolution is marked. We feed the current LR video frame, LR draft cube, and previous output O_{t-1} to the SRNet. $\copyright$ represents the concatenation of images in the channel dimension. $\oplus$ indicates that the feature maps are added.

Our method can be extended by adjusting the number of residual blocks in SRNet. The residual block can fully transfer the extracted feature information backward, so increasing the number of residual blocks can improve the VSR performance of the model while the inference speed will decrease slightly. The inference speed of our model can also be improved by reducing the number of residual blocks. Experiment result shows that our method has good inference speed and VSR performance. Under the condition of the same inference speed, or the same VSR performance, our method outperforms the existing VSR methods (see Fig. 1).

3.4 Loss Functions

To train SRFNet and SRNet more efficiently, we use the Charbonnier loss function [19] in our model, and we use two loss terms to train our model (see Fig. 6). The first loss function is applied to the output of SRNet.

$$L_{sr} = \sqrt{\|H_t^{est} - H_t^{GT}\|^2 + \varepsilon^2} \tag{5}$$

The second loss function calculates the difference between the SR draft and the ground truth HR video frame.

$$L_f = \sqrt{\|W_{t-1} - H_t^{GT}\|^2 + \varepsilon^2} \tag{6}$$

where ε is a constant with value 10^{-6}. H_t^{est} is the estimated HR video frame. H_t^{GT} is the ground truth HR video frame. W_{t-1} is the SR draft. The total loss used for training is $L = L_{sr} + L_f$.

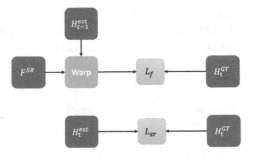

Fig. 6. The loss functions used for training the FFRVSR. We apply a loss on H_t^{est} and a loss on the warped previous SR frame to aid SRFNet.

4 Experiments

4.1 Datasets

In our work, we adopt REDS [23] as our training dataset. REDS is a public dataset for video restoration tasks. It consists of 240 video clips with various motion types. Each video clip contains 100 frames, and the resolution of each frame is 720×1280. During training, we randomly sample patches of size 128×128 from the HR video sequences as target patches and apply Gaussian blur with $\sigma = 1.6$ to the target patches, then perform ×4 downsampling to generate the corresponding LR patches with the size 32×32. We evaluate our model on the Vid4 [20] and assess the SR results with terms of PSNR and SSIM on the Y channel of YCbCr space. For a fair comparison of these VSR networks, PSNR and SSIM are measured excluding the first frame. We also assess the parameters and runtime of each VSR network and use them as two important references.

4.2 Implementation Details

During training, we use the Adam optimizer to optimize the model and set $\beta_1 = 0.9$ and $\beta_2 = 0.999$. The initial learning rate is 5×10^{-5}. The batch size is 8, and each batch contains 10 samples. The total number of epochs is 534. The training data is augmented with random standard flip operations during training. In recent years, the matrix multiplication (MatMul) algorithm has been widely used in convolutional neural networks(CNN) [2,24]. It has been proven that it can save inference time and improve computational efficiency through memory space. Therefore, after training, we use the matrix multiplication algorithm to optimize the model to improve the model's computational efficiency further and reduce the model's inference time.

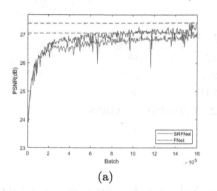
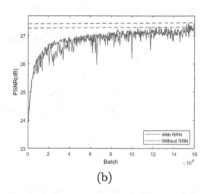

(a)   (b)

Fig. 7. The evaluation of PSNR on the Vid4 [20]. (a) The results of the methods using SRFNet and without SRFNet. (b) The results of methods using RRN and without RRN.

4.3 Ablation Experiment

In this section, we examine the effectiveness of the SRFNet and SRNet in FFRVSR. To test the efficacy of SRFNet, we compare SRFNet and FNet, which used in FRVSR [25]. For a fair comparison, we train both networks on the same training dataset and evaluate them on Vid4 [20]. The evaluation of PSNR on the Vid4 during the training are illustrated in Fig. 7(a). The network using SRFNet achieves notable improvement by 0.59 dB over the network using FNet on Vid4. It proves that the SRFNet can effectively avoid generating error information and improve the accuracy of optical flow estimation of inter-frames. In order to verify the effectiveness of the RRN framework, we trained the network with RRN and the network without RRN, respectively. Their training results are shown in Fig. 7(b). The network using RRN has an improvement of about 0.3 dB on Vid4 compared to the network without RRN. At the same time, we found that the network using RRN needs to be trained more epochs to achieve the optimal VSR performance, which also verifies that VSR networks using the implicit motion compensation methods are more challenging to converge during the training.

We also designed three networks of different sizes to analyze the scalability of our method. These three networks use 6 residual blocks, 12 residual blocks, and 18 residual blocks in SRNet. We trained these three networks on REDS [23]. Their final test results are shown in Table 1. FFRVSR-M has a 0.4 dB improvement on Vid4 compared to FFRVSR-S, and FFRVSR-L has a 0.54 dB improvement compared to FFRVSR-M, the inference speeds of the networks decrease linearly with the increase in the network size.

Table 1. Comparison of different size methods for × 4 VSR. Y denotes the evaluation on luminance channel. Runtime is calculated on an LR frame of size 320×180, and all models run on the same 2080Ti GPU.

Mode	FFRVSR-S	FFRVSR-M	FFRVSR-L
Blocks	6	12	18
Parameter(M)	2.3	2.7	3.2
Runtime(ms)	18	24	30
Vid4(Y)	27.29/0.8369	27.69/0.8480	28.23/0.8580

Table 2. Comparison in existing VSR methods. Y denotes the evaluation on luminance channel. Runtime is calculated on an LR frame of size 320 × 180, and all models run on the same 2080Ti GPU. Red and blue colors indicate the best and the second-best performance.

Mode	Bicubic	SOFVSR	FRVSR	PFNL	EDVR	TGA	RRN	BasicVSR	IconVSR	FFRVSR-L(ours)
Parameter(M)	N/A	1.6	5.1	3.0	20.6	5.8	3.4	6.3	8.7	3.2
Runtime(ms)	N/A	97	129	295	378	375	45	63	70	30
Vid4(Y)	21.82/0.5406	25.31/0.7466	26.48/0.8104	27.16/0.8365	27.85/0.8503	27.59/0.8419	27.65/0.8479	27.91/0.8536	28.04/0.8570	28.23/0.8580

4.4 Comparison with Other Methods

In this section, we compare our method with several famous VSR methods on the standard testing dataset Vid4 [20], including SOFVSR [28], FRVSR [25], PFNL [34], EDVR [29], TGA [15], RRN [16], BasicVSR [3], IconVSR [3]. SOFVSR uses a sliding window approach to achieve VSR. FRVSR and RRN perform VSR in a unidirectional recurrent way. The difference is that FRVSR uses an explicit motion compensation method, while RRN uses an implicit motion compensation method. PFNL uses a non-local method for VSR, EDVR uses deformable convolution for VSR, TGA uses temporal grouping and intra-group fusion methods

Fig. 8. Quantitative comparison on Vid4 [37] for 4× VSR. FFRVSR restore finer details and sharper edges.

Fig. 9. Comparison of VSR performance between ours and RRN. Our method can restore finer details.

to achieve implicit alignment and VSR. BasicVSR and IconVSR use the bidirectional recurrent network for VSR. To make a fair comparison, we reproduced each network, using the same downsampling filter, and tested their VSR performance and runtime on the same device. The comparison is provided in Table 2. Our method outperforms these famous VSR methods in inference speed and VSR performance. FFRVSR-L surpasses RRN by 0.58 dB on Vid4 and reduces runtime by 15 ms. Compared with BasicVSR, FFRVSR-L has a 0.32 dB improvement on Vid4 and reduces the running time by 35 ms. FFRVSR-L also surpasses IconVSR by 0.19 dB on Vid4 but reduces runtime by 40 ms.

The qualitative comparisons are shown in Fig. 8. FFRVSR can more efficiently utilize the relevant information in the video frames and restore higher-quality SR video frames on Vid4. Compared with other VSR networks, FFRVSR can recover finer details and sharper edges. Other VSR networks cannot retrieve the missing information well (e.g., numbers on the calendar) or are prone to some artifacts (e.g., window outlines of the buildings). Compared with RRN, our method can more efficiently utilize the effective information in the previous video frames and can effectively reduce the accumulation of error information in the iterative process (see Fig. 9).

5   Conclusion

In this work, we propose an efficient and scalable VSR network. We fuse frame-recurrent and recurrent-residual nets in our method and design a novel end-to-end trainable VSR network. The network can efficiently extract the spatiotemporal information between video frames and reduce the error accumulation during inference. To avoid misinformation during optical flow estimation, we design an efficient super-resolution flow estimation network to more accurately estimate optical flow between video frames and minimize misinformation generation. We also employ new loss function terms to train our network more efficiently. Experiment result shows that our method has good inference speed and VSR performance. Under the condition of the same inference speed or the same VSR performance, our method outperforms the existing VSR methods. Moreover, our network can be extended to apply to real-time VSR tasks or high-quality VSR tasks.

Acknowledgments. This work was supported by the National Natural Science Foundation of China under Grant 62176161, and the Scientific Research and Development Foundations of Shenzhen under Grant JCYJ20220818100005011.

References

1. Caballero, J., et al.: Real-time video super-resolution with spatio-temporal networks and motion compensation. In: Proceedings of the IEEE Conference on Computer Vision and Pattern Recognition, pp. 4778–4787 (2017)

2. Cao, Y., Wang, C., Song, C., Tang, Y., Li, H.: Real-time super-resolution system of 4k-video based on deep learning. In: 2021 IEEE 32nd International Conference on Application-specific Systems, Architectures and Processors (ASAP), pp. 69–76. IEEE (2021)
3. Chan, K.C., Wang, X., Yu, K., Dong, C., Loy, C.C.: BasicVSR: the search for essential components in video super-resolution and beyond. In: Proceedings of the IEEE/CVF Conference on Computer Vision and Pattern Recognition, pp. 4947–4956 (2021)
4. Chu, M., Xie, Y., Mayer, J., Leal-Taixé, L., Thuerey, N.: Learning temporal coherence via self-supervision for GAN-based video generation. ACM Trans. Graph. (TOG) **39**(4), 1–75 (2020)
5. Dong, C., Loy, C.C., He, K., Tang, X.: Learning a deep convolutional network for image super-resolution. In: Fleet, D., Pajdla, T., Schiele, B., Tuytelaars, T. (eds.) ECCV 2014. LNCS, vol. 8692, pp. 184–199. Springer, Cham (2014). https://doi.org/10.1007/978-3-319-10593-2_13
6. Freedman, G., Fattal, R.: Image and video upscaling from local self-examples. ACM Trans. Graph. (TOG) **30**(2), 1–11 (2011)
7. Freeman, W.T., Jones, T.R., Pasztor, E.C.: Example-based super-resolution. IEEE Comput. Graph. Appl. **22**(2), 56–65 (2002)
8. Fuoli, D., Gu, S., Timofte, R.: Efficient video super-resolution through recurrent latent space propagation. In: 2019 IEEE/CVF International Conference on Computer Vision Workshop (ICCVW), pp. 3476–3485. IEEE (2019)
9. Goto, T., Fukuoka, T., Nagashima, F., Hirano, S., Sakurai, M.: Super-resolution system for 4k-HDTV. In: 2014 22nd International Conference on Pattern Recognition, pp. 4453–4458. IEEE (2014)
10. Gunturk, B.K., Batur, A.U., Altunbasak, Y., Hayes, M.H., Mersereau, R.M.: Eigenface-domain super-resolution for face recognition. IEEE Trans. Image Process. **12**(5), 597–606 (2003)
11. Haris, M., Shakhnarovich, G., Ukita, N.: Recurrent back-projection network for video super-resolution. In: Proceedings of the IEEE/CVF Conference on Computer Vision and Pattern Recognition, pp. 3897–3906 (2019)
12. He, K., Zhang, X., Ren, S., Sun, J.: Identity mappings in deep residual networks. In: Leibe, B., Matas, J., Sebe, N., Welling, M. (eds.) ECCV 2016. LNCS, vol. 9908, pp. 630–645. Springer, Cham (2016). https://doi.org/10.1007/978-3-319-46493-0_38
13. Huang, Y., Wang, W., Wang, L.: Bidirectional recurrent convolutional networks for multi-frame super-resolution. In: Advances in Neural Information Processing Systems 28 (2015)
14. Isobe, T., et al.: Look back and forth: video super-resolution with explicit temporal difference modeling. In: Proceedings of the IEEE/CVF Conference on Computer Vision and Pattern Recognition, pp. 17411–17420 (2022)
15. Isobe, T., et al.: Video super-resolution with temporal group attention. In: Proceedings of the IEEE/CVF Conference on Computer Vision and Pattern Recognition, pp. 8008–8017 (2020)
16. Isobe, T., Zhu, F., Jia, X., Wang, S.: Revisiting temporal modeling for video super-resolution. arXiv preprint arXiv:2008.05765 (2020)
17. Jo, Y., Oh, S.W., Kang, J., Kim, S.J.: Deep video super-resolution network using dynamic upsampling filters without explicit motion compensation. In: Proceedings of the IEEE Conference on Computer Vision and Pattern Recognition, pp. 3224–3232 (2018)
18. Kappeler, A., Yoo, S., Dai, Q., Katsaggelos, A.K.: Video super-resolution with convolutional neural networks. IEEE Trans. Comput. Imaging **2**(2), 109–122 (2016)

19. Lai, W.S., Huang, J.B., Ahuja, N., Yang, M.H.: Fast and accurate image super-resolution with deep Laplacian pyramid networks. IEEE Trans. Pattern Anal. Mach. Intell. **41**(11), 2599–2613 (2018)

20. Liu, C., Sun, D.: On Bayesian adaptive video super resolution. IEEE Trans. Pattern Anal. Mach. Intell. **36**(2), 346–360 (2013)

21. Liu, H., Ruan, Z., Zhao, P., Dong, C., Shang, F., Liu, Y., Yang, L., Timofte, R.: Video super-resolution based on deep learning: a comprehensive survey. Artif. Intell. Rev. **55**(8), 5981–6035 (2022)

22. Mahapatra, D., Bozorgtabar, B., Garnavi, R.: Image super-resolution using progressive generative adversarial networks for medical image analysis. Comput. Med. Imaging Graph. **71**, 30–39 (2019)

23. Nah, S., et al.: NTIRE 2019 challenge on video deblurring and super-resolution: dataset and study. In: Proceedings of the IEEE/CVF Conference on Computer Vision and Pattern Recognition Workshops (2019)

24. Park, G., Park, B., Kwon, S.J., Kim, B., Lee, Y., Lee, D.: nuQmm: quantized MatMul for efficient inference of large-scale generative language models. arXiv preprint arXiv:2206.09557 (2022)

25. Sajjadi, M.S., Vemulapalli, R., Brown, M.: Frame-recurrent video super-resolution. In: Proceedings of the IEEE Conference on Computer Vision and Pattern Recognition, pp. 6626–6634 (2018)

26. Shi, X., Chen, Z., Wang, H., Yeung, D.Y., Wong, W.K., Woo, W.: Convolutional LSTM network: a machine learning approach for precipitation nowcasting. In: Advances in Neural Information Processing Systems, vol. 28 (2015)

27. Tao, X., Gao, H., Liao, R., Wang, J., Jia, J.: Detail-revealing deep video super-resolution. In: Proceedings of the IEEE International Conference on Computer Vision, pp. 4472–4480 (2017)

28. Wang, L., Guo, Y., Lin, Z., Deng, X., An, W.: Learning for video super-resolution through HR optical flow estimation. In: Jawahar, C.V., Li, H., Mori, G., Schindler, K. (eds.) ACCV 2018. LNCS, vol. 11361, pp. 514–529. Springer, Cham (2019). https://doi.org/10.1007/978-3-030-20887-5_32

29. Wang, X., Chan, K.C., Yu, K., Dong, C., Change Loy, C.: EDVR: video restoration with enhanced deformable convolutional networks. In: Proceedings of the IEEE/CVF Conference on Computer Vision and Pattern Recognition Workshops (2019)

30. Wang, Z., et al.: Multi-memory convolutional neural network for video super-resolution. IEEE Trans. Image Process. **28**(5), 2530–2544 (2018)

31. Xue, T., Chen, B., Wu, J., Wei, D., Freeman, W.T.: Video enhancement with task-oriented flow. Int. J. Comput. Vision **127**, 1106–1125 (2019)

32. Yan, B., Lin, C., Tan, W.: Frame and feature-context video super-resolution. In: Proceedings of the AAAI Conference on Artificial Intelligence, vol. 33, pp. 5597–5604 (2019)

33. Yang, C.-Y., Huang, J.-B., Yang, M.-H.: Exploiting self-similarities for single frame super-resolution. In: Kimmel, R., Klette, R., Sugimoto, A. (eds.) ACCV 2010. LNCS, vol. 6494, pp. 497–510. Springer, Heidelberg (2011). https://doi.org/10.1007/978-3-642-19318-7_39

34. Yi, P., Wang, Z., Jiang, K., Jiang, J., Ma, J.: Progressive fusion video super-resolution network via exploiting non-local spatio-temporal correlations. In: Proceedings of the IEEE/CVF International Conference on Computer Vision, pp. 3106–3115 (2019)

35. Yoo, J., Lee, S., Kwak, N.: Image restoration by estimating frequency distribution of local patches. In: Proceedings of the IEEE Conference on Computer Vision and Pattern Recognition, pp. 6684–6692 (2018)
36. Zhang, L., Zhang, H., Shen, H., Li, P.: A super-resolution reconstruction algorithm for surveillance images. Signal Process. **90**(3), 848–859 (2010)

Structural Properties of Associative Knowledge Graphs

Janusz A. Starzyk[1,2]([✉]) [ID], Przemysław Stokłosa[3] [ID], Adrian Horzyk[4] [ID],
and Paweł Raif[5] [ID]

[1] University of Information Technology and Management in Rzeszow, Rzeszow, Poland
starzykj@gmail.com
[2] Ohio University, Athens, OH 45701, USA
[3] Institute of Management and Information Technology, Bielsko-Biała, Poland
[4] AGH University of Krakow, Krakow, Poland
[5] Silesian University of Technology in Gliwice, Gliwice, Poland

Abstract. This paper introduces a novel structural approach to constructing associative knowledge graphs. These graphs are composed of many overlapping scenes, with each scene representing a specific set of objects. In the knowledge graph, each scene is represented as a complete subgraph associating scene objects. Knowledge graph nodes represent various objects present within the scenes. The same object can appear in multiple scenes. The recreation of the stored scenes from the knowledge graph occurs through association with a given context, which includes some of the objects stored in the graph. The memory capacity of the system is determined by the size of the graph and the density of its synaptic connections. Theoretical dependencies are derived to describe both the critical graph density and the memory capacity of scenes stored in such graphs. The critical graph density represents the maximum density at which it is possible to reproduce all elements of the scene without errors.

Keywords: structural associative knowledge graphs · critical graph density · associative memory capacity · context association · associating scene objects

1 Introduction

Developed over decades, associative networks and memories store and retrieve scene info using contextual cues. Recognizing fragments recalls whole scenes through associations. Hopfield networks, a classical associative memory, have limited capacity; the studies presented in [1] found around 14% of the total number of neurons for stored patterns. Modern Hopfield networks were introduced first in 2016 by Krotov and Hopfield in paper [2] and then generalized by Demircigil et al. [3]. The main modification to the classical Hopfield network, made by Krotov and Hopfield, is the introduction of a new energy function. This enhancement leads to an increased storage capacity and improved convergence of the network. As a result, we can obtain associative memories with significantly larger memory capacity. The paper entitled "Hopfield Networks Is All You Need" [4] introduced Dense Associative Memories as generalizations of classical Hopfield Networks.

There are notable similarities between the ideas of transformers and modern Hopfield networks, primarily because both systems utilize the self-attention mechanism as a core component [5]. Transformers were introduced in paper "Attention Is All You Need" [6] in 2017. Models built with transformers require neither recurrent [2] nor convolutional neural networks [7]. The structure is fast and easy to parallelize. Implementations of the transformer mechanism are available in popular deep learning libraries, for example, TensorFlow [8] or PyTorch [9].

The Hierarchical Temporal Memory (HTM) was proposed by Hawkins and George [10]. The method relies on hierarchical structures that take inspiration from the architecture of the human neocortex. One of the most interesting software projects inspired by HTM is nupic.torch – a library that integrates selected neuroscience principles from HTM into the PyTorch deep learning platform [11]. The idea behind this project has been described in the paper "How Can We Be So Dense? The Benefits of Using Highly Sparse Representations" [12].

The Knowledge Graph represents objects together with their relationships. According to Neo4j [13], knowledge graphs can use the underlying data for complex decision-making. Graph Neural Networks (GNNs) [14] are deep learning-based methods that operate on graphs (including knowledge graphs). In machine learning, knowledge graph analysis focuses on tasks such as node classification, link prediction, and clustering. A survey of deep learning methods based on GNNs is presented in [15].

The ANAKG (Active Neuro-Associative Knowledge Graph) [16, 17] represents a form of emergent neural, cognitive architecture that leverages an associative model of neurons. It employs dynamic neurons with relaxation-refraction mechanisms, enabling the construction of associative AI systems. ANAKG algorithms effectively consolidate learning sequences, giving rise to an associative knowledge graph.

The most important contribution of this work is to demonstrate that a specific type of associative knowledge graphs can serve as semantic memories, even when synaptic connections are limited to binary weights (0 and 1). Their construction is rapid, and they hold larger memory capacities than traditional Hopfield networks.

The study thoroughly examines these graphs' structural properties, including associative memory limits and graph density, crucial for error-free scene retrieval. Validation involves experiments on various datasets.

2 Structural Associative Knowledge Graphs

Associative knowledge graphs were developed to rapidly construct semantic memories that store associations between the observed events, actions, and objects or their parts. These graphs serve as the foundation of knowledge about the outside world, making them invaluable for autonomous learning systems acting in open environments. They can recall sequences, develop associations, and create novel knowledge.

In this paper, we employ structural associative knowledge graphs to represent episodes consisting of intricate scenes featuring diverse objects. Unlike the other associative knowledge graphs, structural associative knowledge graphs rely solely on the graph structure itself, disregarding synaptic connection weights. The essential information required about these graphs is whether a synaptic connection exists between any given pair of nodes. It is obvious that additional information regarding such connections, such as synaptic connections' strength or delays, can be added to enhance a memory structure, as discussed in [16]. However, this paper focuses exclusively on exploring the structural properties of associative memories constructed using such graphs.

Associative knowledge graphs, as presented in [17], offer a convenient framework for constructing a structure in which the memory of scenes and associations among their elements is saved. The structure of the knowledge graph develops automatically by progressively entering additional information, such as sentences containing word relationships. Both the synaptic connections between the represented elements and the strengths of these connections are recorded during the creation of the graph.

Each node in the knowledge graph represents a specific object, word, or concept, and as knowledge is inputted into the graph, these nodes become interconnected. However, when a large amount of information is recorded in the knowledge graph with a fixed number of nodes, the synaptic connections may undergo multiple modifications, resulting in a reduction or loss of the resolution of the original information. In this work, we extensively analyze the structural properties of knowledge graphs, pointing to the importance of knowledge graph density. This density is measured by the ratio of the number of used synaptic connections to the total possible synaptic connections among the graph nodes.

In particular, we will demonstrate that the mere structure of the knowledge graph is capable of capturing and restoring information among its associated elements. We will establish the relationships between the sparsity of the graph and the amount of stored information. We will show these relationships using the statistically normal distributions of objects recorded in the graph scenes and establish conditions that enable flawless recreation of these scenes. Since the distributions of objects recorded in real scenes differ from normal distributions, the number of scenes that can be saved and reproduced without error will deviate from the theoretical predictions. However, the most important dependencies between the number of nodes in the graph, the graph density or context size, and the number of saved scenes will be accurately estimated. It will enable us to determine and use the appropriate graph size or the necessary context based on the size and number of scenes stored within the associative knowledge graph memory (referred to as graph memory). The essential contribution of this paper is that the sparsity of the graph memory allows us to achieve a high capacity of the memory.

All the obtained dependencies were experimentally verified by using diverse types of scenes. These included scenes obtained with the use of a random number generator, the Iris dataset, and scenes containing real images of objects from the set COCO-2017 [18]. These objects were recognized using deep neural networks EfficientDet [19]. Both the neural network and the dataset were obtained from the framework FiftyOne [20].

Definition 1. *Structural Associative Knowledge Graph (SAKG) is a graph where all weights of the synaptic connections (represented by the graph edges) have a value of 1. Each scene stored in the SAKG is represented by a complete subgraph (clique) formed by the nodes that describe the objects present in the scene.*

The SAKG (referred to as a knowledge graph or a graph) density becomes a crucial factor influencing memory storage capacity. However, the scenes stored should be sufficiently intricate to enable scene retrieval with a relatively small context (i.e., the number of observed objects in a scene). Additionally, to maintain graph sparsity, a large number of unique objects must be used in the stored scenes.

Definition 2. *Knowledge graph **memory storage capacity** is the maximum number of scenes that can be uniquely recalled without errors.*

Both aspects of this definition are essential. A recalled scene is considered error-free if all elements of the scene are correctly recalled. A recalled scene is deemed unique if no other scene that matches the provided context is recalled.

In the proposed approach to associative memory in which complete subgraphs represent individual scenes, the storage capacity strongly depends on the resulting graph density, which is a function of the number of nodes in the graph (number of objects represented by the graph) as well as the size and quantity of stored scenes. As more scenes are added to the knowledge graph, its density increases. However, once the density surpasses a **critical graph density**, recalling stored memories without error becomes impossible. The objective is to determine this critical graph density since it defines the memory capacity.

In the cases where the number of unique objects used to construct the knowledge graph is insufficient, resulting in a dense graph as observed with recognized objects from the FiftyOne database, virtual objects can be generated by combining the object's symbol with its location in the scene, as demonstrated in Sect. 5.3. Generally, virtual objects can be created using various methods that consider distinguishing characteristics such as color, position, size, and more.

3 Gradual Increase in the Knowledge Graph Density

This section explores the link between knowledge graph density and the number of stored scenes, showing how density rises with more scenes due to factors like average scene size and graph size (the number of vertices). Equation (1) calculates the total number of synaptic connections in a dense graph of n nodes.

$$e_0 = \frac{n * (n-1)}{2}. \tag{1}$$

For the associative scene memory, complete subgraphs spanning nodes representing scene objects are added. Assuming scenes have equal object count n_f, this does not alter key memory properties like capacity and density dependence. As dense subgraphs accumulate, they may share connections, affecting graph density. The effective synaptic connections in a SAKG storing s dense subgraphs, each with n_f nodes, must factor in increased overlap and graph density.

If, at some stage of the graph development, the graph density is d_i, then $(1 - d_i) * \frac{n_f*(n_f-1)}{2}$ new synaptic connections are added with each newly added complete subgraph, and the new SAKG graph density will be as follows

$$d_{i+1} = \frac{d_i * \frac{n*(n-1)}{2} + (1 - d_i) * \frac{n_f*(n_f-1)}{2}}{\frac{n*(n-1)}{2}} \tag{2}$$

where the first term in the numerator represents the number of existing connections and the second term the number of added connections, while the denominator represents the maximum number of connections in the knowledge graph spanned on n nodes. After simplification, we have

$$d_{i+1} = d_i + (1 - d_i) * \frac{n_f * (n_f - 1)}{n * (n - 1)} = d_i * \left(1 - \frac{n_f * (n_f - 1)}{n * (n - 1)}\right) + \frac{n_f * (n_f - 1)}{n * (n - 1)}$$

Let us denote

$$\xi = \frac{n_f * (n_f - 1)}{n * (n - 1)} \tag{3}$$

then

$$d_{i+1} = d_i * (1 - \xi) + \xi. \tag{4}$$

Assuming a uniform distribution of added connections within the knowledge graph, we will get a reliable estimate of the total number of synaptic connections in the knowledge graph using $d_1 = d_0 * (1 - \xi) + \xi$, and, after s steps, we have

$$d_s = d_0 * (1 - \xi)^s + 1 - (1 - \xi)^s = 1 + (d_0 - 1) * (1 - \xi)^s,$$

where s is the number of stored scenes. Before any subgraph is added, the initial graph density d_0 is 0, the final graph density after all scenes were stored is simply related to the number of scenes (s), the number of graph nodes (n), , and the number of scene objects (n_f)

$$d_n = 1 - (1 - \xi)^n = 1 - \left(1 - \frac{n_f * (n_f - 1)}{n * (n - 1)}\right)^s. \tag{5}$$

From (5), we can derive the number of scenes as a function of the graph density

$$s = \frac{\log(1 - d)}{\log\left(1 - \frac{n_f * (n_f - 1)}{n * (n - 1)}\right)} = \frac{\log(1 - d)}{\log(1 - \xi)}. \tag{6}$$

To illustrate this result, let us calculate the number of scenes in a knowledge graph with $n = 3200$ nodes, density $d = 0.412$, and the number of scene objects $n_f = 15$.

Example 1. For the specified graph parameters, we have $\xi = \frac{n_f*(n_f-1)}{n*(n-1)} = \frac{15*14}{3200*3199} = 2.0514 * 10^{-5}$. Using Eq. (6), we get the maximum number of stored scenes.

$$s = \frac{\log 0.5}{\log(1 - \xi)} = \frac{-1}{\log_2(1 - \xi)} = \frac{-1}{-2.9596 * 10^{-5}} = 33789.$$

4   Dependence on the Size of the Context

The associative knowledge graph serves as a context-addressable memory, allowing for the retrieval of scene memories based on a specific context. In practical applications, this context can correspond to identifying a scene by observing its fragments.

Equation (6) establishes the relationship between the density of the graph and the number of scenes stored within it. However, it is important to note that this relationship alone does not guarantee error-free reproduction of these scenes based on the provided context, which is a necessary requirement for associative memory. In the following discussion, we will explore the critical relationships between the size of the knowledge graph, the number of saved scenes, and the feasibility of accurately recreating them within a given context. By examining these relationships, we will demonstrate that the memory capacity of an associative graph increases quadratically with the number of knowledge graph nodes and is highly dependent on the size of the context provided.

Therefore, the memory capacity of the knowledge graph depends not only on the size of the stored scenes but also on the number of observed context elements. To find this dependence, we calculate the probability of memory retrieval without any errors, given the graph density d and the size of the retrieval context n_c.

Lemma 1. *Assuming a random distribution of connections, the probability that any two nodes in the graph are connected is equal to the graph density d.*

The proof of this lemma is trivial and stems from the definition of graph density as the ratio of the number of existing connections to all possible connections.

Lemma 2. *Assuming a random distribution of connections, the probability that all randomly selected context nodes n_c are connected to each other is given by.*

$$p_c = d^{\frac{n_c*(n_c-1)}{2}}. \tag{7}$$

The rationale behind this lemma is as follows: Since the given context is taken from one stored scene, all nodes of the context are connected to each other by construction. To exclude any other pattern from being incorrectly recognized by having the same context as the stored pattern, all $\frac{n_c*(n_c-1)}{2}$ randomly selected pairs of nodes must be connected. This probability is described by (7). We want this probability to be as small as possible to uniquely and accurately retrieve the desired scene without any errors.

4.1   Critical Graph Density

Lemma 3. *If all scenes stored in a knowledge graph are to be retrieved with an error smaller than any small positive number ε by providing a known context n_c, then the **critical graph density** d can be estimated iteratively from.*

$$d_{i+1} \cong \left(-\frac{\xi*\varepsilon}{\log(1-d_i)}\right)^{\frac{1}{\zeta}} i \in [0, \infty) \tag{8}$$

where $\zeta = \frac{n_c*(n_c-1)}{2}$, and $\xi = \frac{n_f*(n_f-1)}{n*(n-1)}$.

Proof. If more scenes are stored in the memory, the likelihood of falsely recognizing one of them increases when considering a given context. The probability of success, in this case, is that none of the scenes was falsely recognized, which is equal to $(1 - p_c)^s > 1-\varepsilon$ for a given small positive ε, where the number of scenes s can be calculated from (6).

For error-free retrieval, we require that all these scenes are retrieved without errors meaning that $(1 - p_c)^s > 1 - \varepsilon$; therefore, by using Lemma 2, we require that

$$\left(1 - d^{\frac{n_c*(n_c-1)}{2}}\right)^s = \left(1 - d^{\frac{n_c*(n_c-1)}{2}}\right)^{\frac{\log(1-d)}{\log(1-\xi)}} > 1 - \varepsilon. \tag{9}$$

then (9) can be replaced by

$$\left(1 - d^\xi\right)^{\frac{\log(1-d)}{\log(1-\xi)}} > 1 - \varepsilon. \tag{10}$$

For small value d^ξ, we can approximate $\left(1 - d^\xi\right)^{\frac{d^\xi}{d^\xi}} \cong e^{-d^\xi}$, so (10) can be replaced by $e^{-\frac{d^\xi*\log(1-d)}{\log(1-\xi)}} > 1 - \varepsilon$. By taking the logarithm of both sides, we will get the following

For small value d^ξ, we can approximate $\left(1 - d^\xi\right)^{\frac{d^\xi}{d^\xi}} \cong e^{-d^\xi}$, so (10) can be replaced by $e^{-\frac{d^\xi*\log(1-d)}{\log(1-\xi)}} > 1 - \varepsilon$. By taking the logarithm of both sides, we will get the following

$$-\frac{d^\xi * \log(1 - d)}{\log(1 - \xi)} > \log(1 - \varepsilon)$$

or

$$d^\xi * \log(1 - d) > -\log(1 - \xi) * \log(1 - \varepsilon) \cong -\xi * \varepsilon. \tag{11}$$

From Eq. (11), we can get the value d^ξ, which allows us to find the critical graph density d for the given values ζ, ξ, and ε using the iterative process:

$$d \cong \left(-\frac{\xi * \varepsilon}{\log(1 - d)}\right)^{\frac{1}{\zeta}}. \tag{12}$$

The iterations converge very quickly, starting from the initial value $d_0 = 0.5$ to obtain the critical graph density.

The following theorem presents the key result of this study, which establishes the memory capacity of an associative graph based on the provided context, the average size of the scene, and the number of graph nodes.

Theorem 1
For the critical graph density for a given context (12) *, the memory capacity can be determined using Eq.* (6). *Assuming that ξ is small, we have*

$$s = \frac{\log(1 - d)}{\log(1 - \xi)} = \frac{\log(1 - d)}{-\xi} = \frac{\log(1 - d) * n * (n - 1)}{-n_f * (n_f - 1)}. \tag{13}$$

4.2 Finding the Maximum Memory Capacity

To determine the maximum memory capacity concerning the number of graph neurons, the number of features, and the size of the context, we selected $n \in \{10^6, 10^7, 10^8, 10^9\}$, change the number of features n_f from 1 to 100, and set the size of the context n_c from $0.1*n_f$ to $1*n_f$. The results shown in Fig. 1 demonstrate that the largest memory capacity is obtained for a relatively small number of features n_f (scene size).

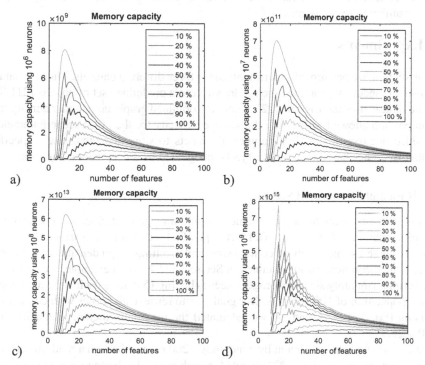

Fig. 1. Memory capacity as a function of the number of features for $n = 10^6, 10^7, 10^8, 10^9$ neurons in figures a), b), c), and d), respectively. The legend shows the size of the context as a percentage of the scene size.

Notably, the optimal context size is approximately $n_c \sim 10 - 12$ regardless of the network size (number of neurons) or the scene size (number of features). Visible noise is the result of rounding up a fraction of the features to determine the size of the context.

4.3 Algorithm for Scene Retrieval

The SAKG graph is created by overlaying complete subgraphs that contain all objects stored in the consecutive scenes stored in the graph. As long as the density of the graph does not exceed the critical graph density, these scenes can be recreated by associating them with the given context o_c. The scene retrieval testing algorithm consists of the following steps:

1. Randomly select a context o_c from each scene o_s stored in the SAKG graph.
2. Determine the set o_m of maximally activated neurons, which are the neurons in the SAKG graph that are connected to all context neurons o_c.
3. The sum of the context neurons and the maximally activated neurons recreates the desired scene $o = o_c \cup o_m$.
4. If the recreated desired scene $o \neq o_s$, then the scene recovery error is increased by one.
5. The scene retrieval error rate is determined by dividing the scene recovery error by the number of tested scenes.

5 Experiments

In this section, we performed experiments using three different datasets. The first dataset consists of randomly generated scenes drawn from a predefined set of objects [21]. The second dataset uses the Iris data [22], where the obtained graph has a higher density than the maximum allowed, surpassing the storage capacity for the given scene parameters. The third one uses a sufficient number of objects to store and recognize a specified number of scenes based on a set of objects recognized by a deep neural network.

5.1 Randomly Generated Scenes

The first set of experiments was performed on randomly generated scenes that contained a specified number of objects drawn from a specified set of objects. Various sizes of the stored scene set, scene numbers, context size, and resulting graph density were tested to validate the theoretical results obtained in Sects. 3 and 4. As an example, we first tested the maximum memory size for storage of scenes having 15 objects each, drawn randomly from a population of 1000 objects. Our goal was to retrieve all scenes with context size 5. Using Eqs. 3, 10, 13, and 14, we calculated the maximum memory capacity to be 1200 scenes.

We conducted an experiment by generating 1200 scenes through a random selection of 15 objects from a pool of 1000 objects for each scene. Each scene was then saved in the knowledge graph as a complete subgraph, distributed across 15 randomly selected nodes within the graph, representing the objects within that scene. We tested the scene retrieval error level with various contexts provided for retrieval using an average of 20 simulations. The error would increase by one each time a wrong scene was extracted from the associative graph memory. Table 1 provides the results.

As we can see, the context of 5 objects was necessary to retrieve all scenes without error. Smaller contexts resulted in increasingly larger error levels, while larger contexts resulted in no error. A similar test performed on 2400 scenes required 7 context objects to retrieve all scenes without an error, as shown in Table 2.

Again, these results fully agree with the theory that predicts the maximum memory capacity of 3197 scenes for the context of 7 objects and 2180 scenes for the context of 6 objects. The use of a smaller context than the one employed to calculate memory capacity results in a considerably larger error in recognizing scenes. This discrepancy is especially apparent when the context size is set to 2. For instance, storing 1200 scenes in the graph's memory already led to an error of over 12%. However, when the number of

Table 1. Errors for various numbers of observed context objects for 1200 scenes with $n_f = 15$.

Number of observed objects	7	6	5	4	3	2
Scene recognition error in %	0	0	0	0.008	0.446	12.30
Testing time in sec	31.0	30.7	31.8	35.0	50.0	120

Table 2. Errors for various numbers of observed context objects for 2400 scenes with $n_f = 15$.

Number of observed objects	7	6	5	4	3	2
Scene recognition error in %	0	0.002	0.002	0.079	1.89	31.8
Testing time in sec	130	135	174	287	534	1102

scenes stored in the same graph structure doubled, the recognition error using the same context more than doubled, reaching almost 40%.

This significant increase in error is primarily attributed to the larger number of scenes that share the same context of two objects in the latter case.

5.2 Iris Data

The second experiment utilized the Iris flower dataset [22], which comprises a total of 150 records, with 50 for each species. This dataset specifies the length and width of sepals and petals in centimeters for three different species of Irises: Setosa, Virginica, and Versicolor.

Scenes were created using the numerical feature values of different Iris samples, along with the iris species class, as distinct object names. Thus, we obtained 150 scenes, each consisting of 5 objects. Objects corresponding to different Iris features (length and width of sepals and petals) were distinguished as having unique values. Finally, all objects were assigned an enumeration from 1 to the sum of the maximum number of different feature values across the various features.

Table 3. Fragment of the original and enumerated features from the Iris dataset

Original feature values	Object ID numbers
5.6 6.2 5.0 4.6 5.2 5.4 7.6	14 20 8 4 10 12 33
3.0 2.8 3.5 3.6 4.1 3.9 3.0	45 43 50 51 56 54 45
4.5 4.8 1.3 1.0 1.5 1.3 6.6	80 83 62 59 64 62 99
1.5 1.8 0.3 0.2 0.1 0.4 2.1	113 116 104 103 102 105 119
2.0 3.0 1.0 1.0 1.0 1.0 3.0	125 126 124 124 124 124 126

To illustrate this process, let us consider the first 7 scenes and their original feature values (as shown in the left part of Table 3). We transformed these feature values into corresponding ID numbers, as shown in the right part of Table 3. The last row in the table indicates the iris class ID, linking the scene to its respective iris species.

A total of 126 different objects were included in various scenes, each with different frequencies. Testing was performed by specifying 4 feature values as a context. The scene with the maximum likelihood was selected as the retrieved scene. All the scenes (all Iris data) were tested using $n_c = 4$, $n_f = 5$, and $n = 126$. The average scene recognition error rate was 9.8% after 20 simulations. The knowledge graph density was 31.25%. In this example, if we use the result of the critical graph density result (12) with $\varepsilon = 0.01$, we will get $d \cong \left(-\frac{\xi*\varepsilon}{\log(0.5)}\right)^{\frac{1}{\xi}} = 16.24\%$, and the maximum number of scenes according to theory (13) $s = \frac{\log(1-d)}{\log(1-\xi)} = 139.43$. Likewise, for $\varepsilon = 0.001$, we will get $d \cong 11.06\%$ and the number of scenes $s = 92.26$.

It is evident that both these critical values, the graph density and the number of scenes, are exceeded in the Irises testing experiment, where we stored 150 scenes. This discrepancy explains the scene recognition error. However, it is important to note that the classification error was 0% since the retrieved scenes were always of the same class as the tested scene.

5.3 Scene Recognition Using Deep Neural Networks

The third experiment involved using deep neural networks to recognize and identify 1,000 scenes each consisting of 25 objects. This experiment aimed to simulate a scenario where video signals are processed, and individual objects are recognized to identify scenes composed of specific objects. To facilitate the storage and retrieval of scenes from semantic memory, symbols representing objects and their corresponding locations in the scene (virtual objects) were used instead of the raw images. A total of 1097 unique virtual objects were used to build the scenes. This approach enables efficient storage and retrieval of scenes based on the provided context. The dataset, using the COCO data format, was generated and is available on the website [21].

If our objective is to recognize these scenes without any errors using 6 context objects, then according to Eqs. 3, 12, and 13, the maximum capacity of the scene memory would be 998 scenes. Therefore, after storing 1000 scenes in the memory, it is expected that some of these scenes will not be recognized correctly. In our test, we encountered two scenes that were not properly recognized. Figure 2 shows the case of two different scenes with the same context of 6 objects. The objects present in both scenes were: 'cat 6', 'cat 13', 'cell phone 10', 'toilet 11', 'zebra 15', and 'zebra 15'.

In this example, objects are identified by the object class and their respective location in the scene.

Figure 3 shows an image that includes only the context from the first scene. Upon observing the second scene, it becomes apparent that the context objects listed are positioned similarly to those in the first scene. It is worth noting that this context contains two objects with the same name ('zebra 15') due to the presence of two zebras at the same grid location. This example not only highlights a critical aspect of memory capacity but

Fig. 2. Two different scenes with the same context of 6 objects located in similar locations.

also demonstrates our ability to handle multiple instances of a single object within a stored scene.

If we increase the context to include 7 elements, the maximum capacity of this associative memory will rise to 1429 scenes. Statistically, this allows for less than one recognition error for 1000 scenes from the same set. Accordingly, all the stored scenes were correctly recognized in this scenario.

Fig. 3. The context provided led to the misclassification of the two scenes.

At the address [21] there is a collection of 1000 scenes that were used in this experiment, along with a sample program that allows you to download any scene.

6 Conclusions

The paper presented a novel structural approach to constructing associative knowledge graphs. Specifically, it demonstrates that these graphs can be utilized to create associative memories by solely utilizing structural information about the synaptic connections, eliminating the need to specify the connection weights. The research reveals that memory capacity is determined by the size of the graph and the density of its synaptic connections. Theoretical dependencies are derived to describe the critical graph density and the memory capacity of the scenes stored in such graphs.

Memory capacity grows quadratically with the number of neurons used to build the knowledge graph. Although smaller graphs do not yield significant scene capacity, larger graphs do. If graphs are much larger than the stored scenes, the scene memory capacity becomes considerable, allowing each neuron to represent multiple scenes. Memory tests were conducted using both randomly generated synthetic data and real-world datasets such as Iris data and a collection of scenes featuring objects recognized through deep neural networks. These tests confirmed the validity of the derived results.

This approach enables the achievement of substantial scene-memory capacities straightforwardly by leveraging the sparsity of the knowledge graph and the relatively modest sizes of recorded scenes. In comparison to traditional associative Hopfield networks, where the number of recorded scenes is proportional to the number of neurons in the network, this method proves more effective thanks to the use of contextual connections in a sparse SAKG graph. Recent advancements in Hopfield networks [2, 3] have also demonstrated enhanced memory capacities; however, they rely on iterative techniques using dense recursive graphs, making it challenging to identify individual stationary points or define regions in the parameter space encapsulating trajectories of cyclic attractors.

Our future work related to this topic will focus on two key aspects. First, we aim to compare the proposed structural associative memory with contemporary associative memory solutions, such as Modern Hopfield Networks [3]. The comparative analysis of these two solutions will form the subject of our upcoming article. Second, we plan to enhance the presented concept by incorporating novel elements such as the use of the utilization of micro-columns.

References

1. Amit, D.J., Gutfreund, H., Sompolinsky, H.: Storing infinite numbers of patterns in a spinglass model of neural networks. Phys. Rev. Lett. **55**, 1530 (1985). https://doi.org/10.1103/PhysRevLett.55.1530
2. Krotov, D., Hopfield, J.J.: Dense Associative Memory for Pattern Recognition (2016). http://arxiv.org/abs/1606.01164
3. Demircigil, M., Heusel, J., Löwe, M., Upgang, S., Vermet, F.: On a model of associative memory with huge storage capacity. J. Stat. Phys. **168**(2), 288–299 (2017). https://doi.org/10.1007/s10955-017-1806-y
4. Ramsauer, H., et al.: Hopfield Networks is All You Need (2008). http://arxiv.org/abs/2008.02217
5. Schlag, I., Irie, K., Schmidhuber, J.: Linear Transformers are Secretly Fast Weight Programmers (2021). https://doi.org/10.48550/ARXIV.2102.11174
6. Vaswani, A., et al.: Attention is All You Need (2022). http://arxiv.org/abs/1706.03762. Accessed 21 Sept 2022
7. Arbib, M.A. (ed.): The Handbook of Brain Theory and Neural Networks, 1 Paperback edn. MIT Press, Cambridge (1998)
8. TensorFlow Transformer. https://www.tensorflow.org/text/tutorials/transformer. Accessed 11 Aug 2022

9. PyTorch Transformer. https://pytorch.org/docs/stable/generated/torch.nn.Transformer.html. Accessed 11 Aug 2022
10. Hawkins, J., George, D.: Hierarchical Temporal Memory: Concepts, Theory, and Terminology, Numenta Inc. Whitepaper (2006)
11. 'NuPIC PyTorch'. https://nupictorch.readthedocs.io/en/latest/. Accessed 11 Aug 2023
12. Ahmad, S., Scheinkman, L.: How Can We Be So Dense? The Benefits of Using Highly Sparse Representations (2023). http://arxiv.org/abs/1903.11257. Accessed 11 Aug 2023
13. Neo4J Graph Data Platform. https://neo4j.com/. Accessed 11 Aug 2023
14. Zhou, J., et al.: Graph Neural Networks: A Review of Methods and Applications (2018). https://doi.org/10.48550/ARXIV.1812.08434
15. Zhang, Z., Cui, P., Zhu, W.: Deep Learning on Graphs: A Survey (2018). http://arxiv.org/abs/1812.04202
16. Horzyk, A., Starzyk, J.A., Graham, J.: Integration of semantic and episodic memories. IEEE Trans. Neural Netw. Learn. Syst. **28**(12), 3084–3095 (2017). https://doi.org/10.1109/TNNLS.2017.2728203
17. Horzyk, A.: How does generalization and creativity come into being in neural associative systems and how does it form human-like knowledge? Neurocomputing **144**, 238–257 (2014). https://doi.org/10.1016/j.neucom.2014.04.046
18. Lin, T.Y., et al.: Microsoft COCO: Common Objects in Context. http://arxiv.org/abs/1405.0312. Accessed: 15 Jan 2023
19. Tan, M., Pang, R., Le, Q.V.: EfficientDet: scalable and efficient object detection (2019). https://doi.org/10.48550/ARXIV.1911.09070
20. Moore, B.E., Corso, J.J.: FiftyOne (2020). GitHub. https://github.com/voxel51/fiftyone
21. Structural Properties of Associative Knowledge Graphs (2023). https://github.com/PrzemyslawStok/Structural-Properties-of-Associative-Knowledge-Graphs.git. Accessed 11 Aug 2023
22. Fisher, R.A.: UCI Machine Learning Repository (1936). https://archive.ics.uci.edu/ml/datasets/Iris. Accessed 11 Aug 2023

Nonlinear NN-Based Perturbation Estimator Designs for Disturbed Unmanned Systems

Xingcheng Tong[1] and Xiaozheng Jin[1,2($\boxtimes$)]

[1] Key Laboratory of Computing Power Network and Information Security, Ministry of Education, Shandong Computer Science Center (National Supercomputer Center in Jinan), Qilu University of Technology (Shandong Academy of Sciences), Jinan, China
[2] Shandong Provincial Key Laboratory of Computer Networks, Shandong Fundamental Research Center for Computer Science, Jinan, China
jin445118@163.com

Abstract. This paper addresses the challenge of estimating perturbations in a classical unmanned system caused by a combination of internal uncertainties within the system and external disturbances. To accurately approximate these hard-to-measure perturbations, a novel nonlinear radial basis function neural network (RBFNN)-based estimator is introduced. This estimator is designed to reconstruct the perturbation structure effectively. The study demonstrates that utilizing RBFNN-based estimator designs, coupled with Lyapunov stability analysis, leads to achieving asymptotic estimation results. The effectiveness of the proposed perturbation estimation approach is validated through simulations conducted on both an unmanned marine system and a quadrotor system.

Keywords: Nonlinear RBFNN-based estimators · perturbation estimation · unmanned systems

1 Introduction

In recent decades, there has been a growing concern in the field of motion control systems, particularly in unmanned systems including robotic systems [1,2], unmanned aerial vehicles [3,4], and unmanned marine systems [5,6]. These systems possess desirable characteristics such as excellent mobility, low risk, high adaptability, and reliable operation, making them highly suitable for practical applications. It should be noted that unmanned systems typically exhibit multiple input and multiple-output, as well as nonlinear dynamics. Consequently, perturbations arising from factors like parametric uncertainties and external disturbances often impact the operation of these systems. In order to achieve precise perturbation information in unmanned systems, it is imperative to explore the

Supported in part by the Taishan Scholars Program under Grant tsqn202211208; in part by the National Natural Science Foundation of China under Grant 62173193; and in part by the Science Education Industry Integration and Innovation Project under Grant 2023PYI001.

development of high-precision estimation techniques that can effectively estimate perturbations.

In order to address the detrimental effects of perturbations and reinforce the robustness of systems, extensive research has been conducted on effective control technologies. These technologies aim to suppress the negative impacts of perturbations or optimize system performance in the presence of perturbations (see, for example, [7–10]). Within the field of perturbation rejection studies, two kinds of advanced control technique have been widely investigated: direct compensation control technique and indirect estimation technique. Direct compensation control technique involves the characteristics of self-regulation and intelligent adjustments to directly eliminate the effects of perturbations. Several classical control approaches fall under this category, including neural network-based control technique [11–13], sliding mode control technique [14,15], adaptive compensation control technique [16–19], among others. On the other hand, indirect estimation approaches rely on observation and estimation techniques to indirectly restructure the dynamics of perturbations. Various disturbance estimators/observers have been designed as part of these approaches, such as sliding mode observers [20], extended state observers [21,22], nonlinear disturbance observers [23], unknown system dynamics estimators [24,25]. By utilizing the observation or estimation information, control schemes are further proposed to effectively handle perturbations. For instance, in the field of fully-actuated marine vehicles, a combination of sliding mode disturbance observers and super twisting control strategy was developed to strengthen the tracking performance [20]. In the case of unmanned surface vessels with lumped disturbances, estimation results obtained from extended state observers were utilized to develop surface control methods and predictive control methods, respectively, to govern trajectory tracking behaviors [21,22]. Furthermore, an event-triggered control scheme based on nonlinear disturbance observation signals was constructed to deal with path-following issues for an unmanned surface vehicle system [23]. Unlike the aforementioned observers, linear filtering operations for available system states were employed in [20–23] to approximate disturbance based on unknown system dynamics estimator designs. The studies mentioned above indicate that accurate disturbance estimations can be achieved, leading to the effective enhancement of system robustness. However, it is important to note that certain limitations, such as assuming that the differential of perturbations is bounded, should be taken into consideration in the aforementioned research.

Motivated by the preceding analysis, we develop a method for designing a nonlinear dynamics estimator to address the perturbation estimation problem in disturbed unmanned systems. This approach involves the utilization of a newly proposed RBFNN-based estimator to accurately reformulate the unknown perturbations. By employing Lyapunov stability theory, we ensure that the estimation results for the unmanned system asymptotically converge to the true values of perturbations, with estimation errors tending to zero as time approaches infinity. To validate the effectiveness of the developed nonlinear estimator design method, we conduct simulations on two unmanned systems, that is, a unmanned marine system and a quadrotor system.

The paper is organized as follows: In Sect. 2, we present the dynamic model of the unmanned system and provide an overview of the RBFNN preliminaries. Section 3 outlines the construction of the nonlinear dynamics estimator for the unmanned system. To demonstrate the superiority of the proposed control method, simulations of perturbation estimation on the unmanned marine system are presented in Sect. 4. Finally, in Sect. 5, the paper concludes with a summary of the findings.

2 Preliminaries

Before proceeding with the experiments, we have to introduce our models.

2.1 Dynamics of the Unmanned System

According to the models of unmanned aerial vehicles, autonomous underwater vehicles, unmanned ground vehicles, unmanned surface vehicles, the dynamic of an unmanned system can be expressed as follows without loss of generality

$$\begin{cases} \dot{x} = W(x)v \\ M\dot{v} + C(v)v + D(v)v = \tau + w, \end{cases} \tag{1}$$

where $\{x, v\} \in R^n$ describe the position and velocity of the unmanned system, respectively; $W(x)$ denotes the rotation matrix of the system, matrices $\{M, C(v), D(v)\}$ denote the inertia, the Coriolis, and the hydrodynamic damping parameters of the system, respectively; $\tau \in R^m$ is the force afforded by the actuators; $w \in R^m$ stands for the external disturbance, which is assumed to be continuous but unknown.

It follows from (1) that

$$\begin{cases} \dot{x} = \zeta_x \\ \dot{\zeta}_x = u + \gamma(v), \end{cases} \tag{2}$$

where $\zeta_x = [\zeta_{x,1}, \zeta_{x,2}, \ldots, \zeta_{x,m}]^{\mathrm{T}} = W(x)v$, $u = W(x)M^{-1}\tau$, $\gamma(v) = [\dot{W}(x) - W(x)M^{-1}C(v) - W(x)M^{-1}D(v)]v + W(x)M^{-1}w$ denotes the perturbation of the unmanned system. From the formulation of $\gamma(v)$, we know that the perturbation is composed by the internal uncertainties of the system and external disturbances.

Assumption 1: The accurate value of $\dot{\zeta}_{x,i}$, $i = 1, 2, \ldots, m$ in (2) shall not be obtained, but its rough value can be measured with acceptable errors.

Assumption 2: There exists a known constant $\bar{w}_i$ such as $|w_i| \leq \bar{w}_i$, $i = 1, 2, \ldots, m$ and the derivative value of perturbation w exists.

Remark 1. Assumption 1 indicates that the acceleration signal of the unmanned system can be roughly measured. It is a reasonable condition because various sensors can sketchily measure the acceleration signal in practice, although the exact acceleration signal is hardly to be obtained. It is obvious that Assumption 2 is less conservative than the bounded condition of the derivative value of perturbation considered in [20–25] (i.e., $|\dot{\gamma}_i| \leq \delta$, where δ is a known constant).

2.2   Radial Basis Function Neural Networks

Considering a nonlinear function $g(z) \in R^m$ and referring to [26], a RBFNN whose structure is formulated as follows

$$R(z) = \theta^{\mathrm{T}} H(z), \tag{3}$$

It can be used to estimate the nonlinear function, where θ is the weight vector to be adjusted, $H(z)$ is the activation function of the network. Thus, the nonlinear term $g(z)$ could be approximated by

$$\hat{g}(z) = \hat{\theta}^{\mathrm{T}} H(z), \tag{4}$$

where $\hat{\theta}$ is the estimate of θ.

Let $\hat{z} = [\hat{z}_1^{\mathrm{T}}, \ldots, \hat{z}_m^{\mathrm{T}}]^{\mathrm{T}}$ be the estimation of the state vector $z = [z_1^{\mathrm{T}}, \ldots, z_m^{\mathrm{T}}]^{\mathrm{T}}$. Then, one can get the expression:

$$\hat{g}(\hat{z}|\hat{\theta}) = \hat{\theta}^{\mathrm{T}} H(\hat{z}). \tag{5}$$

The optimal weight matrix θ^* is denoted as

$$\theta^* := \arg\min\{\|\hat{g}(\hat{z}|\hat{\theta}) - g(z)\|\}. \tag{6}$$

Define the following minimum estimation error ε and the estimation error σ:

$$\varepsilon(t) = g(z) - \hat{g}(\hat{z}|\theta^*), \quad \sigma(t) = g(z) - \hat{g}(\hat{z}|\hat{\theta}), \tag{7}$$

respectively, where ε and σ are respectively limited by $\|\varepsilon\| \leq \bar{\varepsilon}$ and $\|\sigma\| \leq \bar{\sigma}$, where $\bar{\varepsilon} > 0$ and $\bar{\sigma} > 0$.

2.3   Objectives

The main research objective of this paper is to develop a nonlinear perturbation estimation strategy for the unmanned system, which is capable of governing the estimation to rebuild the perturbation asymptotically.

3   Nonlinear NN-Based Dynamic Estimator Designs

The nonlinear estimator for signals ζ_x and u is developed by

$$\begin{cases} \lambda \dot{\zeta}_x^f + \zeta_x^f = \zeta_x, \\ \lambda \dot{u}^f + u^f = u + \mathrm{sgn}(\xi)\bar{\lambda}_u, \end{cases} \tag{8}$$

where ζ_x^f and u^f denote the estimator states, $\lambda > 0$ is the estimation parameter, $\bar{\lambda}_u := [\bar{\lambda}_{u,1}, \ldots, \bar{\lambda}_{u,m}]$ is a RBFNN-based function which will be proposed later; ξ is an virtual signal defined by

$$\xi = \tfrac{1}{\lambda}(\zeta_x - \zeta_x^f) - (u^f + \gamma). \tag{9}$$

From Assumption 1, it is reasonable to assume the value of sgn(ξ) knowing.

Define $g(z_i) = \dot{\gamma}_i$, $z_i = [z_i, \dot{z}_i]^T$, $i = 1, 2, \ldots, m$. Based on the definition of γ under (2), we know $g(z_i)$ is a continuous nonlinear function. According to Sect. 2.2, when the RBFNN output $\hat{g}(z_i, \hat{\theta}_i)$ approaching the function $g(z_i)$, a suboptimal weight $\theta_i^* > 0$ can be achieved by

$$\theta_i^* = \arg\min \left\{ \sup \left| \hat{g}(z_i, \hat{\theta}_i) - g(z_i, \theta_i^*) \right| \right\} \tag{10}$$

where

$$g(z_i, \theta_i^*) = \theta_i^{*T} H(z_i) + \epsilon_i, \tag{11}$$

and

$$\hat{g}(z_i, \theta_i^*) = \hat{\theta}_i^{*T} H(z_i) + \hat{\epsilon}_i^*, \tag{12}$$

where ϵ_i denotes the approximation error bounded by an unknown constant ϵ_i^* such that $|\epsilon_i| \leq \epsilon_i^*$. Signals $\hat{\theta}_i^*$ and $\hat{\epsilon}_i^*$ represent the estimates of θ_i^* and ϵ_i^*, respectively, regulated based on the following adaptive schemes

$$\dot{\hat{\theta}}_i^* = \begin{cases} \alpha_{1,i} |\bar{\xi}_i| H(z_i), & \xi \neq 0 \\ 0, & \xi = 0 \end{cases} , \quad \dot{\hat{\epsilon}}_i^* = \begin{cases} \alpha_{2,i} |\bar{\xi}_i|, & \xi \neq 0 \\ 0, & \xi = 0, \end{cases} \tag{13}$$

where $\{\alpha_{1,i}, \alpha_{2,i}\} > 0$ are constants, and $|\bar{\xi}_i|$ is the upper boundary of $|\xi_i|$.

Lemma 1. *Consider the virtual signal ξ in (9). If the estimator dynamic is modeled in (8), $\bar{\lambda}_{u,i}$ is designed by*

$$\bar{\lambda}_{u,i} = \lambda(\hat{\theta}_i^{*T} H(z_i) + \hat{\epsilon}_i^*), \tag{14}$$

and adaptive schemes are developed in (13), then the signal ξ is bounded and the following condition is satisfied

$$\lim_{t \to \infty} \left[\frac{1}{\lambda} (\zeta_x - \zeta_x^f) - (u^f + \gamma) \right] = 0. \tag{15}$$

Proof. According to (9), the derivative of ξ with respect to time is expressed by

$$\dot{\xi} = \frac{1}{\lambda}(\dot{\zeta}_x - \dot{\zeta}_x^f) - (\dot{u}^f + \dot{\gamma}) = \frac{1}{\lambda}(u + \gamma - \frac{1}{\lambda}(\zeta_x - \zeta_x^f) - (\frac{u - u^f + \text{sgn}(\xi)\bar{\lambda}_u}{\lambda} + \dot{\gamma})$$
$$= -\frac{1}{\lambda}\xi - \frac{1}{\lambda}\text{sgn}(\xi)\bar{\lambda}_u - \dot{\gamma}. \tag{16}$$

Select a candidate Lyapunov function as $V_\xi = \frac{1}{2}\xi^T\xi + \frac{1}{2}\Gamma_1^{-1}\tilde{\theta}^{*T}\tilde{\theta}^* + \frac{1}{2}\Gamma_2^{-1}\tilde{\epsilon}^{*T}\tilde{\epsilon}^*$, where $\tilde{\theta}^* = \hat{\theta}^* - \theta^*$ and $\tilde{\epsilon}^* = \hat{\epsilon}^* - \epsilon^*$. The differential of V_ξ is

$$\dot{V}_\xi = -\frac{1}{\lambda}\|\xi\|^2 - \frac{1}{\lambda}\xi^T\text{sgn}(\xi)\bar{\lambda}_u - \xi^T\dot{\gamma} + \frac{1}{2}\Gamma_1^{-1}\tilde{\theta}^{*T}\dot{\tilde{\theta}}^* + \frac{1}{2}\Gamma_2^{-1}\tilde{\epsilon}^{*T}\dot{\tilde{\epsilon}}^*$$
$$\leq -\frac{1}{\lambda}\|\xi\|^2 - \frac{1}{\lambda}\sum_{i=1}^{m}\xi_i\text{sgn}(\xi_i)\bar{\lambda}_{u,i} + \sum_{i=1}^{m}|\xi_i||\dot{\gamma}_i| + \sum_{i=1}^{m}\frac{1}{2\alpha_{1,i}}\tilde{\theta}_i^{*T}\dot{\tilde{\theta}}_i^* + \sum_{i=1}^{m}\frac{1}{2\alpha_{2,i}}\tilde{\epsilon}_i^*\dot{\tilde{\epsilon}}_i^*. \tag{17}$$

In the light of (11) and Assumption 2, we know that the upper boundary information of $|\xi|$, i.e., $|\bar{\xi}|$ is known. Then (17) can be further rewritten by

$$\dot{V}_\xi \leq -\frac{1}{\lambda}\|\xi\|^2 - \frac{1}{\lambda}\sum_{i=1}^{m}|\bar{\xi}_i|\bar{\lambda}_{u,i} + \sum_{i=1}^{m}|\bar{\xi}_i||\dot{\gamma}_i| + \sum_{i=1}^{m}\frac{1}{2\alpha_{1,i}}\tilde{\theta}_i^{*T}\dot{\tilde{\theta}}_i^* + \sum_{i=1}^{m}\frac{1}{2\alpha_{2,i}}\tilde{\epsilon}_i^*\dot{\tilde{\epsilon}}_i^*. \tag{18}$$

It is followed by $\bar{\lambda}_{u,i}$ in (14) and the adaptive schemes in (13) that

$$
\begin{aligned}
\dot{V}_\xi &\leq -\tfrac{1}{\lambda}\|\xi\|^2 - \sum_{i=1}^m |\bar{\xi}_i|(\hat{\theta}_i^{*\mathrm{T}}H(z_i) + \hat{\epsilon}_i^*) + \sum_{i=1}^m |\bar{\xi}_i||\theta_i^{*\mathrm{T}}H(z_i) + \epsilon_i^*| \\
&\quad + \sum_{i=1}^m \tfrac{1}{2\alpha_{1,i}}\tilde{\theta}_i^{*\mathrm{T}}\dot{\tilde{\theta}}_i^* + \sum_{i=1}^m \tfrac{1}{2\alpha_{2,i}}\tilde{\epsilon}_i^*\dot{\tilde{\epsilon}}_i^* \\
&\leq -\tfrac{1}{\lambda}\|\xi\|^2 - \sum_{i=1}^m |\bar{\xi}_i|\hat{\theta}_i^{*\mathrm{T}}H(z_i) - \sum_{i=1}^m |\bar{\xi}_i|\hat{\epsilon}_i^* + \sum_{i=1}^m |\bar{\xi}_i|\theta_i^{*\mathrm{T}}H(z_i) + \sum_{i=1}^m |\bar{\xi}_i|\epsilon_i^* \\
&\quad + \sum_{i=1}^m \tfrac{1}{2\alpha_{1,i}}\tilde{\theta}_i^{*\mathrm{T}}\dot{\tilde{\theta}}_i^* + \sum_{i=1}^m \tfrac{1}{2\alpha_{2,i}}\tilde{\epsilon}_i^*\dot{\tilde{\epsilon}}_i^* \\
&= -\tfrac{1}{\lambda}\|\xi\|^2 - \sum_{i=1}^m (|\bar{\xi}_i|(\tilde{\theta}_i^{*\mathrm{T}}H(z_i) + \tilde{\epsilon}_i) - \tfrac{1}{2\alpha_{1,i}}\tilde{\theta}_i^{*\mathrm{T}}\dot{\tilde{\theta}}_i^* - \tfrac{1}{2\alpha_{2,i}}\tilde{\epsilon}_i^*\dot{\tilde{\epsilon}}_i^*) \leq -\tfrac{1}{\lambda}\|\xi\|^2.
\end{aligned}
$$

$$(19)$$

Thus employing Barbalat lemma, we can obtain the asymptotic convergence result, and the signal ξ converges to zero when time approaching to infinity.

Note that the signal ξ can display the relationship among the estimator states ζ_x^f, u^f and perturbation γ. Therefore, a perturbation estimation can be developed by

$$
\hat{\gamma} = \lambda^{-1}(\zeta_x - \zeta_x^f) - u^f, \tag{20}
$$

where $\hat{\gamma}$ stands for the estimation of the perturbation. Furthermore, we define the following perturbation estimation error

$$
\tilde{\gamma} = \gamma - \hat{\gamma}. \tag{21}
$$

Theorem 1. *If the estimator dynamic is modeled in (8), $\bar{\lambda}_{u,i}$ is designed in (14) and adaptive schemes are developed in (13), the perturbation estimation error $\tilde{\gamma}$ can reduce to 0 when time approaches to infinity.*

Proof. In terms of the dynamics in (2) and (8), one can obtain

$$
\dot{\zeta}_x^f = \lambda^{-1}(\zeta_x - \zeta_x^f) = u^f + \hat{\gamma}. \tag{22}
$$

From (20), the derivative of $\hat{\gamma}$ with respect to time can be formulated as

$$
\dot{\hat{\gamma}} = \tfrac{1}{\lambda}(u + \gamma - \tfrac{1}{\lambda}(\zeta_x - \zeta_x^f)) - \frac{u - u^f + \mathrm{sgn}(\xi)\bar{\lambda}_u}{\lambda} = \tfrac{1}{\lambda}\tilde{\gamma} - \tfrac{1}{\lambda}\mathrm{sgn}(\xi)\bar{\lambda}_u. \tag{23}
$$

Then it follows from $\tilde{\gamma}$ in (21) that

$$
\dot{\tilde{\gamma}} = -\tfrac{1}{\lambda}\tilde{\gamma} + \dot{\gamma} + \tfrac{1}{\lambda\{\mathrm{sgn}}(\xi)\bar{\lambda}_u. \tag{24}
$$

A candidate Lyapunov function is then selected as

$$
V_1 = \tfrac{1}{2}\tilde{\gamma}^{\mathrm{T}}\tilde{\gamma} + \tfrac{1}{2}\Gamma_1^{-1}\tilde{\theta}^{*\mathrm{T}}\tilde{\theta}^* + \tfrac{1}{2}\Gamma_2^{-1}\tilde{\epsilon}^{*\mathrm{T}}\tilde{\epsilon}^*. \tag{25}
$$

Calculating the differential of (25) yields

$$
\begin{aligned}
\dot{V}_1 &= -\tfrac{1}{\lambda}\tilde{\gamma}^{\mathrm{T}}\tilde{\gamma} + \tilde{\gamma}^{\mathrm{T}}(\dot{\gamma} + \tfrac{1}{\lambda}\mathrm{sgn}(\xi)\bar{\lambda}_u) + \tfrac{1}{2}\Gamma_1^{-1}\tilde{\theta}^{*\mathrm{T}}\dot{\tilde{\theta}}^* + \tfrac{1}{2}\Gamma_2^{-1}\tilde{\epsilon}^{*\mathrm{T}}\dot{\tilde{\epsilon}}^* \\
&= -\tfrac{1}{\lambda}\tilde{\gamma}^{\mathrm{T}}\tilde{\gamma} + \tilde{\gamma}^{\mathrm{T}}\dot{\gamma} + \tfrac{1}{\lambda}\tilde{\gamma}^{\mathrm{T}}\mathrm{sgn}(\xi)\bar{\lambda}_u + \tfrac{1}{2}\Gamma_1^{-1}\tilde{\theta}^{*\mathrm{T}}\dot{\tilde{\theta}}^* + \tfrac{1}{2}\Gamma_2^{-1}\tilde{\epsilon}^{*\mathrm{T}}\dot{\tilde{\epsilon}}^*.
\end{aligned}
$$

$$(26)$$

Based on (9), (20) and (21), we obtain $\tilde{\gamma} = -\xi$ as well as $|\tilde{\gamma}_i| \leq |\bar{\xi}_i|$. Then, one gets

$$
\begin{aligned}
\dot{V}_1 &= -\tfrac{1}{\lambda}\tilde{\gamma}^{\mathrm{T}}\tilde{\gamma} + \tilde{\gamma}^{\mathrm{T}}\dot{\gamma} - \tfrac{1}{\lambda}\xi^{\mathrm{T}}\mathrm{sgn}(\xi)\bar{\lambda}_u + \tfrac{1}{2}\tilde{\theta}^{*\mathrm{T}}\dot{\tilde{\theta}}^* + \tfrac{1}{2\alpha_{2,i}}\tilde{\epsilon}^{*\mathrm{T}}\dot{\tilde{\epsilon}}^* \\
&\leq -\tfrac{1}{\lambda}\tilde{\gamma}^{\mathrm{T}}\tilde{\gamma} + \sum_{i=1}^{m}|\tilde{\gamma}_i||\dot{\gamma}_i| - \tfrac{1}{\lambda}\sum_{i=1}^{m}\xi_i\mathrm{sgn}(\xi_i)\bar{\lambda}_{u,i} + \sum_{i=1}^{m}\tfrac{1}{2\alpha_{1,i}}\tilde{\theta}_i^{*\mathrm{T}}\dot{\tilde{\theta}}_i^* + \sum_{i=1}^{m}\tfrac{1}{2\alpha_{2,i}}\tilde{\epsilon}_i^*\dot{\tilde{\epsilon}}_i^* \\
&= -\tfrac{1}{\lambda}\tilde{\gamma}^{\mathrm{T}}\tilde{\gamma} + \sum_{i=1}^{m}|\bar{\xi}_i||\dot{\gamma}_i| - \tfrac{1}{\lambda}\sum_{i=1}^{m}|\bar{\xi}_i|\bar{\lambda}_{u,i} + \sum_{i=1}^{m}\tfrac{1}{2\alpha_{1,i}}\tilde{\theta}_i^{*\mathrm{T}}\dot{\tilde{\theta}}_i^* + \sum_{i=1}^{m}\tfrac{1}{2\alpha_{2,i}}\tilde{\epsilon}_i^*\dot{\tilde{\epsilon}}_i^*.
\end{aligned}
\tag{27}
$$

From $\bar{\lambda}_{u,i}$ in (14) and adaptive schemes in (13), the above inequality can be further formulated by

$$
\begin{aligned}
\dot{V}_1 &\leq -\tfrac{1}{\lambda}\tilde{\gamma}^{\mathrm{T}}\tilde{\gamma} - \sum_{i=1}^{m}|\bar{\xi}_i|(\hat{\theta}_i^{*\mathrm{T}}H(z_i) + \hat{\epsilon}_i^*) + \sum_{i=1}^{m}|\bar{\xi}_i||\theta_i^{*\mathrm{T}}H(z_i) + \epsilon_i^*| \\
&\quad + \sum_{i=1}^{m}\tfrac{1}{2\alpha_{1,i}}\tilde{\theta}_i^{*\mathrm{T}}\dot{\tilde{\theta}}_i^* + \sum_{i=1}^{m}\tfrac{1}{2\alpha_{2,i}}\tilde{\epsilon}_i^*\dot{\tilde{\epsilon}}_i^* \\
&= -\tfrac{1}{\lambda}\|\tilde{\gamma}\|^2 - \sum_{i=1}^{m}|\bar{\xi}_i|(H(z_i)\tilde{\theta}_i^* + \tilde{\epsilon}_i^*) + \sum_{i=1}^{m}\tfrac{1}{2\alpha_{1,i}}\tilde{\theta}_i^{*\mathrm{T}}\dot{\tilde{\theta}}_i^* + \sum_{i=1}^{m}\tfrac{1}{2\alpha_{2,i}}\tilde{\epsilon}_i^*\dot{\tilde{\epsilon}}_i^* \\
&\leq -\tfrac{1}{\lambda}\|\tilde{\gamma}\|^2.
\end{aligned}
\tag{28}
$$

Then employing Barbalat lemma, the conclusion of the asymptotic estimation result is obtained when time approaches to infinity.

4   Simulation Examples

For the sake of evaluating the efficiency of the proposed nonlinear RBFNN estimator design method, two unmanned systems, i.e., a unmanned marine system and a quadrotor system borrowed by [22] and [27], respectively, are used to derive the simulation results.

Note that in both of simulations, the RBFNN activation function of is chosen as $H(z; b, d) = 1/(1 + e^{(-(bz+d))})$, where b and d represent the center and the width of the activation functions. There are 20 hidden layer neurons in the network, and the parameters b and d are randomly selected within the interval $[-4, 4]$ and $[0, 2]$, respectively.

4.1   Unmanned Marine Systems

For the unmanned marine system, we define the position and velocity states as

$$
x := (x_s, y_s, \psi_s)^{\mathrm{T}}, \quad v = [u_s, v_s, r_s]^{\mathrm{T}},
\tag{29}
$$

respectively, where (x_s, y_s) denote the plane position and ψ_s represents its yaw angle; u_s, v_s, and r_s represent the velocities of surge, sway, and yaw, respectively. Based on [22], the dynamic of the unmanned marine system can be illustrated in (2) and the parameters of the system are given by

$$
C = \begin{bmatrix} 0 & 0 & a_{13} \\ 0 & 0 & a_{23} \\ -a_{13} & -a_{23} & 0 \end{bmatrix}, \quad D = \begin{bmatrix} b_{11} & 0 & 0 \\ 0 & b_{22} & 0 \\ 0 & 0 & b_{33} \end{bmatrix}, \quad M = \begin{bmatrix} 25.8 & 0 & 0 \\ 0 & 33.8 & 1.0948 \\ 0 & 1.0948 & 2.76 \end{bmatrix},
$$

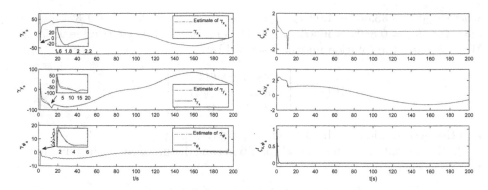

Fig. 1. Estimations of perturbations by using nonlinear RBFNN-based estimator.

Fig. 2. Estimator states $\{\zeta^f_{x,x_s}, \zeta^f_{x,y_s}, \zeta^f_{x,\phi_s}\}$.

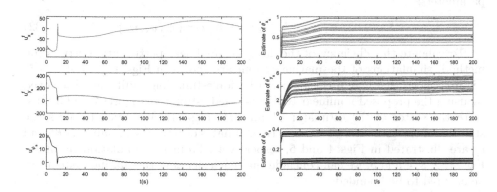

Fig. 3. Estimator states $\{u^f_{x_s}, u^f_{y_s}, u^f_{\phi_s}\}$.

Fig. 4. Adaptive estimates of $\hat{\theta}^*_i$, $i \in \{x_s, y_s, \phi_s\}$.

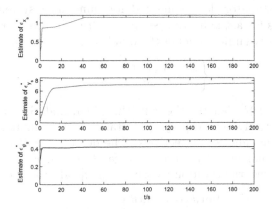

Fig. 5. Adaptive estimations of $\hat{\epsilon}^*_i$, $i \in \{x_s, y_s, \phi_s\}$.

where $a_{13} = -33.8v_s - 1.098r_s$, $a_{23} = 25.8u_s$, $b_{11} = 0.7225 + 1.3274|u_s| + 5.8864u_s^2$, $b_{22} = 0.8896 + 36.4728|v_s| + 0.805|r_s|$, $b_{33} = 1.9 - 0.08|v_s| + 0.75|r_s|$. $W(x)$ denotes the rotation matrix described by

$$W(x) = \begin{bmatrix} \cos(\psi_s) & -\sin(\psi_s) & 0 \\ \sin(\psi_s) & \cos(\psi_s) & 0 \\ 0 & 0 & 1 \end{bmatrix},$$

and the perturbation can be defined by $\gamma = [\gamma_{x_s}, \gamma_{y_s}, \gamma_{\psi_s}]^{\mathrm{T}}$.

In the simulation, system initial conditions and disturbances of the unmanned marine system are:

$$[x_s(t_0), y_s(t_0), \psi_s(t_0)]^{\mathrm{T}} = [0.8, 1.6, 0]^{\mathrm{T}}, \tag{30}$$

$$[w_{x_s}(t), w_{y_s}(t), w_{\psi_s}(t)]^{\mathrm{T}} = [0.4, 0.4 \times \cos(0.3 \times t), -0.6 \times \cos(0.3 \times t)]^{\mathrm{T}}, \tag{31}$$

respectively. And the unmanned marine system is simulated under the following parameters:

$$\lambda = 0.05, \ \alpha_{1,i} = 0.5, \ \alpha_{2,i} = 0.5, \ i = \{x_s, u_s, \psi_s\}. \tag{32}$$

To validate the efficiency of the designed nonlinear perturbation estimation method, simulations of the marine system are shown in Figs. 1, 2, 3, 4 and 5. Figure 1 shows that high precision perturbation estimation result can be achieved by using the proposed nonlinear RBFNN-based estimator. Figures 2 and 3 illustrate the states of the estimator ζ_x^f and u^f, respectively, which exhibits the stability of the estimator. The adaptive estimates of RBFNN parameters θ_i^* and ϵ_i^* are illustrated in Figs. 4 and 5, respectively. From the simulation results of Figs. 1, 2, 3, 4 and 5, we can conclude that each signal in the marine system can be ensured to be bounded.

4.2 A Quadrotor System

To further embody the efficiency and superiority for the developed nonlinear RBFNN-based estimator design method for unmanned systems, comparative simulation results for a quadrotor system is utilized in this subsection. Refer to [27], the attitude dynamic model of the quadrotor can be expressed by

$$\begin{cases} \dot{x} = W(x)v \\ J\dot{v} = -v^\times Jv + \tau + w, \end{cases} \tag{33}$$

where $x := (\phi, \theta, \psi)^{\mathrm{T}}$ denotes the roll, the pitch, and the yaw angles of the quadrotor, respectively; $v := (v_\phi, v_\theta, v_\psi)^{\mathrm{T}}$ stands for the angular velocities of the roll, pitch, and yaw angles, respectively. The definite positive diagonal matrix J stands for the moment of inertia; $w = (w_\phi(t), w_\theta(t), w_\psi(t))^{\mathrm{T}}$ is the external disturbance. Besides, the rotation matrix of the quadrotor is defined as

$$W(x) = \begin{bmatrix} 1 & \sin\phi\tan\theta & \cos\phi\tan\theta \\ 0 & \cos\phi & -\sin\phi \\ 0 & \sin\phi\sec\theta & \cos\phi\sec\theta \end{bmatrix}, \quad v^\times = \begin{bmatrix} 0 & -v_\psi & v_\theta \\ v_\psi & 0 & -v_\phi \\ -v_\theta & v_\phi & 0 \end{bmatrix}.$$

According to (33) and defining $u = J^{-1}W(x)\tau$ and $\gamma = [\gamma_\phi, \gamma_\theta, \gamma_\psi]^{\mathrm{T}} = \dot{W}(x)v + W(x)J^{-1}(-v^\times Jv + w)$, the above dynamic model can be simplified as

$$\begin{cases} \dot{x} = \zeta_x \\ \dot{\zeta}_x = u + \gamma, \end{cases} \tag{34}$$

where $\zeta_x = W(x)v$.

In the simulation, the initial conditions of the quadrotor are:

$$[\phi(t_0), \theta(t_0), \psi(t_0)]^{\mathrm{T}} = [1.8, 0.2, 0.5]^{\mathrm{T}}. \tag{35}$$

The disturbances for the quadrotor are considered by

$$[w_\phi(t), w_\theta(t), w_\psi(t)]^{\mathrm{T}} = [-0.6, 0.6 \times \sin(0.4 \times t), -0.5 \times \sin(5 \times t)] \cdot \tag{36}$$

And the quadrotor is simulated under the following parameters:

$$\lambda = 0.01, \ \alpha_{1,i} = 0.5, \ \alpha_{2,i} = 0.5, \ i = \{\phi, \theta, \psi\}.$$

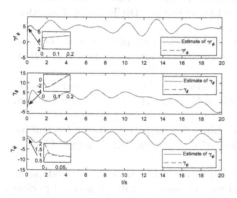

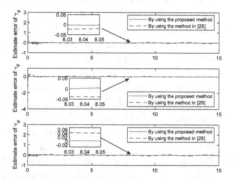

Fig. 6. Estimations of perturbations by using nonlinear estimators.

Fig. 7. Comparative results of estimation errors $\{\tilde{\gamma}_\psi, \tilde{\gamma}_\theta, \tilde{\gamma}_\phi\}$ via developed approach and approach in [25].

Considering the above mentioned cases, Fig. 6 is provided to show the satisfied perturbation estimation result for the quadrotor system. Note that the response curves of the estimator states ζ_x^f and u^f and adaptive estimations of RBFNN parameters θ_i^* and ϵ_i^* are similar with the curves demonstrated in Figs. 4 and 5, so that they are omitted here. For the sake of confirming the advantages of the developed nonlinear estimator design approach, comparative results with the existing linear estimators, that is, the unknown system dynamics estimator proposed in [25] are demonstrated in Fig. 7. From Fig. 7, we can easily see that the perturbation estimation errors based on the developed approach are smaller than the errors by utilizing the approach designed in [25]. Thus, the comparative results validate the advantages of the developed estimation approach.

5   Conclusion

In this paper, a nonlinear RBFNN-based estimator design method has been investigated to deal with the perturbation estimation problem for perturbed unmanned systems. The nonlinear RBFNN-based estimator has approximated the unknown perturbation online by showing the relationship between the estimation operation and the perturbation. Asymptotic perturbation estimation results have been obtained by analyzing Lyapunov functions. Finally, simulation results for two unmanned systems have also provided to confirm the effectiveness of the developed nonlinear RBFNN estimator design method.

References

1. Jin, X.Z., Che, W.W., Wu, Z.G., Zhao, Z.: Adaptive consensus and circuital implementation of a class of faulty multiagent systems. IEEE Trans. Syst. Man Cybern. Syst. **52**(1), 226–237 (2022)
2. Wang, H., et al.: Precise discrete-time steering control for robotic fish based on data-assisted technique and super-twisting-like algorithm. IEEE Trans. Ind. Electron. **67**(12), 10587–10599 (2020)
3. Jin, X.Z., He, T., Wu, X.M., Wang, H., Chi, J.: Robust adaptive neural network-based compensation control of a class of quadrotor aircrafts. J. Frankl. Inst. **357**(17), 12241–12263 (2020)
4. Jin, X., Che, W.-W., Wu, Z.-G., Deng, C.: Robust adaptive general formation control of a class of networked quadrotor aircraft. IEEE Trans. Syst. Man Cybern. Syst. **52**(12), 7714–7726 (2022)
5. Fossen, T.I., Pettersen, K.Y., Galeazzi, R.: Line-of-sight path following for Dubins paths with adaptive sideslip compensation of drift forces. IEEE Trans. Control Syst. Technol. **23**(2), 820–827 (2015)
6. Jin, X., Jiang, J., Wang, H., Deng, C.: Nonlinear ELM estimator-based path-following control for perturbed unmanned marine systems with prescribed performance. Neural. Comput. Appl. (2023). https://doi.org/10.1007/s00521-023-08653-3
7. Gao, H., Li, Z., Yu, X., Qiu, J.: Hierarchical multiobjective heuristic for PCB assembly optimization in a beam-head surface mounter. IEEE Trans. Cybern. **52**(7), 6911–6924 (2022)
8. Jin, X., Jiang, C., Qin, J., Zheng, W.X.: Robust pinning constrained control and adaptive regulation of coupled Chua's circuit networks. IEEE Trans. Circ. Syst. I Reg. Papers **66**(10), 3928–3940 (2019)
9. Liu, Z., Lin, W., Yu, X., Rodriguez-Andina, J.J., Gao, H.: Approximation-free robust synchronization control for dual-linear-motors-driven systems with uncertainties and disturbances. IEEE Trans. Ind. Electron. **69**(10), 10500–10509 (2022)
10. Jin, X.Z., Yang, G.H.: Robust synchronization control for complex networks with disturbed sampling couplings. Commun. Nonlinear Sci. Numer. Simul. **19**(6), 1985–1995 (2014)
11. Jin, X., Lü, S., Yu, J.: Adaptive NN-based consensus for a class of nonlinear multi-agent systems with actuator faults and faulty networks. IEEE Trans. Neural Netw. Learn. Syst. **33**(8), 3474–3486 (2022)

12. Jin, X., Lü, S., Qin, J., Zheng, W.X., Liu, Q.: Adaptive ELM-based security control for a class of nonlinear interconnected systems with DoS attacks. IEEE Trans. Cybern. **53**(8), 5000–5012 (2023)

13. Jin, X., Zhao, X., Yu, J., Wu, X., Chi, J.: Adaptive fault-tolerant consensus for a class of leader-following systems using neural network learning strategy. Neural Netw. **121**, 474–483 (2020)

14. Hu, Y., Wang, H., Yazdani, A., Man, Z.: Adaptive full order sliding mode control for electronic throttle valve system with fixed time convergence using extreme learning machine. Neural Comput. Appl. **34**, 5241–5253 (2022)

15. Wang, H., et al.: Adaptive integral terminal sliding mode control for automotive electronic throttle via an uncertainty observer and experimental validation. IEEE Trans. Veh. Technol. **67**(9), 8129–8143 (2018)

16. Jin, X., Wang, S., Qin, J., Zheng, W.X., Kang, Y.: Adaptive fault-tolerant consensus for a class of uncertain nonlinear second-order multi-agent systems with circuit implementation. IEEE Trans. Circ. Syst. I Reg. Papers **5**(7), 2243–2255 (2018)

17. Ma, Y., Che, W., Deng, C., Wu, Z.: Distributed model-free adaptive control for learning nonlinear MASs under DoS attacks. IEEE Trans. Neural Netw. Learn. Syst. **34**(3), 1146–1155 (2023)

18. Jin, X.Z., Yang, G.H.: Distributed adaptive robust tracking and model matching control with actuator faults and interconnection failures. Int. J. Control Autom. Syst. **7**, 702–710 (2009)

19. Jin, X., Che, W.-W., Wu, Z.-G., Wang, H.: Analog control circuit designs for a class of continuous-time adaptive fault-tolerant control systems. IEEE Trans. Cybern. **52**(6), 4209–4220 (2020)

20. Karl, D., Helen, E., Henninger, C.: A higher order sliding mode controller-observer for marine vehicles. IFAC-PapersOnLine **52**(21), 341–346 (2019)

21. Cao, H., Xu, R., Zhao, S., Li, M., Song, X., Dai, H.: Robust trajectory tracking for fully-input-bounded actuated unmanned surface vessel with stochastic disturbances: an approach by the homogeneous nonlinear extended state observer and dynamic surface control. Ocean Eng. **243**, 110113 (2022)

22. Lü, G., Peng, Z., Wang, H., Liu, L., Wang, D., Li, T.: Extended-state-observer-based distributed model predictive formation control of under-actuated unmanned surface vehicles with collision avoidance. Ocean Eng. **238**, 109587 (2021)

23. Feng, N., Wu, D., Yu, H., Yamashita, A.S., Huang, Y.: Predictive compensator based event-triggered model predictive control with nonlinear disturbance observer for unmanned surface vehicle under cyber-attacks. Ocean Eng. **259**, 111868 (2022)

24. Na, J., Yang, J., Shu, S.B.: Unknown dynamics estimator-based output-feedback control for nonlinear pure-feedback systems. IEEE Trans. Syst. Man Cybern. Syst. **51**(6), 3832–3843 (2021)

25. Huang, Y.B., Wu, J.D., Na, J., Han, S., Gao, G.: Unknown system dynamics estimator for active vehicle suspension control systems with time-varying delay. IEEE Trans. Cybern. **52**(8), 8504–8514 (2022)

26. Buhmann, M.D.: Radial Basis Functions. Theory and Implementations. Cambridge University (2003)

27. Tian, B.L., Cui, J., Lu, H.C.: Adaptive finite-time attitude tracking of quadrotors with experiments and comparisons. IEEE Trans. Ind. Electron. **66**(12), 9428–9438 (2019)

DOS Dataset: A Novel Indoor Deformable Object Segmentation Dataset for Sweeping Robots

Zehan Tan[1], Weidong Yang[1,2(✉)], and Zhiwei Zhang[3]

[1] School of Computer Science, Fudan University, Shanghai, China
{18110240062,wdyang}@fudan.edu.cn
[2] Zhuhai Fudan Innovation Institute, Hengqin New Area, Zhuhai, Guangdong, China
[3] Gree Electric Appliances, INC.Of Zhuhai, Zhuhai, China

Abstract. Path planning for sweeping robots requires avoiding specific obstacles, particularly deformable objects such as socks, ropes, faeces, and plastic bags. These objects can cause secondary pollution or hinder the robot's cleaning capabilities. However, there is a lack of specific datasets for deformable obstacles in indoor environments. Existing datasets either focus on outdoor scenes or lack semantic segmentation annotations for deformable objects. In this paper, we introduce the first dataset for detecting and segmenting deformable objects in indoor sweeping robot scenarios, DOS Dataset. We believe that DOS will catalyze research in semantic segmentation of deformable objects for indoor robot obstacle avoidance applications.

Keyword: Computer Vision, Semantic Segmentation, Indoor Deformable Object, Sweeping Robot

1 Introduction

Service robots are becoming increasingly prevalent in various application scenarios, making robot perception crucial for improving service quality and expanding service content. A typical application of service robots indoors is sweeping robots. The first prerequisite for an indoor sweeping robot to move autonomously is localization and mapping. Once localization and mapping are completed, path planning is carried out. In the process of moving, for safety reasons, the robot must be able to sense the state of obstacles on the map. The recognition of certain obstacles in the indoor environment is therefore particularly important. Current 2D-based target detection and segmentation algorithms can be categorized into feature-based matching and deep learning models. Relevant indoor datasets [1, 8, 15, 21, 31, 33, 37] mainly focus on detecting and segmenting objects with fixed shapes, such as furniture, electrical appliances, and decorative objects (discussed in Sect. 2). Unfortunately, some obstacles of particular interest, such as plastic bags that may tangle the robot roller or pet faeces causing significant secondary pollution (see Fig. 1), lack fixed shapes. We define such objects as deformable objects, and detecting them in 2D images is our primary concern. We aim to collect images of indoor deformable objects, design an effective labeling format for them, and develop an efficient semantic segmentation algorithm.

Left: Example of a robot roller tangled. **Right:** Example of secondary pollution.

Fig. 1. Sweeping robot fails to avoid deformable obstacles.

Detecting and segmenting deformable objects in indoor environments can be more challenging than identifying objects with fixed shapes. To the best of our knowledge, there are no publicly available datasets for semantic segmentation of deformable objects for indoor sweeping robots. Moreover, few relevant datasets either provide indoor object detection without semantic segmentation or cater exclusively to outdoor scenes. Therefore, we introduce the first openly available indoor deformable object detection dataset, DOS, designed specifically for detecting and segmenting deformable obstacles encountered by indoor sweeping robots. We envision that our open-access dataset will enable researchers to develop more robust and data-efficient algorithms for deformable object detection and other related problems, such as robot grasping.

The remainder of this paper is organized as follows: In Sect. 2, we summarize past studies on commonly used databases for image segmentation and their corresponding algorithms, primarily focusing on deep learning. In Sect. 3, we describe the proposed DOS dataset in detail. In Sect. 4, we evaluate several state-of-the-art semantic segmentation methods based on deep learning using the proposed dataset. Finally, in Sect. 5, we conclude this paper.

2 Related Work

With the rapid development of artificial intelligence technology, indoor sweeping robots have become widely applied in daily human life. Sweeping robots encompass three core elements: navigation, obstacle avoidance, and cleaning. Obstacle avoidance poses a challenge in accurately identifying obstacles encountered by the sweeping robot. Once the robot detects an obstacle, it must analyze it in real-time, using vision sensors to detect the obstacle's three-dimensional information and assess its size, then choose a suitable obstacle avoidance strategy to plan an appropriate navigation route [18].

A high-quality dataset of a large scale (i.e., clear images, balanced classes, and precise annotations) aids a model in learning the abstract representation of a problem from generalizable features [26,40]. Nevertheless, the paucity of waste image datasets poses a considerable barrier to the progress of intelligent waste recognition. This section will provide a succinct overview of recent works on indoor deformable object classification, detection, and segmentation, which closely align with ours.

2.1 Detection and Segmentation Datasets

Many datasets for image segmentation have been proposed to densely recognize general objects and "stuff" in image scenes like waste images [3,14,20,25, 30,32,39], natural scenes [4,10,22,50], and indoor spaces [1,8,15,21,31,33,37]. However, few have been designed for the challenging vision task required by sweeping robots to avoid deformable obstacles. We found that related datasets for natural scenes and indoor spaces focus predominantly on indoor furniture, appliances, daily supplies, and decorations, rather than on garbage or obstacles on the floor. For example, InteriorNet [21] contains 22 million indoor scenes, and Scene [50] includes over 20,000 images, but these indoor spaces datasets barely classify ground trash, let alone study deformable objects. Although waste images datasets [14,25,32] have explored ground trash recognition, and the paper [14] specifically proposed a method to distinguish cleanable and non-cleanable objects, these datasets and corresponding methods focus only on object detection and do not involve ground obstacle segmentation. The application scenarios of datasets [3,20] are conveyor belts, making them unsuitable for indoor sweeping robot obstacle detection. Deformable Linear Objects (DLOs) [17] are a class of objects that includes ropes, cables, threads, sutures, and wires, but DLOs only use traditional skeletonization, and contour extraction methods to detect objects, and no dataset has been provided. MJU-Waste [39] formulates the problem as a two-class (waste vs. background) semantic segmentation for waste object segmentation, with 2475 images. Perhaps the most similar dataset to ours is the TACO [30] of 1500 densely annotated images containing objects of 60 litter classes. However, TACO primarily captures outdoor scenes and lacks object classification for socks and pet faeces, rendering it less practical for indoor sweeping robot obstacle avoidance scenarios. In contrast, our DOS dataset comprises nearly three times more annotated images than TACO, collected from indoor environments in a manner that simulates the visual acquisition of a sweeping robot. This approach makes our dataset more closely aligned with real applications, such as sweeping robot obstacle avoidance or indoor robot grasping.

2.2 Detection and Segmentation Methods

Semantic segmentation, which aims to partition an image into visually significant regions, plays a vital role in numerous applications such as scene understanding, medical image analysis, robot perception, and satellite image segmentation [27,28]. This domain has seen significant progress, with contemporary models demonstrating leading-edge performance in recognizing diverse object and stuff classes in natural scene images [5,6,19,23,38,41,43,44,46,48,49]. Initially pioneered by Long, Shelhamer, and Darrell [24], a host of architectures and techniques have been developed to augment Fully Convolutional Networks (FCN), capture multi-scale context, or refine boundaries of objects [29,36,45]. Further explorations have delved into the realms of encoder-decoder structures [2,6,34] and self-attention mechanism-based contextual dependencies [13,16,42]. Noteworthy advances include the Pyramid Scene Parsing Network (PSPNet) [48], which utilizes a pyramid pooling module to amalgamate multi-scale representations. KNet [47] leverages a group of learnable kernels, each tasked with generating a mask for a potential instance or a stuff class. SegNeXt [12], on the other hand, prompts spatial attention via multi-scale convolutional features. In recent years, the landscape of semantic segmentation has been heavily influenced by transformer-based models, largely attributed to their adeptness in spatial information encoding through self-attention. For instance, Segmenter [38] expands the applications of the recently developed Vision Transformer (ViT) [9] to encompass semantic segmentation.

3 DOS Dateset

In this section, we summarize the data we collected using our framework to establish the DOS dataset. This dataset is a snapshot of available data from roughly one month of data acquisition by 17 users. We used the open-source LabelMe [35] annotation toolkit, whice provides an object refinement yielding close-to-exact per pixel label (pixel boundaries), to manually collect the polygon annotations of deformable objects of four types: faeces, socks, plastic bag, and rope. Here, we report some statistics for an initial 3,056 images of the DOS dataset, which are summarized in Table 1.

However, TACO contains an intentional collection of outdoor scenes with a small number of deformable objects. For instance, the number of deformable dangerous objects available for indoor obstacle avoidance in this dataset is 850 labeled plastic bags & wrappers and 29 labeled ropes & strings, but no faeces and socks label. This makes the model trained on TACO less practical for indoor sweeping robot obstacle avoidance scenarios. Obviously, there are more classes of deformable objects in the DOS than TACO, and the number of images is also greater than TACO, with all collected scenes being indoor scenes. This enables the model trained on DOS to be effectively applied to indoor sweeping robot obstacle avoidance scenarios. For a comparison, please refer to Table 1 and Fig. 2.

Table 1. Summary statistics for DOS compared to the most similar existing dataset (TACO [30]). DOS has significantly more data than TACO, with over twice as many deformable objects and more than 2,900 annotated object instances.

Statistic	TACO	DOS
Images	1500	3056
Objects Per Image	3.19	2.52
Deformable Objects(for sweeping robot)	879	7687
Labeled Objects	4784	7687
Classes	60	4
Collected Scenes	outdoor	indoor

Left: Examples of the existing waste detection and segmentation dataset (TACO). **Right:** Examples of our deformable object detection and segmentation dataset (DOS).

Fig. 2. Cropped example images from TACO and DOS datasets and their groundtruths.

3.1 Data Collection

We utilized mobile phones for capturing images in our dataset and saved them using a uniform data format (JPEG). To guarantee the dataset's diversity, we collected images from various household settings (over 45 households), encompassing different interior design styles, room configurations, and flooring materials. Identifying appropriate target categories within the scenes is crucial for dataset collection. We examined and analyzed prevalent target categories that sweeping robots need to avoid during their operation, ultimately selecting four objects: faeces, rope, plastic bag, and socks. Sweeping robots struggle to clean feces efficiently, resulting in secondary pollution, while ropes (e.g., data cables, plastic ropes, and wires), plastic bags, and socks tend to impede the robot's motion mechanism. Consequently, these four objects are essential in the dataset, as the sweeping robot must detect and evade these targets while operating.

To emulate the sweeping robot's viewpoint, we required a horizontal photo angle, with the camera placed approximately 20–30 cm above the floor and at a distance of roughly 0.5–3 m from the target object. Moreover, we randomly altered the target objects' shapes during the image collection process and took photos under diverse lighting conditions to accommodate the actual working environment of the sweeping robot. To collect more objects, we substituted some of the feces with play-doh.

3.2   Data Annotation

The annotation process was carried out by individuals experienced in semantic segmentation research who had also undergone annotation training. We used the LabelMe tool [35] to annotate the DOS dataset by drawing polygon bounding boxes to represent specific category instances within the images. The annotated categories comprise faeces, ropes, plastic bags, and socks, which are labeled as foreground, while the remaining areas are designated as background. A polygon region must encompass all foreground object pixels and may contain a minimal amount of background pixels. If an object is partially occluded, we only annotate the segmentation mask for the visible part. For the rope category, since ropes often become entangled, we focus on depicting the entire bundle's outer contour rather than adhering strictly to the rope's linear features. To simplify the validation of deep learning algorithms and ensure the dataset's adaptability, we converted the dataset annotation files into COCO [22] and PASCAL VOC [10] data formats.

3.3   Dataset Description

Our dataset comprises 3,056 images, which have been randomly partitioned into training and validation sets at 8:1 ratio. The training set contains 6,800 semantic segmentation labels, while the validation set includes 887 annotated labels. We examined the categories present within the dataset, as depicted in Fig. 3(a). Faeces is the most prevalent category, accounting for 3,257 instances (42.37%). Interestingly, socks have the lowest occurrence rate, with merely 704 labels, representing 9.16% of the total instances.

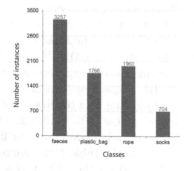

(a) Number of annotations per category.

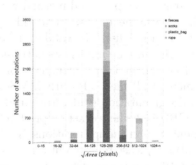

(b) Histogram of bounding box sizes per category.

Fig. 3. Overview of Dataset: Annotation Categories and Bounding Box Sizes.

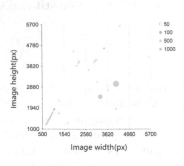

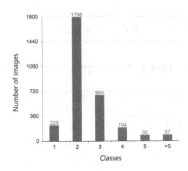

(a) Distribution of image resolutions.   (b) Instances per image.

Fig. 4. Dataset Details: Image Resolution Distribution and Foreground Instances.

Figure 3(b) illustrates the dimensions of the labels for each category. As observed, the majority of faeces areas range from 64×64 to 256×256 pixels. The area occupied by socks is marginally larger than that of faeces, typically spanning between 128×128 and 512×512 pixels. Plastic bag labels exhibit the most substantial area, predominantly between 128×128 and 1024×1024 pixels. The size distribution of ropes is relatively consistent, which correlates with the deformability of ropes. Based on the aforementioned analysis, our constructed dataset aligns with real-world scenarios. The dataset encompasses high-resolution images with various size combinations, as demonstrated in Fig. 4(a). Figure 4(b) presents the statistics for the number of foreground instances in each frame within the training and validation sets. As can be seen, the majority of images feature two foreground targets.

4   Experiments

In this section, we provided baseline results for our proposed DOS dataset. We compared the proposed method with state-of-the-art semantic segmentation baselines with the dataset randomly partitioned into training and validation sets at 8:1 ratio where the validation set and the test set uses the same data. In this experiment, the models were trained with Intel Core i7-10700 CPU, Nvidia GTX 1080 Ti, and 32G memory. For semantic segmentation experiments, we implemented all methods with MMSegmentation [7]. Some key parameters in the MMSegmentation configuration file are as the following: the batch size of a single GPU is 2, the worker to pre-fetch data for each single GPU is also 2, and the number of training iterations is 160K. For some detailed configuration file descriptions, please refer to our open-source code on GitHub. Table 2 summarises the quantitative performance evaluation results obtained on the DOS test set. The performance of the following baseline methods is reported for this dataset:

– DeepLabv3 [5] utilises the Atrous Spatial Pyramid Pooling (ASPP) module, enabling the capture of long-range contexts. Specifically, ASPP employs parallel dilated convolutions with variable atrous rates to encode features from different sized receptive fields.

- PSPNet [48] introduces the pyramid pooling module for multi-scale context aggregation. This module concatenates the last layer features of the conv4 block with the same features applied with 1×1, 2×2, 3×3, and 6×6 average pooling and upsampling, resulting in the harvesting of multi-scale contexts.
- PointRend [19] frames image segmentation as a rendering problem. PointRend conducts point-based segmentation predictions at adaptively selected locations, leveraging an iterative subdivision algorithm. It can be conveniently applied to both instance and semantic segmentation tasks.
- OCRNet [42] tackles the context aggregation problem in semantic segmentation, proposing an effective object-contextual representations approach. This approach characterises a pixel by utilising the representation of the corresponding object class.
- ISANet [13] factorizes the dense affinity matrix as the product of two sparse affinity matrices, introducing an interlaced sparse self-attention approach to enhance the self-attention mechanism's efficiency in semantic segmentation. Two attention modules each estimate a sparse affinity matrix. These modules are designed to allow every position to receive information from all other positions.
- KNet [47] employs a set of learnable kernels to segment both instances and semantic categories. Each kernel is responsible for generating a mask for a potential instance or a stuff class. KNet introduces a kernel update strategy that enables each kernel to be dynamic and conditional on its meaningful group in the input image.
- Segmenter [38] introduces a transformer model for semantic segmentation. Unlike convolution-based methods, Segmenter allows global context modelling from the first layer and throughout the network. It builds on the Vision Transformer (ViT) [9], extending it to semantic segmentation. Segmenter relies on output embeddings corresponding to image patches, obtaining class labels from these embeddings using a point-wise linear decoder or a mask transformer decoder.
- SegFormer [41] designs a series of Mix Transformer encoders (MiT) that output multi-scale features, obviating the need for positional encoding. SegFormer eschews complex decoders, opting for a lightweight multi-layer perceptron (MLP) decoder that aggregates information from different layers. It combines local and global attention to produce potent representations.
- SegNeXt [12] stacks building blocks to create a convolutional encoder (MSCAN) that invokes spatial attention via multi-scale convolutional features for semantic segmentation. SegNeXt also designs a novel convolutional attention network using inexpensive convolutional operations. It incorporates a decomposition-based Hamburger module [11] for global information extraction.

For the quantitative evaluation, we report the performance of baseline methods and the proposed method by three metrics: mean intersection over union

($mIoU$), mean Accuracy ($mAcc$), and all Accuracy ($aAcc$). Let TP, FP, and FN denote the total number of true positive, false positive, and false negative pixels, respectively. The Intersection over Union (IoU) is calculated as follows:

$$IoU_i = \frac{GT_i \cap \text{Pred}_i}{GT_i \cup \text{Pred}_i} \tag{1}$$

$$mIoU = \frac{1}{n} \sum_i^n IoU_i \tag{2}$$

The accuracy metric, denoted as Acc, is computed as follows:

$$Acc = \frac{TP + TN}{TP + FN + FP + FN} \tag{3}$$

$$mAcc = \frac{1}{n} \sum_i^n Acc_i \tag{4}$$

$$aAcc = \frac{Num_{\text{True}}}{Num} \tag{5}$$

where GT stands for ground truth, i denotes the semantic categories, and n symbolizes the total number of classes. Specifically, the average accuracy ($aAcc$) is determined by dividing the count of correctly classified pixels within the prediction map by the total pixel count.

Table 2. Performance comparisons on the test set of DOS. For each method, we report the mean intersection over union ($mIoU$), mean Accuracy ($mAcc$), and all Accuracy ($aAcc$).

Method	Backbone	aAcc	mIoU	mAcc
DeepLabv3	ResNet-101	99.29	76.36	84.90
PSPNet	ResNet-101	99.35	78.34	84.57
PointRend	ResNet-101	99.42	81.50	87.37
OCRNet	ResNet-101	99.32	76.91	85.64
ISANet	ResNet-101	99.36	78.96	86.18
KNet	ResNet-101	99.40	80.35	88.50
Segmenter	ViT	99.42	81.91	88.03
SegFormer	MiT	99.43	82.63	89.61
SegNeXt	MSCAN	**99.54**	**86.06**	**91.90**

Table 2 results demonstrate that our DOS dataset poses a challenging task for detecting and segmenting deformable objects in indoor sweeping robots. We note that in the experimental results: the worst $aAcc$, $mIoU$, and $mAcc$ are 99.29 % 76.36%, and 84.57%; the best $aAcc$, $mIoU$, and $mAcc$ are 99.54%, 86.06%, and 91.90%. This means that the accuracy of all three metrics is good, and all methods are using the default parameters of the model, which further indicates that our dataset has a good distribution and annotations. To ensure a robust, reproducible metric of these models in the experiment, we ran each model five times and averaged $aAcc$, $mIoU$, and $mAcc$ with a random test set.

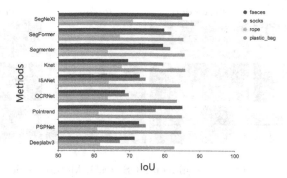

Fig. 5. Comparison of IoU performance for different categories on the DOS test set. For each method, we report the IoU for each category on the test set of DOS.

Figure 5 shows the IoU of the four categories of deformable objects with different methods. From the perspective of area (see Fig. 3(b)) and number (see Fig. 3(a)) of each category, we find that plastic bags have the largest area and number, resulting in the best IoU. They are followed by faeces, which have the highest number, and Fig. 3(b) shows that most faeces have areas concentrated in the intervals 64–128 and 128–256. Ropes and socks perform worse, as shown in the analysis of the confusion matrices in Fig. 6.

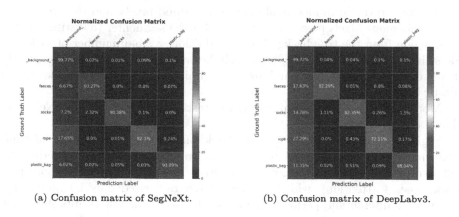

(a) Confusion matrix of SegNeXt.   (b) Confusion matrix of DeepLabv3.

Fig. 6. Confusion Matrices of SegNeXt and DeepLabv3.

The confusion matrices in Fig. 6(a) (best $mIoU$) and Fig. 6(b) (worst $mIoU$) show that each model achieves good segmentation accuracy in our dataset. However, it is notable that the higher probability of each deformable object being segmented as the background is related to the higher percentage of background pixel regions in the original dataset. Specifically, the highest probability of ropes being identified as background may be related to the way our rope category is labeled. The random stacking form of ropes and the way the outer contours are

Image	Prediction	Image	Prediction

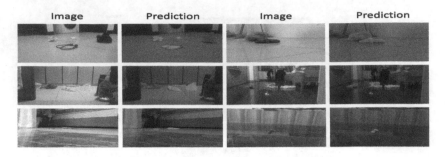

Fig. 7. Segmentation results on DOS (test). Method is SegNeXt.

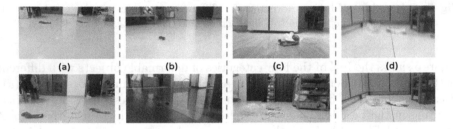

(a)	(b)	(c)	(d)

Fig. 8. Examples of Dataset Variability.

labeled lead to the introduction of many background pixels, and the presence of these ambiguous pixels presents a challenge for the segmenter to distinguish the rope region from the true background. In addition, a small number of socks are identified as faeces and plastic bags, likely because socks appear less frequently in the dataset (see Fig. 3(a)) and are more similar in appearance to faeces and plastic bags.

More example results obtained on the test set of DOS with SegNeXt are shown in Fig. 7.

SegNeXt (characterized by a strong backbone network, multi-scale information interaction, and spatial attention) performs the best in the experiments, followed by SegFormer and Segmenter (characterized by a strong backbone network, spatial attention), while DeepLabv3 (characterized only by multi-scale information interaction) performs the worst. The experimental results show that a good semantic segmentation model should have a strong backbone network, multi-scale information interaction, and spatial attention.

In order to showcase the diversity of the collected images, Fig. 8 provides examples of objects with occlusions, varying lighting conditions, and varying distances within the dataset. Specifically, Fig. 8(a) displays images with objects at different distances, Fig. 8(b) provides images under different illumination conditions, Fig. 8(c) presents images with occluded objects, and Fig. 8(d) offers images with both blurred and clear quality.

5   Conclusion

Following a comprehensive examination of related works, we present a novel dataset, designated as DOS, designed specifically to serve as a benchmark for semantic segmentation of deformable objects within the context of obstacle avoidance in indoor robotic sweeping scenarios. Performance evaluation of leading object detection algorithms on this dataset reveals that convolutional networks employing convolutional attention mechanisms exhibit superior performance in the semantic segmentation of deformable objects.

It is our anticipation that this contribution will catalyze further exploration into vision-based indoor robotic perception, particularly within complex real-world scenarios. Future research directions could include investigations into point cloud segmentation, with a focus on handling multiple datasets presenting occluded objects and variations in camera perspectives.

Our dataset, code, configuration files, and trained model weights used in this work are publicly available at: https://github.com/zehantan6970/DOS_Dataset.

Acknowledgements. This work is supported by Anonymity.

References

1. Adhikari, B., Peltomaki, J., Puura, J., Huttunen, H.: Faster bounding box annotation for object detection in indoor scenes. In: 2018 7th European Workshop on Visual Information Processing (EUVIP), pp. 1–6. IEEE (2018). https://doi.org/10.1109/EUVIP.2018.8611732
2. Badrinarayanan, V., Kendall, A., Cipolla, R.: SegNet: a deep convolutional encoder-decoder architecture for image segmentation. IEEE Trans. Pattern Anal. Mach. Intell. **39**(12), 2481–2495 (2017). https://doi.org/10.1109/TPAMI.2016.2644615
3. Bashkirova, D., et al.: ZeroWaste dataset: towards deformable object segmentation in cluttered scenes. In: Proceedings of the IEEE/CVF Conference on Computer Vision and Pattern Recognition, pp. 21147–21157 (2022). https://doi.org/10.48550/arXiv.2106.02740
4. Caesar, H., Uijlings, J., Ferrari, V.: COCO-stuff: thing and stuff classes in context. In: Proceedings of the IEEE Conference on Computer Vision and Pattern Recognition, pp. 1209–1218 (2018). https://doi.org/10.48550/arXiv.1612.03716
5. Chen, L.C., Papandreou, G., Schroff, F., Adam, H.: Rethinking atrous convolution for semantic image segmentation. arXiv preprint arXiv:1706.05587 (2017). https://doi.org/10.48550/arXiv.1706.05587
6. Chen, L.C., Zhu, Y., Papandreou, G., Schroff, F., Adam, H.: Encoder-decoder with atrous separable convolution for semantic image segmentation. In: Proceedings of the European Conference on Computer Vision (ECCV), pp. 801–818 (2018). https://doi.org/10.1007/978-3-030-01234-2_49
7. Contributors, M.: MMSegmentation: Open MMLab semantic segmentation toolbox and benchmark (2020). https://github.com/open-mmlab/mmsegmentation

8. Dai, A., Chang, A.X., Savva, M., Halber, M., Funkhouser, T., Nießner, M.: Scan-Net: richly-annotated 3D reconstructions of indoor scenes. In: Proceedings of the IEEE Conference on Computer Vision and Pattern Recognition, pp. 5828–5839 (2017). https://doi.org/10.48550/arXiv.1702.04405

9. Dosovitskiy, A., et al.: An image is worth 16x16 words: transformers for image recognition at scale. arXiv preprint arXiv:2010.11929 (2020). https://doi.org/10.48550/arXiv.2010.11929

10. Everingham, M., Van Gool, L., Williams, C.K., Winn, J., Zisserman, A.: The pascal visual object classes (VOC) challenge. Int. J. Comput. Vis. **88**, 303–308 (2009). https://doi.org/10.1007/s11263-009-0275-4

11. Geng, Z., Guo, M.H., Chen, H., Li, X., Wei, K., Lin, Z.: Is attention better than matrix decomposition? arXiv preprint arXiv:2109.04553 (2021). https://doi.org/10.48550/arXiv.2109.04553

12. Guo, M.H., Lu, C.Z., Hou, Q., Liu, Z., Cheng, M.M., Hu, S.M.: SegNeXt: rethinking convolutional attention design for semantic segmentation. arXiv preprint arXiv:2209.08575 (2022). https://doi.org/10.48550/arXiv.2209.08575

13. Huang, L., Yuan, Y., Guo, J., Zhang, C., Chen, X., Wang, J.: Interlaced sparse self-attention for semantic segmentation. arXiv preprint arXiv:1907.12273 (2019). https://doi.org/10.48550/arXiv.1907.12273

14. Huang, Q.: Weight-quantized SqueezeNet for resource-constrained robot vacuums for indoor obstacle classification. AI **3**(1), 180–193 (2022). https://doi.org/10.3390/ai3010011

15. Huang, X., Sanket, K., Ayyad, A., Naeini, F.B., Makris, D., Zweir, Y.: A neuromorphic dataset for object segmentation in indoor cluttered environment. arXiv preprint arXiv:2302.06301 (2023). https://doi.org/10.48550/arXiv.2302.06301

16. Huang, Z., Wang, X., Huang, L., Huang, C., Wei, Y., Liu, W.: CCNet: criss-cross attention for semantic segmentation. In: Proceedings of the IEEE/CVF International Conference on Computer Vision, pp. 603–612 (2019). https://doi.org/10.1109/ICCV.2019.00069

17. Keipour, A., Mousaei, M., Bandari, M., Schaal, S., Scherer, S.: Detection and physical interaction with deformable linear objects. arXiv preprint arXiv:2205.08041 (2022). https://doi.org/10.48550/arXiv.2205.08041

18. Kim, W., Seok, J.: Indoor semantic segmentation for robot navigating on mobile. In: 2018 Tenth International Conference on Ubiquitous and Future Networks (ICUFN), pp. 22–25. IEEE (2018). https://doi.org/10.1109/ICUFN.2018.8436956

19. Kirillov, A., Wu, Y., He, K., Girshick, R.: PointRend: image segmentation as rendering. In: Proceedings of the IEEE/CVF Conference on Computer Vision and Pattern Recognition, pp. 9799–9808 (2020). https://doi.org/10.48550/arXiv.1912.08193

20. Koskinopoulou, M., Raptopoulos, F., Papadopoulos, G., Mavrakis, N., Maniadakis, M.: Robotic waste sorting technology: toward a vision-based categorization system for the industrial robotic separation of recyclable waste. IEEE Robot. Autom. Mag. **28**(2), 50–60 (2021). https://doi.org/10.1109/MRA.2021.3066040

21. Li, W., et al.: InteriorNet: mega-scale multi-sensor photo-realistic indoor scenes dataset. arXiv preprint arXiv:1809.00716 (2018). https://doi.org/10.48550/arXiv.1809.00716

22. Lin, T.-Y., et al.: Microsoft COCO: common objects in context. In: Fleet, D., Pajdla, T., Schiele, B., Tuytelaars, T. (eds.) ECCV 2014. LNCS, vol. 8693, pp. 740–755. Springer, Cham (2014). https://doi.org/10.1007/978-3-319-10602-1_48

23. Liu, Z., et al.: Swin Transformer: hierarchical vision transformer using shifted windows. In: Proceedings of the IEEE/CVF International Conference on Computer Vision, pp. 10012–10022 (2021). https://doi.org/10.48550/arXiv.2103.14030

24. Long, J., Shelhamer, E., Darrell, T.: Fully convolutional networks for semantic segmentation. In: Proceedings of the IEEE Conference on Computer Vision and Pattern Recognition, pp. 3431–3440 (2015). https://doi.org/10.1109/CVPR.2015.7298965

25. Lv, Y., Fang, Y., Chi, W., Chen, G., Sun, L.: Object detection for sweeping robots in home scenes (ODSR-IHS): a novel benchmark dataset. IEEE Access **9**, 17820–17828 (2021). https://doi.org/10.1109/ACCESS.2021.3053546

26. Majchrowska, S., et al.: Deep learning-based waste detection in natural and urban environments. Waste Manage. **138**, 274–284 (2022). https://doi.org/10.1016/j.wasman.2021.12.001

27. Minaee, S., Boykov, Y.Y., Porikli, F., Plaza, A.J., Kehtarnavaz, N., Terzopoulos, D.: Image segmentation using deep learning: a survey. IEEE Trans. Pattern Anal. Mach. Intell. **44**, 3523–3542 (2021). https://doi.org/10.48550/arXiv.1809.00716

28. Mo, Y., Wu, Y., Yang, X., Liu, F., Liao, Y.: Review the state-of-the-art technologies of semantic segmentation based on deep learning. Neurocomputing **493**, 626–646 (2022). https://doi.org/10.1016/j.neucom.2022.01.005

29. Pohlen, T., Hermans, A., Mathias, M., Leibe, B.: Full-resolution residual networks for semantic segmentation in street scenes. In: Proceedings of the IEEE Conference on Computer Vision and Pattern Recognition, pp. 4151–4160 (2017). https://doi.org/10.48550/arXiv.1611.08323

30. Proença, P.F., Simoes, P.: TACO: trash annotations in context for litter detection. arXiv preprint arXiv:2003.06975 (2020). https://doi.org/10.48550/arXiv.2003.06975

31. Rafique, A.A., Jalal, A., Kim, K.: Statistical multi-objects segmentation for indoor/outdoor scene detection and classification via depth images. In: 2020 17th International Bhurban Conference on Applied Sciences and Technology (IBCAST), pp. 271–276. IEEE (2020). https://doi.org/10.1109/IBCAST47879.2020.9044576

32. Rao, J., Bian, H., Xu, X., Chen, J.: Autonomous visual navigation system based on a single camera for floor-sweeping robot. Appl. Sci. **13**(3), 1562 (2023). https://doi.org/10.3390/app13031562

33. Richtsfeld, A., Mörwald, T., Prankl, J., Zillich, M., Vincze, M.: Segmentation of unknown objects in indoor environments. In: 2012 IEEE/RSJ International Conference on Intelligent Robots and Systems, pp. 4791–4796. IEEE (2012). https://doi.org/10.1109/IROS.2012.6385661

34. Ronneberger, O., Fischer, P., Brox, T.: U-Net: convolutional networks for biomedical image segmentation. In: Navab, N., Hornegger, J., Wells, W.M., Frangi, A.F. (eds.) MICCAI 2015. LNCS, vol. 9351, pp. 234–241. Springer, Cham (2015). https://doi.org/10.1007/978-3-319-24574-4_28

35. Russell, B.C., Torralba, A., Murphy, K.P., Freeman, W.T.: LabelMe: a database and web-based tool for image. Int. J. Comput. Vis. **77**(1), 157–173 (2008). https://doi.org/10.1007/s11263-007-0090-8

36. Shelhamer, E., Long, J., Darrell, T.: Fully convolutional networks for semantic segmentation. IEEE Trans. Pattern Anal. Mach. Intell. **39**(4), 640–651 (2017). https://doi.org/10.1109/TPAMI.2016.2572683

37. Silberman, N., Hoiem, D., Kohli, P., Fergus, R.: Indoor segmentation and support inference from RGBD images. ECCV **5**(7576), 746–760 (2012). https://doi.org/10.1007/978-3-642-33715-4_54

38. Strudel, R., Garcia, R., Laptev, I., Schmid, C.: Segmenter: transformer for semantic segmentation. In: Proceedings of the IEEE/CVF International Conference on Computer Vision, pp. 7262–7272 (2021). https://doi.org/10.48550/arXiv.2105.05633

39. Wang, T., Cai, Y., Liang, L., Ye, D.: A multi-level approach to waste object segmentation. Sensors **20**(14), 3816 (2020). https://doi.org/10.3390/s20143816

40. Wu, T.W., Zhang, H., Peng, W., Lü, F., He, P.J.: Applications of convolutional neural networks for intelligent waste identification and recycling: A review. Resour. Conserv. Recycl. **190**, 106813 (2023). https://doi.org/10.1016/j.resconrec.2022.106813

41. Xie, E., Wang, W., Yu, Z., Anandkumar, A., Alvarez, J.M., Luo, P.: SegFormer: simple and efficient design for semantic segmentation with transformers. In: Advances in Neural Information Processing Systems, vol. 34, pp. 12077–12090 (2021). https://doi.org/10.48550/arXiv.2105.15203

42. Yuan, Y., Chen, X., Chen, X., Wang, J.: Segmentation transformer: object-contextual representations for semantic segmentation. arXiv preprint arXiv:1909.11065 (2019). https://doi.org/10.1007/978-3-030-58539-6_11

43. Yuan, Y., Chen, X., Wang, J.: Object-contextual representations for semantic segmentation. In: Vedaldi, A., Bischof, H., Brox, T., Frahm, J.-M. (eds.) ECCV 2020. LNCS, vol. 12351, pp. 173–190. Springer, Cham (2020). https://doi.org/10.1007/978-3-030-58539-6_11

44. Yuan, Y., Huang, L., Guo, J., Zhang, C., Chen, X., Wang, J.: OCNet: object context for semantic segmentation. Int. J. Comput. Vis. **129**(8), 2375–2398 (2021). https://doi.org/10.1007/s11263-021-01465-9

45. Zhang, H., et al.: Context encoding for semantic segmentation. In: Proceedings of the IEEE Conference on Computer Vision and Pattern Recognition, pp. 7151–7160 (2018). https://doi.org/10.48550/arXiv.1803.08904

46. Zhang, W., Pang, J., Chen, K., Loy, C.C.: K-Net: towards unified image segmentation. In: Advances in Neural Information Processing Systems, vol. 34, pp. 10326–10338 (2021). https://doi.org/10.48550/arXiv.2106.14855

47. Zhang, W., Pang, J., Chen, K., Loy, C.C.: K-Net: towards unified image segmentation. In: NeurIPS (2021). https://doi.org/10.48550/arXiv.2106.148550

48. Zhao, H., Shi, J., Qi, X., Wang, X., Jia, J.: Pyramid scene parsing network. In: Proceedings of the IEEE Conference on Computer Vision and Pattern Recognition, pp. 2881–2890 (2017). https://doi.org/10.1109/CVPR.2017.660

49. Zhao, H., et al.: PSANet: point-wise spatial attention network for scene parsing. In: Proceedings of the European Conference on Computer Vision (ECCV), pp. 267–283 (2018). https://doi.org/10.1007/978-3-030-01240-3_17

50. Zhou, B., Zhao, H., Puig, X., Fidler, S., Barriuso, A., Torralba, A.: Scene parsing through ade20k dataset. In: Proceedings of the IEEE Conference on Computer Vision and Pattern Recognition, pp. 633–641 (2017). https://doi.org/10.1109/CVPR.2017.544

Leveraging Sound Local and Global Features for Language-Queried Target Sound Extraction

Xinmeng Xu[1], Yiqun Zhang[1], Yuhong Yang[1,3], and Weiping Tu[1,2,3(✉)]

[1] National Engineering Research Center for Multimedia Software,
School of Computer Science, Wuhan University, Wuhan, China
{xuxinmeng,2018302110188,yangyuhong,tuweiping}@whu.edu.cn
[2] Hubei Luojia Laboratory, Wuhan, China
[3] Hubei Key Laboratory of Multimedia and Network Communication Engineering,
Wuhan University, Wuhan, China

Abstract. Language-queried target sound extraction is a fundamental audio-language task that aims to estimate the audio signal of the target sound event class by a natural language expression in a sound mixture. One of the key challenges of this task is leveraging the language expression to highlight the target sound features in the noisy mixture interpretably. In this paper, we leverage language expression to guide the model to extract the most informative features of the target sound event by adaptively using local and global features, and we present a novel language-aware synergic attention network (LASA-Net) for language-queried target sound extraction, as the first attempt to leverage local and global operations using language representation to extract target sound in single or multiple sound source environments. In particular, language-aware synergic attention consists of a local operation submodule, a global operation submodule, and an interaction submodule, in which local and global operation submodules extract sound local and global features while the interaction submodule adaptively selects the most discriminative features with the guidance of linguistic features. In addition, we introduce a linguistic-acoustic fusion module that leverages the well-proven correlation modeling power of self-attention for excavating helpful multi-modal contexts. Extensive experiments demonstrate that our proposed LASA-Net is able to achieve state-of-the-art performance while maintaining an attractive computational complexity.

Keywords: Language-queried Target Sound Extraction ·
Language-aware synergic attention · Local and global operation ·
Linguistic-acoustic fusion module

1 Introduction

Given a sound mixture and a text description of the target sound event class, language-queried target sound extraction aims at extracting the sound of the

© The Author(s), under exclusive license to Springer Nature Singapore Pte Ltd. 2024
B. Luo et al. (Eds.): ICONIP 2023, LNCS 14450, pp. 367–379, 2024.
https://doi.org/10.1007/978-981-99-8070-3_28

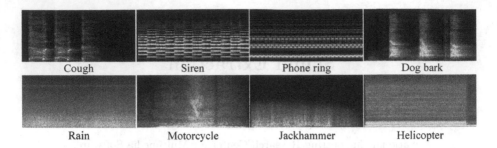

| Cough | Siren | Phone ring | Dog bark |

| Rain | Motorcycle | Jackhammer | Helicopter |

Fig. 1. Demonstration of patterns in non-mechanical sounds (first row) and mechanical sounds (second row).

target event class [2]. It yields great value applications such as language-based human-robot interaction, dynamic editing of sound, and video recording. In contrast with traditional single-modality target sound extraction tasks based on fixed category conditions [12,13,25,26], language-queried target sound extraction deals with the much larger vocabularies and syntactic varieties of human natural languages.

Benefiting from the huge success of deep learning, many deep learning-based methods have been developed to solve this problem [9,12]. Most focus on local processing by adopting convolutional layers and non-linear operations. These methods mainly address sound local time-frequency dynamics, such as pitch shifting and format patterns. However, the long-term structures, such as variability and recurrence properties, also carry rich discriminant acoustic information. For example, mechanical sounds, which are generated by reciprocating or rotational motions, usually exhibit homogeneous repetitive patterns that should be exploited for sound extraction. We demonstrate the property in Fig. 1, in which 8 types of sound events are presented: 4 non-mechanical sounds and 4 mechanical sounds. Although some strategies, such as encoder-decoder architecture [18,19] or dilated convolutions [4,13], attempt to aggregate more useful sound non-local features from border receptive field, the global features are still unable to explore.

Self-attention calculates response at a given time-frequency bin by a weighted sum of all other positions and thus has been explored in deep neural networks for various speech and sound signal processing tasks [1,25,26]. Benefiting from the advantage of global processing, self-attention achieves a significant performance boost over convolutional operations. However, the self-similarity calculation of self-attention based on time-frequency bins is more likely influenced by noisy signals, thus being biased and unreliable. In addition, local processing is an important strategy in acoustic signal processing tasks and has been studied in many sound source separation and extraction tasks. However, the main drawback of applying local operations to target sound extraction is that the receptive field is relatively small, which is unfavorable to the use of global information, such as variability and recurrence properties.

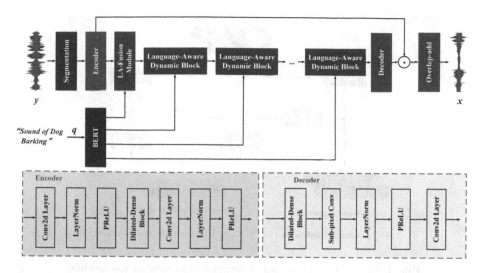

Fig. 2. Framework of our proposed Language-aware Synergic Attention Network (LASA-Net). LASA-Net mainly consists of the encoder, decoder, and several Language-aware Dynamic Blocks (LADBs). The second row further shows the detailed architecture of the encoder and decoder.

According to the analysis above, local and global operations have unique drawbacks but can complement each other in many aspects. To tackle this issue, we leverage local and global operations by involving linguistic features that provide information about the target sound event to guide the local and global feature selection adaptively. This paper proposes a novel language-aware synergic attention network (LASA-Net) for language-queried target sound extraction. In particular, local and global operations extract sound local and global features while the interaction sub-module adaptively selects the most discriminative features with the guidance of linguistic features. In addition, a linguistic-acoustic fusion module that leverages the well-proven correlation modeling power of self-attention is proposed for excavating helpful multi-modal contexts.

2   Proposed Method

2.1   Framework

This paper proposes a Language-aware Synergic Attention Network (LASA-Net) for language-queried target sound extraction. As shown in Fig. 2, LASA-Net consists of a segmentation operation, encoder, separator, masking module, decoder, and overlap-add operation. The input audio mixture $\mathbf{y} \in \mathbb{R}^{1 \times L}$ is firstly segmented into frames of length F with hop size H by segmentation operation. Then all the frames are stacked to form a 3 dimension tensor $Y \in \mathbb{R}^{1 \times N \times F}$, in which N represents the number of frames and is expressed as:

$$F = \lceil (L - F)/(F - H) + 1 \rceil, \tag{1}$$

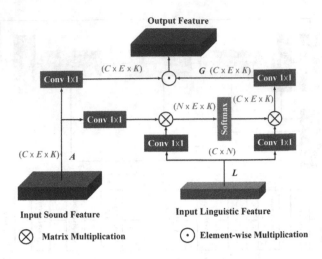

Fig. 3. Diagram of the proposed linguistic-acoustic fusion module.

In addition, the overlap-add operation is an inverse of the segmentation operation to merge the frames for the enhanced speech waveform reconstruction.

Next, we feed Y to the encoder part that plays the role of feature extractor [10, 27] and contains two convolutional layers, of which the first one is increasing the number of channels to 64 using convolution with a kernel size of $(1, 1)$ and the second one halves the dimension of frame size using a kernel size of $(1, 3)$ with a stride of $(1, 2)$, in which a dilated-dense block [14] by using four dilated convolutions collaborates between them to obtain the audio mixture latent space representation. Then, the encoded feature is fed to the separation part, which is composed of a linguistic-acoustic (LA) fusion module to learn more effective cross-modal alignments for deeply leveraging linguistic and acoustic features, and several language-aware dynamic blocks (LADBs). More details of the separation part will be given in the next subsection.

The separation part outputs a mask for sound extraction by the element-wise multiplication between the mask and the output of the encoder part. The final masked encoder feature is then fed into the decoder part that is responsible for the feature reconstruction and contains a dilated-dense block, a sub-pixel convolution [17], where followed by a layer normalization, PReLU, and a 2D convolutional layer with a kernel size of $(1, 1)$ for the channel dimension recovery of enhanced speech feature into 1. Finally, the overlap-add method is used as the inverse operation of segmentation for recovering enhanced waveform. The goal of training LASA-Net is to minimize the negative scale-dependent signal-to-ratio (SNR) loss function between target sound and estimated sound, which can be defined as:

$$\mathcal{L} = -10 \log_{10}(||x||^2/||x - \hat{x}||^2), \tag{2}$$

where x and $\hat{x}$ are the sample index of the target and estimated sounds, respectively.

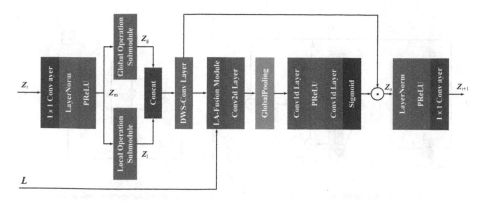

Fig. 4. The architecture of proposed language-aware dynamic block.

In our design, LASA-Net extracts the target sound described by the natural language expressions, and we use BERT [6] as the query network to extract the language query embedding from the input keywords in SA-Net. We use the pre-trained BERT [20] consisting of 4 Transformer encoder blocks, each with 4 heads and 256 hidden dimensions, respectively.

2.2 Linguistic-Acoustic (LA) Fusion Module

In order to extract the target sound signal from a noisy mixture, it is important to align the acoustic and linguistic representations of the object across modalities. To this end, we propose a novel LA-fusion module that is schematically illustrated in Fig. 3. Given the input acoustic features $A \in \mathbb{R}^{C \times E \times K}$ and linguistic features $L \in \mathbb{R}^{C \times N}$, LA-fusion module performs multi-modal fusion in two steps, as introduced in the following.

First, the LA-fusion module aggregates the linguistic features L across the word dimension to generate a position-specific, sentence-level feature vector, which collects linguistic information most relevant to the current local nearing features. Through this step, we obtain the attention map, $G \in \mathbb{R}^{C \times E \times K}$, as follow,

$$G = \mathcal{H}_c^1(\text{softmax}(\mathcal{H}_c^2(A)^\top \mathcal{H}_c^3(L))\mathcal{H}_c^4(L)), \tag{3}$$

where $\mathcal{H}_c^1(\cdot)$, $\mathcal{H}_c^2(\cdot)$, $\mathcal{H}_c^3(\cdot)$, and $\mathcal{H}_c^4(\cdot)$ are four 1×1 convolutional layers. The Eq. 3 implements the scaled dot-product attention [21] using acoustic features, A, as query and linguistic features as the key and value, with instance normalization after linear transformation in query projection function, $\mathcal{H}_c^2(\cdot)$, and the output projection function, $\mathcal{H}_c^4(\cdot)$.

Second, after obtaining G that has the same shape as A, we combine them to produce a set of multi-modal feature maps, F, via element-wise multiplication. Consequently, The step is described as follows:

$$F = \mathcal{H}_c^6(\mathcal{H}_c^5(A) \odot G), \tag{4}$$

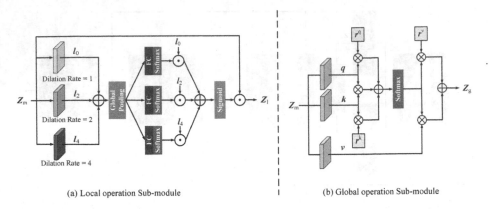

(a) Local operation Sub-module   (b) Global operation Sub-module

Fig. 5. The architecture of (a) local operation submodule and (b) global operation submodule.

where "$\odot$" represents element-wise multiplication, and $\mathcal{H}_c^5(\cdot)$ and $\mathcal{H}_c^6(\cdot)$ are an acoustic projection and a multi-modal projection, respectively. Each of the two functions is implemented as a 1×1 convolution.

2.3 Language-Aware Dynamic Block (LADB)

The role of LADB is to make a trade-off between local and global operations and maximize their performance by involving linguistic features. As shown in Fig. 4, the LADB consists of three main components: local operation submodule, global operation submodule, and fusion submodule. Given the input feature, $\mathbf{Z}_i$, an encoder module containing a 1×1 convolution, layer normalization, and PReLU, is adopted to obtain an intermediate feature, $\mathbf{Z}_m$, and the local and global operation submodules, respectively, process $\mathbf{Z}_m$ and output global feature, $\mathbf{Z}_g$, and local feature, $\mathbf{Z}_l$, in a paralleled manner:

$$\begin{cases} \mathbf{Z}_g = \mathcal{H}_g(\mathbf{Z}_m), \\ \mathbf{Z}_l = \mathcal{H}_l(\mathbf{Z}_m), \end{cases} \tag{5}$$

where $\mathcal{H}_g$ and $\mathcal{H}_l$ refer to the functions of the local and global operation submodules. The architecture of these two sub-modules will be elaborated in the following subsections.

After getting $\mathbf{Z}_g$ and $\mathbf{Z}_l$, we introduce a fusion sub-module to obtain the output feature, $\mathbf{Z}_o$. The fusion sub-module is designed to select the most informative features from $\mathbf{Z}_g$ and $\mathbf{Z}_l$ with the guidance of linguistic features, L. Concretely, the fusion sub-module first concatenates $\mathbf{Z}_g$ and $\mathbf{Z}_l$ and feeds them into depth-wise separable (DWS) convolution [3] to model each channel of the concatenation. Then, the output is fed into a channel selection module, which contains an LA-fusion module with a convolutional layer to involve linguistic features in the selection module, and then a global pooling operation and two 1-D convolutional layers take the features containing linguistic features as input and adaptively select the channels of output feature of DSC to get a residual representation. In this way, the linguistic features of the target sound event are injected into the selection module, and the linguistic modality is only used for the feature selection rather than fusion without interpretability.

2.4 Local and Global Operation Submodules

In this section local and global operation submodules are demonstrated as follow:

Local Operation Submodule. The local operation is designed by employing the channel-wise attention mechanism [8] to perform channel selection with multiple receptive field sizes. The architecture of the proposed local operation submodule is shown in Fig. 5 (a). In this submodule, we adopt three branches to carry different dilation rates of convolutional layers to generate feature maps with different receptive field sizes. The channel-wise attention is performed on the outputs of these three branches, and results are added together.

Global Operation Submodule. Inspired by [24,25], we develop the global operation submodule with positional sensitivity to capture long-range interactions with precise positional information at a reasonable computation cost. The detailed architecture of the developed global operation submodule is shown in Fig. 4. Conventionally, given the input $\mathbf{Z}_m$, the output $\mathbf{Z}_g'$ is defined as:

$$\mathbf{Z}_g' = \sum_{p \in \mathcal{N}} \text{Softmax}(\mathbf{q}_c^\top \mathbf{k}_p)\mathbf{v}_p, \tag{6}$$

where the query $\mathbf{q} = \mathbf{W}_q\mathbf{Z}_m$, key $\mathbf{k} = \mathbf{W}_k\mathbf{Z}_m$, and value $\mathbf{v} = \mathbf{W}_v\mathbf{Z}_m$ are all linear projections of $\mathbf{Z}_m$. In addition, $c = (i,j)$ indicates the position of the output, and $\mathcal{N}$ represents the whole location lattices. The conventional global attention pools value $\mathbf{v}$ globally based on affinities $(\mathbf{Z}_{m,c}^\top \mathbf{W}_q^\top \mathbf{W}_k \mathbf{Z}_{m,p})$, resulting computational complexity $\mathcal{O}(n^2 f^2)$.

Instead, the positional-sensitive global attention defines the relative distances of $c = (i,j)$ to the position $p = (a,b)$, in which $c, p \in \mathcal{N}$ has resulted in the offset $p - c$ [15]. Consequently, the output of positional-sensitive global attention, $\mathbf{Z}_g$, is designed to involve relative positional encoding of query, key, and value, which is expressed as:

$$\mathbf{Z}_g = \sum_{p \in \mathcal{N}_{1 \times m}(o)} \text{Softmax}(\mathbf{q}_c^\top \mathbf{k}_p + \mathbf{q}_c^\top \mathbf{r}_{p-c}^q + \mathbf{k}_p^\top \mathbf{r}_{p-c}^k) \times (\mathbf{v}_p + \mathbf{r}_{p-c}^v), \tag{7}$$

where the $\mathbf{r}_{p-c}^q$, $\mathbf{r}_{p-c}^k$, and $\mathbf{r}_{p-c}^v$ are the learnable positional encoding of query, key, and value. The positional-sensitive global attention reduces the computational complexity to $\mathcal{O}(nfm)$, which enables the global receptive field achieved by setting the span m directly to the whole feature input.

3 Experimental Setup

3.1 Dataset

We create the dataset for SA-Net based on the FSD Kaggle 2018 dataset [5]. FSD Kaggle 2018 is a set of sound event and class label pairs, including 41 different sound classes that are a subset of the Audioset dataset [11]. The audio caption generation is based on the AudioCaps approach [7]. In addition, the synthetic sound mixture dataset contains 50,000 training samples, 5,000 validation samples, and 10,000 test samples. The interference sound sources are collected from TAU Urban Acoustic Scenes 2019 dataset [16]. We mix the sound sources by using Scaper toolkit [16], each with 3–5 foreground sounds randomly sampled from the FSD Kaggle 2018 dataset and a background sound randomly sampled from the TAU Urban Acoustic Scenes 2019 dataset. Sound mixtures are established by sampling 3–5 seconds crops from each foreground sound and then pasting them on a 6 s background sound. The foreground sounds' signal-to-noise ratios (SNRs) are randomly chosen between 15 and 25 dB relative to the background sound.

Particularly, training and validation sets are collected from the development splits of the FSD Kaggle 2018 and TAU Urban Acoustic Scenes 2019 datasets, and the test set is collected from the test splits of these two datasets. The choices of the target foreground sounds are fixed after generating the validation and test sets to ensure evaluations are reproducible. Since we mainly consider human listening applications for streaming target sound extraction, we run our experiments at a 44.1 kHz sampling rate to cover the full audible range.

3.2   Training Parameters

We set the stride of the initial convolution of SA-Net, L, to 32, which is about 0.73 ms at 44.1 kHz. We set the batch size to 16 and used Adam optimizer for model training with an initial learning rate of 1×10^{-3} and a weight decay of 1×10^{-5}. The whole training epochs are set to 50 using one NVIDIA Geforce RTX 3060 GPU.

3.3   Evaluation Metrics

In this paper, three commonly used evaluation metrics in sound source separation are adopted for model performance evaluation [23], which are (1) scale-invariant source-to-distortion ratio (SI-SDR), (2) source-to-inferences ratio (SIR), and (3) sources to artifact ratio (SAR). For all metrics, a higher number indicates a better performance. In addition, to accurately measure these models' computational complexities, we also evaluate the model in terms of the number of parameters (Param.) and real-time factor (RTF).

4   Results and Analysis

4.1   Model Comparison

In this study, three baselines are selected for model comparison: (1) **Conv-TasNet** [10], originally proposed for speech separation, can also be used for target sound extraction, (2) **Waveformer** [22], a real-time and streaming target sound extraction system, in which the separator contains 10 encoder layers and

Table 1. Performance comparison on single-target and multi-target universal sound extraction task.

Num. of Target	1			2			3			Param.	RTF
Metrics	SI-SDR	SIR	SAR	SI-SDR	SIR	SAR	SI-SDR	SIR	SAR		
Unprocessed	0.24	10.96	0.26	0.19	9.43	0.22	0.20	9.21	0.21	–	–
Conv-TasNet	5.93	16.64	4.98	2.94	13.75	2.38	1.21	11.89	1.06	4.57	0.68
LASS-Net	7.76	17.70	7.03	3.74	13.68	5.81	1.99	12.98	1.82	4.34	0.72
Waveformer	9.84	17.76	9.24	4.86	14.53	5.96	3.32	14.03	2.25	3.88	0.43
LASA-Net (AudioCaps)	11.24	18.34	9.36	5.17	16.65	6.14	2.85	15.62	3.78	3.01	0.46
LASA-Net (Human)	9.85	17.29	9.04	4.56	15.69	5.79	2.67	15.06	2.92	3.01	0.46
LASA-Net (One-hot)	12.03	18.96	9.76	5.38	16.01	6.24	3.18	16.12	3.11	3.01	0.46

1 decoder layer, and (3) **LASS** [9], a language-queried audio source separation system. In addition, we use two types of methods to generate language queries, i.e., the AudioCaps approach and human description. Note that we average the evaluation results of five language queries based on the human description for each target audio source. Furthermore, the one-hot label is also evaluated in our experiment.

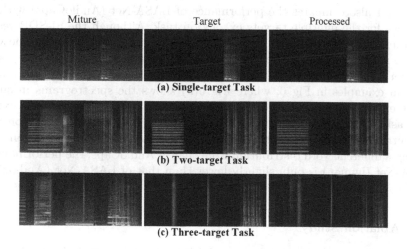

(a) Single-target Task

(b) Two-target Task

(c) Three-target Task

Fig. 6. Visualization of the spectrograms in (a) single-target extraction task, (b) two-target extraction task, and (c) three-target extraction task.

Table 1 shows the comparison results in the single-target extraction task. LASA-Net achieves SI-SDR, SIR, and SAR improvement of 11.24 dB, 18.34 dB, and 9.36 dB, respectively, compared with the original mixtures. Compared with the **LASS** baseline system, LASA-Net (AudioCaps) achieves SI-SDR, SIR, and SAR improvement of 2.26 dB, 0.33 dB, and 2.38 dB, respectively. These results indicate the effectiveness of the proposed SA-Net. Compared with **Waveformer**,

LASA-Net (AudioCaps) shows superior performance while using a close number of trainable parameters and showing a similar real-time factor.

Table 2. Ablation study on single-target sound extraction task.

Case Index	0 (default)	1	2	3	4	5	6
Local operation	✓	✓	×	✓	✓	✓	×
Global operation	✓	×	✓	✓	✓	×	✓
Self-attention	×	×	×	×	×	✓	×
Conventional Conv	×	×	×	×	×	×	✓
LA-Fusion Module	✓	✓	✓	×	✓	✓	✓
Fusion Module	✓	✓	✓	✓	×	✓	✓
Fusion Module (w/o LA-Fusion Module)	×	×	×	×	✓	×	×
Param. (M)	3.01	1.96	2.56	2.94	2.76	2.81	2.09
SI-SDR (dB)	11.24	6.24	8.88	7.21	9.54	8.89	6.11

Table 1 also compares the performance of LASA-Net (AudioCaps) with the baselines for the multiple targets extraction task. Although the SI-SDR results are lower in the 3-target extraction task since there is greater similarity between the input mixture and the target signal, LASA-Net (AudioCaps) achieves the best performance in all conditions in terms of SI-SDR. We show the sound spectrogram examples in Fig. 6, where Fig. 6(a) shows the spectrograms in single-target extraction task, Fig. 6(b) shows the spectrograms in two-target extraction task, and Fig. 6(c) shows the spectrograms in three-target extraction task. Furthermore, we observe that although the human-annotated descriptions containing a different word distribution from that in AudioCaps, the performance of LASA-Net (human) is only slightly lower than that of LASA-Net (AudioCaps) in all conditions.

4.2   Ablation Study

In this section, we explore the impact of different components in LASA-Net. The results of the ablation study are shown in Table 2. Note that **Case 0** represents the basic LASA-Net with the default setting. In the ablation study, we compare our method with several models: **Case 1**: we keep local operation submodule only in LADB, **Case 2**: we keep global operation submodule only in LADB, **Case 3**: we remove the LA-fusion module from the main body of LASA-Net, **Case 4**: we remove the LA-fusion module from LADB, **Case 5**: we replace global operation submodule with self-attention, and **Case 6**: we replace the local operation submodule with conventional convolutional layer.

According to the comparison results in Table 2, four observations can be gathered: (1) compare **Case 0** with **Case 2**, the local operation contributes

0.14 dB SI-SDR improvement and the comparison between **Case 0** with **Case 6** shows that proposed local operation outperforms conventional convolution, which proves the effectiveness of local operation submodule, (2) compare **Case 0** and **Case 1**, the global attention contributes 2.17 dB SI-SDR improvement, and compare **Case 0** with **Case 5**, proposed global attention has the similar performance with self-attention but with much lower RTF score, which indicates that the positional-sensitive global attention effectively captures long-range feature dependencies with lower computation complexity, (3) compare **Case 0** with **Case 4**, one can observe the necessity of fusion module for balancing local and global operations by involving linguistic features, and (4) according to the comparison between **Case 0** with **Case 3**, we observe the efficiency of proposed LA-fusion module.

5 Conclusion

This paper proposes a novel Language-aware Synergic Attention Network, dubbed LASA-Net, by incorporating local and global attention mechanisms by using the guidance of linguistic features of the target sound class for target sound extraction tasks. LASA-Net is the first attempt to integrate local and global attention operations adaptively involving linguistic features and to enable extracting the target sound source by incorporating. In addition, we introduce a linguistic-acoustic fusion module that leverages the well-proven correlation modeling power of self-attention for excavating helpful multi-modal contexts. Experimental results show that the proposed LASA-Net achieves state-of-the-art results while maintaining an attractive computational complexity.

Acknowledgements. This paper is partly supported by the National Nature Science Foundation of China (No. 62071342, No.62171326), the Special Fund of Hubei Luojia Laboratory (No. 220100019), the Hubei Province Technological Innovation Major Project (No. 2021BAA034) and the Fundamental Research Funds for the Central Universities (No.2042023kf1033).

References

1. Chen, K., Du, X., Zhu, B., Ma, Z., Berg-Kirkpatrick, T., Dubnov, S.: Zero-shot audio source separation through query-based learning from weakly-labeled data. In: Proceedings of the AAAI Conference on Artificial Intelligence, vol. 36, pp. 4441–4449 (2022)
2. Cherry, E.C.: Some experiments on the recognition of speech, with one and with two ears. J. Acoust. Soc. Am. **25**(5), 975–979 (1953)
3. Chollet, F.: Xception: deep learning with depthwise separable convolutions. In: Proceedings of the IEEE Conference on Computer Vision and Pattern Recognition, pp. 1251–1258 (2017)
4. Delcroix, M., Vázquez, J.B., Ochiai, T., Kinoshita, K., Araki, S.: Few-shot learning of new sound classes for target sound extraction. In: Proceedings of Interspeech 2021, pp. 3500–3504 (2021). https://doi.org/10.21437/Interspeech.2021-1369

5. Gemmeke, J.F., et al.: Audio set: an ontology and human-labeled dataset for audio events. In: 2017 IEEE International Conference on Acoustics, Speech and Signal Processing (ICASSP), pp. 776–780. IEEE (2017)
6. Kenton, J.D.M.W.C., Toutanova, L.K.: BERT: pre-training of deep bidirectional transformers for language understanding. In: Proceedings of NAACL-HLT, pp. 4171–4186 (2019)
7. Kim, C.D., Kim, B., Lee, H., Kim, G.: Audiocaps: generating captions for audios in the wild. In: Proceedings of the 2019 Conference of the North American Chapter of the Association for Computational Linguistics: Human Language Technologies, vol. 1 (Long and Short Papers), pp. 119–132 (2019)
8. Li, X., Wang, W., Hu, X., Yang, J.: Selective kernel networks. In: Proceedings of the IEEE/CVF Conference on Computer Vision and Pattern Recognition, pp. 510–519 (2019)
9. Liu, X., et al.: Separate what you describe: language-queried audio source separation. In: Proceedings of Interspeech 2022, pp. 1801–1805 (2022). https://doi.org/10.21437/Interspeech.2022-10894
10. Luo, Y., Mesgarani, N.: Conv-tasnet: surpassing ideal time-frequency magnitude masking for speech separation. IEEE/ACM Trans. Audio Speech Lang. Process. **27**(8), 1256–1266 (2019)
11. Mesaros, A., Heittola, T., Virtanen, T.: A multi-device dataset for urban acoustic scene classification. In: Scenes and Events 2018 Workshop (DCASE 2018), p. 9 (2018)
12. Ohishi, Y., et al.: Conceptbeam: concept driven target speech extraction. In: Proceedings of the 30th ACM International Conference on Multimedia, pp. 4252–4260 (2022)
13. Pan, Z., Ge, M., Li, H.: USEV: universal speaker extraction with visual cue. IEEE/ACM Trans. Audio Speech Lang. Process. **30**, 3032–3045 (2022)
14. Pandey, A., Wang, D.: Densely connected neural network with dilated convolutions for real-time speech enhancement in the time domain. In: ICASSP 2020–2020 IEEE International Conference on Acoustics, Speech and Signal Processing (ICASSP), pp. 6629–6633. IEEE (2020)
15. Ramachandran, P., Parmar, N., Vaswani, A., Bello, I., Levskaya, A., Shlens, J.: Stand-alone self-attention in vision models. Adv. Neural Inf. Process. Syst. **32** (2019)
16. Salamon, J., MacConnell, D., Cartwright, M., Li, P., Bello, J.P.: Scaper: a library for soundscape synthesis and augmentation. In: 2017 IEEE Workshop on Applications of Signal Processing to Audio and Acoustics (WASPAA), pp. 344–348. IEEE (2017)
17. Shi, W., et al.: Real-time single image and video super-resolution using an efficient sub-pixel convolutional neural network. In: Proceedings of the IEEE Conference on Computer Vision and Pattern Recognition, pp. 1874–1883 (2016)
18. Slizovskaia, O., Kim, L., Haro, G., Gomez, E.: End-to-end sound source separation conditioned on instrument labels. In: ICASSP 2019–2019 IEEE International Conference on Acoustics, Speech and Signal Processing (ICASSP), pp. 306–310. IEEE (2019)
19. Stoller, D., Ewert, S., Dixon, S.: Wave-u-net: a multi-scale neural network for end-to-end audio source separation. arXiv preprint arXiv:1806.03185 (2018)
20. Turc, I., Chang, M.W., Lee, K., Toutanova, K.: Well-read students learn better: on the importance of pre-training compact models. arXiv preprint arXiv:1908.08962 (2019)

21. Vaswani, A., et al.: Attention is all you need. Adv. Neural Inf. Process. Syst. **30** (2017)

22. Veluri, B., Chan, J., Itani, M., Chen, T., Yoshioka, T., Gollakota, S.: Real-time target sound extraction. In: ICASSP 2023–2023 IEEE International Conference on Acoustics, Speech and Signal Processing (ICASSP), pp. 1–5. IEEE (2023)

23. Vincent, E., Gribonval, R., Févotte, C.: Performance measurement in blind audio source separation. IEEE Trans. Audio Speech Lang. Process. **14**(4), 1462–1469 (2006)

24. Wang, H., Zhu, Y., Green, B., Adam, H., Yuille, A., Chen, L.-C.: Axial-DeepLab: stand-alone axial-attention for panoptic segmentation. In: Vedaldi, A., Bischof, H., Brox, T., Frahm, J.-M. (eds.) ECCV 2020. LNCS, vol. 12349, pp. 108–126. Springer, Cham (2020). https://doi.org/10.1007/978-3-030-58548-8_7

25. Xu, X., Hao, J.: U-Former: improving monaural speech enhancement with multi-head self and cross attention. In: 2022 26th International Conference on Pattern Recognition (ICPR), pp. 663–369 (2022). https://doi.org/10.1109/ICPR56361.2022.9956638

26. Xu, X., Tu, W., Yang, Y.: Selector-enhancer: learning dynamic selection of local and non-local attention operation for speech enhancement. In: Proceedings of the AAAI Conference on Artificial Intelligence (2023)

27. Yu, G., Li, A., Zheng, C., Guo, Y., Wang, Y., Wang, H.: Dual-branch attention-in-attention transformer for single-channel speech enhancement. In: ICASSP 2022–2022 IEEE International Conference on Acoustics, Speech and Signal Processing (ICASSP), pp. 7847–7851. IEEE (2022)

PEVLR: A New Privacy-Preserving and Efficient Approach for Vertical Logistic Regression

Sihan Mao[1,2]($\boxtimes$), Xiaolin Zheng[1], Jianguang Zhang[2], and Xiaodong Hu[2]

[1] Zhejiang University, Hangzhou, China
sihanmao0317@163.com, xlzheng@zju.edu.cn
[2] Hangzhou Finance and Investment Group, Hangzhou, China
{zhangjianguang,huxiaodong}@hzfi.cn

Abstract. In our paper, we consider logistic regression in vertical federated learning. A new algorithm called PEVLR (**P**rivacy-preserving and **E**fficient **V**ertical **L**ogistic **R**egression) is proposed to efficiently solve vertical logistic regression with privacy preservation. To enhance the communication and computational efficiency, we design a novel local-update and global-update scheme for party $\mathcal{A}$ and party $\mathcal{B}$, respectively. For the local update, we utilize hybrid SGD rather than vanilla SGD to mitigate the variance resulted from stochastic gradients. For the global update, full gradient is adopted to update the parameter of party $\mathcal{B}$, which leads to faster convergence rate and fewer communication rounds. Furthermore, we design a simple but efficient plan to exchange intermediate information with privacy-preserving guarantee. Specifically, random matrix sketch and random selected permutations are utilized to ensure the security of original data, label information and parameters under honest-but-curious assumption. The experiment results show the advantages of PEVLR in terms of convergence rate, accuracy and efficiency, compared with other related models.

Keywords: Vertical federated learning · Logistic regression · Privacy preservation · Hybrid SGD · Selected permutation · Matrix sketch

1 Introduction

Machine learning models have been traditionally trained using a centralized dataset, which can be assembled by a single entity or multiple data providers. This paradigm is efficient when the data is owned by a single provider or when data from multiple sources can be legally shared and integrated. However, in recent years, as the significance of data has become more pronounced, data transmission between multiple parties across domains has faced severe issues, such as privacy leaks and loss of data value. Federated learning (FL) offers a potential solution to address these challenges. FL can be broadly classified into three categories based on the characteristics of data distribution: Horizontal

B. Luo et al. (Eds.): ICONIP 2023, LNCS 14450, pp. 380–392, 2024.
https://doi.org/10.1007/978-981-99-8070-3_29

Federated Learning (HFL), Vertical Federated Learning (VFL), and Federated Transfer Learning (FTL). VFL, also known as feature-based federated learning, is applicable for scenarios where a dataset owned by different parties share the same sample ID space but differs in feature space. In our paper, we focus on the application of logistic regression model within the privacy-preserving and efficient framework for vertical federated learning.

Logistic Regression (LR) is a widely used machine learning model for classification tasks, known for its simplicity, efficiency, and interpretability. When two parties collaborate to train a LR model and the dataset is vertically partitioned, both parties lack complete feature information. Traditional joint modeling methods are no longer viable. Moreover, the users' data is often sensitive, private, and valuable, making direct exchange impossible. Thus, researching privacy-preserving and efficient vertical logistic regression (VLR) is crucial in these scenarios. Despite the existence of several proposed methods for implementing VLR, there are still some issues that need to be addressed. Mohassel and Zhang [18] proposed a scheme for implementing VLR by encrypting the original data and distributing it to each party for gradient calculation and model learning. However, this method suffers from significant efficiency loss because of the frequent data encryption and decryption process. Hardy et al. [11] proposed a gradient sharing scheme that relies on Taylor series expansion and only requires intermediate information for computation, but this approximation of the loss function can lead to a decrease in accuracy.

Besides, the above solutions for VLR involve a three-party architecture, comprising active party $\mathcal{A}$, passive party $\mathcal{B}$ and party $\mathcal{C}$ acting as a coordinator, responsible for coordinating the collaboration. It increases the risk of data leakage by involving a third party, as it is difficult to find a completely trustworthy intermediary that is accepted by both party $\mathcal{A}$ and $\mathcal{B}$. The work [30] originally proposed a model training strategy based on homomorphic encryption that does not require a third party, but it has limitations in performance due to strict training constraints and may lead to potential privacy leakage. Furthermore, this approach incurs high communication costs due to the need for frequent gradient exchange. To address excessive communication costs, Wei et al. [27] propose an approach that involves the data label holder doing more local updates of parameters. The results of the experiments show an improvement in the efficiency of model training. However, [27] adopts stochastic gradient descent (SGD) to achieve the local update, which results in a low cost per iteration, but also leads to a high variance in stochastic gradients because of the random sampling involved. Our work proposes a new model training algorithm, which combines hybrid SGD method [7,25] and local updates [16,27] to enhance the stability of the training process and improve the efficiency of the entire system.

In terms of privacy protection, Sun et al. [24] point out that the methods [27,30] have the potential for privacy leaks. When the number of iterations is too large or the data matrix held by the data label holder has sparse properties, such as an upper or lower triangular matrix, there is a risk of data leakage. Therefore, [24] makes improvements to the encryption scheme to ensure the security of the entire protocol. In our work, we point out that the privacy analysis in [24,27,29] is not accurate. The specific discussion will be presented in Sect. 4.2.

Overall, the contributions of our work can be summarized as follows:

- Firstly, we design a new local-update scheme to improve the convergence rate compared with previous baseline models. The combination of momentum-based SGD and SARAH [19], which is the most advanced variance-reduced method, can alleviate the instability incurred by large variance from SGD.
- In order to decrease the communication rounds and improve the communication efficiency, we not only add the local update in party $\mathcal{A}$, but also utilize the full gradient to update the parameter in party $\mathcal{B}$. The idea comes from SVRG [13] and SARAH. These two algorithms consist one inner loop and one outer loop, which correspond to local update and global update in party $\mathcal{A}$ and $\mathcal{B}$, respectively.
- For the privacy protection, previous related works exist flaws in the security analysis and we modify them. Especially, our design adopts random matrix sketch and random selected permutations, not homomorphic encryption, to protect the privacy information, which guarantees the privacy security and significantly improves the efficiency of the complete system.

2 Related Work

We give a brief review of related works in this section. To prevent privacy leakage from the intermediate results, the application of privacy-enhancing techniques, including Homomorphic Encryption (HE) [20,21], Secure Multi-Party Computation (MPC) [15,31], and Trusted Execution Environment (TEE) [23] can be incorporated into the VFL protocol to safeguard vital information from both internal and external threats. The most prevalent methodology to train the VFL model is gradient descent [26], which achieves the transmission of local model outputs and their respective gradients instead of the local data among parties. However, gradient descent requires heavy computational cost to deal with large-scale machine learning problems. To tackle this problem, SGD [3] and its variance reduction variants [8,13,14,19] have been dominant algorithms in the training process. We will give a specific introduction of these methods in Sect. 3.2. FDML [12] is the first work of asynchronous distributed SGD algorithm for VFL. Zhang et al. propose FD-SVRG [32] for VFL, which is the first work of privacy-preservation and variance-reduced method. Gu et al. [9] propose an asynchronous federated stochastic gradient descent (AFSGD-VP) algorithm for multiparty VFL, which can support SVRG and SAGA in addition to SGD.

Our work studies the logistic regression model for VFL. Unlike the traditional centralized distributed manners in FL, we consider the scenario without the third coordinator, which reduces the cost of communication and alleviates the risk of privacy leakage. The schemes of non-3rd party have also been studied in [24,27,30,34]. In order to save the communication cost and improve the efficiency in VFL, local-update scheme is adopted in our work. Local updates allow the parties to perform multiple local parameter updates during each iteration.

Wei et al. [27] use SGD as the algorithm of local updates. Liu et al. [16] introduce a federated stochastic block coordinate descent algorithm, named FedBCD, which enables each party to carry out multiple client updates prior to each communication, thereby reducing the number of synchronizations and alleviating the communication overhead.

3   Preliminaries

3.1   Logistic Regression

The logistic regression model has been widely used in many fields. It is a binary classification model, using the sigmoid function to map the independent variable x to $(0, 1)$ interval. The formula of logistic regression can be written as

$$\mathbb{P}\{y = 1|x; \theta\} = \text{Sig}(x\theta) = \frac{1}{1 + e^{-x\theta}},$$

where θ is the model parameter.

Assume the dataset $\mathcal{D} = \{\boldsymbol{X}, \boldsymbol{y}\}$, where $\boldsymbol{X} = (\boldsymbol{x}_1^\mathsf{T}, \cdots, \boldsymbol{x}_n^\mathsf{T})$ is a data matrix with dimensions $n \times p$, n is the number of samples, p represents the number of features, and $\boldsymbol{x}_i \in \mathbb{R}^p$ is the feature of the i-th sample. The vector $\boldsymbol{y} = (y_1, \cdots, y_n)^\mathsf{T}$ is a label vector, where $y_i = \{0, 1\}$ is the label of the i-th sample. The cross-entropy loss function is as follows:

$$\mathcal{L}(\boldsymbol{\theta}) = -\frac{1}{n}\sum_{i=1}^{n}\left[y_i\boldsymbol{x}_i^\mathsf{T}\boldsymbol{\theta} - \log\left(1 + e^{\boldsymbol{x}_i^\mathsf{T}\boldsymbol{\theta}}\right)\right], \tag{1}$$

where $\boldsymbol{\theta} \in \mathbb{R}^p$ is the parameter vector. The corresponding gradient vector is

$$\nabla_{\boldsymbol{\theta}}\mathcal{L} = -\frac{1}{n}\sum_{i=1}^{n}(y_i - \frac{1}{1 + e^{-\boldsymbol{x}_i^\mathsf{T}\boldsymbol{\theta}}})\boldsymbol{x}_i = -\frac{1}{n}\sum_{i=1}^{n}(y_i - \text{Sig}(\boldsymbol{x}_i^\mathsf{T}\boldsymbol{\theta}))\boldsymbol{x}_i.$$

3.2   SGD and Variance-Reduced Variants

As previously mentioned, SGD-based algorithms are widely used for solving large-scale machine learning problems [3]. Here, we provide a brief overview of SGD and its variance-reduced variants. Consider the following optimization problem:

$$\min_{\boldsymbol{\theta} \in \mathbb{R}^p} \mathbb{E}\left[f(\boldsymbol{\theta}; \boldsymbol{\xi})\right],$$

where $\boldsymbol{\xi} = (\boldsymbol{x}, \boldsymbol{y})$ represents a random variable satisfying a distribution $\mathcal{P}$. SGD leverages an unbiased noisy gradient estimator $g(\boldsymbol{\theta}^t; \boldsymbol{\xi}^t) = \nabla f(\boldsymbol{\theta}^t; \boldsymbol{\xi}^t)$ to update the parameter. Note that $\boldsymbol{\xi}^t = (\boldsymbol{x}_i, y_i)$ is used to denote a single training example independently sampled from $\mathcal{P}$ at the t-th iteration. The randomness of SGD makes each iteration inexpensive, but it also results in the large variance produced by the stochastic gradient due to random sampling. Thus vanilla SGD methods tend to converge slowly.

Recently, several algorithms have been proposed to reduce the variance of the stochastic gradients and speed up the convergence rate, such as SAG [22], SAGA [8], SVRG [13], SARAH [19], and SCSG [14]. In our work, we adopt a hybrid stochastic gradient descent (HSGD) method [7,25] that combines SARAH and SGD, with the momentum parameter $\beta \in [0,1]$ as the weighting parameter. The formula of the gradient estimator is as follows:

$$v^t := (1 - \beta) \cdot g(\theta^t; \xi^t) + \beta \left(v^{t-1} + g(\theta^t; \xi^t) - g(\theta^{t-1}; \xi^t) \right). \tag{2}$$

The first term in (2) is the stochastic gradient estimated at θ^t, and the second term is the SARAH estimator, which modifies the previous step direction v^{t-1} by adding the difference between two stochastic gradient estimators. When $\beta = 0$, (2) reduces to SGD; and when $\beta = 1$, it is equivalent to SARAH estimator.

The advantages of HSGD are mainly in two aspects. On one hand, HSGD is a single-loop algorithm, which is more simplified and easier to implement than the double-loop methods of SVRG-like methods; on the other hand, due to the sensitivity of the convergence of SGD, a decreasing step size is usually required, which needs to be hand-picked. However, HSGD can ensure the stability and convergence of the algorithm by using a fixed step size.

3.3   Random Matrix Sketch

Random matrix sketch [10,17,28] is a technique used in linear algebra and data analysis to approximate a large matrix by a smaller one, while preserving certain key properties of the original matrix. The basic idea is to multiply the original matrix by a random matrix to obtain a smaller, compressed version of the original matrix that can be used for various tasks such as dimensional reduction, low-rank approximation, and compressed sensing. Hence random matrix sketching is a powerful tool for reducing the computational cost and storage requirements of large-scale data analysis problems while maintaining the accuracy of the results.

CountSketch matrix [4,28,33] is classical in random matrix sketch. It can be constructed as follows. Denote $S \in \mathbb{R}^{p \times s}$. Each row of S has only one nonzero element and the value is uniformly sampled from $\{+1, -1\}$. The index of nonzero entry is also uniformly sampled from $\{1, 2, \cdots, s\}$. Here we present a 5×3 CountSketch matrix as an example:

$$S^\mathsf{T} = \begin{bmatrix} 1 & 0 & -1 & 0 & 0 \\ 0 & 0 & 0 & 1 & 0 \\ 0 & 1 & 0 & 0 & -1 \end{bmatrix}.$$

Due to the sparsity of S, matrix multiplication via CountSketch is more computational efficiency than the standard matrix multiplication. Moreover, the theoretical guarantee in [4,28] shows that $\mathbb{E}\left[SS^\mathsf{T} \right] = I$ and $\left\| SS^\mathsf{T} - I \right\|$ is concentrated, which is the vital property of CountSketch to sketch high-dimensional matrix and keep the results accurately.

4   Solution

In this section, we firstly establish the problem setup of logistical regression for vertical federated learning. Then a new algorithm called PEVLR will be proposed to solve the problem. We give a specific discussion to compare PEVLR with other methods proposed in related works to highlight its advantages. Besides, we conduct the security analysis to point out the guarantee of data privacy in our model.

4.1   Problem Formulation

We propose the formulation of the logistic regression model in the vertical federated learning with two participants as an example. This framework can be naturally extended to the multiple-party scenario. Assume that party $\mathcal{A}$ has a part of dataset $X_1 \in \mathbb{R}^{n \times p_1}$ and the labels for the whole samples $y \in \mathbb{R}^n$. Party $\mathcal{B}$ has additional dataset $X_2 \in \mathbb{R}^{n \times p_2}$. The parameters corresponding to party $\mathcal{A}$ and $\mathcal{B}$ are $\theta_1 \in \mathbb{R}^{p_1}$ and $\theta_2 \in \mathbb{R}^{p_2}$, where $p_1 + p_2 = p$. Before training process, the two parties have employed private set intersection (PSI) techniques [5] to establish and synchronize their shared samples without disclosing the samples that do not intersect.

In our work, we assume that each party is semi-honest [2]. Concretely, they will utilize correct raw data and not intentionally tamper with the training data or disturb the training process to carry out inference attacks. But they still keep curiosity, which means they may use intermediate results obtained during the process to infer information such as features, labels, or model parameters of others. Thus our goal is to train a vertical LR model without exposing the private information of any party.

4.2   Algorithm

We present a new vertical federated learning algorithm for logistic regression, which is called PEVLR shown in Algorithm 1. Compared with related works [24, 27, 30], it not only satisfies the privacy security mentioned above, but also improves the efficiency and accuracy of the model. Figure 1 illustrates the framework of a two-party vertical logistic regression with PEVLR.

Local-Update Scheme. Inspired by local-update methods in [16, 27], we design a inner loop to update the parameter θ_1 locally in party $\mathcal{A}$ as shown in STEP $4 - 9$. Wei et al. [27] use SGD to increase the local training rounds. Their experimental results illustrate that the loss curve is unstable due to the large variance incurred by stochastic gradient estimators. As described in Sect. 3.2, we apply HSGD to achieve the local updates. The number of local training rounds m is a hyper-parameter to be adjusted to improve the efficiency of training. STEP 4 states that the batch size of HSGD is free to select and the samples are randomly drawn. STEP 8 shows the gradient estimator of HSGD, where β is the momentum parameter.

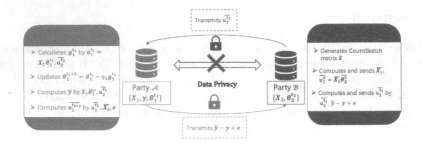

Fig. 1. The framework of a two-party vertical logistic regression with PEVLR

Privacy Security. As pointed out in [24], the protocols proposed in [30] exist the risk of privacy leakage. Since [30] use full gradient information to update parameters, party $\mathcal{B}$ transmits $X_2\theta_2^t$ to party $\mathcal{A}$ and party $\mathcal{A}$ obtains n equations and $(n+1)\times p$ unknowns at the t-th iteration. Within r iterations, party $\mathcal{A}$ collects $n \times r$ equations and $(n+r)\times p$ unknowns. In order to ensure party $\mathcal{A}$ cannot solve an overdetermined system to derive the information of X_2 and $\{\theta_2^t\}$, it should be satisfied that $r \le \frac{np}{n-p}$, where $n > p$. (We consider the underparameterized regime.) When $n \gg p$ and p is quite small, the number of iterations r will be restricted to a very small range, which severely affects the training accuracy of the model. However, their analysis is inaccurate since there always exits an orthogonal matrix Q to obey $X_2QQ^\mathsf{T}\theta_2^t = X_2\theta_2^t$. Moreover, we adopt Countsketch to further reduce the dimension of raw data and improve the efficiency of system. To be concrete, Countsketch matrix is generated by party $\mathcal{B}$ and used to compress X_2 and θ_2^t into $\widetilde{X}_2$ and $\widetilde{\theta}_2^t$, the intermediate results $\widetilde{X}_2$ and $\widetilde{u}_2^t$ are secure and party $\mathcal{A}$ is unable to infer X_2 and θ_2 exactly no matter how many iterations achieve.

Another place in [27,30] may cause privacy leakage is that party $\mathcal{B}$ obtains the decrypted gradient estimator

$$g_2^t = \frac{1}{b}X_2^{\ell\,\mathsf{T}}(\widehat{y}_\ell^t - y_\ell) = \frac{1}{b}X_2^{\ell\,\mathsf{T}}\Delta y_\ell^t$$

at t-th iteration. Note that $X_2^\ell \in \mathbb{R}^{b \times p_2}$ and $\Delta y_\ell^t \in \mathbb{R}^b$. Party $\mathcal{B}$ obtains p_2 equations and b unknowns in each iteration. To ensure the information of label y safe, the batch size b should be always chosen larger than p_2. It is impossible to deal with high-dimensional scenario, that is p_2 is quite large. Besides, previous works all adopt HE to guarantee the security of label information. But the computational complexity of HE severely affects the efficiency of training. To tackle these issues, we adopt the noisy global update in party $\mathcal{B}$, which directly updates $\widehat{u}_2^{t_2+1}$ by using noisy full gradient $g_2^t := \frac{1}{n}\widetilde{X}_2\widetilde{X}_2^\mathsf{T}(\widetilde{y} - y + e)$ as stated in STEP 12. Note that the residuals $\widetilde{y} - y$ are masked by the noise vector e and the value should be dominated by e to hide the true information of y. Since $y_i = \{0, +1\}$ and $\widetilde{y}_i - y_i = o(1)$, e_i can be drawn from $N(0, O(1))$ to guarantee the domination of $\widetilde{y}_i - y_i + e_i$. In the following experiments, we will sample e_i from $N(0, 100)$.

Algorithm 1. PEVLR

Input: $\theta_1^0, \theta_2^0, X_1, X_2, y, n, p_1, p_2, b, l, \beta, \eta_1, \eta_2$

$\mathcal{B}$ generates CountSketch matrix S, computes and sends $\widetilde{u_2}^0 = X_2 S S^\top \theta_2^0, \widetilde{X}_2 = X_2 S$ to $\mathcal{A}$.

for $t_2 = 0, 1, \cdots, T$ **do**

    1. $\mathcal{A}$ randomly selects index set ℓ and calculates $u_\ell = X_1^\ell \theta_1^0 + \widetilde{u_{2\ell}}^{t_2}$, where $|\ell| = b$.

    2. $\mathcal{A}$ computes $\widehat{y}_\ell = \text{Sig}(u_\ell)$, $\Delta y = \widehat{y}_\ell - y_\ell$ and $g_1^0 = \frac{1}{b} X_1^{\ell\top} \Delta y_\ell$.

    3. $\mathcal{A}$ updates $\theta_1^1 = \theta_1^0 - \eta_1 g_1^0$.

    for $t_1 = 1, 2, \cdots, m - 1$ **do**

        4. $\mathcal{A}$ randomly selects index set i from $\{1, \cdots, n\}$.

        5. $\mathcal{A}$ computes $u_1^{t_1-1} = X_1^i \theta_1^{t_1-1}$, $u^{t_1-1} = u_1^{t_1-1} + u_2^{t_2}$

          and $u_1^{t_1} = X_1^i \theta_1^{t_1}$, $u^{t_1} = u_1^{t_1} + u_2^{t_2}$.

        6. $\mathcal{A}$ computes $\widehat{y}_i^{t_1-1} = \text{Sig}(u^{t_1-1})$ and $\widehat{y}_i^{t_1} = \text{Sig}(u^{t_1})$.

        7. $\mathcal{A}$ computes $v^{t_1-1} = \frac{1}{b} X_1^{i\top} \left(\widehat{y}_i^{t_1-1} - y_i \right)$ and $v^{t_1} = \frac{1}{b} X_1^{i\top} \left(\widehat{y}_i^{t_1} - y_i \right)$.

        8. $\mathcal{A}$ computes $g_1^{t_1} = \beta v^{t_1} + (1 - \beta) \left(v^{t_1} - v^{t_1-1} + g_1^{t_1-1} \right)$.

        9. $\mathcal{A}$ updates $\theta_1^{t_1+1} = \theta_1^{t_1} - \eta_1 g_1^{t_1}$.

    end for

    10. $\mathcal{A}$ computes $\widetilde{u} = X_1 \theta_1^m + \widetilde{u_2}^{t_2}$ and sets $\theta_1^0 = \theta_1^m$.

    11. $\mathcal{A}$ computes $\widetilde{y} = \text{Sig}(\widetilde{u})$, generates a noisy vector e and sends $\widetilde{y} - y + e$ to $\mathcal{B}$.

    12. $\mathcal{B}$ computes $\widehat{u}_2^{t_2+1} = \widetilde{u_2}^{t_2} - \frac{\eta_2}{n} \widetilde{X}_2 \widetilde{X}_2^\top (\widetilde{y} - y + e)$ and sends $\widehat{u}_2^{t_2+1}$ to $\mathcal{A}$.

    13. $\mathcal{A}$ computes $\widetilde{u}_2^{t_2+1} = \widehat{u}_2^{t_2+1} + \frac{\eta_2}{n} \widetilde{X}_2 \widetilde{X}_2^\top e$.

end for

The careful readers may notice that the parameter θ_2 is not available for party $\mathcal{B}$ during the training process, then the prediction for the test data can not proceed. In fact, party $\mathcal{B}$ can obtain $\widetilde{u}_2$ from party $\mathcal{A}$ after training and θ_2 can be calculated by solving an overdetermined linear system. Thus the sketch parameter θ_2 can be utilized in the inference stage without significant test accuracy loss.

5 Experiments

In this section, we first illustrate the experimental settings, then evaluate the performance of PEVLR through a series of experiments and make comparison with previous related protocols.

5.1 Experimental Settings and Evaluation Metric

We conduct the experiments on three datasets: GiveMeSomeCredit[1] MNIST[2] and Default-Credit[3]. Through preprocessing of the datasets, including missing value handling, data normalization, and feature engineering, we summarize the details of three datasets, as shown in Table 1.

---

[1] https://www.kaggle.com/c/GiveMeSomeCredit.

[2] http://yann.lecun.com/exdb/mnist.

[3] https://archive.ics.uci.edu/ml/datasets/default+of+credit+card+clients.

Table 1. The details of datasets.

Dataset	#Training	#Test	#Features		Is Dense
			Party $\mathcal{A}$	Party $\mathcal{B}$	
GMSC	112500	3750	5	5	No
MNIST	56000	14000	392	392	Yes
Default-Credit	22500	7500	10	13	Yes

For all experiments, PEVLR adopts a fixed learning rate $\eta_1 = 0.01$ for local updates in party $\mathcal{A}$ and $\eta_2 = 0.5$ in party $\mathcal{B}$. For the momentum parameter, we select $\beta = 0.9$. The batch sizes for MNIST and Default-Credit are fixed to be 64. For GMSC, the batch size is set to be 128. The sketch size s in PEVLR is set to be $p_2/2$ for MNIST and p_2 for GMSC and Default-Credit. The experiments are executed from Python 3.8 on a macOS machine with multi-core Intel CPU at 2.3 GHz CPU and 16 GB RAM.

The performance metrics we use for all datasets include the training loss, the Area Under Curve (AUC) and maximum Kolmogorov-Smirnov (KS). The KS statistic is commonly employed to assess model discrimination in risk management. Typically, a higher KS value indicates that the model has a stronger ability to rank risks.

5.2   Results and Discussion

We first present the curves of training loss and training AUC of PEVLR and other three related models in Fig. 2. It can be seen that PEVLR achieves the fastest convergence rate and the lowest training loss among all methods in three datasets. Besides, training AUC is the highest derived by PEVLR in all three datasets. The method proposed by Wei et al. performs slightly worse than ours, although they also introduced a strategy of local updates. However, we improved their strategy by using HSGD and used full-gradient updates instead of stochastic gradient updates at party $\mathcal{B}$, which results in superior performance in experiments.

As shown in Table 2, the test AUCs computed by our models are most closed to original LR and are highest among other related models. Our solutions exhibit superior performance compared to Hardy's approach, as it does not require the implementation of Taylor Expansion on the loss function, leading to significantly improved results. Compared with the protocols proposed in Sun and Wei, PEVLR also performs well in terms of AUC and KS.

Moreover, we particularly conduct the experiment to investigate the impact of local-update rounds m in PEVLR. In the following trial, we select $m = 5, 10, 50, 100$ to evaluate the relationship between the convergence rate and the local iteration. The other settings are the same as before. As illustrated in Fig. 3, the convergence rate of PEVLR speeds up with the increase of local iteration m. But it seems that there is no evident improvement between $m = 50$ and

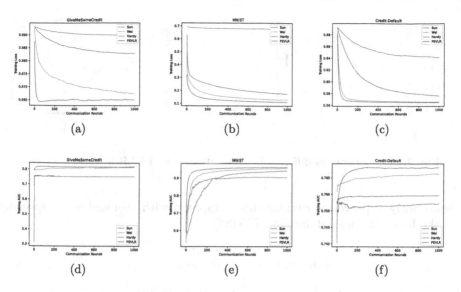

Fig. 2. Comparisons of training loss and training AUC in three datasets with different methods.

Table 2. Results on the AUC for different models

Datasets	Original LR		Hardy [11]		Sun [24]		Wei [27]		PEVLR	
	AUC	KS	AUC	KS	AUC	KS	AUC	KS	AUC	KS
GMSC	0.8102	0.4897	0.8010	0.4899	0.8089	0.4899	0.8072	0.4900	**0.8098**	**0.4909**
MNIST	0.9698	0.8432	0.9261	0.7169	0.9411	0.7485	0.9587	0.8078	**0.9632**	**0.8188**
Default-Credit	0.7839	0.4180	0.7774	0.4207	0.7786	0.4271	0.7773	0.4127	**0.7814**	**0.4160**

$m = 100$, thus the best selection of the number of m to balance the accuracy and efficiency is another interesting work in future.

Finally, we evaluate the efficiency of PEVLR among three datasets. From the previous experiments, we can find out that the performance of Wei's and Sun's models are closed to ours and Wei also adopted the local-update scheme to improve the efficiency and accuracy of the model. Hence we make comparison of running time among Wei, Sun and PEVLR. To proceed the HE technique utilized in Wei and Sun, we use Tenseal library [1] in Python based on CKSS scheme [6]. For each dataset, we repeat the trails for five times and record the average running time. As the recommendation in [27], the local-update round of Wei is set to be 10. In our models, we choose m to be 50 as discussed in the second experiment. The number of iterations is chosen to be 1000 for Sun and Wei. As shown in Fig. 2, the training losses (AUCs) of PEVLR is almost the lowest (highest) among other three models within 100 iterations. Hence we choose the number of iterations of PEVLR to be 100 in this experiment. Besides, other settings are the same as before. Table 3 shows the average running time for Wei, Sun, and PEVLR among three datasets. It is clear to see that our protocols

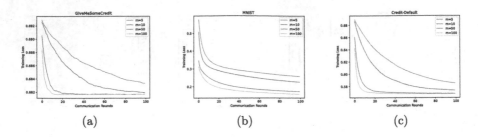

Fig. 3. Comparisons of different local iterations on PEVLR in three datasets.

significantly improve the running time compared with Wei and Sun, especially for the high-dimensional datasets, MNIST.

Table 3. Results on the running time

Datasets	Sun [24]	Wei [27]	PEVLR
GMSC	327.44 s	328.01 s	**1.44 s**
MNIST	188.61 s	192.74 s	**13.86 s**
Default-Credit	161.25 s	161.16 s	**1.02 s**

6   Conclusion

In this paper, a new algorithm called PEVLR is proposed to solve the logistic regression in the vertical federated learning framework. We design a novel local-update and global-update scheme to improve the convergence rate and computational efficiency. For the local update, we utilize hybrid SGD, which is the combination of momentum-based SGD and SARAH to stabilize the large variance incurred by vanilla SGD. For the global update, we adopt the full gradient to update the parameter in party $\mathcal{B}$, which greatly speeds up the training process and decreases the communication rounds. In terms of privacy security, we thoroughly design the plan to exchange intermediate results without the privacy leakage and skillfully avoids the use of homomorphic encryption through random matrix sketch and selected permutation. Moreover, several experiments are conducted to demonstrate the performance of PEVLR. By comparing with other related works, our method displays the superiority in terms of efficiency and accuracy. In future work, we will expect to generalize our algorithms to multi-party scenario and other machine learning models, such as neural network.

References

1. Benaissa, A., Retiat, B., Cebere, B., Belfedhal, A.E.: TenSeal: a library for encrypted tensor operations using homomorphic encryption (2021)
2. Bonawitz, K., et al.: Practical secure aggregation for privacy-preserving machine learning. In: Proceedings of the 2017 ACM SIGSAC Conference on Computer and Communications Security, pp. 1175–1191 (2017)
3. Bottou, L., Curtis, F.E., Nocedal, J.: Optimization methods for large-scale machine learning. SIAM Rev. **60**(2), 223–311 (2018)
4. Charikar, M., Chen, K., Farach-Colton, M.: Finding frequent items in data streams. In: Widmayer, P., Eidenbenz, S., Triguero, F., Morales, R., Conejo, R., Hennessy, M. (eds.) ICALP 2002. LNCS, vol. 2380, pp. 693–703. Springer, Heidelberg (2002). https://doi.org/10.1007/3-540-45465-9_59
5. Chen, H., Laine, K., Rindal, P.: Fast private set intersection from homomorphic encryption. In: Proceedings of the 2017 ACM SIGSAC Conference on Computer and Communications Security, pp. 1243–1255 (2017)
6. Cheon, J.H., Kim, A., Kim, M., Song, Y.: Homomorphic encryption for arithmetic of approximate numbers. In: Takagi, T., Peyrin, T. (eds.) ASIACRYPT 2017. LNCS, vol. 10624, pp. 409–437. Springer, Cham (2017). https://doi.org/10.1007/978-3-319-70694-8_15
7. Cutkosky, A., Orabona, F.: Momentum-based variance reduction in non-convex SGD. IN: Advances in Neural Information Processing Systems, vol. 32 (2019)
8. Defazio, A., Bach, F., Lacoste-Julien, S.: SAGA: a fast incremental gradient method with support for non-strongly convex composite objectives. In: Advances in Neural Information Processing Systems, vol. 27 (2014)
9. Gu, B., Xu, A., Huo, Z., Deng, C., Huang, H.: Privacy-preserving asynchronous vertical federated learning algorithms for multiparty collaborative learning. IEEE Trans. Neural Netw. Learn. Syst. (2021)
10. Halko, N., Martinsson, P.G., Tropp, J.A.: Finding structure with randomness: probabilistic algorithms for constructing approximate matrix decompositions. SIAM Rev. **53**(2), 217–288 (2011)
11. Hardy, S., et al.: Private federated learning on vertically partitioned data via entity resolution and additively homomorphic encryption. arXiv preprint arXiv:1711.10677 (2017)
12. Hu, Y., Niu, D., Yang, J., Zhou, S.: FDML: a collaborative machine learning framework for distributed features. In: Proceedings of the 25th ACM SIGKDD International Conference on Knowledge Discovery and Data Mining, pp. 2232–2240 (2019)
13. Johnson, R., Zhang, T.: Accelerating stochastic gradient descent using predictive variance reduction. In: Advances in Neural Information Processing Systems, vol. 26 (2013)
14. Lei, L., Jordan, M.: Less than a single pass: stochastically controlled stochastic gradient. In: Artificial Intelligence and Statistics, pp. 148–156. PMLR (2017)
15. Lindell, Y.: Secure multiparty computation for privacy preserving data mining. In: Encyclopedia of Data Warehousing and Mining, pp. 1005–1009. IGI global (2005)
16. Liu, Y., et al.: FedBCD: a communication-efficient collaborative learning framework for distributed features. IEEE Trans. Sig. Process. **70**, 4277–4290 (2022)
17. Mahoney, M.W., et al.: Randomized algorithms for matrices and data. Found. Trends® Mach. Learn. **3**(2), 123–224 (2011)

392   S. Mao et al.

18. Mohassel, P., Zhang, Y.: SecureML: a system for scalable privacy-preserving machine learning. In: 2017 IEEE Symposium on Security and Privacy (SP), pp. 19–38. IEEE (2017)
19. Nguyen, L.M., Liu, J., Scheinberg, K., Takáč, M.: SARAH: a novel method for machine learning problems using stochastic recursive gradient. In: International Conference on Machine Learning, pp. 2613–2621. PMLR (2017)
20. Paillier, P.: Public-key cryptosystems based on composite degree residuosity classes. In: Stern, J. (ed.) EUROCRYPT 1999. LNCS, vol. 1592, pp. 223–238. Springer, Heidelberg (1999). https://doi.org/10.1007/3-540-48910-X_16
21. Rivest, R.L., Adleman, L., Dertouzos, M.L., et al.: On data banks and privacy homomorphisms. Found. Secure Comput. 4(11), 169–180 (1978)
22. Roux, N., Schmidt, M., Bach, F.: A stochastic gradient method with an exponential convergence rate for finite training sets. In: Advances in Neural Information Processing Systems, vol. 25 (2012)
23. Sabt, M., Achemlal, M., Bouabdallah, A.: Trusted execution environment: what it is, and what it is not. In: 2015 IEEE Trustcom/BigDataSE/ISPA, vol. 1, pp. 57–64. IEEE (2015)
24. Sun, H., Wang, Z., Huang, Y., Ye, J.: Privacy-preserving vertical federated logistic regression without trusted third-party coordinator. In: 2022 The 6th International Conference on Machine Learning and Soft Computing, pp. 132–138 (2022)
25. Tran-Dinh, Q., Pham, N.H., Phan, D.T., Nguyen, L.M.: Hybrid stochastic gradient descent algorithms for stochastic nonconvex optimization. arXiv preprint arXiv:1905.05920 (2019)
26. Wan, L., Ng, W.K., Han, S., Lee, V.C.: Privacy-preservation for gradient descent methods. In: Proceedings of the 13th ACM SIGKDD International Conference on Knowledge Discovery and Data Mining, pp. 775–783 (2007)
27. Wei, Q., Li, Q., Zhou, Z., Ge, Z., Zhang, Y.: Privacy-preserving two-parties logistic regression on vertically partitioned data using asynchronous gradient sharing. Peer-to-Peer Network. Appl. 14(3), 1379–1387 (2021)
28. Woodruff, D.P., et al.: Sketching as a tool for numerical linear algebra. Found. Trends® Theor. Comput. Sci. 10(1–2), 1–157 (2014)
29. Yang, Q., Liu, Y., Chen, T., Tong, Y.: Federated machine learning: concept and applications. ACM Trans. Intell. Syst. Technol. (TIST) 10(2), 1–19 (2019)
30. Yang, S., Ren, B., Zhou, X., Liu, L.: Parallel distributed logistic regression for vertical federated learning without third-party coordinator. arXiv preprint arXiv:1911.09824 (2019)
31. Yao, A.C.: Protocols for secure computations. In: 23rd Annual Symposium on Foundations of Computer Science (SFCS 1982), pp. 160–164. IEEE (1982)
32. Zhang, G.D., Zhao, S.Y., Gao, H., Li, W.J.: Feature-distributed SVRG for high-dimensional linear classification. arXiv preprint arXiv:1802.03604 (2018)
33. Zhang, M., Wang, S.: Matrix sketching for secure collaborative machine learning. In: International Conference on Machine Learning, pp. 12589–12599. PMLR (2021)
34. Zhao, D., Yao, M., Wang, W., He, H., Jin, X.: NTP-VFL-A new scheme for non-3rd party vertical federated learning. In: 2022 14th International Conference on Machine Learning and Computing (ICMLC), pp. 134–139 (2022)

Semantic-Pixel Associative Information Improving Loop Closure Detection and Experience Map Building for Efficient Visual Representation

Yufei Deng[1], Rong Xiao[1,2(✉)], Jiaxin Li[1,3], and Jiancheng Lv[1,2]

[1] College of Computer Science, Sichuan University, Chengdu, China
{dengyufei,lijiaxin1993}@stu.scu.edu.cn, {rxiao,lvjiancheng}@scu.edu.cn
[2] Engineering Research Center of Machine Learning and Industry Intelligence,
Ministry of Education, Chengdu, China
[3] PetroChina Southwest Oil and Gasfield Company, Chengdu, China

Abstract. RatSLAM is a brain-inspired simultaneous localization and mapping (SLAM) system based on the rodent hippocampus model, which is used to construct the experience map for environments. However, the map it constructs has the problems of low mapping accuracy and poor adaptability to changing lighting environments due to the simple visual processing method. In this paper, we present a novel RatSLAM system by using more complex semantic object information for loop closure detection (LCD) and experience map building, inspired by the effectiveness of semantic information for scene recognition in the biological brain. Specifically, we calculate the similarity between current and previous scenes in LCD based on the pixel information computed by the sum of absolute differences (SAD) and the semantic information extracted by the YOLOv2 network. Then we build an enhanced experience map with object-level information, where the 3D model segmentation technology is used to perform instance semantic segmentation on the recognized objects. By fusing complex semantic information in visual representation, the proposed model can successfully mitigate the impact of illumination and fully express the multi-dimensional information in the environment. Experimental results on the Oxford New College, City Center, and Lab datasets demonstrate its superior LCD accuracy and mapping performance, especially for environments with changing illumination.

Keywords: Brain-inspired simultaneous localization and mapping ·
Loop closure detection · Semantic information

1 Introduction

Simultaneous Localization and Mapping (SLAM) localizes a robot and builds a map as it explores an unknown environment [1]. This map defines the robot's orientation by pose (position and orientation). Recently, the literature has grown up around introducing mathematical probability filtering into SLAM algorithms, such as Kalman Filter (KF), Particle Filter (PF), and their improved algorithms [2]. Since only considering the recursive relationship before the data, it is easy to cause error accumulation problems and map inconsistencies. In large-scale and complex environments, it is a big challenge for the mapping performance of the algorithm. To settle this problem, brain-inspired SLAM models have become a promising alternative method for robot spatial cognition development by transforming neuroscience research into engineering solutions [3–5]. Among them, the rodent-inspired simultaneous localization and mapping algorithm (RatSLAM) has achieved promising performance by emulating the spatial awareness of the hippocampal system [6]. RatSLAM works indoors and outdoors and requires only a monocular vision sensor. However, multiple problems still need to be solved before the RatSLAM model could be used as a practical solution in complex environments. Specifically, Loop closure detection (LCD) is the critical process for robots to relocalize themselves and correct accumulative errors. The cognitive map constructed by RatSLAM is composed of many connected experiences, which describe the spatial structure of the environment. Recently, many different methods have been proposed for detecting correct loop closure and building experience map accurately [7–11]. Although the existing methods can achieve effective LCD and experience map building in RatSLAM systems, they often result in relatively low efficiency and poor robustness to lighting changes by only using pixel information in template matching.

An emerging trend seeks help from semantic information for more accurate LCD. Image semantic information with pixel information not only increases the computational depth of the LCD but also enhances the brain-inspired feature of the entire system. People identify the place where they are not only spatially, but also conceptually [12]. Semantic information is a higher-level concept obtained through vision, which can successfully provide landmarks for the map.

Although semantic information enables scene recognition and human-intelligible map building, it is still lacking in LCD and experience map building in RatSLAM. Further research on them may help us encode information perceived by humans and refine brain-inspired SLAM.

In this paper, inspired by the effectiveness of semantic information in the brain, semantics is added to LCD and experience map building in RatSLAM. The view template records the experienced scene. We use YOLOv2 to extract the object information from the vision, and store them in the view template together with the pixel information. For template matching in LCD, we calculate the template similarity by fusing the semantic similarity computed by the number and position of recognized objects with the pixel similarity computed by SAD. Comparing the template similarity with the threshold detects whether a closed loop is generated, the detected closed loop will correct the experience map. To improve the legibility of the map, we project the recognized objects to the corresponding positions on the experience map through 3D model segmentation technology, thus constructing a map with object-level information. By detecting semantic information, the proposed model can successfully mitigate the impact of illumination and fully express the multi-dimensional information in the environment. The experiment results show that with the fusion of semantic information, our system has higher LCD accuracy and more stable mapping performance, especially for environments with changing illumination. The main contributions of this paper can be summarized as:

- For increasing the representation capacity of the system, we store the object information recognized from the visual scene into the view templates, and fuse it with the pixel information for template matching in LCD;
- For improving the map display capability and establishing a multi-dimensional experience map, we project objects identified in the scene onto the map through 3D model segmentation technology;
- We also have verified the advantages of the semantic-pixel associative representation method and comprehensively compared the impact of various parameters and lighting conditions on performance.

2 Related Work

In traditional RatSLAM system, LCD usually downscales and vectorizes camera image vectors and then compares them. Current research on LCD can be divided into two categories: improvement of LCD detection and visual processing. Gu et al. proposed to replace the templates search module in RatSLAM with the multi-index hashing-based loop closure detection (MILD) [7] algorithm, building a more accurate experience map [8]. Xu et al. used the Bag of Words (BoW) model and the dynamic island mechanism to achieve quick and efficient image retrieval [13]. In contrast, another type of LCD innovation lies in the processing of visual information. Zhou et al. used ORB as the feature when matching [9]. Accordingly, Kazmi et al. used the Gist descriptor as the feature matching method, which can also reduce the matching error [10]. To be closer to the biological model, Wu et al. improved the visual space of the image, and deal with the characteristics of image brightness and saturation from the perspective of the visual model [11]. It has gradually become a trend to extract visual information close to biological models. But it still has a long way to brain-inspired SLAM, and the application of semantics in it needs to be further studied.

Besides, there is only a single topology environment display method for map in RatSLAM. In traditional SLAM, multidimensional representation using semantics has become a hotspot. Nakajima et al. integrated object information to associate geometric map points with semantic labels [14]. Iqbal et al. inserted complete object instances into the map instead of classifying already obtained map elements [15]. Yet the semantic map for RatSLAM is still being explored.

3 Methodology

In this section, we will first review the RatSLAM and its limitations. The proposed novel SLAM system is then introduced. Finally, we will display how to integrate semantics in LCD and build a multi-dimensional experience map.

3.1 Revisiting RatSLAM

RatSLAM is composed of three main modules: pose cell, local view cell, and experience map. The local view cells store the view template in it [16]. If a new visual scene is encountered, a new local view cell would be created, and a link would also be built to connect this local view cell with the active pose cell. When a familiar scene is sensed, the corresponding local view cell would be activated and inject energy to the connected pose cell. We detect familiar scenes by calculating the similarity between the visual templates in LCD. The pose cells are represented the robot's perception of its current pose (x', y', θ'). Their dynamics are regulated by the path integration and the energy from local view cells. Finally, the experience map groups the information from local view cells and pose cells to represent the robot pose uniquely. Due to the alone visual information in LCD, conventional RatSLAM, despite being simple and effective, struggles with low precision and poor stability.

3.2 Semantic-Pixel Associative RatSLAM

The overview of the proposed semantic-pixel association RatSLAM is presented in Fig. 1, which includes two major components: LCD and experience map building. Our system will use semantic information and pixel information in LCD to eliminate trajectory offset caused by accumulated errors. Here, YOLOv2 network is introduced to obtain semantic information. We then estimate the pose through pose cells, build an experience map fused with semantic information. 3D models will be projected in the corresponding position according to semantic information, making a comprehensive representation of the environment.

3.3 Integrate Semantic Information in LCD

To improve the accuracy of LCD, our system adds semantic information in conventional template similarity calculation. When a visual scene is an input, it is annotated by YOLOv2 and stored in the view template. In the LCD, semantic information is firstly used for rough matching, and templates that are less similar to the current scene are quickly excluded to reduce unnecessary calculations. Afterwards, we calculate the similarity between the current scene and the remaining templates and treat the template with the highest similarity as the matching template. In addition, we set it not to match with the recently saved templates to reduce the false closure caused by the short interval between two images. Our proposed LCD method mainly consists of two parts: (1) Semantic annotation part. We identify objects and store the semantic information and pixel information in the view template. (2) Template matching part. We compute template similarity using multiple levels of information. The architecture of the proposed method is shown in the LCD module in Fig. 1.

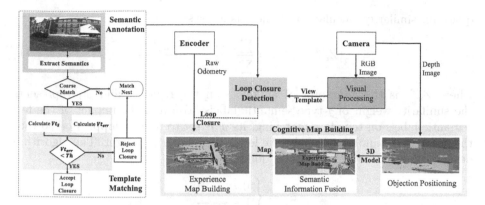

Fig. 1. The architecture of our semantic-pixel association RatSLAM system.

Semantic Annotation. To perform stable and accurate LCD in complex scenes, it is necessary to perform object detection on RGB images and find out images' discriminative landmarks, including their categories and positions. YOLOv2 is a target algorithm introduced in the YOLO family. Its detection framework can be roughly divided into two parts: feature extraction network and action network. The feature extraction network is modified based on the DarkNet-19 network. The detection network consists of 4 convolutional layers, a transfer layer, and a detection layer [17]. In contrast to other YOLO detectors, YOLOv2 is relatively basic, but with better performance and a simple structure. It is proved by experiments that even with the basic detector, we can have better performance than traditional RatSLAM. By using the YOLOv2 network on the images, we obtain a set of labeled semantic regions $S = \{s_1, s_2, \ldots, s_n\}$, $s_i = [type_i, x_i, y_i, w_i, h_i]$, where $type, x, y, w, h$ are the label category, center point coordinates, width, and height values respectively. The semantic information s_i of each image is stored in the label set S_k. To facilitate the use of semantic information in LCD, S_k of the image is added to the original visual template V_k.

Template Matching. After semantic annotation, an image is abstracted into a semantic label set S. Taking the center of the image as the origin, we divide an image into four quadrants: upper left, lower left, upper right, and lower right. Any label s_i in S is classified into one quadrant through its center coordinates. Count the number of all regions according to the type of semantic label, and construct a scene descriptor $U = [Z^{lu}; Z^{ld}; Z^{ru}; Z^{rd}]$, $Z = [z_1, z_2, \ldots, z_n]$, where Z^{lu}, Z^{ld}, Z^{ru} and Z^{rd} represent the descriptor information of the four quadrants of a picture respectively, n is the number of semantic label categories, and z_i represents the number of a certain semantic tag in a quadrant.

After obtaining U, the algorithm calculates the similarity between the current image and the four quadrants of the template. For the current image a and the template b to be matched, according to its scene descriptors, we define the quadrant similarity calculation method as follows:

$$\zeta(Z_i^{(a)}, Z_i^{(b)}) = \frac{\sum \omega \cdot |Z_i^{(a)} - Z_i^{(b)}|}{\sum \omega \cdot |Z_i^{(a)} + Z_i^{(b)}|} \tag{1}$$

where $Z_i^{(a)}$ is the descriptor of an image in the region i and w_j represents the similarity weight of j-type semantic labels. To avoid the impact of the few semantic labels on template matching, its w is set to 0. w is suppressed so that the greater the number of labels, the lower the similarity weight. The algorithm designs label similarity weight is as follows:

$$\omega_j = \begin{cases} 0, c_j \to 0 \\ 1 - \frac{c_j}{\sum_j^n c_j} \end{cases} \tag{2}$$

where c_j is the total number of labels of j-type. Sum the similarities of the four quadrants to obtain the semantic similarity between the image and the template:

$$Vt_s(a,b) = \sum_{i=1}^{4} \zeta(Z_i^{(a)}, Z_i^{(b)}) \tag{3}$$

Combined with the Vt_s with Vt_p calculated by the SAD algorithm, the template similarity Vt_{err} between the image and the template is defined as follows:

$$Vt_{err} = \alpha \cdot Vt_s + \beta \cdot Vt_p \tag{4}$$

where α and β are the weight parameters of semantic similarity and pixel similarity, respectively. The most similar template to the current image, which with the highest similarity is preferentially selected as the best-matched closed-loop template. If a closed loop is detected, the system will correct the cognitive map, otherwise, a new template will be created.

3.4   Experience Map Building with Semantics

In this part, we project the recognized objects onto the map, displaying semantic information in the environment. Then we construct a topology map, 1displaying the physical information of the environment.

Objection Positioning. When the objects are detected, we perform object instance segmentation by adopting primitive 3D shape priors. Then we reconstruct and find all 3D points inside the detected box, where the primitive model of the objects is fitted. The Euclidean region growing segmentation technique [18] is introduced as the clustering technique adopted in the shape model fitting, returning the centroid and respective convex hull dimensions of all classes except doors. Whenever doors are recognized, we fit a planar patch using RANSAC [19] for estimating the pose. The projected pose of each object, denoted by $y \in R^3$, is then represented by the 2D projected centroid from the camera coordinate system to the global map coordinate system and its orientation.

Experience Map Building. Experience map in RatSLAM system is a semi-topology map composed of many experiences. When the system is running, the subsequent experiences are calculated based on the position of the previous experience and the robot's motion. When the closure is detected, the update procedure for experiences will be executed. Then the active pose cell and the local view cell will be reset to the matched experience. The map will be adjusted by the matched experience and a new experience will be created.

4 Experimental Analysis

In this section, we verify the proposed method from the following four perspectives. Firstly, LCD with semantic annotation is compared with traditional LCD in terms of Precision and Recall. Next, we discuss how the weight parameters of Vt_s and Vt_p affect the performance of our proposed method on specific datasets. We then evaluate the performance of the proposed LCD method facing noisy data. Finally, for showing the effectiveness of the entire RatSLAM system with our proposed method, the experiment is performed in real environment datasets.

4.1 Datasets

Experimental verification is performed using two datasets that are widely used to evaluate LCD algorithms because these datasets provide ground truth loop closures, which is convenient to measure the correctness of the results. Lab datasets are recorded to verify the effectiveness of multidimensional cognitive map construction with LCD. The datasets used in experiments are listed:

- Oxford New College (NC) dataset [20]. The left and right cameras each capture 1073 images, with 423 and 430 real closed loops respectively.
- City Center (CiC) dataset [20]. The left and right cameras each capture 1237 images, each with 561 real closed loops.
- Recorded Lab datasets. The two indoor datasets are gathered from a laboratory with images recorded by the RGB-D sensor, which is controlled to do semi-automatic environmental exploration. The sensor respectively captures 2893 and 1784 image pairs (RGB and depth).

4.2 Performance Comparison of LCD

Generally, the Precision-Recall (PR) curve is used to evaluate the performance of the LCD algorithm. Precision reflects the ability of the algorithm to detect correctly, and Recall reflects the ability to detect comprehensively. By changing the threshold of the similarity, different Precision and Recall can be obtained, thereby obtaining the PR curve. Figure 2 shows the comparison of the PR curves obtained by our proposed method and the original Ratslam, DBoW2 [21], and SeqSLAM [22] algorithms for LCD on four datasets, including NC's, and CiC's left and right image datasets. For NC, we set the label type total is 17, while

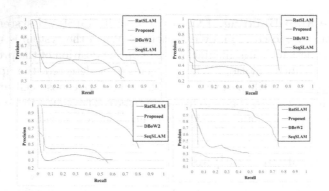

Fig. 2. Comparison results of the proposed LCD method, RatSLAM, DBoW2 and SeqSLAM on different datasets.

40 for CiC. α is set to 0.9, 0.8 0.9 0.8 for NC's, and CiC's left and right image datasets respectively, and β is set to 0.09 0.01 0.05 0.09 respectively.

Fig. 2 shows that as the Recall increases, the Precision of the method we propose drops slowly, while the others drop rapidly. When the Recall is 0.5, the Precision of our method can still maintain 0.9. This is because in these datasets, dynamic objects such as vehicles and pedestrians have great interference with traditional LCD algorithms, and their robustness is relatively poor in the face of similar structures and local changes. The proposed algorithm calculate the pixel and semantic similarity of different quadrants, which increases the amount of matching information, performs stricter similarity calculations according to areas, and improves the accuracy of the LCD.

4.3   Performance Trade-Offs for LCD

To express the proposed method more clearly, we discuss the relationship between different α and β under the same Th. 9 values of α and β are tested in the NC's left and CiC's right camera image dataset respectively. We set $\alpha \in [0.1, 0.9]$ with an interval of 0.1, and set $\beta \in [0.01, 0.09]$ with an interval of 0.01. Under the same α condition, Fig. 3 indicates that the smaller the β value, the lower the Precision, and the higher the Recall. This is because the single semantics will generate a large number of false matches when the use of pixels in the image-matching process is greatly weakened. And it also shows that the greater the β, the increase of the Precision, and the decrease of the Recall. Since the SAD is sensitive to noise, it will match the image and the template too strictly. Therefore, when the use of pixels is greatly enhanced, closed loops that should be matched are not matched. When α is larger, the average performance of LCD is better, indicating that more semantic information is used in the process of template comparison, and better LCD results; when α is less than 0.5, the average performance of LCD crosses, and the difference not much, indicating that when the α is less than 0.5, its influence on the comprehensive similarity is lower than that of the pixel similarity, and the matching is mainly determined by the SAD. Therefore, the selection of parameters α and β requires a trade-off between Precision and Recall for LCD.

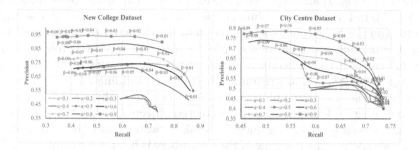

Fig. 3. PR curves under different similarity parameters.

4.4 Brightness Transformation on LCD

In this section, we conduct experiments under different lighting conditions to demonstrate the robustness of the proposed algorithm under the influence of lighting. We find the closed-loop matching images from the CiC's right image dataset, and select serial 38 even frames from 0304-0368 as the test dataset. The brightness of the 38 frames is increased by 10 and decreased by 10 respectively to obtain the test dataset S1 and the test dataset S2.

Table 1 shows the original algorithm is more obviously affected when the brightness changes. Compared with the original result, the matching template has a more obvious change, and more than 90% of the changed templates are wrong matches. However, our method is less affected by brightness changes, and the proportion of false matches in templates with matching changes is less than 65%. These results show that our proposed algorithm is still more robust than the original under changing lighting conditions.

Table 1. Closed-loop detection effect under brightness change

Method	brightness	changed matched images	rate of change	match errors	error rate
Original	+10	14	36.84%	13	92.86%
algorithm	−10	21	55.26%	21	100.00%
Proposed	+10	6	15.79%	2	33.33%
algorithm	−10	14	36.84%	9	64.29%

4.5   Experience Map Building on Lab Dataset

We compare the original algorithm and our method on the recorded Lab datasets
to verify the effectiveness of fusing semantic information for improving mapping
accuracy (see Fig. 4) and expressing the experience map with object information
(see Fig. 5). In Fig. 4 the left column shows the maps constructed by the original
system, and the right column shows the maps constructed by our system. Obviously, the original system cannot correct the map because LCD relying on pure
pixel information has difficulty in correcting the cumulative error. In contrast,
the proposed RatSLAM algorithm has better map-building effects. Through 3D
modeling of the recognized objects, the map shows the physical environment of
the robot in multiple dimensions. The detected objects are shown in green, white,
blue, and brown representing "door", "desk", "water" and "box" respectively.
Figure 5 shows that the mapping of the laboratory is built more comprehensively,
and gives people a better understanding of the environment.

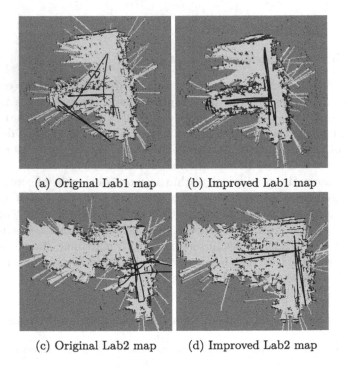

(a) Original Lab1 map   (b) Improved Lab1 map

(c) Original Lab2 map   (d) Improved Lab2 map

Fig. 4. The cognitive map generated from the Lab datasets by the original RatSLAM
algorithm and our proposed method.

(a) Lab1   (b) Lab2

Fig. 5. The multi-dimensional maps generated from the Lab datasets.

5   Conclusion

In this paper, an improved RatSLAM model incorporating semantic information has been proposed to improve the performance of LCD and construct an experience map with semantic information. In this model, we use YOLOv2 to extract semantic labels in images, integrate semantic information and pixel information to create visual templates, calculate the similarity between visual templates based on pixel level and object level, improve the accuracy of LCD, and reduce the interference of lightness. We use 3D model segmentation technology to project the objects extracted from the image to the corresponding positions on the map, and build an experience map with semantic information. Experiments have proved that the performance of LCD has been significantly enhanced in the RatSLAM system, and the experience map constructed has better accuracy and comprehensively reflects the environment structure and semantics.

Acknowledgements. This work was supported by the National Natural Science Foundation of China (Grant No. 62206188), the China Postdoctoral Science Foundation (Grant No. 2022M712237), Sichuan Province Innovative Talent Funding Project for Postdoctoral Fellows and the 111 Project under grant B21044.

References

1. Hamid Taheri and Zhao Chun Xia: SLAM; definition and evolution. Eng. Appl. Artif. Intell. **97**, 104032 (2021)
2. Davison, A.J., Reid, I.D., Molton, N.D., Stasse, O.: MonoSLAM: real-time single camera slam. IEEE Trans. Pattern Anal. Mach. Intell. **29**(6), 1052–1067 (2007)
3. Sharma, S., Sur, C., Shukla, A., Tiwari, R.: CBDF based cooperative multi robot target searching and tracking using BA. In: Jain, L.C., Behera, H.S., Mandal, J.K., Mohapatra, D.P. (eds.) Computational Intelligence in Data Mining - Volume 3. SIST, vol. 33, pp. 373–384. Springer, New Delhi (2015). https://doi.org/10.1007/978-81-322-2202-6_34
4. Calvo, R., de Oliveira, J.R., Figueiredo, M., Romero, R.A.F.: A distributed, bioinspired coordination strategy for multiple agent systems applied to surveillance tasks in unknown environments. In: The 2011 International Joint Conference on Neural Networks, pp. 3248–3255. IEEE (2011)

5. Silva, G., Costa, J., MagalhÃes, T., et al.: Cyberrescue: a pheromone approach to multi-agent rescue simulations. In 5th Iberian Conference on Information Systems and Technologies, pp. 1–6. IEEE (2010)
6. Ball, D., Heath, S., Wiles, J., Wyeth, G., Corke, P., Milford, M.: OpenratSLAM: an open source brain-based slam system. Auton. Robot. **34**(3), 149–176 (2013)
7. Han, L., Fang, L.: Mild: multi-index hashing for loop closure detection. arXiv preprint arXiv:1702.08780 (2017)
8. Gu, T., Yan, R.: An improved loop closure detection for ratSLAM. In: 2019 5th International Conference on Control, Automation and Robotics (ICCAR), pp. 884–888. IEEE (2019)
9. Zhou, S.-C., Yan, R., Li, J.-X., Chen, Y.-K., Tang, H.: A brain-inspired slam system based on ORB features. Int. J. Autom. Comput. **14**(5), 564–575 (2017)
10. Kazmi, S.M.A.M., Mertsching, B.: GIST+ ratSLAM: an incremental bio-inspired place recognition front-end for ratSLAM. EAI Endors. Trans. Creat. Technol. **3**(8), e3–e3 (2016)
11. Chong, W., Shumei, Yu., Chen, L., Sun, R.: An environmental-adaptability-improved ratSLAM method based on a biological vision model. Machines **10**(4), 259 (2022)
12. Crespo, J., Castillo, J.C., Mozos, O.M., Barber, R.: Semantic information for robot navigation: a survey. Appl. Sci. **10**(2), 497 (2020)
13. Xu, J., Yan, N., Tang, F.: An improvement of loop closure detection based on bow for ratSLAM. In: 2022 37th Youth Academic Annual Conference of Chinese Association of Automation (YAC), pp. 634–639. IEEE (2022)
14. Hempel, T., Al-Hamadi, A.: An online semantic mapping system for extending and enhancing visual SLAM. Eng. Appl. Artif. Intell. **111**, 104830 (2022)
15. Iqbal, A., Gans, N.R.: Data association and localization of classified objects in visual slam. J. Intell. Robot. Syst. **100**(1), 113–130 (2020)
16. Wyeth, G., Milford, M.: Spatial cognition for robots. IEEE Robot. Autom. Mag. **16**(3), 24–32 (2009)
17. Redmon, J., Farhadi, A.: Yolo9000: better, faster, stronger. In: Proceedings of the IEEE Conference on Computer Vision and Pattern Recognition, pp. 7263–7271 (2017)
18. Rusu, R.B., Cousins, S.: 3D is here: Point cloud library (PCL). In: 2011 IEEE International Conference on Robotics and Automation, pp. 1–4. IEEE (2011)
19. Raguram, R., Frahm, J.-M., Pollefeys, M.: A comparative analysis of RANSAC techniques leading to adaptive real-time random sample consensus. In: Forsyth, D., Torr, P., Zisserman, A. (eds.) ECCV 2008. LNCS, vol. 5303, pp. 500–513. Springer, Heidelberg (2008). https://doi.org/10.1007/978-3-540-88688-4_37
20. Cummins, M., Newman, P.: FAB-MAP: probabilistic localization and mapping in the space of appearance. Int. J. Robot. Res. **27**(6), 647–665 (2008)
21. Gálvez-López, D., Tardos, J.D.: Bags of binary words for fast place recognition in image sequences. IEEE Trans. Robot. **28**(5), 1188–1197 (2012)
22. Milford, M.J., Wyeth, G.F.: SeqSLAM: visual route-based navigation for sunny summer days and stormy winter nights. In: 2012 IEEE International Conference on Robotics and Automation, Saint Paul, MN, USA, pp. 1643–1649 (2012). https://doi.org/10.1109/ICRA.2012.6224623

Knowledge Distillation via Information Matching

Honglin Zhu[1], Ning Jiang[1,3]([✉]), Jialiang Tang[2], and Xinlei Huang[1,3]

[1] School of Computer Science and Technology, Southwest University of Science and Technology, Mianyang 621000, Sichuan, China
jiangning@swust.edu.cn
[2] School of Computer Science and Engineering, Nanjing University of Science and Technology, Nanjing 210094, Jiangsu, China
[3] Jiangxi Qiushi Academy for Advanced Studies, Nanchang 330036, Jiangxi, China

Abstract. Knowledge distillation can enhance network generalization by guiding a smaller student network to learn from a more complex teacher network. The challenge lies in maximizing the performance of the student network under the supervision of the teacher network. Currently, the feature-based distillation approach utilizes the middle-layer features of the teacher network to improve the performance of the student network. However, this approach lacks a measure to evaluate the content of the information present in the intermediate layers of both the teacher and student networks, which leads to a distillation mismatch of features and damages the student's performance. In this study, we propose a new feature distillation method to solve this problem. We measure the information content in the intermediate layers of the teacher and student networks based on the receptive fields of corresponding features. Subsequently, the suitable number and locations of transmission features are decided based on information content, effectively alleviating the risk of information mismatch during distillation. Our experimental results demonstrate that the proposed method significantly improves the performance of the student network.

Keywords: Model Compression · Knowledge Distillation · Receptive Field · Feature Distillation

1 Introduction

Deep neural networks (DNNs) have been widely used in computer vision for tasks such as object detection [16] and semantic segmentation [7]. However, these excellent DNNs with a large number of parameters often require huge computing consumption and memory occupation, which make it difficult to deploy the model on resource-constrained devices. Many model compression methods have been proposed to solve this problem, including network quantization [9], network pruning [6,11,12,20], lightweight network design [14,28], and knowledge distillation (KD) [1,8,17,26,29]. Among them, the knowledge distillation method has attracted much attention due to its effectiveness and simplicity.

Knowledge distillation refers to using the pre-training teacher network to transfer knowledge to a small student network to improve the performance of the student network. Generally speaking, there are different types of knowledge, such as soft labels [8] and middle hidden layer information [17]. Compared with the category dependency information in soft labels, there is rich texture information contained in the features of the middle hidden layer, which can be taken as useful supervision signals for the student network. In general, as shown in Fig. 2a and 2b, most of the existing methods artificially select feature transfer locations and lack measurement methods to quantify the information content in different feature locations between the teacher and the student. For example, a series of methods [2,3,17,25,26] propose to match the middle-layer features of the teacher network and the student network one-to-one. In fact, due to the structural differences between the teacher and the student, the student network needs to mimic information suitable for itself. Furthermore, with the change of network parameters during training, the conventional one-to-one feature matching may cause an information mismatch, which inevitably damages the performance of the student network.

In deep learning, the receptive field [13] is an important tool for understanding how the network works, which is defined as a region where a specific feature is affected by the input space. The receptive field can quantify the size of an image captured by a feature. Intuitively, if a feature can capture a large size for an image, it contains much information content about the image. Therefore, we calculate the receptive fields of the features of the teacher and student to quantify their information content, which can be utilized to precisely match the transferring features from teacher to student.

Based on the above analysis, as shown in Fig. 2, we propose a distillation framework called **I**nformation **M**atching **K**nowledge **D**istillation (IMKD) to efficiently align the teacher and the student. Firstly, the receptive field computation is introduced to automatically compute the information difference between student and teacher in the middle layer. Secondly, we propose a simple feature-matching method to obtain suitable teacher-student layer pairs according to the information difference. The student network mimics the information of the suitable feature locations to obtain higher performance and stronger generalization. Our method has mainly two advantages for distillation over previous feature-based distillation: 1) By measuring the information difference between the features of the teacher and the student, we can properly assign features of the teacher network to transfer to the student network; 2) we can adaptively match the target features according to the performance improvement of the student network. We conduct experiments on different combinations of teacher-student networks. The experimental results in Sect. 4 show that our method significantly improves different neural networks. For example, the proposed method improves the performance of ShuffleNetV2 [14] from 71.82% to 78.13% on CIFAR100. To sum up, the main contributions of this paper are as follows:

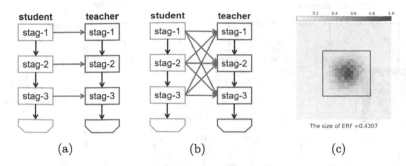

Fig. 1. (a) One-to-one feature matching. (b) Fully connected feature matching. (c) Effective receptive field. The area in the red box is a high contribution area. The effective receptive field size is obtained by calculating the ratio of the high contribution area to the entire area. Visualization method derived from work [5]

- We propose a novel information-matching knowledge distillation (IMKD) to adaptively match target features, which can effectively alleviate the problem of information mismatch in feature distillation.
- We analyze the importance of the receptive field for feature distillation and introduce it as a new metric to quantify the difference between teacher-student layer pairs.
- Our proposed method can adjust the learning features for the student network according to the performance change of the student network in distillation to further improve the performance of the student network.

2   Related Work

2.1   Knowledge Distillation

Knowledge distillation aims to train an excellent, compact student network by learning the predictions of a cumbersome teacher network. Hinton *et al.* [8] propose the pioneering method, where the student network learns the soften output probabilities of the teacher network. Subsequent works explore rich knowledge, *e.g.*, intermediate layer responses [17], attention maps [26], sample relations composed of sample similarity matrices [15], and representation learning [21]. In addition, cross-layer feature distillation methods [2,3,31] propose to make the student network fully mimic the multi-layer features of the teacher network and achieve surprising results.

2.2   Receptive Field

The receptive field is a concept that helps to understand and analyze DNNs, which can measure the degree of association between the feature output of the network and the input. In 2016, Luo *et al.* [13] proposed the concept of the

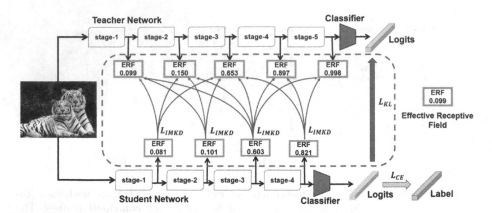

Fig. 2. Overview of our proposed distillation method (IMKD). First, the features and effective receptive fields of each layer are obtained through forward propagation. Second, the difference between each teacher-student layer pair is computed, and the proposed IMKD automatically matches suitable teacher-student layer pairs. Finally, the student mimics the teacher's predictions via L_{total}.

effective receptive field (ERF) and visualized the ERF of the network. The ERF of the network is much smaller than the theoretical receptive field, which presents a Gaussian distribution in the center of the image. Recently, Ding *et al.* [5] greatly improved the performance of the network by increasing the ERF. In this study, we measure the information difference between the teacher and the student by introducing the ERF into distillation training. We propose a matching algorithm to enable the student network to obtain better feature representation capabilities by matching appropriate feature information.

3 Method

In this section, we first introduce the necessary concepts of knowledge distillation and basic notations in Sect. 3.1. Then, effective receptive field calculation is described in Sect. 3.2. Finally, we propose the novel matching mechanism and main formulations in Sect. 3.3.

3.1 Preliminaries

Formally, we define the training dataset as $\mathcal{D} = \{(\mathbf{x}_i, y_i)\}_{i=1}^{n}$, where n denotes the number of examples in $\mathcal{D}$ and $\mathcal{D}$ contains c categories. The fixed pre-trained teacher network and the randomly initialized student network are denoted as T and S, respectively. For an image $\mathbf{x} \in \mathcal{D}$ inputting into the teacher network T and the student network S, the logits are obtained as of the student network S and the teacher network T as $\mathbf{g}_S = S(\mathbf{x})$ and $\mathbf{g}_T = T(\mathbf{x})$, respectively. The features of the student network and the teacher network are denoted as $F_{S_l} \in \mathbb{R}^{C_{S_l} \times H_{S_l} \times W_{S_l}}$ and $F_{T_l} \in \mathbb{R}^{C_{T_l} \times H_{T_l} \times W_{T_l}}$, where $S_l \in \{S_1, ..., S_L\}$ and $T_l \in \{T_1, ..., T_L\}$. C, H, and W represent the channel number, height, and width of the corresponding feature, respectively. S_L and T_L are the total number of features of the teacher and the student, where S_L and T_L are not correlated. $\sigma(\cdot)$ denotes as the softmax function, and l represents the index of layers in the corresponding model. The symbol ω is a balance parameter that is used to reasonably control the size of the difference.

The core idea of knowledge distillation is to improve network performance by transferring the knowledge of the teacher network to the student network. In classical KD [8], distillation training is achieved by a minidiscrepancy between the soft labels of the teacher network and student network:

$$L_{KD} = L_{CE}(y, \sigma(\mathbf{g}_S)) + \tau^2 L_{KL}(\sigma(\frac{\mathbf{g}_S}{\tau}), \sigma(\frac{\mathbf{g}_T}{\tau})), \tag{1}$$

where $L_{CE}(\cdot, \cdot)$ represents the cross-entropy loss, which is used to calculate the difference between the truth label y and the prediction $\sigma(g_S)$, L_{KL} is the KL divergence, τ is a temperature parameter to control the soften degree of the soft labels.

3.2   Effective Receptive Field Calculation

In our IMKD, we introduce ERF to quantify the gap of features between the teacher and the student in preparation for selecting the appropriate feature pairs.

To obtain the effective receptive field of a feature, we first compute the gradient of the feature. Suppose $\mathbf{x}^{j,u,v}$ is denoted as a pixel of the input $\mathbf{x}$, where $\mathbf{x} \in \mathbb{R}^{3 \times H_{\mathbf{x}} \times W_{\mathbf{x}}}$. First, the feature $F_{S_l} \in \mathbb{R}^{C_{S_l} \times H_{S_l} \times W_{S_l}}$ are converted to $\tilde{F}_{S_l} \in \mathbb{R}^{H_{S_l} \times W_{S_l}}$ by summation operation. Following the work [13], the partial derivative $\partial \tilde{F}_{S_l}^{1,1} / \partial \mathbf{x}^{j,u,v}$ can measure the importance of $\mathbf{x}^{j,u,v}$ with respect to $\tilde{F}_{S_l}^{1,1}$. Similarly, the importance of an input $\mathbf{x}$ with respect to $\partial \tilde{F}_{S_l}^{1,1}$ can be expressed as $G_{S_l}^{1,1} = \sum_{j=1}^{3} \sum_{u=1}^{H_{\mathbf{x}}} \sum_{v=1}^{W_{\mathbf{x}}} \frac{\partial \tilde{F}_{S_l}^{1,1}}{\partial \mathbf{x}^{j,u,v}}$. Then, we can obtain the gradient G_{S_l} of the feature F_{S_l}, where the gradient $G_{S_l} = [G_{S_l}^{1,1}, ..., G_{S_l}^{H_{S_l}, W_{S_l}}] \in \mathbb{R}^{H_{S_l} \times W_{S_l}}$. We show the effective receptive field by visualizing this gradient in Fig. 2d. Moreover, the high contribution region containing 99% of non-zero gradient values is selected, and its height and width are both K_{S_l}. We obtain the size R_{S_l} of the ERF through the area ratio of the high contribution area to the entire area:

$$R_{S_l} = \frac{K_{S_l} \times K_{S_l}}{H_{S_l} \times W_{S_l}}. \tag{2}$$

In this study, the size R of the ERF at each layer of the teacher network and the student network can be expressed as $C = \{(R_{S_l}, R_{T_l}) | \forall R_{S_l} \in [R_{S_1}, \cdots, R_{S_L}], \forall R_{T_l} \in [R_{T_1}, \cdots, R_{T_L}]\}$.

3.3   Feature Matching Distillation

The key point of our method is to measure the information difference and match target features suitable for the student network to learn. Overall, our method can be divided into three steps: 1) The same examples are input into the teacher and the student to obtain effective receptive fields; 2) through the calculated effective receptive fields, the student network selects suitable target features; and 3) knowledge from the teacher network is transferred to the student network based on the selected features.

First, we can use the following formula to get the difference in the receptive field of each layer:

$$\alpha_{(T_l, S_l)} = |R_{T_l} - R_{S_l}|. \tag{3}$$

We need to obtain target feature maps matching the student network to avoid negative effects. Therefore, we classify the difference factor α as follows: 1) If α is smaller than ω, we keep such a teacher-student layer pair; 2) otherwise, we regard such a feature pair as a semantic mismatch and discard it. The specific formula is as follows:

$$W_{(T_l, S_l)} = \begin{cases} 1, & if \quad \alpha_{(T_l, S_l)} < \omega \\ 0, & \text{otherwise}, \end{cases} \tag{4}$$

where the reasonable value range of ω is shown in Sect. 4.4. Then, the mean square error (MSE) is used to minimize the distance between selected target features from the teacher network and those from the student network. Distillation loss L_{IMKD} is as follows:

$$L_{IMKD} = \sum_{S_l=1}^{S_L} \sum_{T_l=1}^{T_L} W_{(T_l, S_l)} MSE(r(F_{S_l}), r(F_{T_l})), \tag{5}$$

where $r(\cdot)$ is an adaptive layer composed of 1×1 convolution and an adaptive pooling layer, which is used to align the dimension of features.

Finally, we use the following total loss to train the student network:

$$L_{TOTAL} = L_{CE} + \tau^2 L_{KL} + \beta L_{IMKD}, \tag{6}$$

where β is a hyperparameter used to balance the loss function.

4   Experiments

The training details and datasets for all experiments are in Sect. 4.1. The effectiveness of the proposed distillation method has been verified on different datasets. In Sect. 4.2, we show experimental results of teacher-student networks with the same structures and different structures for distillation on CIFAR-100, STL10 and Tiny-ImageNet. To further verify the effectiveness of the method, we provide the results of the comparison of different distillation methods. Moreover, Sect. 4.3 provides some visualizations of our approach. Finally, the results of ablation experiments are presented in Sect. 4.4.

4.1   Datasets and Experiments Configuration

Three basic image classification datasets, including CIFAR100 [10], Tiny-ImageNet [22], and STL10 [4], are used to verify the effectiveness of the proposed method.

Following previous work [2,21], we employ a stochastic gradient descent optimizer with momentum set to 0.9. For all experiments, the weight decay is set to 5×10^{-4}, and the batch size is set to 64. For CIFAR100, the initial learning rate is set to 0.05, and the learning rate decreases by 10 at 150, 180, and 210 epochs until 240 epochs. The configuration of the Tiny-ImageNet dataset is consistent with that of CIFAR100. For STL10, the initial learning rate is set to 0.05 and is decayed by 0.1 every 30 epochs. We set the L_{CE} weight and temperature parameters τ as 1 and 4, respectively.

Table 1. Top-1 test accuracy of different distillation approaches on CIFAR100.

| Teacher | ResNet32x4 | ResNet32x4 | ResNet-110x2 | ResNet32x4 | WRN-40-2 |
Student	ShuffleNetV1	ShuffleNetV2	ResNet-116	resnet8x4	ShuffleNetV1
Teacher	79.42%	79.42%	78.18%	79.42%	75.61%
Student	70.50%	71.82%	74.46%	72.50%	70.50%
KD [8]	74.07%	74.45%	76.14%	73.33%	74.83%
FitNet [17]	73.59%	73.54%	76.20%	73.44%	73.73%
AT [26]	71.73%	72.73%	76.84%	72.94%	73.32%
CRD [21]	75.11%	75.65%	76.83%	75.51%	76.05%
SRRL [24]	75.18%	76.19%	77.19%	75.39%	76.30%
SemCKD [2]	76.31%	77.62%	76.69%	76.23%	76.83%
DKD [30]	76.45%	77.07%	77.08%	76.32%	76.70%
Ours	**77.17 %**	**78.13%**	**78.05%**	**76.89%**	**77.31%**

4.2   Compared to Different Distillation Methods

Table 1 show the top-1 test accuracy of multiple methods with different network combinations on CIFAR100. We compare eight different distillation methods, among which FitNet [17], AT [26], and SemCKD [2] are feature-based distillation methods. From Table 1, we can see that the proposed method significantly improves network accuracy compared to other methods on CIFAR100. For example, in the network combination "ResNet32x4 [23]/ShuffleNetV1 [28]", the test accuracy of our method achieves a performance improvement of 6.51% compared to the baseline. The higher performance of the network reflects that the proposed method alleviates the information mismatch problem of feature distillation.

Table 2 shows the experimental results of different teacher-student architectures on Tiny-ImageNet. From Table 2, we can see that IMKD outperforms other feature distillation methods. Especially, the best example is in the setting of "ResNet32x4 [23]/ShuffleNetV2 [14]" where the student's performance is even better than the teacher's performance. The reason for the excellent performance of our proposed approach is to select connections suitable for training.

Table 3 shows the comparison results of different distillation methods on STL10. Our method significantly improves the test accuracy of different networks on the STL10 dataset. For example, for the network combinations "WRN-40-2 [27]/ShuffleNetV1 [28]" and "WRN-40-2 [27]/MobileNetV2 [18]", IMKD outperforms semckd by 2.05% and 3.03%. Furthermore, IMKD significantly outperforms KD methods in multiple network combinations. These excellent test results show that our method can select the target features suitable for distillation, and the training of IMKD is stable and robust.

Table 2. Top-1 test accuracy of different distillation approaches on Tiny-ImageNet.

Teacher	WRN-40-2	ResNet110	ResNet32x4
Student	WRN-16-2	ResNet20	ShuffleNetV2
Teacher	62.53%	60.36%	65.28%
Student	58.60%	54.03%	62.40%
KD [8]	60.92%	55.75%	67.95%
FitNet [17]	58.63%	53.20%	67.82%
AT [26]	60.20%	55.86%	66.90%
RKD [15]	58.72%	54.22%	68.28%
SemCKD [2]	60.82%	56.52%	68.13%
Ours	**61.67%**	**57.28%**	**68.87%**

Table 3. Top-1 test accuracy of different distillation approaches on STL10.

Teacher	VGG13	WRN-40-2	WRN-40-2	ResNet32x4
Student	VGG8	MobileNetV2	ShuffleNetV1	MobileNetV2
Teacher	71.58%	71.26%	71.26%	69.67%
Student	69.30%	53.30%	59.60%	53.30%
KD [8]	71.65%	58.07%	61.33%	58.06%
FitNet [17]	71.81%	63.79%	68.13%	63.26%
AT [26]	70.41%	58.43%	69.86%	60.51%
SRRL [24]	69.76%	53.83%	65.86%	57.09%
SemCKD [2]	72.61%	66.58%	70.33%	61.99%
DKD [30]	69.73%	58.04%	63.58%	58.89%
Ours	**73.30%**	**69.61%**	**72.38%**	**65.01%**

4.3   Visualization

To verify the advantage of our proposed IMKD, we provide the visualization of the penultimate layer of the student network. In Fig. 3, IMKD has stronger

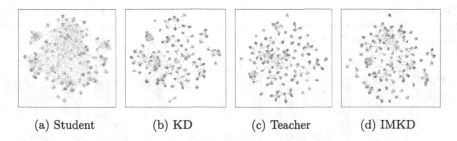

|(a) Student|(b) KD|(c) Teacher|(d) IMKD|

Fig. 3. The penultimate layer visualization of ShuffleNetV1 (student) with KD, IMKD and the teacher (ResNet32x4) on CIFAR-100.

Table 4. The ablation experiments with different feature matching approachs.

| Teacher | ResNet32x4 | WRN-40-2 | ResNet32x4 |
Student	ShuffleNetV1	ShuffleNetV1	ResNet8x4
Teacher	79.42%	75.61%	79.42%
Student	70.50%	70.50%	72.50%
one-to-one feature matching	76.09%	75.85%	75.18%
Fully connected feature matching	76.10%	76.20%	75.54%
Ours(IMKD)	**77.17%**	**77.31%**	**76.68%**

representation ability compared to the student network. Moreover, our approach has more scattered embeddings than KD.

To provide visual interpretations of IMKD, we use Grad-CAM [19] to highlight regions that the model values in Table 5. Three images labeled "Espresso", "Labrador" and "Orange" are randomly selected on Tiny-ImageNet. In Table 5, the Grad-CAM maps of IMKD pays more attention to important regions than KD and baseline. As shown in the visualization in the fourth row of Table 5, our method can focus on the location of the orange.

4.4   Ablation

We set up three feature matching methods to verify the effectiveness of the proposed information matching method in Table 4. One-to-one feature matching is set as the same stage of information transfer, as shown in Fig. 2a. The fully connected feature matching approach is shown in Fig. 2b. Different from other feature distillation frameworks, our IMKD performs automatic feature matching based on information differences. According to the data in Table 4, suitable information transfer can significantly improve the performance of the student network. In particular, the other feature matching methods are significantly lower than the information matching distillation method, which illustrates the suboptimal performance of hand-crafted feature matching methods. The information matching distillation can effectively obtain excellent target features.

Table 5. Grad-CAM visualization of WRN-16-2 (student) with KD, IMKD and the teacher (WRN-40-2) on Tiny-ImageNet. The labels of the three randomly selected photos are "Espresso", "Labrador" and "Orange".

Table 6. The ablation of β and ω on CIFAR100.

β	40	60	80	100
Top-1	77.02%	**78.13%**	77.54%	77.61%
ω	0.3	0.5	0.7	0.9
Top-1	77.46%	77.65%	**78.13%**	77.27%

Table 6 shows the test accuracy of the network under different settings of β and ω, where β and ω belong to Eq. 6 and Eq. 4, respectively. The network combination is set to "ResNet32x4 [23]/ShuffleNetV2 [14]". From Table 6, the experimental results show that matching suitable target features can effectively improve performance ($\omega = 0.7$). Furthermore, boost is best when β is equal to 60. Combining the information in Table 6 and Fig. 2, when the locations of multiple target features are set to be similar to the stage of the student network, the performance of the student network is significantly improved.

5 Conclusion

This academic paper highlighted a long-neglected aspect of feature map distillation-based methods: the absence of effective metrics and appropriate matching techniques for training student networks. To address this issue, we propose a novel approach called IMKD. Our IMKD utilizes the receptive field computation to calculate the difference in information between the teacher network and student network. To ensure that the features between teacher and student are appropriately matched during the distillation process, we introduce a feature matching distillation technique that adapts to the information differences, rather than relying on manual selection. This approach enables the student network to acquire the most optimal matching information continuously and iteratively throughout the distillation training process. Our extensive experiments demonstrate that IMKD outperforms previous distillation methods.

Acknowledgement. This research is supported by Sichuan Science and Technology Program (No. 2022YFG0324), SWUST Doctoral Research Foundation under Grant 19zx7102.

References

1. Chen, D., Mei, J.P., Zhang, H., Wang, C., Feng, Y., Chen, C.: Knowledge distillation with the reused teacher classifier. In: Proceedings of the IEEE/CVF Conference on Computer Vision and Pattern Recognition, pp. 11933–11942 (2022)
2. Chen, D., Mei, J.P., Zhang, Y., Wang, C., Wang, Z., Feng, Y., Chen, C.: Cross-layer distillation with semantic calibration. In: Proceedings of the AAAI Conference on Artificial Intelligence, vol. 35, pp. 7028–7036 (2021)
3. Chen, P., Liu, S., Zhao, H., Jia, J.: Distilling knowledge via knowledge review. In: Proceedings of the IEEE/CVF Conference on Computer Vision and Pattern Recognition, pp. 5008–5017 (2021)
4. Coates, A., Ng, A., Lee, H.: An analysis of single-layer networks in unsupervised feature learning. In: Proceedings of the Fourteenth International Conference on Artificial Intelligence and Statistics, pp. 215–223. JMLR Workshop and Conference Proceedings (2011)
5. Ding, X., Zhang, X., Han, J., Ding, G.: Scaling up your kernels to 31x31: Revisiting large kernel design in cnns. In: Proceedings of the IEEE/CVF Conference on Computer Vision and Pattern Recognition. pp. 11963–11975 (2022)
6. Ghosh, S., Srinivasa, S.K., Amon, P., Hutter, A., Kaup, A.: Deep network pruning for object detection. In: 2019 IEEE International Conference on Image Processing (ICIP), pp. 3915–3919. IEEE (2019)
7. He, K., Gkioxari, G., Dollár, P., Girshick, R.: Mask r-cnn. In: Proceedings of the IEEE International Conference on Computer Vision, pp. 2961–2969 (2017)
8. Hinton, G., Vinyals, O., Dean, J.: Distilling the knowledge in a neural network. arXiv preprint arXiv:1503.02531 (2015)
9. Jacob, B., et al.: Quantization and training of neural networks for efficient integer-arithmetic-only inference. In: Proceedings of the IEEE Conference on Computer Vision and Pattern Recognition, pp. 2704–2713 (2018)
10. Krizhevsky, A., Hinton, G., et al.: Learning multiple layers of features from tiny images (2009)
11. Liu, Z., Sun, M., Zhou, T., Huang, G., Darrell, T.: Rethinking the value of network pruning. arXiv preprint arXiv:1810.05270 (2018)
12. Luo, J.H., Wu, J., Lin, W.: Thinet: a filter level pruning method for deep neural network compression. In: Proceedings of the IEEE International Conference on Computer Vision, pp. 5058–5066 (2017)

13. Luo, W., Li, Y., Urtasun, R., Zemel, R.: Understanding the effective receptive field in deep convolutional neural networks. Advances in neural information processing systems 29 (2016)

14. Ma, N., Zhang, X., Zheng, H.T., Sun, J.: Shufflenet v2: Practical guidelines for efficient cnn architecture design. In: Proceedings of the European Conference on Computer Vision (ECCV), pp. 116–131 (2018)

15. Park, W., Kim, D., Lu, Y., Cho, M.: Relational knowledge distillation. In: Proceedings of the IEEE/CVF Conference on Computer Vision and Pattern Recognition, pp. 3967–3976 (2019)

16. Ren, S., He, K., Girshick, R., Sun, J.: Faster r-cnn: towards real-time object detection with region proposal networks. Advances in neural information processing systems 28 (2015)

17. Romero, A., Ballas, N., Kahou, S.E., Chassang, A., Gatta, C., Bengio, Y.: Fitnets: Hints for thin deep nets. arXiv preprint arXiv:1412.6550 (2014)

18. Sandler, M., Howard, A., Zhu, M., Zhmoginov, A., Chen, L.C.: Mobilenetv 2: Inverted residuals and linear bottlenecks. In: Proceedings of the IEEE Conference on Computer Vision and Pattern Recognition, pp. 4510–4520 (2018)

19. Selvaraju, R.R., Cogswell, M., Das, A., Vedantam, R., Parikh, D., Batra, D.: Grad-cam: Visual explanations from deep networks via gradient-based localization. In: Proceedings of the IEEE International Conference on Computer Vision, pp. 618–626 (2017)

20. Tang, J., Liu, M., Jiang, N., Cai, H., Yu, W., Zhou, J.: Data-free network pruning for model compression. In: 2021 IEEE International Symposium on Circuits and Systems (ISCAS), pp. 1–5. IEEE (2021)

21. Tian, Y., Krishnan, D., Isola, P.: Contrastive representation distillation. arXiv preprint arXiv:1910.10699 (2019)

22. Torralba, A., Fergus, R., Freeman, W.T.: 80 million tiny images: a large data set for nonparametric object and scene recognition. IEEE Trans. Pattern Anal. Mach. Intell. **30**(11), 1958–1970 (2008)

23. Xie, S., Girshick, R., Dollár, P., Tu, Z., He, K.: Aggregated residual transformations for deep neural networks. In: Proceedings of the IEEE Conference on Computer Vision and Pattern Recognition, pp. 1492–1500 (2017)

24. Yang, J., Martinez, B., Bulat, A., Tzimiropoulos, G., et al.: Knowledge distillation via softmax regression representation learning. International Conference on Learning Representations (ICLR) (2021)

25. Yim, J., Joo, D., Bae, J., Kim, J.: A gift from knowledge distillation: Fast optimization, network minimization and transfer learning. In: Proceedings of the IEEE Conference on Computer Vision and Pattern Recognition, pp. 4133–4141 (2017)

26. Zagoruyko, S., Komodakis, N.: Paying more attention to attention: Improving the performance of convolutional neural networks via attention transfer. arXiv preprint arXiv:1612.03928 (2016)

27. Zagoruyko, S., Komodakis, N.: Wide residual networks. arXiv preprint arXiv:1605.07146 (2016)

28. Zhang, X., Zhou, X., Lin, M., Sun, J.: Shufflenet: an extremely efficient convolutional neural network for mobile devices. In: Proceedings of the IEEE Conference on Computer Vision and Pattern Recognition, pp. 6848–6856 (2018)

29. Zhang, Y., Xiang, T., Hospedales, T.M., Lu, H.: Deep mutual learning. In: Proceedings of the IEEE Conference on Computer Vision and Pattern Recognition, pp. 4320–4328 (2018)

30. Zhao, B., Cui, Q., Song, R., Qiu, Y., Liang, J.: Decoupled knowledge distillation. In: Proceedings of the IEEE/CVF Conference on Computer Vision and Pattern Recognition, pp. 11953–11962 (2022)
31. Zhu, H., Jiang, N., Tang, J., Huang, X., Qing, H., Wu, W., Zhang, P.: Cross-layer fusion for feature distillation. In: Neural Information Processing: 29th International Conference, ICONIP 2022, Virtual Event, November 22–26, 2022, Proceedings, Part IV. pp. 433–445. Springer (2023)

CenAD: Collaborative Embedding Network for Anomaly Detection with Leveraging Partially Observed Anomalies

Li Cheng[1], Bin Li[2(✉)], Renjie He[1], and Feng Yao[1]

[1] College of System Engineering, National University of Defense Technology, Changsha, China
{chengli09,herenjie,yaofeng}@nudt.edu.cn
[2] Intelligent Game and Decision Lab (IGDL), Beijing, China
libin_bill@126.com

Abstract. Leveraging observed anomalies in anomaly detection can significantly improve detection accuracy. Assuming that observed anomalies cover all anomaly distributions, existing methods commonly learn the anomaly distributions from these observed anomalies and assign each object an anomaly score according to the similarities between it and observed anomalies. However, these observed anomalies may partially cover anomaly distributions, which severely restrains the performance in detecting uncovered anomalies. To address this issue, we propose a novel collaborative embedding network for this task, named CenAD. By leveraging partially observed anomalies, the collaborative learning derives a loss with maximum neighbor dispersion and minimum volume estimation as guidance to make anomalies more dispersed. Each object is assigned to an anomaly score by its contributions to data dispersion, which distinguishes these anomalies from the entire data effectively. To investigate the effectiveness of CenAD with partially observed anomalies, we conduct extensive results on several datasets to validate the superiority of our method, in which we obtain average improvement up to 13.92% in AUC-ROC and 29.44% in AUC-PR compared with previous methods.

Keywords: Anomaly Detection · Partially Observed Anomalies · Neighbor Dispersion · Embedding Network

1 Introduction

Anomaly detection focuses on identifying the unexpected patterns from usual behaviors in a dataset, which has been widely and successfully used in broad domains [4], such as fraud detection [21], intrusion detection [6], and health monitoring [22]. In practical settings, observed anomalies could be leveraged as the prior information to help learn anomaly-informed detection models, reducing the false positive rate and obtaining a better performance [1]. For example, since detected abnormal data flow in the intrusion detection system can be effectively

B. Luo et al. (Eds.): ICONIP 2023, LNCS 14450, pp. 418–432, 2024.
https://doi.org/10.1007/978-981-99-8070-3_32

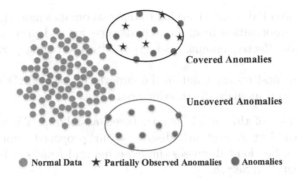

Fig. 1. Blue and red dots represent normal data and anomalies, respectively. Black stars are partially observed anomalies that only cover the anomaly distribution in the black circle. In contrast, these anomalies included in the red circle cannot be covered by partially observed anomalies, and the distribution of them cannot be directly learned from partially observed anomalies. (Color figure online)

leveraged to help learn the characteristics of anomalies, the accuracy of the intrusion detection system could significantly be improved.

However, since anomalies often demonstrate different anomalous behaviors and are difficult to be obtained, observed anomalies hardly cover all anomaly distributions. We call them partially observed anomalies. As shown in Fig. 1, these partially observed anomalies only locate in the anomaly distribution at the upper right. Anomalies whose distributions covered and uncovered by partially observed anomalies are called **Covered Anomalies** and **Uncovered Anomalies**, respectively. Most of the existing methods in this field learn the distribution of observed anomalies and assign each object an anomaly score according to similarities between it and observed anomalies [13,15,18,23]. But it overlooks the potential problem of incomplete coverage of observed anomalies, leading to ineffective detection of uncovered anomalies.

To address aforementioned issues, this paper introduces a **C**ollaborative embedding **n**etwork for **A**nomaly **D**etection with leveraging partially observed anomalies, termed **CenAD**. It fulfills an end-to-end anomaly detection in collaborative learning with maximum neighbor dispersion and minimum volume estimation in the embedding space as guidance. On the one hand, CenAD maximizes the average distance of each object with their random nearest neighbors. On the other hand, it minimizes the volume of a data-enclosing hypersphere with a global center. The collaborative learning makes anomalies more dispersed than normal data in the embedding space and assigns each object an anomaly score by contributions it makes to the data dispersion. With augmenting the effects of partially observed anomalies on data dispersion, CenAD detects anomalies effectively with higher anomaly scores. The contributions of this work are summarized as below:

- We focus on a rarely explored problem of anomaly detection with partially observed anomalies. And we define the covered and uncovered anomalies according to whether their distributions are represented by partially observed anomalies.

- We derive the collaborative learning to make anomalies more dispersed, which exploits the information from data characteristics and prior knowledge, and is tailored for effective anomaly detection with leveraging partially observed anomalies.
- Our experimental results validate the superiority of CenAD when observed anomalies cover anomaly distributions partially.

The remainder of this article is structured as follows. We will discuss the related work in Sect. 2. Section 3 discusses our proposed anomaly detection method CenAD. Section 4 discusses the experimental results, while the conclusions are presented in Sect. 5.

2   Related Work

In this section, we summarize related works on anomaly detection with observed anomalies, which can be divided into two main categories: traditional anomaly detection with observed anomalies and deep anomaly detection with observed anomalies. These studies show that the anomaly detection accuracy can be improved significantly with prior anomaly-information. But how to detect anomalies effectively with partially observed anomalies is still an open problem.

2.1   Traditional Anomaly Detection with Observed Anomalies

As for traditional anomaly detection methods, observed anomalies were incorporated into a belief propagation process to achieve more reliable anomaly scoring in [11,20], but they were more suitable for the graph data. Zhang proposed a cluster-based method ADOA [23], which got potential anomalies and reliable normal instances based on the isolation scores and similarity scores by clustering observed anomalies. Because of the curse of dimensionality, it is difficult to apply the traditional anomaly detection methods, like ADOA, to high-dimensional data. Moreover, due to the incomplete coverage of partially observed anomalies, uncovered anomalies cannot be effectively detected by clustering observed data to compute anomaly scores.

2.2   Deep Anomaly Detection with Observed Anomalies

Recent studies detect anomalies effectively in high dimensional data by resorting to the deep neural network. Ruff presented DeepSAD [18], a semi-supervised version of [17], to detect anomalies by compacting the unlabeled data to a center from a global view based on the majority of data are normal data. DevNet [15] adopted a few observed anomalies and a prior probability to enforce statistically significant deviations of the anomaly scores of covered anomalies from that of other data in the upper tail. It tries to maximize the difference between covered anomalies and the rest in unlabeled data. Although the result shows its effective performance in detecting covered anomalies, it is difficult for uncovered anomalies to be distinguished by these methods due to the incomplete coverage of partially observed anomalies. PRIOR [14] first discovered the problem that observed anomalies hardly cover all anomaly distributions. It formulated the problem as a pairwise relation learning task and assigned each instance pairs with specified scalars of three different classes. However, since the specified scalars of classes severely affect the accuracy of anomaly detection, it is difficult to determine proper scalars to achieve desirable detection results.

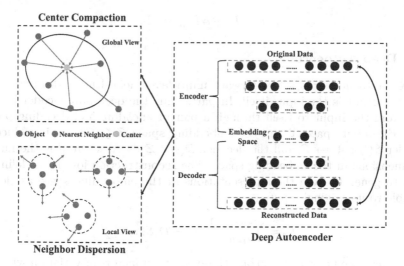

Fig. 2. Overview of CenAD

3   CenAD

3.1   Problem Statement

Given a set of $n+m$ training data objects $\mathcal{X} = \{x_0, \cdots, x_{n-1}, x_n, \cdots, x_{n+m-1}\}$ with $m \ll n$ and $x_i \in \mathbb{R}^D$ where D is the number of dimensions, in which $\{x_0, x_1 \cdots, x_{n-1}\}$ are n unlabeled data and $\{x_n, x_{n+1}, \cdots, x_{n+m-1}\}$ are m partially observed anomalies. $y_i \in \{0, 1\}$ is the label of each object where $y_i = 1$ and $y_i = 0$ are denoted for partially observed anomalies and unlabeled data, respectively.

3.2   Overview

CenAD defines a collaborative embedding network that synthesizes a deep autoencoer, random nearest neighbors, data-enclosing hypersphere, and a new loss function to train a deep anomaly detector in an end-to-end fashion, which is illustrated in Fig. 2. CenAD works as follows: **Deep Autoencoder** is adopted to reconstruct the original data, with preserving the data characteristics of the input data in the embedding space. To detect uncovered and covered anomalies effectively, **Neighbor Dispersion** and **Center Compaction** are responsible for manipulating the embedding space in collaborative learning, which makes anomalies more dispersed than normal data from both the local and global view. The objective is defined as Eq. 1, where L_r, L_{nn} and L_c are reconstruction loss, neighbor loss, and center loss, respectively. α, $\beta > 0$ are coefficients that control the degree of distorting embedding space.

$$L = L_r + \alpha L_{nn} + \beta L_c \tag{1}$$

3.3   Deep Autoencoder

A deep autoencoder is a feed-forward multi-layer neural network in which the desired output is the input itself. In particular, the deep autoencoder learns a map from the input to itself through a pair of encoding and decoding phases. With the input space $\mathcal{X}$ and the embedding space $\mathcal{Z} \subseteq \mathbb{R}^d$, we introduce the encoder $\mathcal{E}(\cdot) : \mathcal{X} \mapsto \mathcal{Z}$ and the decoder $\mathcal{D}(\cdot) : \mathcal{Z} \mapsto \mathcal{X}$ where d is the number of dimensions in the embedding space. The reconstruction loss L_r is defined as Eq. 2 to generate from the representations in the embedding space as close as possible to its original input.

$$L_r = \sum_{i=0}^{n+m-1} \|x_i - \mathcal{D}(\mathcal{E}(x_i))\|_2 \tag{2}$$

It forces the network to extract those common factors of variation which are most stable within a dataset. Intuitively, if it could make the differences between original data and their reconstructions low, the embedding representations could better preserve the key data characteristics of input samples.

3.4   Collaborative Learning

Having obtained the embedding features in deep autoencoder, the collaborative learning guides the optimization with maximum neighbor dispersion and minimum volume estimation in the embedding space. It exploits the information from data characteristics and prior knowledge in the local and the global view effectively, which makes anomalies more dispersed.

Neighbor Dispersion. As inspired by [16], anomalies distribute far away from normal data in the feature space and thus have a greater distance to their nearest neighbors in small random samples of a dataset. We define the nearest neighbors in small random samples of a dataset as random nearest neighbors in the following:

Definition 1. *The random nearest neighbor nn_i^j of each object x_i in the j-th ensemble is computed from random samples Q_i^j based on the l_2-norm distance:*

$$nn_i^j = \arg\min_{q \in Q_i^j} \|x_i - q\|_2 \tag{3}$$

In terms of the local information, CenAD trains random nearest neighbors of each object by maximizing the degree of dispersion in an ensemble way as shown in Eq. 4.

$$L_{nn} = \sum\nolimits_{i=0}^{n+m-1} -\frac{1}{\gamma^{y_i} N} \sum\nolimits_{j=0}^{N-1} \left\| \mathcal{E}(x_i) - \mathcal{E}(nn_i^j) \right\|_2 \tag{4}$$

The size of ensembles and each random sample are denoted as N and K, respectively. In each ensemble, different K objects are randomly selected to find the nearest neighbor. L_{nn} pushes the nearest neighbors away from each trained object x_i by increasing the average distance between them in the embedding space. Moreover, L_{nn} pushes the neighbors of partially observed anomalies further away with a coefficient of γ ($\gamma > 1$) to augment the degree of dispersion for observed anomalies. There exist three potential possibilities for random neighbors, which all make anomalies more dispersed.

A. *The nearest neighbor of a (an) normal object (anomaly) is an (a) anomaly (normal object).*
B. *The nearest neighbor of a normal object is a normal object.*
C. *The nearest neighbor of an anomaly is an anomaly.*

For A, L_{nn} guides the neighbors away from the trained object, leading to increasing the distance between them. For B and C, since anomalies are irregular and dissimilar, it is more difficult to separate two normal objects (similar) than two anomalies (dissimilar) with optimizing proper neural-network parameters. Thus, according to the above analysis, the distances of anomalies to their neighbors are trained to become greater than those of normal data, leading to anomalies more dispersed from a local view.

Center Compaction. In terms of the global information, CenAD focuses on compacting the entire data into a global updating center c in Eq. 5 by minimizing the volume of a data-enclosing hypersphere.

$$L_c = \sum\nolimits_{i=0}^{n+m-1} \gamma^{y_i} \left(\| \mathcal{E}(x_i) - \mathcal{E}(c) \|_2 \right)^{-2y_i+1} \tag{5}$$

Specifically, the center c is initialized as the mean of the embedding representations of all unlabeled data. To map data points to the updating center c, the neural network penalizes the mean l_2-norm distance over all unlabeled data. The normal data will become more compact than anomalies in unlabeled data, which is consistent with the assumption in the real applications that the majority of training data are normal data. In contrast, L_c penalizes the inverse of the distances for the partially observed anomalies and augments their penalization with γ such that they must be mapped further away from the global updating center.

Consequently, from both the local and global view, collaborative learning exploits the data characteristics and prior information with partially observed anomalies effectively, which makes anomalies more dispersed than normal data.

Algorithm 1. Training CenAD

Input: $\mathcal{B}$ - a batch of training data objects $\mathcal{X}$; y - training labels of the batch data.
Output: the optimized neural network parameters Θ^* and the optimized hypersphere center c^*.

1: Initialize neural network parameters Θ and the center c.
2: **for** $j = 0$ to N-1 **do**
3:     Randomly sample a handful of neighbors $Q_{\mathcal{X}}^j$.
4:     Find the nearest neighbors $nn_{\mathcal{X}}^j$ of the entire data $\mathcal{X}$.
5: **end for**
6: **for** epoch in n_epochs **do**
7:     **for** batch $\mathcal{B}$ in $n\_batches$ **do**
8:         Encode $\mathcal{B}$, $nn_{\mathcal{B}}^j$, c to obtain $\mathcal{E}(\mathcal{B})$, $\mathcal{E}(nn_{\mathcal{B}}^j)$, $\mathcal{E}(c)$.
9:         Compute each loss as Eq.2, Eq.4, Eq.5 to get L in Eq.1.
10:         Perform a gradient descent w.r.t. Θ and c.
11:     **end for**
12: **end for**
13: **return** Θ^*, c^*

3.5   Anomaly Detection Using CenAD

Training Stage. Algorithm 1 presents the procedure of CenAD. Random weights are initialized in Step 1. Then, CenAD finds the nearest neighbors of data $\mathcal{X}$ from Step 2 to Step 5. From Step 6 to Step 9, CenAD gets each loss item and performs the forward propagation of the collaborative embedding network. Step 10 performs stochastic gradient descent-based optimization to learn the neural network parameter Θ and the center c.

Testing Stage. At the testing stage, CenAD uses the function ϕ with optimized parameters Θ^* and center c^* to produce an anomaly score for every test object depending on the contributions that they make to data dispersion, which is shown in Eq. 6.

$$\phi(x_i; \Theta^*, c^*) = score_i^{nn} + score_i^c$$
$$score_i^{nn} = \frac{1}{N} \sum_{j=0}^{N-1} \left\| \mathcal{E}(x_i) - \mathcal{E}(nn_i^j) \right\|_2 \tag{6}$$
$$score_i^c = \left\| \mathcal{E}(x_i) - \mathcal{E}(c^*) \right\|_2$$

In the local view, the average distance to the nearest neighbor $score_i^{nn}$ is adopted to represent the contributions that each object makes to data dispersion. In the global view, the radius of a hypersphere (the distance of an object to the center c) $score_i^c$ is adopted to represent the contributions that each object makes to data dispersion. The more contributions it makes to data dispersion, the greater the anomaly score is, the more likely the object is an anomaly.

Table 1. We create four setups for USPS and MNIST, three setups for HAR, and one setup for UNSW-NB15. In USPS, MNIST, and HAR, the numbers denote the index of classes we select to represent normal distributions and anomaly distributions. In UNSW-NB15, we use the class name to represent the corresponding distributions.

Dataset	Setups (Normal v.s. Anomalies)	
USPS	{7 v.s. 1,2,3,4,5,6}	{8 v.s. 1,2,3,4,5,6}
	{9 v.s. 1,2,3,4,5,6}	{10 v.s. 1,2,3,4,5,6}
MNIST	{6 v.s. 0,1,2,3,4,5}	{7 v.s. 0,1,2,3,4,5}
	{8 v.s. 0,1,2,3,4,5}	{9 v.s. 0,1,2,3,4,5}
HAR	{1 v.s. 2,4,6} {3 v.s. 2,4,6} {5 v.s. 2,4,6}	
UNSW-NB15	{Normal v.s. DoS, Reco, Anal, Back, Shell}	

4 Experiments

4.1 Datasets

Three publicly available real-world classification datasets USPS [8] (10 classes, 256 dimensions), MNIST [5] (10 classes, 784 dimensions), HAR [2] (6 classes, 561 dimensions) and a network intrusion detection dataset UNSW-NB15 [12] (10 classes, 41 dimensions) are used in our experiments. Especially, we use the one-hot method to change categorical features of UNSW-NB15 to numerical features so that the number of its dimensions increases to 194.

We create 12 setups by selecting different classes to represent the normal distribution and anomaly distributions. In every setup, for USPS, MNIST, and HAR, we fix several classes as anomaly distributions and randomly sample 5% objects in each of them to be anomalies. The remaining classes in each dataset represent normal distributions and only one of them is set to be the normal data in each setup. For UNSW-NB15, the entire data in five anomaly classes, *DoS, Reco (Reconnaissance), Anal (Analysis), Back (Backdoor)*, and *Shell (Shellcode)* are regarded as anomalies and the data in *Normal* class are regarded as normal data. The created setups are shown in Table 1. We repeat experiments on each setup with 10 independent runs. In each run, according to the number of covered distributions, we select anomaly classes as covered anomaly distributions randomly. Furthermore, we preprocess all these data with normalization for each object. These datasets are split into two subsets, with 70% data as the train set and the other 30% data as the test set.

4.2 Comparative Methods

CenAD is compared with state-of-the-art methods, including anomaly detection methods with leveraging observed anomalies, DevNet [15], DeepSAD [18], ADOA [23] and unsupervised anomaly detection methods AE (autoencoder)

Table 2. The specific average AUC-ROC results with standard deviation (mean±std) in USPS, MNIST, HAR, and UNSW-NB15 when an anomaly distribution (∗-1) and three anomaly distributions (∗-3) are covered by observed anomalies. UNSW-NB15 is denoted as UNSW here.

Dataset	AUC-ROC (mean±std)					
	CenAD	DeepSAD	DevNet	iForest	AE	ADOA
USPS-1	**0.907±0.061**	0.892±0.032	0.638±0.012	0.880±0.058	0.850±0.060	0.882±0.016
MNIST-1	**0.900±0.036**	0.850±0.035	0.646±0.019	0.796±0.089	0.827±0.077	0.723±0.108
HAR-1	0.850±0.032	0.801±0.031	0.705±0.041	0.847±0.057	0.730±0.099	**0.888±0.025**
UNSW-1	**0.945±0.066**	0.896±0.122	0.925±0.076	0.526±0.017	0.627±0.037	0.521±0.035
Average	**0.894**	0.856	0.681	0.814	0.793	0.800
USPS-3	**0.939±0.043**	0.933±0.030	0.775±0.041	0.880±0.058	0.850±0.060	0.891±0.014
MNIST-3	**0.936±0.029**	0.928±0.018	0.795±0.020	0.796±0.089	0.827±0.077	0.813±0.018
HAR-3	0.975±0.030	0.947±0.020	**0.977±0.032**	0.847±0.057	0.730±0.099	0.889±0.012
UNSW-3	**0.981±0.006**	0.972±0.009	0.964±0.010	0.526±0.017	0.627±0.037	0.601±0.040
Average	**0.950**	0.938	0.848	0.814	0.793	0.840

Table 3. The specific average AUC-PR results with standard deviation (mean±std) in USPS, MNIST, HAR, and UNSW-NB15 when an anomaly distribution (∗-1) and three anomaly distributions (∗-3) are covered by observed anomalies. UNSW-NB15 is denoted as UNSW here.

Dataset	AUC-PR (mean±std)					
	CenAD	DeepSAD	DevNet	iForest	AE	ADOA
USPS-1	0.798±0.116	**0.809±0.071**	0.545±0.051	0.696±0.124	0.727±0.095	0.617±0.039
MNIST-1	**0.761±0.085**	0.708±0.050	0.531±0.030	0.569±0.132	0.677±0.142	0.462±0.099
HAR-1	**0.656±0.071**	0.599±0.039	0.575±0.088	0.500±0.031	0.375±0.201	0.534±0.049
UNSW-1	**0.831±0.094**	0.759±0.171	0.786±0.153	0.158±0.007	0.426±0.027	0.139±0.065
Average	**0.753**	0.719	0.568	0.560	0.597	0.505
USPS-3	0.865±0.089	**0.879±0.049**	0.724±0.075	0.696±0.124	0.727±0.095	0.656±0.035
MNIST-3	**0.846±0.058**	0.833±0.041	0.742±0.039	0.569±0.132	0.677±0.142	0.560±0.033
HAR-3	0.935±0.059	0.835±0.037	**0.945±0.066**	0.500±0.031	0.375±0.201	0.664±0.055
UNSW-3	**0.898±0.038**	0.888±0.043	0.843±0.045	0.158±0.007	0.426±0.027	0.181±0.011
Average	**0.879**	0.854	0.795	0.560	0.597	0.586

[19], iForest [9]. These methods are chosen because they show effective performance in relevant areas, i.e., DevNet in deep anomaly detection by learning anomaly scores, DeepSAD in deep anomaly detection by learning feature representations, ADOA in non-deep anomaly detection, autoencoder in deep unsupervised anomaly detection, and iForest in non-deep unsupervised anomaly detection.

4.3   Parameters Settings

For deep methods, since our experiments focus on unordered multi-dimensional data, the encoder network is set as a fully connected four-layer perceptron, running with 256-128-64-10 for USPS, 784-256-64-10 for MNIST, 561-256-64-10 for HAR, and 194-100-50-10 for UNSW-NB15. The decoder network is a mirror of the encoder. All internal layers are activated by LeakyReLU non-linear function

[10]. According to [3], we use a normalization constraint on representations during encoding training to make data more separable for different distributions and more compact for the same distribution. Other deep methods with learning feature representations (AE and DeepSAD) are trained with the same architecture in CenAD , while one hidden layer structure is adopted with 20 neurons by default for DevNet. Additionally, the root mean square propagation (RMSprop) optimizer [7] is adopted in deep methods to perform gradient descents. Due to different sizes of datasets, these methods are trained using 100 epochs, with 64, 512, 128, and 1024 mini-batches for USPS, MNIST, HAR, and UNSW-NB15, respectively. The learning rate is set to 0.001 and adjusted to 0.0001 after 50 epochs for fine-tuning. For non-deep methods, the subsampling size and ensemble size of iForest are set to 256 and 100. As for ADOA, the contamination is set 0.2 and the trade-off parameter between the isolation score and the similarity score is 1.

Moreover, due to page limit, we set $\gamma = 10$, $\alpha = 0.1$, and $\beta = 10$ because that they render desirable results in CenAD. As for the neighbor set size N and sample size K, we set $N = 50$ and $K = 16$ as the recommended settings from [16].

4.4 Metrics

First, we use two widely adopted complementary performance metrics, the Area Under Receiver Operating Characteristic Curve (AUC-ROC) and the Area Under Precision-Recall Curve (AUC-PR) to show the performance of anomaly detection. AUC-ROC summarizes the ROC curve of true positives against false positives, while AUC-PR is a summarization of the curve of precision against recalls.

Second, to measure the performance of each method on uncovered anomalies and covered anomalies, we define $UCDR$ and CDR satisfying $UCDR = \frac{Detected_{uncovered}}{All_{uncovered}}$ and $CDR = \frac{Detected_{covered}}{All_{covered}}$. Given a dataset $\mathcal{X}$ with M anomalies, All_* shows the number of *-anomalies in the entire data. $Detected_*$ shows the number of *-anomalies in the ranked highest M objects w.r.t anomaly scores.

4.5 Effectiveness Tests w.r.t. Numbers of Covered Anomaly Distributions

Experimental Settings. This section examines the performance of CenAD when partially observed anomalies cover different numbers of anomaly distributions. The consistent settings, partially observed anomalies accounting for 2% of the entire data in each setup, are used across all setups to gain the performance of each method. We compare CenAD with state-of-the-art methods introduced in Sect. 4.2. Unsupervised methods are plotted in dash lines as the baseline.

Results. CenAD *achieves consistently higher detection accuracy than other methods in most data, especially when partially observed anomalies only cover an anomaly distribution.* In Table 2 and Table 3, when partially observed anomalies only cover an anomaly distribution, CenAD obtains substantially better average

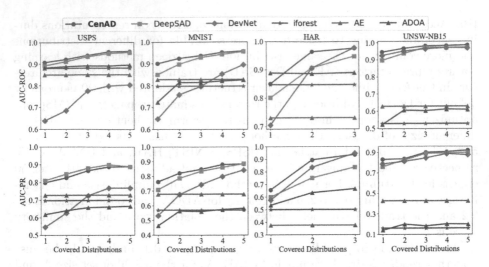

Fig. 3. Changes of AUC-ROC and AUC-PR w.r.t. the Number of Covered Anomaly Distributions

improvement than DeepSAD (4.42%, 4.79%), DevNet (31.16%, 32.61%), iForest (9.72%, 34.47%), AE (12.62%, 26.08%) and ADOA (11.65%, 49.25%) in terms of AUC-ROC and AUC-PR, respectively. When the number of covered anomaly distributions increases to 3, the average performance of CenAD are still superior than DeepSAD (1.34%, 2.96%), DevNet (12.08%, 10.53%), ADOA (13.08%, 49.88%), iForest (16.7%, 56.94%), and AE (19.78%, 47.14%) in terms of AUC-ROC and AUC-PR, respectively.

As we can see from Fig. 3, with the number of covered anomaly distributions increasing, CenAD performs the best in most datasets and it performs comparably well to the best performer on a few datasets where it ranks in second. CenAD keeps a relatively moderate growth trend, while DeepSAD and DevNet show obvious growth and the performance of unsupervised methods keeps level with the increasing number of covered anomaly distributions. These results are due to the reason that CenAD efficiently leverages the data characteristics and prior knowledge to make anomalies more dispersed, resulting in high-quality anomaly rankings, while most contenders overlook the incomplete coverage of observed anomalies, leading to weak capability of distinguishing uncovered anomalies.

4.6 Effectiveness Tests w.r.t. Proportions of Partially Observed Anomalies

Experimental Settings. This section examines the performance of CenAD with different proportions of partially observed anomalies, which is a critical factor due to the difficulties in obtaining anomalies in real applications. The proportion of partially observed anomalies occupying all data ranges from 0.1% to 5% for setups in MNIST and UNSW-NB15. And it ranges from 0.5% to 5%

for setups in USPS and HAR since they have fewer data. In each setup, the number of random covered anomaly distributions is fixed as 3. The competing methods are adopted as the former experiments.

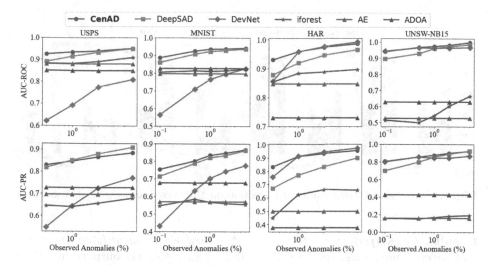

Fig. 4. Changes of AUC-ROC and AUC-PR w.r.t. the Proportions of Partially Observed Anomalies

Results. *CenADis the most data-efficient method and needs less observed anomalies to achieve comparable performance in four datasets.* Figure 4 shows the performance of each method w.r.t. different proportions of partially observed anomalies. CenAD obtains a better or comparable performance than other methods in all datasets, especially when partially observed anomalies are extremely limited (0.1% in MNIST and UNSW-NB15, 0.5% in USPS and HAR). Since more observed information generally helps train the model better, AUC-ROC and AUC-PR show a similar increasing tendency, first obvious growth and then flattening gradually, except that ADOA shows fluctuations due to its ineffectiveness in dealing with high dimensional data. The CenAD's superiority is due to exploiting both the local information and global information, which helps to leverage the observed anomalies much more sufficiently than other methods.

4.7 Effectiveness Tests w.r.t. Uncovered Anomalies and Covered Anomalies

Experimental Settings. This section examines the performance of CenAD in detecting uncovered and covered anomalies by adopting $UCDR$ and CDR as measures. $UCDR$ and CDR show the detection performance of each method on uncovered and covered anomalies directly, instead of comprehensive detection results like AUC-ROC and AUC-PR. The proportion of partially observed anomalies occupying the entire data is 2% and the number of random covered anomaly distribution is fixed as 1. Since the concept of covered anomaly distributions is not introduced in unsupervised methods, we only compare CenAD with DeepSAD, DevNet, and ADOA here.

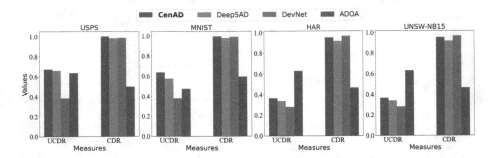

Fig. 5. $UCDR$ and CDR Results of Detection on Uncovered and Covered Anomalies

Results. CenAD *performs better than other methods in identifying uncovered and covered anomalies.* As we can see from Fig. 5, CenAD outperforms than other methods in most datasets in terms of $UCDR$ and CDR. It is because existing methods separate anomalies and normal data based on the partially observed anomalies, while CenAD leverages the partially observed anomalies as auxiliary data and exploits the useful information from data characteristics and prior anomaly knowledge to make anomalies more dispersed.

4.8 Ablation Study

Experimental Settings. This section examines the importance of the key components of CenAD by comparing CenAD to its two variants in Table 4. The first variant is CenAD-nn that removes the center loss by only considering the local information. The second variant is CenAD-c that removes the neighbor loss by only considering the global information. The proportion of partially observed anomalies occupying the entire data is fixed as 2% and the number of random covered anomaly distribution is 1.

Results. *The collaborative learning has major contributions to the superior performance of* CenAD. Table 4 shows the performance of CenAD and its two variants in AUC-ROC and AUC-PR. On average, due to comprehensive information considered in CenAD, CenAD performs slightly better than CenAD-c (5.6% and 7.0%) and substantially better than CenAD-nn (151.6% and 295.2%) in terms of AUC-ROC and AUC-PR, which indicates a significant role of the collaborative learning in CenAD.

Table 4. AUC-ROC and AUC-PR Results of CenAD and Its Variants.

Dataset	AUC-ROC			AUC-PR		
	CenAD-nn	CenAD-c	CenAD	CenAD-nn	CenAD-c	CenAD
USPS	0.320	0.851	**0.907**	0.203	0.726	**0.798**
MNIST	0.382	0.835	**0.900**	0.181	0.706	**0.761**
HAR	0.341	0.829	**0.850**	0.207	0.645	**0.656**
UNSW-NB15	0.431	0.920	**0.945**	0.129	0.778	**0.831**
Average	0.355	0.846	**0.894**	0.191	0.703	**0.753**

5 Conclusion

This paper proposes a collaborative embedding network for anomaly detection with leveraging partially observed anomalies, which jointly performs anomaly detection on embedding representations. The collaborative learning makes anomalies show a more dispersed distribution and assigns each object an anomaly score according to its contributions to data dispersion. Experimental results in real-world datasets show the superiority of our method, which in turn demonstrates the effectiveness of collaborative learning in anomaly detection with partially observed anomalies.

References

1. Aggarwal, C.C.: Supervised outlier detection. In: Outlier Analysis, pp. 219–248. Springer, Cham (2017). https://doi.org/10.1007/978-3-319-47578-3_7
2. Anguita, D., Ghio, A., Oneto, L., Parra, X., Reyes-Ortiz, J.L.: A public domain dataset for human activity recognition using smartphones. In: Esann (2013)
3. Aytekin, C., Ni, X., Cricri, F., Aksu, E.: Clustering and unsupervised anomaly detection with l 2 normalized deep auto-encoder representations. In: 2018 International Joint Conference on Neural Networks (IJCNN), pp. 1–6. IEEE (2018)
4. Chandola, V., Banerjee, A., Kumar, V.: Anomaly detection: a survey. ACM Comput. Surv. (CSUR) **41**(3), 15 (2009)
5. Deng, L.: The mnist database of handwritten digit images for machine learning research [best of the web]. IEEE Signal Process. Mag. **29**(6), 141–142 (2012)
6. Eskin, E., Arnold, A., Prerau, M., Portnoy, L., Stolfo, S.: A geometric framework for unsupervised anomaly detection. In: Applications of Data Mining in Computer Security, pp. 77–101. Springer, Boston (2002). https://doi.org/10.1007/978-1-4615-0953-0_4
7. Hinton, G., Srivastava, N., Swersky, K.: Overview of mini-batch gradient descent. Neural Networks for Machine Learning 575 (2012)
8. Hull, J.J.: A database for handwritten text recognition research. IEEE Trans. Pattern Anal. Mach. Intell. **16**(5), 550–554 (1994)
9. Liu, F.T., Ting, K.M., Zhou, Z.H.: Isolation forest. In: 2008 Eighth IEEE International Conference on Data Mining, pp. 413–422. IEEE (2008)
10. Maas, A.L., Hannun, A.Y., Ng, A.Y.: Rectifier nonlinearities improve neural network acoustic models. In: Proceedings of ICML, vol. 30, p. 3 (2013)

11. McGlohon, M., Bay, S., Anderle, M.G., Steier, D.M., Faloutsos, C.: Snare: a link analytic system for graph labeling and risk detection. In: Proceedings of the 15th ACM SIGKDD International Conference on Knowledge Discovery and Data Mining, pp. 1265–1274. ACM (2009)

12. Moustafa, N., Slay, J.: Unsw-nb15: a comprehensive data set for network intrusion detection systems (unsw-nb15 network data set). In: 2015 Military Communications and Information Systems Conference (MilCIS), pp. 1–6. IEEE (2015)

13. Pang, G., Cao, L., Chen, L., Liu, H.: Learning representations of ultrahigh-dimensional data for random distance-based outlier detection. In: Proceedings of the 24th ACM SIGKDD International Conference on Knowledge Discovery & Data Mining, pp. 2041–2050. ACM (2018)

14. Pang, G., Hengel, A.v.d., Shen, C.: Weakly-supervised deep anomaly detection with pairwise relation learning. arXiv preprint arXiv:1910.13601 (2019)

15. Pang, G., Shen, C., van den Hengel, A.: Deep anomaly detection with deviation networks. In: Proceedings of the 25th ACM SIGKDD International Conference on Knowledge Discovery & Data Mining, pp. 353–362. ACM (2019)

16. Pang, G., Ting, K.M., Albrecht, D.: Lesinn: detecting anomalies by identifying least similar nearest neighbours. In: 2015 IEEE International Conference on Data Mining Workshop (ICDMW), pp. 623–630. IEEE (2015)

17. Ruff, L., et al.: Deep one-class classification. In: International Conference on Machine Learning, pp. 4393–4402 (2018)

18. Ruff, L., et al.: Deep semi-supervised anomaly detection. arXiv preprint arXiv:1906.02694 (2019)

19. Sakurada, M., Yairi, T.: Anomaly detection using autoencoders with nonlinear dimensionality reduction. In: Proceedings of the MLSDA 2014 2nd Workshop on Machine Learning for Sensory Data Analysis, p. 4. ACM (2014)

20. Tamersoy, A., Roundy, K., Chau, D.H.: Guilt by association: large scale malware detection by mining file-relation graphs. In: Proceedings of the 20th ACM SIGKDD International Conference on Knowledge Discovery and Data Mining, pp. 1524–1533. ACM (2014)

21. Wang, C.D., Deng, Z.H., Lai, J.H., Philip, S.Y.: Serendipitous recommendation in e-commerce using innovator-based collaborative filtering. IEEE Trans. Cybern. **49**(7), 2678–2692 (2018)

22. Wang, H., Yang, Y., Liu, B., Fujita, H.: A study of graph-based system for multi-view clustering. Knowl.-Based Syst. **163**, 1009–1019 (2019)

23. Zhang, Y.L., Li, L., Zhou, J., Li, X., Zhou, Z.H.: Anomaly detection with partially observed anomalies. In: Companion Proceedings of the Web Conference 2018, pp. 639–646. International World Wide Web Conferences Steering Committee (2018)

PAG: Protecting Artworks from Personalizing Image Generative Models

Zhaorui Tan[1], Siyuan Wang[1], Xi Yang[1(✉)], and Kaizhu Huang[2]

[1] Xi'an Jiaotong-Liverpool University, Suzhou, China
`Zhaorui.Tan21@student.xjtlu.edu.cn`, `Siyuan.Wang21@alumni.xjtlu.edu.cn`,
`Xi.Yang01@xjtlu.edu.cn`
[2] Duke Kunshan University, Suzhou, China
`kaizhu.huang@dukekunshan.edu.cn`

Abstract. Recent advances in conditional image generation have led to powerful personalized generation models that generate high-resolution artistic images based on simple text descriptions through tuning. However, the abuse of personalized generation models may also increase the risk of plagiarism and the misuse of artists' painting styles. In this paper, we propose a novel method called **P**rotecting **A**rtworks from Personalizing Image **G**enerative Models framework (PAG) to safeguard artistic images from the malicious use of generative models. By injecting learned target perturbations into the original artistic images, we aim to disrupt the tuning process and introduce the distortions that protect the authenticity and integrity of the artist's style. Furthermore, human evaluations suggest that our PAG model offers a feasible and effective way to protect artworks, preventing the personalized generation models from generating similar images to the given artworks.

Keywords: Image protection · Conditional image generation · Personalizing generation models

1 Introduction

The rapid growth of conditional generative models, such as Stable Diffusion [10] and VQGAN [2], has revolutionized large-scale image generation. These models demonstrate an unprecedented ability to generate highly realistic images by leveraging text conditions. Built upon them, personalizing large generative models with a few images by fine-tuning presents a popular downstream task. For instance, LoRA [12], a tuning method of Stable Diffusion, significantly reduces tuning time and memory requirements; VQGAN-CLIP [1] leverages CLIP's text embeddings [8] as test-time guidance to get expected images via iterative optimization of input noise. As a result, any user can flexibly and easily generate

---

Z. Tan and S. Wang—Equal contribution.

© The Author(s), under exclusive license to Springer Nature Singapore Pte Ltd. 2024
B. Luo et al. (Eds.): ICONIP 2023, LNCS 14450, pp. 433–444, 2024.
https://doi.org/10.1007/978-981-99-8070-3_33

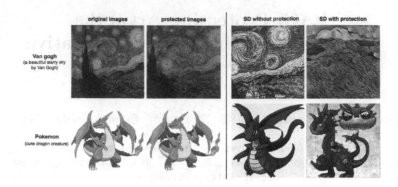

Fig. 1. PAG limits the ability of generative models to learn art styles. The left images are the inputs for fine-tuning the pre-trained Stable Diffusion, while the results on the right are the outputs generated by the personalized Stable Diffusion.

special-style images by adapting large-scale generation models with a few reference images in just a few hours through tuning (as shown in the third column of Fig. 1).

Unfortunately, the convenience and accessibility of personalized models may also pose risks to the artists, as their artworks may be maliciously mimicked. For example, artists may consume thousands of hours to create a painting, while the aforementioned models only require a few reference images to learn the style of artistic images in several hours. Consequently, if artists share their artistic images on social media platforms without protection, it becomes easier for malicious users to download and replicate the artists' styles (as shown in Fig. 2). This presents a significant threat that popular and powerful generation models may potentially plagiarize creative ideas and painting styles of artistic images when exploited by malicious users, which devalues the artworks.

Various image protection methods have been developed to counter this abuse of personalizing models. For example, Anti-DreamBooth [15] is a defense method specifically designed to safeguard adversarially only face-related images from diffusion-based generative models and requires the knowledge of prompts and models used by attackers. GLAZE [14] employs a pre-trained style transfer model to transfer the artistic images to another certain style, trying to break the semantic alignment in the mimicking models. However, the style transfer model is pre-trained on other artworks, threatening the style learning of some other artists. The limitations of current methods make it challenging for protectors to ensure the effectiveness of their protection models when they lack prior knowledge about malicious users' operations.

To alleviate these problems, we propose the **P**rotect **A**rtworks from Personalizing Image **G**enerative Models framework (PAG) to safeguard artistic images (see Fig. 2). Different from previous methods, PAG provides general protection for all images, instead of only safeguarding specific contents locally and individually against various potential models without knowledge of them (i.e., black-box protection). Additionally, PAG does not require a pre-trained style transfer model, which may already harm artists or face the unknown risks of uploading artistic works online. Specifically, PAG aims to prevent generative models from learning artistic images by an offline generative protection model which generates semantically abnormal and visually normal samples. These abnormal samples can disturb the visual distribution of the mimicking model when using them for tuning and then protecting the original artistic images. Importantly, the entire abnormal perturbation injection process is self-constrained and requires no additional knowledge, making PAG highly suitable for real-world protection scenarios.

Our main contributions can be summarized as follows:

- We propose a novel black-box image protection framework, called PAG, through invisible semantically abnormal perturbation injection, which requires no knowledge of mimicking tuning, including the model structure and prompts.
- We theoretically show that semantically abnormal perturbation injection introduces bias into the mimicking models, leading them to learn an abnormal visual distribution.
- We develop a black-box abnormal sample generation network with learnable conditions to generate abnormal samples and mix them with original images.
- We test our proposed PAG on two SOTA generative models, VQGAN-CLIP and Stable Diffusion via human evaluation. The results suggest that our PGA produces protected images with high acceptance and effectiveness.

2   Literature Review

Large-Scale Pretrianed Image Generative Models: Recently, large-scale image generative models have demonstrated an unprecedented power to generate semantic and high-realistic images, such as VQGAN [2], Stable diffusion model [10], and DALL·E2 [9]. VQGAN uses an autoregressive transformer to learn a context-rich codebook to synthesize high-resolution images. Stable Diffusion is a state-of-the-art generative model trained in a latent space with powerful pre-trained autoencoders. However, those image generative models are trained on extremely large-scale datasets, where famous artistic works like Van Goghs' artworks are also contained in the training sets. With the convenience of generative models, more and more malicious users can easily generate artistic images by generative models.

Personalization: Furthermore, personalized text-to-image generation models have the ability to generate specific and realistic images by fine-tuning large-scale image generative models with just a few reference images. VQGAN-CLIP, for instance, utilizes the pre-trained CLIP [8] text-image encoder to steer the image generative model VQGAN [2]. DreamBooth [11], on the other hand, fine-tunes large-scale diffusion models using a reconstruction loss and a prior preservation loss to address over fitting and language shift issues. Additionally, the other focus of fine-tuning the large-scale text-to-image generation models is the cost of the memory and the time. LoRA is another lightweight tool to generate personalized images with few-shot image sets [5]. All these methods offer malicious users one easy way to learn the special style of artistic images from a few reference images.

Image Protection: With the unpredicted growth speed of generation models, for artists, it is essential to protect their artistic images from being mimicked by generative models. For naive mimicry, in which the mocked artists are well-known and have a large number of artistic images online. Image mimicry can easily mock the style of famous artists' works by the large generative models which have already seen those famous artistic images. For individual artists, who publish a few artworks on social media, it is also a threat in that their works may be downloaded to tune a large-scale generation model by malicious users. Recently, there have been some protection models to protect images from generative models. Anti-DreamBooth is the first diffusion-based face protection model to protect human face images by generating some adversarial noise to prevent generative models from reconstructing images. However, in most successful settings, Anti-DreamBooth requires the generative models or the input prompts of the attackers used [15]. GLAZE proposes a style cloak method to shift the feature representations of artistic images in the style transfer generative models. Unfortunately, its pre-trained style transfer network also learns a lot of styles of other artists' artworks, which cannot completely protect artistic images [14]. To alleviate these shortcomings, we propose an abnormal perturbation injection method, an adversary training process to learn an abnormal target data distribution through a random text-based generative model which is offline used and unrelated to the generation models used by attackers.

3   Theoretical Motivations

In this section, we introduce some basic terminologies in conditional image generation and present our theoretical motivations.

Preliminaries: Unconditional and conditional image generative models aim to learn image distribution. We take conditional image generation as an example, where unconditional cases can be considered as a special case of conditional by taking an unchanged condition. Considering an i.i.d dataset $\mathcal{X} = \{X_k\}_{k=1}^K$ where each example $X_k = \{i_k, y_k\}$ contains an image i_k and its corresponding label y_k. The conditional generative model G with parameters α aims to maximize the likelihood:

$$\arg \max_{\alpha} \mathbb{E}_{X_k \sim \mathcal{X}} [\log \int_{-\infty}^{\infty} p_\alpha(I, Y) \, dy_k], \tag{1}$$

$$p_\alpha(I, Y) = \prod_{k=1}^K p_\alpha(i_k \mid y_k) \, p(y_k). \tag{2}$$

where $I = \{i_k\}_{k=1}^N$, $Y = \{y_k\}_{k=1}^N$. In the mimicry case, y_k is an explicit or implicit label (e.g., a word/description or semantic embedding) that can anchor the mimicking target.

Previous Protection Methods: Equation (2) suggests that the alignment between x_i, y_i is vital in conditional image generation. Therefore, most current image protection methods, such as GLAZE [14], try to undermine the alignment between i_k and y_k by adding minimal cloaks ϵ_k to i_k. Specifically, GLAZE tends to lead the model to mimic another specific target $y'_k \neq y_k$ by adding minimal ϵ_k, which needs a pre-trained model that has already learned the styles of some artists. It violates the original purpose of protecting all victims. Some other methods like Anti-DreamBooth [15] focus on obtaining an $i_k + \epsilon_k = i'_k \neq i_k$ where ϵ_k adversarially disturbs the alignment learning, and requires access to the mimicking model. However, the mimicking model is not always available.

Our Proposed Method: Distinct from GLAZE, which focuses on aligning x_i to a certain known y'_i, and Anti-DreamBooth, which needs access to the mimicking model, our method aims to learn a mapping ϕ with parameters θ to obtain $I' = \{\phi_\theta(i_k)\}_{k=1}^K$ whose semantics are most anomaly with minimal visual changes, misleading the mimicking model to I' without any acknowledgement to the mimicking model:

$$\arg\min_\theta \sum_{k=1}^K D(\phi_\theta(i_k), i_k), \ \arg\max_\theta \sum_{k=1}^K S(\phi_\theta(i_k), y_k), \tag{3}$$

where $D(\cdot, \cdot)$ denotes a visual distance measurement and $S(\cdot, \cdot)$ denotes a semantic distance measurement. To engage with victims, hyperparameters are introduced for them to control the mix between the original image and the semantic abnormal image: $i''_k = a \cdot i_k + (1-a)i'_k, a \leq c$, where c controls the minimal protection that should be applied. The semantically abnormal samples mislead G to learn an abnormal visual semantic distribution while tuning, thus failing to mimic artistic images.

4 Main Method

In order to protect artworks, we propose PAG to locally transfer artistic images into semantically abnormal images but visually similar to the normal images. Our PAG requires no access to mimicking models, tuning details, and pre-trained style transfer models. Under the circumstances, PAG-protected images work for most generative models without sacrificing any artists' styles. Specifically, to achieve the objective mentioned in Sect. 3, we propose a semantic abnormal transfer network that can inject abnormal visual semantics into artistic images as protection.

4.1 Overall Framework

The overview of the PAG structure is presented as Fig. 2. First, artists can locally train their protection model ϕ without the threat of uploading their original artistic images to social websites or online protection models. Then, they can

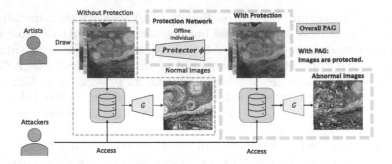

Fig. 2. Illustration of PAG: the method of protecting artworks from generative models.

offline adjust the protection effectiveness and the fidelity of each protected image i_k'' before they upload it to social websites. At the bottom of Fig. 2, we assume that there are some malicious users, who want to download the artistic images to tune the generative models G and fake their styles. With our PAG protection method, the inputs of the generator are all the protected images, which contain invisible perturbations leading the generator to learn an abnormal visual distribution. Finally, the PAG protection method effectively protects artistic images by preventing generative models from learning the artwork styles.

4.2 Generate Semantically Abnormal Samples

The proposed semantically abnormal transfer network structure can be seen in Fig. 3. To achieve Eq. 3, we adopt a randomly selected word and its CLIP embedding $\mathbf{Y}$ with a mapping ψ with parameters β as an anchor. After training, the optimized $\psi(\mathbf{Y})$ should locate where it is semantically abnormal but with minimal visual changes. As such, protected artistic images that have been transferred to $\psi(\mathbf{Y})$ are semantically abnormal but barely visually changed. Empirically, we weigh maximizing semantically abnormal more than minimizing visual changes, providing more choices for artists so that they can customize the protection strength. Note that we adopt CLIP [8] semantic space as our standard semantic space, and we take the generator of CLIPStyler [6] as our ϕ for the generation. Specifically, ϕ's parameters are denoted as θ.

Maximize Semantic Difference: We leverage a semantic loss to maximize the style difference between the original images and generated abnormal samples. In detail, we aim to learn a semantically abnormal target $\psi(\mathbf{Y})$ by minimizing its cosine distance with the semantics of generated image samples $E_I(\phi(i_k))$, while maximizing the semantic cosine distance of the semantically abnormal target with the original normal image. We optimize the objective by the semantic loss as shown in Eq. 4:

$$L_{sem}(\theta, \beta) = \lambda_{sem} \frac{D_{cos}(\psi_\beta(\mathbf{Y}), E_I(\phi_\theta(i_k)))}{D_{cos}(\psi_\beta(\mathbf{Y}), E_I(i_k))}, \tag{4}$$

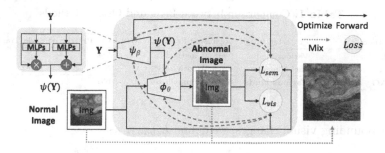

Fig. 3. Structure of our proposed Semantically Abnormal Transfer Network.

where $\psi_\beta(\mathbf{Y})$ is the abnormal condition; i_k describes an artistic image; E_I is an image encoder that maps images into semantic space, and we use CLIP [8] in practice; D_{cos} denotes the cosine distance between two semantic embeddings z^i, z^t: $D_{cos}(\cdot, \cdot) = 1 - (z^i z^t)/(\|z^i\| \|z^t\|)$. In practice, our ψ structure can be seen in Fig. 3 at the left top. With several MLPs that learn weight and bias, $\mathbf{Y}$ is mapped to an abnormal semantic target by multiplying weight and adding bias.

Producing Semantically Abnormal Images: In order to generate semantically abnormal images with minimal visual changes, the loss used for training ϕ is designed as:

$$L_{vis}(\theta, \beta) = L_{transfer}(\phi_\theta(i_k), \psi_\beta(\mathbf{Y})) + L_{change}(\phi_\theta(i_k), i_k), \qquad (5)$$

where $L_{transfer}$ is used to transfer the image to the learnable abnormal condition $\psi(\mathbf{Y})$; L_{change} bounds the visual changes between transferred images and original images. We assume that image content changes will lead to most visual changes. Therefore, we alter CILPStyler [6] that exploits a pre-trained locked VGG-19 [4] $(V(\cdot))$ to extract only content-level features as our basic backbone. Same as CILPStyler, we split the original images into M patches $\{i_k^m\}_{m=1}^M$. Following CILPStyler and StyleGAN-NADA [3], we define $L_{transfer}$ that forces generated images to obtain abnormal semantics as:

$$L_{transfer}(\theta, \beta) = \lambda_d \underbrace{D_{cos}(E_I(\phi_\theta(i_k)) - E_I(i_k), \psi_\beta(\mathbf{Y}) - \mathbf{Y})}_{\text{Global Level Abnormal Semantic Matching}} +$$

$$\lambda_{patch} \underbrace{\frac{1}{M} \sum_{m=1}^M D_{cos}(E_I(aug(\phi_\theta(i_k^m))) - E_I(i_k^m), \psi_\beta(\mathbf{Y}) - \mathbf{Y})}_{\text{Patch Level Abnormal Semantic Matching}}, \qquad (6)$$

where $aug(\cdot)$ is the random perspective augmentation for patch-level images. λ_* are hyperparameter weights for losses.

Table 1. Comparison between PAG and Other Methods

Method	Protected Image	Againsted Personalization	Use Pretrianed Style Transfer
GLAZE	Various kinds	Various methods	Yes
Anti-DreamBooth	Facial images	DreamBooth only	No
PAG	**Various kinds**	**Various methods**	**No**

For bounding visual changes, we define L_{change} as:

$$L_{change}(\theta, \beta) = \underbrace{\lambda_{cont}MSE(V(\phi_\theta(i_k)), V(i_k))}_{\text{Feature-Level Visual Bounding}} + \underbrace{\lambda_{recon}\frac{1}{M}\sum_{m=1}^{M}MSE(\phi_\theta(i_k^m) - i_k^m)}_{\text{Pixel-Level Visual Bounding}}.$$

$$(7)$$

Equation 7 means the feature-level visual bounding as feature-level content invariance loss and the pixel-level visual bounding as a reconstruction loss.

Interactivate with Artists: Increasing the strength of the abnormal perturbation injection enhances the protection performance. However, the visibility of mixed artworks impacts the details, such as the lightness, darkness, and noise. In order to alleviate the trade-off problem, we propose a mix strategy, in which the artists can adjust the mix strength of the original art and the added abnormal perturbation. We use discrete Fourier transform to transfer the original images from the spatial domain to the frequency domain, which is helpful to preserve the low-frequency signal from images. Additionally, we mix the original images i_k and the generated abnormal images i_k' to reduce the visible difference between original and abnormal images. The mixed strategy is shown as follows:

$$i_k'' = (1 - m) \times i_k + (1 - \lambda_p) \times m \times i_k + \lambda_p \times m \times i_k', \ \lambda_p \geq 25\%, \quad (8)$$

where i_k'' is the final protected image and λ_p is mix strength ratio. m describes a customized mask for artists to choose the areas in which they want to remain the same as the original. In our experiments, the mask is randomly generated as 20% at the pixel level.

Compare to Previous Methods: Key differences between PAG and previous methods are highlighted in Table 1. Unlike GLAZE, which requires a pretrained style transfer model, our PAG protects images without transferring images to any known artistic style. Our PAG also protects various kinds of images from various personalizing methods, while Anti-DreamBooth is specially designed for facial protection from DreamBooth.

5 Experiments

Experimental Setup: The experimental procedure is arranged as follows. For each set of artworks, the abnormal image generation network is trained individually at first. Afterward, artists protect each work by mixing the generated

Fig. 4. Our PAG protection results on two datasets of Stable Diffusion with LoRA.

abnormal images and the originals. We use protected images to tune SOTA image generative models to examine our protection effectiveness. Finally, we evaluate the protection acceptance (i.e., if the protected image is visually acceptable) and its effectiveness through human evaluations. To test the feasibility of our PAG on different kinds of works, we choose two distinct typical styles: Impressionist oil painting and anime-style painting. For the former, we take the works of Van Gogh from WikiArt datasets [13]; for the latter, we leverage the Pokemon dataset [7] with its BLIP captions. We set $\lambda_{sem}, \lambda_{recon} = 0.1$ while other hyperparameters follow CLIPstyler. Additionally, the learning rate for the abnormal generative model is initialized as 1e-4.

Tuning: We use LoRA with Stable Diffusion version 1.4 and VQGAN-CLIP with VQGAN as mimicking models. In the LoRA fine-tuning process, we use protected images to fine-tune the Stable Diffusion model for 50 epochs with a constant learning rate of 1e-4. Specifically for Van Gogh's works, we use "a drawing by Van Gogh" as a prompt for Van Gogh's artworks, while we adopt BLIP captions of the Pokemon dataset as prompts for each image. For VQGAN-CLIP whose tuning is testing time iterations that optimize image input tensor, we use the protected image as the initial input image and semantic target. All tuning tests are conducted by using one Nvidia Quadro RTX 8000.

Human Evaluation: To evaluate how much protection strength is acceptable (*Acceptance*) and the effectiveness of the protection to mimicking model (*Effectiveness*), we conduct human evaluations on our proposed PAG. Participants are required to complete two evaluation tasks: one focuses on the acceptance of visual changes between the original and protected images; the other examines the effectiveness of PAG and how much it can protect artistic from mimicking. We provide images with different protection strengths for each task: $\lambda_p = [0.75, 0.5, 0.25]$, respectively. For both Acceptance and Effectiveness, participants can choose from five levels (1 for "not acceptable at all" and "have no effectiveness" to 5 for "totally acceptable" and "have strong effectiveness"). We received total 168 groups of anonymous human evaluation results in Table 2.

5.1 Human Evaluation Results

Human evaluation results are shown in Table 2. As protection strength increases, the acceptance of protected images would be decreased. Participants show a greater preference for images with a 25% protection strength. Surprisingly,

Table 2. Human evaluated Acceptance and Effectiveness of PAG to mimicking models.

Datasets	PAG $\lambda_p = 75\%$		PAG $\lambda_p = 50\%$		PAG $\lambda_p = 25\%$	
	Acceptance	Effectiveness	Acceptation	Effectiveness	Acceptation	Effectiveness
Van Gogh	2.5	3.4	3.6	3.5	3.9	3.6
Pokemon	3.1	4.1	3.4	4.2	4.1	4.2

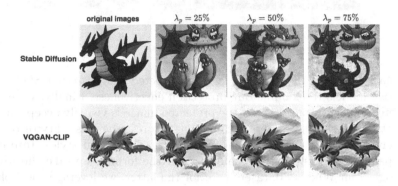

Fig. 5. Our PAG protection results on two generative models, Stable Diffusion with LoRA and VQGAN-CLIP.

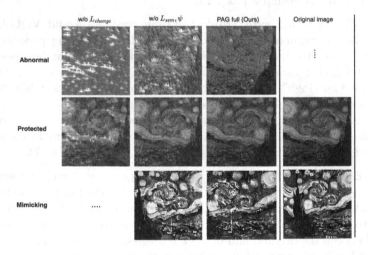

Fig. 6. Ablation results of PAG-protected images for Van Gogh's paintings.

although the protection strength is decreased, the effectiveness of our PAG still ranges at 3.5 for Van Gogh datasets and 4.1 for Pokemon datasets. Based on these results, our PAG can trade off acceptance and effectiveness by introducing the least disruptive abnormal perturbations to protect the original image while preserving its visual characteristics and artistic value.

5.2   Qualitative Protection Performance

Our qualitative results can be seen in Fig. 4 and Fig. 5. For Stable Diffusion with LoRA (as shown in Fig. 5), it can be observed that mimicking artistic images succeeds when the attackers are accessible to the artistics' unprotected artworks. Meanwhile, for VQGAN-CLIP, though its generation quality is compromised due to its testing time training, it still maintains the basic styles of the original artworks. Furthermore, in the first and fifth columns of Fig. 4, the mimicking ability of models does not vary when different datasets are given for Stable Diffusion with LoRA. Meanwhile, our PAG protects artistic images by mixing them with their abnormally transferred samples, hindering the mimicking models from generating high-quality images. As shown in Fig. 4 and Fig. 5, the mimicked images with 75% protection strength have a visible difference from the original art images, while the protection is the most efficient. With 50% protection strength, the protection is almost invisible, while the mimicking images cannot reconstruct the style of the original images. Nevertheless, with only 25% protection strength, the protection is invisible with barely compromising protection strength.

5.3   Ablation Study

For the ablation study, we evaluate the effect of different losses in Sect. 4.2. As shown in Fig. 6, without L_{change}, ϕ can generate abnormal images, however with significant visible changes in protected images. Without using $\psi(\mathbf{Y})$ to a semantically abnormal target with minimal visual changes, protected images still suffer from obvious visible changes. Our PAG can offer the most acceptable images and effective protection with all these losses.

6   Conclusion

In this paper, we present Protecting Artworks from Personalizing Image Generative Models framework (PAG) to protect artistic images through a mixture between semantically abnormal images and originals from maliciously learned by generative models. Specifically, we adopt a learnable mapping that transfers the semantic anchor to a semantically abnormal one. By taking this as the semantic target, the original images can be transferred to abnormal images. We show that, tuning with the mixed abnormally protected images, mimicking models fail to generate high-quality images compared with using the originals. Importantly, PAG is customized for each artist without any pre-trained style-transfer model or knowledge of possible mimicking models. In the future, we aim to minimize the visible difference between the original artworks and protected images while maximizing the protection effectiveness.

Acknowledgement. This research was funded by the National Natural Science Foundation of China under no. 62206225; Jiangsu Science and Technology Programme (Natural Science Foundation of Jiangsu Province) under no. BE2020006-4; Natural

Science Foundation of the Jiangsu Higher Education Institutions of China under no. 22KJB520039; Xi'an Jiaotong-Liverpool University's Research Development Fund in XJTLU under no. RDF-19-01-21.

References

1. Crowson, K., et al.: VQGAN-CLIP: open domain image generation and editing with natural language guidance. http://arxiv.org/abs/2204.08583
2. Esser, P., Rombach, R., Ommer, B.: Taming transformers for high-resolution image synthesis. In: Proceedings of the IEEE/CVF Conference on Computer Vision and Pattern Recognition (CVPR), pp. 12873–12883 (2021)
3. Gal, R., Patashnik, O., Maron, H., Bermano, A.H., Chechik, G., Cohen-Or, D.: StyleGAN-NADA: clip-guided domain adaptation of image generators. ACM Trans. Graph. (TOG) **41**(4), 1–13 (2022)
4. Gatys, L.A., Ecker, A.S., Bethge, M.: Image style transfer using convolutional neural networks. In: Proceedings of the IEEE Conference on Computer Vision and Pattern Recognition (CVPR), pp. 2414–2423 (2016)
5. Hu, E.J., et al.: LoRA: low-rank adaptation of large language models. http://arxiv.org/abs/2106.09685
6. Kwon, G., Ye, J.C.: CLIPstyler: image style transfer with a single text condition. http://arxiv.org/abs/2112.00374
7. Pinkney, J.N.M.: Pokemon blip captions. http://huggingface.co/datasets/lambdalabs/pokemon-blip-captions/ (2022)
8. Radford, A., et al.: Learning transferable visual models from natural language supervision. In: International Conference on Machine Learning, pp. 8748–8763. PMLR (2021)
9. Ramesh, A., et al.: Zero-shot text-to-image generation. http://arxiv.org/abs/2102.12092
10. Rombach, R., Blattmann, A., Lorenz, D., Esser, P., Ommer, B.: High-resolution image synthesis with latent diffusion models. In: 2022 IEEE/CVF Conference on Computer Vision and Pattern Recognition (CVPR), pp. 10674–10685. IEEE (2022). https://doi.org/10.1109/CVPR52688.2022.01042, http://ieeexplore.ieee.org/document/9878449/
11. Ruiz, N., Li, Y., Jampani, V., Pritch, Y., Rubinstein, M., Aberman, K.: Dream-Booth: fine tuning text-to-image diffusion models for subject-driven generation. http://arxiv.org/abs/2208.12242
12. Ryu, S.: Low-rank adaptation for fast text-to-image diffusion fine-tuning. http://github.com/cloneofsimo/lora, original-date: 2022–12-08T00:09:05Z
13. Saleh, B., Elgammal, A.: Large-scale classification of fine-art paintings: learning the right metric on the right feature. http://arxiv.org/abs/1505.00855
14. Shan, S., Cryan, J., Wenger, E., Zheng, H., Hanocka, R., Zhao, B.Y.: GLAZE: protecting artists from style mimicry by text-to-image models. http://arxiv.org/abs/2302.04222
15. Van Le, T., Phung, H., Nguyen, T.H., Dao, Q., Tran, N., Tran, A.: Anti-DreamBooth: protecting users from personalized text-to-image synthesis. http://arxiv.org/abs/2303.15433, version: 1

Attention Based Spatial-Temporal Dynamic Interact Network for Traffic Flow Forecasting

Junwei Xie, Liang Ge[✉], Haifeng Li, and Yiping Lin

College of Computer Science, Chongqing University, Chongqing 400030, China
{xiejunwei,lihaifeng}@stu.cqu.edu.cn, {geliang,yplin}@cqu.edu.cn

Abstract. The prediction of spatio-temporal traffic flow data is challenging due to the complex dynamics among different roads. Existing approaches often focused on capturing traffic patterns at a single temporal granularity, disregarding spatio-temporal interactions and relying heavily on prior knowledge. However, this limits the generality of the models and their ability to adapt to dynamic changes in traffic patterns. We argue that traffic flow changes co-occur in the road network's temporal and spatial dimensions, which leads to commonalities and regularities in the data across these dimensions, with their dynamic changes depending on the temporal granularity. In this research, we propose an attention based spatio-temporal dynamic interaction network consisting of a spatio-temporal interaction filtering module and a spatio-temporal dynamic perception module. The interaction filtering module captures commonalities and regularities from a global perspective, ensuring adherence to the temporal and spatial dimensions of the road network structure. The dynamic perception module incorporates a sliding window attention mechanism to capture local dynamic correlations between the temporal and spatial dimensions at different time granularities. To address the issue of time series span, we design a more adaptive time-aware attention mechanism that effectively captures the impact of time intervals. Extensive experiments on four real-world datasets demonstrate that our approach achieves state-of-the-art performance and consistently outperforms other baseline methods. The source code is available at https://github.com/JunweiXie/ASTDIN.

Keywords: Traffic flow prediction · Interaction filtering · Multi-granularity

1 Introduction

Traffic flow prediction constitutes a fundamental element of Intelligent Transportation Systems (ITS) [1,2], employing traffic sensors, such as cameras, radar detectors, and infrared detectors, strategically placed along urban roads. These sensors collect historical traffic data, which is subsequently used to predict future traffic flow patterns.

B. Luo et al. (Eds.): ICONIP 2023, LNCS 14450, pp. 445–457, 2024.
https://doi.org/10.1007/978-981-99-8070-3_34

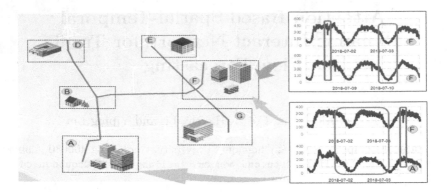

Fig. 1. An example of the traffic flow system. (Color figure online)

Traffic flow data is considered a typical example of spatio-temporal data as it captures changes occurring in both the temporal and spatial dimensions [3,4], the changes in traffic flow data coincide in both the temporal and spatial dimensions of the entire traffic network structure and these dimensions mutually affect and interact with each other in real-time [5]. Consequently, the temporal and spatial dimensions exhibit certain commonalities, which dynamically evolve with different temporal granularities. As illustrated in Fig. 1, when considering coarse spatio-temporal granularity, the commonality of spatio-temporal interaction in the temporal dimension is evident through macroscopic periodicity. For instance, sensor F exhibits periodic hourly characteristics, as indicated by similar traffic patterns (red round rectangle) observed during specific hourly periods daily. The similarity in traffic patterns within neighborhoods sharing similar characteristics demonstrates the commonality in the spatial dimension. For instance, sensor A and sensor F, situated in different commercial areas, display similar traffic patterns (red round rectangle). At a finer spatio-temporal granularity, the commonality in both the temporal and spatial dimensions of spatio-temporal interaction manifests through short-term micro fluctuations. These fluctuations can be observed as temporary reductions in traffic flow due to unexpected events such as extreme weather conditions or traffic accidents (green round rectangle).

To effectively model spatio-temporal data, it is essential to consider the impact of commonalities resulting from intricate spatio-temporal interactions on the temporal and spatial dimensions and the dynamic nature of these interactions across different temporal granularities. Therefore, models need to possess the following capabilities:

(1) Capture the commonalities and regularities observed in spatio-temporal interaction patterns in the temporal and spatial dimensions.

(2) Dynamically capture the correlations between the temporal and spatial dimensions, which can vary based on the temporal granularity.

However, previous studies directly exploit commonalities and regularities in modeling, such as introducing a priori knowledge, such as random variables. [6] uses a random variable to capture the time-varying model parameters at a specific location to achieve spatio-temporal aware prediction, and such methods cannot appropriately adapt to dynamic changes in data distribution because they ignore spatio-temporal interactions, and rely heavily on domain knowledge.

Furthermore, previous studies have primarily focused on modeling temporal and spatial dimensional correlations. [7] employs separate components to capture temporal and spatial correlations, respectively. They input the spatial feature output into the temporal component to indirectly capture spatio-temporal correlations. In contrast, [8,9] directly capture spatio-temporal correlations by constructing local spatio-temporal graphs. However, most of these approaches utilize shared parameters for single-time granularity prediction, which fail to ensure the complementary integration of temporal and spatial patterns grounded in distinct temporal granularities, thus impacting the robustness of the prediction model.

This paper introduces a novel two-stage model for traffic flow prediction. The first stage of the module incorporates two gates to collaboratively capture temporal and spatial dependencies, forming temporal and spatial specific partitions. It then simulates spatio-temporal interaction patterns and filters out redundant features. In the second stage, we aim to enable the model to dynamically capture detailed features of the temporal and spatial dimensions without relying on the same parameter space. To achieve this, we adopt a sliding window mechanism to facilitate rich interactions between adjacent windows, ensuring the incorporation of rich information context. We partition the data into three distinct temporal granularities, allowing us to extract dynamic features from both the temporal and spatial dimensions. Moreover, we have designed mechanisms in the temporal dimension to ensure adaptive time intervals, further enhancing the model's flexibility. Extensive experiments are conducted on four real datasets, and the results consistently demonstrate that our model outperforms all baseline methods.

2   Related Works

There are two main categories of traffic flow prediction models: those based on statistical theory and those based on deep learning. Early models based on statistical theory treated traffic flow data as simple time series. They utilized historical data, other relevant factors, and statistical methods to identify potential patterns and predict the future traffic flow [10–12]. However, these models heavily relied on the assumption of smoothness correlation, which did not align with the characteristics of traffic flow data.

Deep learning methods have recently been proposed to capture the complex spatio-temporal correlations in traffic flow data [13–15]. For example, [7] represented the spatial correlation between sensor pairs as a directed graph, modeling it using a bidirectional random walk, and introduced a GRU [16] structure to capture the temporal correlation. Furthermore, local spatio-temporal graphs were extended from the spatial graph structure of adjacent time steps, as seen in models like [8,9,17], which directly utilized graph convolution methods for capturing spatio-temporal correlations. However, these methods employed shared parameters across different time series at different locations, failing to capture the dynamic variability of temporal and spatial patterns over time, which is often observed in real-world scenarios. In addition to deep learning models, approaches incorporating prior knowledge on top of these models have achieved impressive results [18–20]. For example, [6] introduced random variables in the time series encoding to generate location-specific time-varying parameters, enhancing the capture of spatio-temporal dynamic correlations. However, these methods relying on prior knowledge heavily depend on domain expertise and predefined designs, needing more adaptability to dynamic changes in the data, ultimately limiting their performance.

3   Methodology

3.1   Preliminaries

Traffic flow data is a typical spatio-temporal graph data consisting of a set of N sensors deployed in the traffic network structure, along with their adjacency relationships. We use $E \in R^{N \times N}$ to represent the adjacency matrix, where $E_{ij} = 1$ indicates the existence of a connection between node i and j Each sensor periodically records multiple attributes, including traffic flow and traffic speed. In our case, we only consider traffic flow as the input feature. Therefore, we use $X \in R^{T \times N \times din}$ to represent the traffic flow data of all N sensors for T time steps, where din equals 1. The goal of the traffic flow prediction problem is to learn a mapping function F based on the historical data of T time steps and the adjacency matrix in order to predict the traffic flow data for the future T' time steps:

$$F([X_{t-T+1}, \cdots , X_{t-1}, X_t] ; E)=([X_{t+1}, \cdots , X_{t+T'-1}, X_{t+T'}]) \qquad (1)$$

where, $X_t \in R^{N \times din}$.

3.2   Attention Based Spatial-Temporal Dynamic Interact Network

In this section, we introduce the ASTDIN model, as illustrated in Fig. 2. The model comprises two stages: (1) The first stage focuses on simulating the spatio-temporal interactions and decoding them to extract location-specific temporal and spatial patterns from a global perspective, achieved through graph convolution and spatio-temporal interaction filtering modules. (2) In the second stage, a dimensional mapping operation is performed on the input from a local perspective. Subsequently, the decoded output from the first stage is embedded into the attention module, which is based on sliding window spatio-temporal dynamic perception and allows the model to capture implicit spatio-temporal features. Finally, several traffic patterns with different temporal granularity are fused to predict future traffic flow.

Spatio-Temporal Interaction Filtering. To simulate data interaction in the temporal and spatial dimensions, we perform dimensional compression on the input. This compression step ensures that the temporal and spatial dimensions

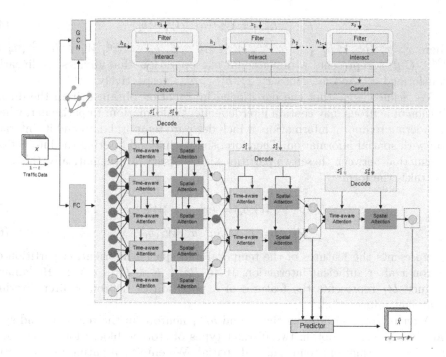

Fig. 2. The overall architecture of ASTDIN. The model comprises two stages: (1) The first focuses on simulating the spatio-temporal interactions. (2) The second stage is based on sliding window spatio-temporal dynamic perception. FC represents a fully connected layer. Considering an input time step of 12. The first layer has a window size of 4 and a step size of 2, resulting in 5 windows. The second layer has a window size of 3 and a step size of 2, leading to 2 windows. The third layer has a window size of 2 and 1 window.

information is fully interconnected in the subsequent spatio-temporal interaction module. By employing the conventional GCNs, we aggregate each node's and its neighbors' information, aiming to capture the static dependencies within the network. The specific calculation formula is as follows:

$$X^{in} = (ReLu(Squeeze(X)W_{a1} + b_{a1})W_{a2} + b_{a2})E \qquad (2)$$

where $X^{in} \in R^{T \times N}$, $Squeeze(\cdot)$ represents feature dimensional compression operations, E represents the connectivity of the urban road network structure, $W_{a1} \in R^{N \times C}$, $b_{a1} \in R^{C}$, $W_{a2} \in R^{C \times N}$, $b_{a2} \in R^{N}$, $ReLu(\cdot)$ represents the rectified linear unit activation function.

First, we divide the neuron storage information into two parts. Specifically, we will divide the current moment input neuron x_t^{in} and the last moment hidden layer state h_{t-1} into two partitions, including temporal information partition and spatial information partition. The specific partitioning formula is as follows:

$$x_t^t, x_t^s = Cut(x_t^{in}W_c + b_c) \qquad (3)$$

$$h_{t-1}^t, h_{t-1}^s = Cut(h_{t-1}W_l + b_l) \tag{4}$$

where x_t^t, x_t^s, h_{t-1}^t, $h_{t-1}^s \in R^{N \times d}$, $W_c \in R^{N \times 2d}$, $b_c \in R^{2d}$, $W_l \in R^{d \times 2d}$, $b_l \in R^{2d}$, $Cut(\cdot)$ represents a dimensional cut operation that divides its linearly transformed x_t^{in} and h_{t-1} into x_t^t, x_t^s, h_{t-1}^t, h_{t-1}^s respectively.

We acknowledge that the partitioned information obtained from the direct division of neurons may contain interdependent information. In particular, when considering temporal information, it includes pure temporal data and its interaction with spatial information. Therefore, simulating and filtering the interaction information between these two partitions is necessary. Mathematically, this process takes the form:

$$\mathcal{U}_i(x^t, x^s, W_i, b_i) = (x^t \odot x^s)W_i + b_i \tag{5}$$

$$\mathcal{U}_f(x) = x - \mathcal{U}_i(x^t, x^s, W_i, b_i) \tag{6}$$

$\mathcal{U}_i$ represents the features of the temporal and spatial information partition of neuron x after sufficient interaction, $W_i \in R^{d \times d}$, $b_i \in R^d$, $\odot$ is the Hadamard product. $\mathcal{U}_f$ represents the features of a partition of neuron x after filtering redundant information.

After interaction filtering the x_t and h_{t-1} neurons in the temporal and spatial dimensions, we obtain two distinct types of storage blocks for current and historical information: temporal and spatial. We employ a gating unit controls the flow of information at different time steps, regulating the inflow and outflow of information. This gating mechanism is implemented as follows:

$$\mathcal{U}_g(x_t, h_{t-1}, W_g, b_g) = \sigma((\mathcal{U}_f(x_t) \parallel \mathcal{U}_f(h_{t-1}))W_g + b_g) + x_t \tag{7}$$

$\mathcal{U}_g$ represents the flow of information between neuron x at time t and h_{t-1} at time $t - 1$. $W_g \in R^{2d \times d}$, $b_g \in R^d$, $(\cdot \parallel \cdot)$ means concatenation according to the feature dimension.

Next, we must calculate the hidden layer state h_t passed between steps. In calculating spatio-temporal information, we still use the gating mechanism to control the temporal information partitioning, spatial information partitioning and the flow of spatio-temporal interaction information at the current moment t. The formula for calculating the hidden layer state h_t is as follows:

$$h_t = \mathcal{U}_i(x^t, x^s, W_i, b_i) \odot tanh((x_t^t + x_t^s)W_h + b_h) + tanh((s_t^t + s_t^s)W_o + b_o) \tag{8}$$

where W_h, $W_o \in R^{d \times d}$, b_h, $b_o \in R^d$. s_t^t and s_t^s represent the outputs of temporal and spatial partitions at time t in method $\mathcal{U}_g$.

Spatio-Temporal Interaction Decoding Module. The spatio-temporal interaction decoding module aims to decode the exclusive temporal and spatial patterns resulting from the filtered interactions. This decoding process enables the generation of temporal-specific and spatial-specific parameter spaces for subsequent modules. We concatenate the temporal and spatial information outputs from each moment of the spatio-temporal interaction filtering module to learn these specific temporal and spatial parameter spaces. We then utilize Multi-layer Perceptrons (MLPs) to model and learn these parameter spaces, namely P_w^t, P_b^t, P_w^s and P_b^s. P_w^t, P_b^t represent the parameter space generated by decoding the temporal dimension. P_w^s, P_b^s represent the parameter space generated by decoding the spatial dimension. P_w^t, $P_w^s \in R^{T^s \times d_c \times d_c}$. P_b^t, $P_b^s \in R^{T^s \times d_c}$.

Attention Module for Spatio-Temporal Dynamic Perception Based on Sliding Window. To predict future traffic flow accurately, we propose a spatio-temporal dynamic perceptual attention module based on sliding windows. This module leverages the temporal and spatial parameter spaces obtained from the previous step and incorporates them into the temporal and spatial attention mechanisms of the each sliding window to capture local spatio-temporal dynamic correlations and incorporates the effect of the time span in the time-aware attention mechanism.

Time-Aware Attention Mechanism. Before capturing the spatio-temporal correlations, the dimensions are mapped using a fully connected layer. The specific equation for the fully connected layer is as follows:

$$X^* = XW_m + b_m \tag{9}$$

where $X^* \in R^{T \times N \times d_c}$, $W_m \in R^{1 \times d_c}$, $b_m \in R^{d_c}$, We embed the commonalities exhibited by the temporal dimensions generated by the decoding module into the keys and values in the attention module. This embedding process is performed using the following formula:

$$Q^t = p, \ K^t = (p \parallel H_i^l) \otimes P_w^t + P_b^t, \ V^t = (p \parallel H_i^l) \otimes P_{w\prime}^t + P_{b\prime}^t \tag{10}$$

where H_i^l represents the $i - th$ window of layer l. When $l = 0$, $H^0 = X^*$. $(\cdot \parallel \cdot)$ means concatenation according to the time dimension. P_w^t, P_b^t and $P_{w\prime}^t$, $P_{b\prime}^t$ representing the two sets of parameters generated by the spatio-temporal interaction decoding module, respectively. $Q^t \in R^{N \times d^P \times d_c}$, $K^t \in R^{N \times T^s \times d_c}$, $V^t \in R^{N \times T^s \times d_c}$.

To ensure the diversity of features, we use a multi-headed mechanism to slice and dice Q^t, K^t, V^t, and compute the attention weights for each sub-feature.

$$H^t = MLP^t(Concat(head_1^t, head_2^t, \cdots, head_h^t)) \tag{11}$$

$$head_i^t = Attention(Q_i^t, K_i^t, V_i^t) = Softmax(Score_i^t V_i^t) \tag{12}$$

where $H^t \in R^{T^s \times N \times d_c}$ is the output of the time-aware attention mechanism, $Softmax(\cdot)$ is the normalized exponential function. Compared with the traditional attention, we build a more adaptive time-aware mechanism to flexibly understand the effect of time intervals on each time step.

$$Score_i^t = \tanh\left(\frac{sig\left(\frac{Q_i^t K_i^{t^T}}{\sqrt{d_c}}\right)}{1 + sig\left(sig\left(\frac{Q_i^t K_i^{t^T}}{\sqrt{d_c}}\right) * sig(\beta) * \Delta t\right)}\right) \tag{13}$$

where $Score_i^t \in R^{N \times T^s \times T^s}$, β is the learnable parameter for the interval feature to control the effect of time interval on the corresponding feature, Δt is the time interval of the latest time.

Spatial Attention Mechanism. To enhance the capture of spatio-temporal dependencies, we utilize the temporal features obtained from the temporal attention module as the input for the spatial attention module. This approach allows us to leverage the captured temporal patterns and incorporate them into the spatial attention mechanism. The specific formula is as follows:

$$Q^s = H^t, \ K^s = H^t \otimes P_w^s + P_b^s, \ V^s = H^t \otimes P_{w\prime}^s + P_{b\prime}^s \qquad (14)$$

Using traditional attention mechanism to calculate implicit temporal dependencies and capturing diversity using a multi-head mechanism [21], only the $Score^s$ are calculated differently compared to the time-aware attention mechanism, which is calculated as follows:

$$Score_i^s = \frac{Q_i^s K_i^{s^T}}{\sqrt{d_k}} \qquad (15)$$

$$H^s = MLP^s(Concat(head_1^s, head_2^s, \cdots, head_h^s)) \qquad (16)$$

where $H^s \in R^{T^s \times N \times d_c}$.

Finally, we aggregate each window output in the temporal dimension as input for subsequent spatio-temporal correlation capture at different time granularities, as follows:

$$H^l = (Agg(MLP^l(H_i^s) * H_i^s) \ \| \cdots \| \ Agg(MLP^l(H_{d_s\prime}^s) * H_{d_s\prime}^s)) \qquad (17)$$

where, $Agg(\cdot)$ represents summation by time dimension, $H^l \in R^{d_s\prime \times N \times d_c}$, and $d_s\prime$ represents the next layer window size.

3.3   Prediction Layer and Loss Function

We fuse the spatio-temporal features obtained from multiple temporal granularities by a multilayer perceptron and finally predict the output.

$$\hat{X} = MLP^o(\sum_{l=1}^{L} H^l W^l + b^l) \qquad (18)$$

where L represents the number of layers and $\hat{X}$ is the final predicted output. We complete the learning of the model by optimizing the following loss function:

$$\mathcal{L} = \begin{cases} \frac{1}{2*\delta}(X - \hat{X})^2 & |X - \hat{X}| \leq \delta \\ |X - \hat{X}| - \frac{1}{2}\delta & otherwise \end{cases} \qquad (19)$$

where δ is a threshold parameter to control the squared error loss.

4   Experiments

4.1   Experimental Setup

Datasets. We select four widely-used large public real-world datasets collected by Caltrans, namely PEMS03, PEMS04, PEMS07 and PEMS08. Each consists of tens of thousands of time steps and hundreds of sensors. These datasets also include structural information about the traffic road network. All four datasets are traffic flow datasets, where the primary variable of interest is the number of passing vehicles.

Implementation Details. We conducted the experiments using the default parameters specified in the baseline paper. We evaluated the results using three commonly employed evaluation metrics for traffic flow prediction tasks: Mean Absolute Error (MAE), Mean Absolute Percentage Error (MAPE), and Root Mean Square Error (RMSE).

We implemented ASTDIN on an NVIDIA 3060 GPU using Python 3.8.13 and PyTorch 1.11.0. We used Adam as the optimizer, the total number of training rounds was set to 200, and an early stop strategy with 30 patients was used. The number of layers of the model was set to 3, and the window size of each layer was 4, 3, and 2. The learning rate was 0.0005, the batch size was 32 using the PEMS04 dataset, and we set $T = T' = 12$. the hidden dimension d_c was set to 128 and the agent dimension p was set to 1 in this paper. The experimental parameters were established based on previous research and empirical values.

4.2   Experimental Results

For a fair comparison, we follow existing studies and divide the dataset into 60% for training, 20% for validation, and 20% for testing in chronological order. We used the first 12 time steps (1 h) to predict the subsequent 12 time steps (1 h), and the experimental results are shown in Table 1.

The experimental results demonstrate that ASTDIN outperforms all other methods regarding the Mean Absolute Error (MAE) evaluation metric across all datasets. Notably, it achieves the best performance in both the PEMS04 and PEMS08 datasets, highlighting the effectiveness of our proposed model. We observed that the statistical theory-based methods performed poorly. These methods heavily rely on the assumption of smoothness correlation and require extensive parameter tuning. On the other hand, approaches like [7] capture spatio-temporal correlation through stacking, and methods like [8] capture spatio-temporal correlation by constructing local spatio-temporal maps. However, none of these methods can guarantee the dynamic changes in spatio-temporal correlation with fine-grained time granularity. In contrast, [6] captures location-specific time-varying patterns by introducing random variables. However, it overlooks the spatio-temporal interaction process and lacks the adaptability to changes in data distribution. In summary, the results presented in the table validate the superiority of ASTDIN compared to the other methods considered in this study.

Table 1. Traffic flow forecasting on the PEMS03, PEMS04, PEMS07 and PEMS08.

Methods	PEMS03			PEMS04		
	MAE	$MAPE(\%)$	RMSE	MAE	$MAPE(\%)$	RMSE
VAR [22]	23.65	24.51	38.26	23.75	18.09	36.66
DCRNN [7]	18.18	18.91	30.31	24.7	17.12	38.12
STSGCN [8]	17.48	16.78	29.21	21.19	13.90	33.65
ASTGNN [23]	15.07	15.8	26.88	19.26	12.65	13.02
STFGNN [9]	16.77	16.3	28.34	19.83	13.02	31.88
ST-WA [6]	15.17	15.83	**26.63**	19.06	12.52	31.02
ASTDIN	**15.00**	**14.77**	26.97	**18.83**	**12.23**	**30.49**
Methods	PEMS07			PEMS08		
	MAE	$MAPE(\%)$	RMSE	MAE	$MAPE(\%)$	RMSE
VAR [22]	75.63	32.22	115.24	23.46	15.42	36.33
DCRNN [7]	25.30	11.66	38.58	17.86	11.45	27.83
STSGCN [8]	24.26	10.21	39.03	17.13	10.96	26.8
ASTGNN [23]	22.23	9.25	35.95	15.98	9.97	25.67
STFGNN [9]	22.07	9.21	35.80	16.64	10.60	26.22
ST-WA [6]	20.74	**8.77**	34.05	15.41	9.94	24.62
ASTDIN	**20.67**	8.87	**33.83**	**15.38**	**9.88**	**24.45**

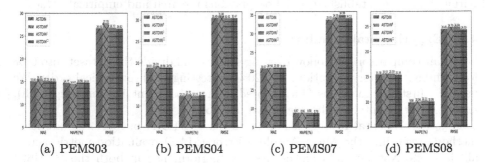

(a) PEMS03   (b) PEMS04   (c) PEMS07   (d) PEMS08

Fig. 3. The ablation study results on the four public datasets.

4.3 Ablation Experiments

In this section, we will conduct ablation studies in three aspects to demonstrate the validity of our work.

ASTDIN[‡]: To assess the significance of spatio-temporal interaction modeling, we introduced the ASTDIN[‡] model, which omits the spatio-temporal interaction process and instead employs random parameters. Figure 3 demonstrates that when the spatio-temporal interaction process is disregarded, the model's performance experiences a significant decline. This result underscores the importance of incorporating the spatio-temporal interaction process.

ASTDIN$^\diamond$: To assess the importance of adaptive time intervals, we introduced the ASTDIN$^\diamond$ model, which disregards the traditional approach of capturing temporal correlations without considering the impact of time intervals. Figure 3 clearly illustrates that incorporating time intervals into capturing temporal correlations significantly improves model performance. This outcome substantiates the significance of accounting for adaptive time intervals in traffic flow prediction.

ASTDIN$^\square$: To examine the significance of fusing multiple temporal granularities, we used ASTDIN$^\square$ by solely considering a single temporal granularity. As depicted in Fig. 3, modeling with only one temporal granularity yields better results than incorporating multiple temporal granularities for datasets with relatively small nodes, such as PEMS08. This finding suggests that the flexibility to choose the appropriate temporal granularity for modeling is crucial, particularly considering the size of the dataset.

5   Conclusion

This study introduces an attention based spatio-temporal dynamic interaction network, ASTDIN, for traffic flow prediction. Our proposed model effectively simulates the process of spatio-temporal interaction within the structure of traffic road networks. Additionally, ASTDIN captures spatio-temporal correlations from multiple temporal granularities, considering the impact of time intervals on capturing temporal correlations. The results on four datasets demonstrate that ASTDIN achieves highly competitive performance in traffic flow prediction. In future work, we plan to extend the application of ASTDIN to handle other data with temporal and spatial attributes, such as meteorological forecasting, population forecasting, and other scenarios.

References

1. Lv, Y., Duan, Y., Kang, W., Li, Z., Wang, F.Y.: Traffic flow prediction with big data: a deep learning approach. IEEE Trans. Intell. Transp. Syst. **16**(2), 865–873 (2014)
2. Ryu, U., Wang, J., Kim, T., Kwak, S., Juhyok, U.: Construction of traffic state vector using mutual information for short-term traffic flow prediction. Transp. Res. Part C: Emerg. Technol. **96**, 55–71 (2018)
3. Pavlyuk, D.: Feature selection and extraction in spatiotemporal traffic forecasting: a systematic literature review. Eur. Transp. Res. Rev. **11**(1), 6 (2019)
4. Huang, X., Ye, Y., Wang, C., Yang, X., Xiong, L.: A multi-mode traffic flow prediction method with clustering based attention convolution LSTM. Appl. Intell., 1–14 (2021)

5. Duan, H., Xiao, X., Xiao, Q.: An inertia grey discrete model and its application in short-term traffic flow prediction and state determination. Neural Comput. Appl. **32**, 8617–8633 (2020)

6. Cirstea, R.G., Yang, B., Guo, C., Kieu, T., Pan, S.: Towards spatio-temporal aware traffic time series forecasting. In: 2022 IEEE 38th International Conference on Data Engineering (ICDE), pp. 2900–2913. IEEE (2022)

7. Li, Y., Yu, R., Shahabi, C., Liu, Y.: Diffusion convolutional recurrent neural network: data-driven traffic forecasting. In: International Conference on Learning Representations (2018)

8. Song, C., Lin, Y., Guo, S., Wan, H.: Spatial-temporal synchronous graph convolutional networks: a new framework for spatial-temporal network data forecasting. In: Proceedings of the AAAI Conference on Artificial Intelligence, vol. 34, pp. 914–921 (2020)

9. Li, M., Zhu, Z.: Spatial-temporal fusion graph neural networks for traffic flow forecasting. In: Proceedings of the AAAI Conference on Artificial Intelligence, vol. 35, pp. 4189–4196 (2021)

10. Drucker, H., Burges, C.J., Kaufman, L., Smola, A., Vapnik, V.: Support vector regression machines. In: Advances in Neural Information Processing Systems 9 (1996)

11. Williams, B.M., Hoel, L.A.: Modeling and forecasting vehicular traffic flow as a seasonal ARIMA process: theoretical basis and empirical results. J. Transp. Eng. **129**(6), 664–672 (2003)

12. Emami, A., Sarvi, M., Asadi Bagloee, S.: Using Kalman filter algorithm for short-term traffic flow prediction in a connected vehicle environment. J. Mod. Transp. **27**, 222–232 (2019). https://doi.org/10.1007/s40534-019-0193-2

13. Choi, J., Choi, H., Hwang, J., Park, N.: Graph neural controlled differential equations for traffic forecasting. In: Proceedings of the AAAI Conference on Artificial Intelligence, vol. 36, pp. 6367–6374 (2022)

14. Yu, L., Du, B., Hu, X., Sun, L., Han, L., Lv, W.: Deep spatio-temporal graph convolutional network for traffic accident prediction. Neurocomputing **423**, 135–147 (2021)

15. Bai, L., Yao, L., Kanhere, S.S., Wang, X., Sheng, Q.Z.: STG2Seq: spatial-temporal graph to sequence model for multi-step passenger demand forecasting. In: 28th International Joint Conference on Artificial Intelligence, IJCAI 2019, pp. 1981–1987. International Joint Conferences on Artificial Intelligence (2019)

16. Cho, K., et al.: Learning phrase representations using RNN encoder-decoder for statistical machine translation. arXiv preprint arXiv:1406.1078 (2014)

17. Kong, X., Zhang, J., Wei, X., Xing, W., Lu, W.: Adaptive spatial-temporal graph attention networks for traffic flow forecasting. Appl. Intell., 1–17 (2022)

18. Guo, S., Lin, Y., Feng, N., Song, C., Wan, H.: Attention based spatial-temporal graph convolutional networks for traffic flow forecasting. In: Proceedings of the AAAI Conference on Artificial Intelligence, vol. 33, pp. 922–929 (2019)

19. Duan, Y., Chen, N., Shen, S., Zhang, P., Qu, Y., Yu, S.: FDSA-STG: fully dynamic self-attention spatio-temporal graph networks for intelligent traffic flow prediction. IEEE Trans. Veh. Technol. **71**(9), 9250–9260 (2022)

20. Belhadi, A., Djenouri, Y., Djenouri, D., Lin, J.C.W.: A recurrent neural network for urban long-term traffic flow forecasting. Appl. Intell. **50**, 3252–3265 (2020)
21. Vaswani, A., et al.: Attention is all you need. In: Advances in Neural Information Processing Systems 30 (2017)
22. Hamilton, J.D.: Regime-switching models (2005)
23. Guo, S., Lin, Y., Wan, H., Li, X., Cong, G.: Learning dynamics and heterogeneity of spatial-temporal graph data for traffic forecasting. IEEE Trans. Knowl. Data Eng. **34**(11), 5415–5428 (2022)

Staged Long Text Generation
with Progressive Task-Oriented Prompts

Xingjin Wang[1,2], Linjing Li[1,2(✉)], and Daniel Zeng[1,2]

[1] School of Artificial Intelligence, University of Chinese Academy of Sciences, Beijing,
China
[2] State Key Laboratory of Multimodal Artificial Intelligence Systems, Institute of
Automation, Chinese Academy of Sciences, Beijing, China
{wangxingjin2021,Dajun.Zeng,linjing.li}@ia.ac.cn

Abstract. Generating coherent and consistent long text remains a challenge for artificial intelligence. The state-of-the-art paradigm partitions the whole generating process into successive stages, however, the content plan applied in each stage may be error-prone and fine tuning large-scale language models, one for each stage, is resource-consuming. In this paper, we follow the above paradigm and devise three stages: keyphrase decompression, transition paraphrase, and text generation. We leverage task-oriented prompts to direct the producing of text in each stage which improves the quality of the generated text. Further, we propose a new content plan representation with elastic mask tokens to reduce model bias and irregular words. Moreover, we introduce length control and commonsense knowledge prompts to increase the adaptability of the proposed model. Extensive experiments conducted on two challenging tasks demonstrated that our model outperforms strong baselines significantly, and it is able to generate longer high quality texts with fewer parameters.

Keywords: Long Text Generation · Content Plan · Prompt Based
Learning

1 Introduction

Text generation has attracted great attention from both the academia and industry in recent years, because of its broad applications such as report creating, opinion essay generating, and story writing. Generating coherent and cohesive long-form text is more challenging compared with short text generation, since it needs to capture global context, plan content, and produce coherent words. Leveraging multiple large pre-trained language models (PLMs) such as GPT-2 [24] and BART [15] has achieved promising results, but these methods are not only resource consuming, but also lack of coherence.

Generating long text usually requires to couple a content planning module with a text generation module, where the content planning module selects and orders content to form a high level intermediate representation, and the text generation module refines the content plan into the final text. There are two types of text planning methods in the literature. The first kind of methods utilizes an one-stage paradigm, which represents the content plan by latent variables, and the final text is generated in an end-to-end way. However, the content plan form applied is difficult to control as the latent representation is invisible. The other kind utilizes multi-stage paradigm, which trains models to plan and generate intermediate output at each stage [13]. A typical three-stage model is consisted of decompression, transduction, and creation [16]. Nevertheless, the partition of stages and the forms of content plans may introduce cascading errors. Meanwhile, this kind of methods inevitably requires huge number of computing resources to train multiple large-scale language models.

Prompts can be employed to guide the tuning of large-scale models to generate text with relatively less resource. However, existing prompt-based methods mainly aim at attribute-controllable text generation, such as sentiment, topic, etc. [18]. In this paper, we further leverage task-oriented prompts to direct the planing and generating at each stage. We partition the whole generation process into three stages: keyphrase decompression, transition paraphrase, and text generation. First, the keyphrase decompression stage aims to decompress a set of keyphrases and forms an initial sketch consisting of key-phrase assignments and **mask** token. We propose the usage of elastic **mask** tokens to separate the keyphrases and design a length prompt to direct the model adapting different length text. Second, the transition paraphrase stage empowers the model to generate some transitional words in proper positions according to the word importance. In order to improve coherence and reduce repeat and disorder, the prompt of this stage is trained to convert a manually constructed set of negative samples into correct text with proper order. Finally, the text generation stage trains prompts to convert the content plan, including elastic **mask** tokens, to a completed text output. We design two prompts for this stage, the first is trained on a concept commonsense dataset to help the employed low-level PLMs adapt to the specific generation task, the other is trained on long text datasets [27]. These prompts assist the model adapting the task more quickly and improving the ability of generalization between different datasets.

We evaluate our model on two challenging long text generation tasks: argument generation and story generation. Automated evaluations illustrated that our progressive model outperforms the existing content plan methods significantly and generate much longer text with less parameters. Human evaluations further indicated that our model understands the guidance keyphrases better and generates higher quality output. Our main contributions can be summarized as:

- We propose a novel progressive long text generation model with task-oriented prompts, the whole generating process is partitioned into three stage: decompression, paraphrase, and generation. The task-oriented prompts are leveraged to complete different tasks at different stages.

- We utilize keyphrases with elastic **mask** tokens to form the content plan of each stage, this new form of representation can reduce model bias and irregular words.
- Length control prompt and commonsense knowledge prompt are trained to better adapt the overall generation task.

2   Related Work

Content Planning for Neural Generation. Content planning is a crucial step for high-quality, well-organized text generation. Recent work [21, 28] incorporates text planning into neural systems to enhance content relevance, and planning methods can be divided into two types. The first kind of method is one-stage models [8, 11] which leverage the latent variable to represent the high-level information to plan and generate text in an end-to-end way. However, these models are hard to control plan form because of the invisible latent representation. The second line is two-stage or multi-stage models, which train multi models independently to plan high-level sketches and then generate texts from sketches. The form of middle plans could be plots in Semantic Role Labelling format [7], template with keyphrases assignment and positions [13], and plan in form of binary tree structuring [1]. ProGen [26] proposed multi-stage method that progressively generates multi sketches in form of word list according to word importance. But these methods inevitably require lots of computing resources to train multi large models. Compared with prior methods, our method splits the generation into three stages: uncompression, paraphrase and creation, and we leverage prompt-based learning to reduce the model parameters.

Prompt-Based Text Generation. Prompt-based learning [3] is a new paradigm, which freezes the PLM's parameters and optimizes a continuous task-specific vector to complete correspond task [14, 20]. Some works [4, 18] propose prefix tuning for controllable text generation, and use the control prefix to combine with general prefix to improve text quality. Recent works [9, 23, 27] demonstrate that using multiple prompts can guide the model to generate text with multi attributes. PTG [16] presents a prompt-based transfer learning approach for text generation and proves the effectiveness of prompt to solve different generation task. We follow this idea and leverage the task-oriented prompt to complete uncompression, paraphrase and creation task, and we incorporate the multi-prompt method into each stage.

Commonsense Knowledge. Recent studies have demonstrated that incorporating external commonsense knowledge significantly improved the coherence and informativeness for multi generation tasks, such as dialog generation, storyending generation, essay generation, and argument generation. Previous work [10] leverage the ATOMIC [25] and Concept [17] commonsense dataset to fine-tune model firstly, and then fine-tuning a second time on the target corpus. We also firstly train a knowledge prompt using commonsense dataset and then train on target dataset combined with general prompt.

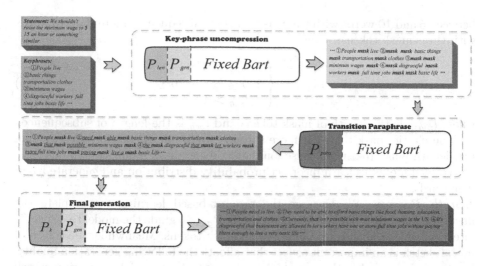

Fig. 1. Overview of our framework. The keyphrases uncompression stage takes a state-ment and a set of keyphrases as input and generates sketch with keyphrases and elastic mask tokens using the length prompt and general prompt. The transition paraphrase stage adds some transitional words and corrects coherence problem in the initial sketch. Finally, the generation stage leverages the knowledge prompt trained by commonsense knowledge and general prompt to convert the template to text.

3 Methodology

3.1 Framework Overview

Task Description. We follow the previous work [13] and model the long text generation task with the input of (1) a statement set as x which could be a proposition for argument generation or a title for story generation, and (2) a set of keyphrases set as $\mathbf{m} = \{m1, m2, ..., m_i\}$ related to the statement as topical guidance signal. The system aims to generate an opinion text or a story set as Y consisting of multiple sentences and properly reflecting the keyphrases in a coherent way.

3.2 Keyphrases Uncompression Stage

Given a set of keyphrases, the uncompression stage aims to generate an initial sketch with proper keyphrases assignment and mask tokens, as shown in the yellow block of Figure 1.
Elastic Mask Tokens. The keyphrases form the main body of sketch and the mask tokens separate the keyphrases and represent context information. The number of mask tokens indicates the sentence length information between the different keyphrases. Different from replacing the words not in keyphrases set with same number mask tokens, we leverage the elastic mask tokens which replaces the subsequences less than 5 words with one mask token, subsequences

between 5 and 10 words with two mask tokens, subsequences more than 10 words with three mask tokens:

$$l_{mask} = \begin{cases} 1 & l_{seq} < 5 \\ 2 & 5 \le l_{seq} \le 10 \\ 3 & l_{seq} > 10 \end{cases} \tag{1}$$

where l_{mask} is the length of mask tokens and l_{seq} is the length of subsequence.

The elastic mask tokens is more suitable for whole generation process than real length mask tokens, because long and repeat mask sequences will lead the pre-trained model to learn incorrect probability distribution and generate irregular words.

Length Prompt. We leverage the prompt-based learning to complete the uncompression task. We use a general prefix P_{gen} to guide the model to a specific direction which expands keyphrases with mask tokens. Meanwhile, the length of initial sketch is an important aspect that affects the following generation quality, hence we design a length control prefix P_{len} to assist model generating sketch with proper length. For each input, we design an artificial sentence with length number as the prompt input and use an extra embedding layer to embed the length information into vector as input of prompt generator. Both length and general prompt use the same prompt generator to achieve parameters efficient.

Then we combine the general prompt and length prompt to generate the reasonable sketch:

$$S = \arg\max_{s} P_{LM}(s \mid P_{len}; P_{gen}; x, m) \tag{2}$$

where LM is the pre-trained model combined with prompt, and P_{len}, P_{gen} represent length and general prompt, respectively.

3.3   Transition Paraphrase Stage

Based on sketch with keyphrases and mask tokens, the transition paraphrase stage aims to produce some transitional words inserted in proper positions and correct some semantic problems to improve coherence.

Transition Words. We follow [26] to define the importance scores of words, which first computes the average TF-IDF scores of each words and ranks all words to get the top frequency words as intermediate vocabulary V.

$$V = importance(D, rate) \tag{3}$$

where D is all documents, and $rate$ is the threshold ratio. Adding transition words could refine the sketch to contains fine-grained information. As shown in Fig. 1, the underlined words in orange block is the transition words.

Negative Sample. Transition paraphrase stage also improves text coherence by recovering the incorrect order and eliminating the repeat sentence span in the above sketch. We manually construct the negative samples to train the prompt of this stage. In particular, two kinds of negative samples are created based on the following strategies: 1) *SHUFFLE*, where we randomly select some text spans less than 10 words and shuffle the order of tokens in these spans. 2) *REPEAT*, where we randomly select some spans to copy and directly concatenate the copy text in the back of original spans. The labels of negative samples are same as original, and we train the model with augmented dataset.

3.4 Final Generation Stage

The final generation stage leverage the transitional template with keyphrases and elastic mask tokens to generate final output.

Knowledge Prompt. Commonsense knowledge can facilitate language comprehension and generation, and we post-train a commonsense prompt to distill the knowledge. We use the ConceptNet[1] [17] and ATOMIC[2] [25] datasets consisting of commonsense triples and transform these triples into readable natural language sentences. In order to adapt the target generation task better, we replace the sentences words not in the triples with elastic mask tokens in the same way mentioned in Sect. 3.2. For example, the original triple is *(telephone, UsedFor, communication)* and the transformed sentence is *"telephone is used for communication"*, and the masked sentence is *"telephone* **mask** *communication"*. The knowledge prompt are trained to teach the model to recover the original sentence from the input of mask sentence.

$$P_k = P_{LM}(Data_{Common}) \tag{4}$$

where P_k is knowledge prompt, and $Data_{Common}$ is the commonsense data.

The knowledge prompt could fill the mask tokens with coherent commonsense words, which assists the model adapting to the target dataset more quickly. We also train a general prompt P_{gen} (yellow arrows) to concatenate with commonsense knowledge prompt P_k (green arrows) to assist the model generating final output:

$$Y = arg \max_y P_{LM}(y \mid P_k; P_{gen}; t) \tag{5}$$

where P_k, P_{gen} represent the knowledge prompt and general prompt respectively, and t is the above template.

4   Experimental Setups

4.1   Tasks and Datasets

We evaluate on two different long-form generation for argument and story generation.

---

[1] https://www.conceptnet.io/.
[2] https://homes.cs.washington.edu/m̃sap/atomic/.

Argument Generation [12] This dataset contains pairs of claims and counter-arguments extracted from titles and responses from the Reddit ChangeMyView (CMV) forum. The original poster title is regarded as the statement, and the high-quality argument replies are regarded as the target. We use 42k samples for training, 6k samples for validation, and 7k samples for testing.

Writing Prompt [6] which is a large collection of user generated stories along with their associated prompts in Reddit. Each story contains lots of sentences, which provides long and complicated story texts. We sample 3k documents for training, 4k documents for validation and 5k documents for testing. In both datasets, the noun phrases and verb phrases that contains at least one topic signature word are extracted to form the guidance keyphrases.

4.2 Baselines and Comparisons

For both long text generation tasks, we evaluate the following baselines:

Seq2Seq: where we directly fine-tune BART with keyphrases concatenated to the statement as input and the whole text as output.

PAIR: [13] which is a two-stage model. The first stage fine-tunes a BART as the planner to generate the position of each pre-defined keyword and forms a template. The second stage fine-tunes another BART as the generator to produce final outputs based on the statement and generated template.

ProGen: [26] which is a multi-stage model and generates text in a progressive manner. The method first produces domain-specific content keywords according to the word importance and then fine-tune multi BARTs to progressively refines keywords into complete passages.

5   Results and Analysis

5.1   Automatic Evaluation

We adopt the following automatic metrics to evaluate the performance of our method: **(1) BLEU-2** [22]: We use BLEU-2 to evaluate *2-gram* overlap between generated texts and target texts. **(2) ROUGE-2** [19] which measures recall of the *2-gram* common subsequences. **(3) METEOR** [5] which account for paraphrase. **(4) MS-Jaccard (MSJ-4)** [2] which is a lexical-based metric and measures the similarity of *4-grams* frequencies between two sets of texts with Jaccard index. The results are shown in Table 1.

Our model consistently outperforms all baseline methods. Compared with Seq2Seq which is trained based on BART with same statement and keyphrases as input, the substantial improvements indicate the effectiveness of using middle

Table 1. Experimental results on argument generation (ArgGen) and story generation (WritingPrmopt). Ours$_{w/o}CM$ is our model variant without elastic mask and Ours$_{w/o}KP$ is our model variant without knowledge prompt. We report BLEU-2, ROUGE-2, METEOR and MSJ-4.

System	ArgGen				WritingPrompt			
	BLEU-2	ROUGE-2	METEOR	MSJ-4	BLEU-2	ROUGE-2	METEOR	MSJ-4
Seq2Seq	38.17	32.03	50.22	56.45	28.28	23.44	36.76	47.63
PAIR	36.75	31.85	47.99	52.81	22.95	25.04	38.30	49.80
ProGen	34.89	30.67	48.03	53.34	21.02	21.36	28.67	34.18
Ours	**40.66**	**35.93**	**52.80**	**56.98**	**28.52**	**27.23**	**43.04**	**50.37**
Ours$_{w/o CM}$	27.30	30.64	42.97	47.73	22.36	22.31	35.93	43.29
Ours$_{w/o KP}$	39.79	34.94	49.71	56.77	26.53	25.35	42.33	49.31

content planning. Meanwhile, the improvements over PAIR show that generating keyphrases plan with elastic mask token autoregressively is better than predicting the position of keyphrases. Furthermore, the lead over ProGen confirms that adding mask tokens in the keyphrases plan is helpful for model to understand the plan in a consistent manner. The mask tokens represents the internal relations between the adjacent keyphrases and provides more effect information for model.

We also evaluate the variants of our model. First, we replace the curtate *mask* tokens with the real length *mask* tokens in all stages, which leads to a significant decrease. This demonstrates the elastic mask tokens is more suitable to separate the keyphrases and represent the internal relations between these words.

Long and repeat subsequences filled with mask tokens will lead the pretrained model to learn incorrect features and generate irregular mask tokens. Besides, removing the knowledge prompt also results in the decrease of performance. Knowledge prompt is trained with commonsense knowledge dataset and integrating external commonsense into a short length prompt. The knowledge prompt could independently complete the mask filling task with some commonsense words, so it could assist model better adapting to the target datasets.

5.2 Human Evaluation

For human evaluation, we hire three proficient English speakers to give a performance of article in terms of (1) **Coherence**, measuring if the text is logical and cohesive; (2) **Relevance**, which measures whether the text is relevant and consistent to the input statement; (3) **Content Richness**, measuring whether exists enough specific details in the output texts. Each aspect is rated on a scale of 1 (worst) to 5 (best). We randomly select 50 samples from the test sets for both tasks.

The results are shown in Table 2. Our model achieve better results than other baseline, which indicates the effectiveness of our progressive task-oriented

Table 2. Human evaluation on coherence (COh.), relatedness (Rel.), content richness (Rich.).

Task	Model	Coh.	Rel.	Rich.
ArgGen	Seq2Seq	3,14	3.34	2.87
	Ours$_{w/0\,Para}$	3.18	3.73	2.83
	Ours	**3.21**	**3.82**	**2.91**
WritingPrompt	Seq2Seq	2.32	2.98	3.27
	Ours$_{w/0\,Para}$	2.45	3.22	3.33
	Ours	**2.48**	**3.27**	**3.35**

prompt-based generation method. Furthermore, the better performance of model with transition paraphrase stage demonstrates the transitional words and self-correct process could improve the coherence of output text. Overall, the human results verify the capability of our progressive content plan stage and task-oriented prompt to complete the long-form text generation.

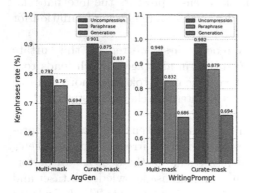

Fig. 2. Percentage of keyphrases in all three stages of elastic mask and multi mask in both datasets.

Fig. 3. BLEU-2 score of validation set of model with and without knowledge prompt in training stage.

5.3   Sketch with Elastic Mask Tokens

We study the influence of elastic mask tokens and real length mask tokens on model performance. The above experimental results show that real length mask tokens plan form leads to a significant performance degradation compared to elastic mask tokens. The reason is that the output of multi-mask generate longer mask tokens subsequences, but it easily causes keyphrases generation confusion and early generation termination results in missing parts of keyphrases.

Table 3. Sample outputs on dataset. The underline words in the paraphrase is generated transitional words. The keyphrases are highlighted in the final text.

Statement: *Nuclear Deterrence is bullshit.* **Keyphrases:** ① difference skirmish Falklands blown war countries fight brink collapse surrenders ② nukes handy ③ nuke moment surrender entire world gunpoint.
Uncompression: ① *mask* difference *mask* skirmish *mask* Falklands *mask* full blown *mask* countries *mask* fight *mask* *mask* brink *mask* collapse *mask* surrenders ② *mask* nukes *mask* handy ③ *mask* nuke *mask* moment *mask* surrender *mask* entire world *mask* gunpoint **Paraphrase:** ① *mask* a difference *mask* a skirmish *mask* the Falklands *mask* a full blown out war *mask* countries *mask* fight until one *mask* the brink *mask* collapse *mask* surrenders ② *mask* nukes *mask* handy ③ Using a nuke *mask* the moment *mask* surrender *mask* holding the entire world *mask* gunpoint **Generation:** ① There's a difference between a skirmish like the Falklands and a full blown - out war where countries continue to fight until one is on the brink of collapse and surrenders. ② That's when nukes come in handy. ③ Using a nuke at the moment of surrender is like holding theentire world at gunpoint.

We further study how keyphrases are utilized in the two forms of content plan. We compute the percentage of keyphrases mentioned in output of these two forms in both stages. As shown in Fig. 2, multi-mask format uses less keyphrases in both stages, and the decrement rate of elastic mask is less than multi mask, which indicates the weakness of multi mask plan form.

5.4   Knowledge Prompt

Knowledge prompt is trained with commonsense knowledge dataset and integrating external commonsense into a short length prompt. We directly adapt the trained knowledge prompt to the target dataset Arggen and WritingPrompt and the trained knowledge prompt could independently complete the mask filling task with some commonsense words and generate a original draft which reasonably reflects the keyphrases into the whole sentence.

We test the BLEU-2 of the validation set of both model based on knowledge prompt and model without commonsense knowledge. As shown in Fig. 3, the performance of model with knowledge prompt is better than another model. The model with knowledge prompt has a higher initial point and a faster ascent speed in both tasks, which indicates that the knowledge prompt assist the model better adapting the target dataset and generate better text.

5.5   Sample Outputs

We show two sample outputs of all stages on both tasks in Table 3. Our model effectively reflects the keyphrases into content plan and forms a coherent long text. The uncompression stage use the mask tokens to separate the keyphrases and form an initial sketch. The paraphrase add some transitional word and correct some semantic problems to improve the coherence. The final stage leverage the template to generate the final long text.

6 Conclusion

We present a novel progressive generation framework which splits the process into three stages and leverages the task-oriented prompt to complete each stage task. The keyphrases uncompression stage use the combination of length prompt and general prompt to uncompress a set of keyphrases to a content sketch consisting of keyphrases and elastic mask tokens. The transition paraphrase stage adds some transition words and corrects some semantic problems to improve coherence. The final generation stage utilizes the knowledge prompt trained with commonsense knowledge to better adapt to target datasets and generate final long text. Experiment results on two text generation tasks demonstrate that our model can reasonably reflect the given keyphrases and generate high-quality long text outputs.

Acknowledgements. This work was supported in part by the Strategic Priority Research Program of Chinese Academy of Sciences under Grant XDA27030100 and the National Natural Science Foundation of China under Grants 72293573 and 72293575.

References

1. Adewoyin, R., Dutta, R., He, Y.: RSTGen: imbuing fine-grained interpretable control into long-FormText generators. In: Proceedings of the 2022 Conference of the North American Chapter of the Association for Computational Linguistics: Human Language Technologies, pp. 1822–1835. Association for Computational Linguistics, Seattle, United States, July 2022. https://doi.org/10.18653/v1/2022.naacl-main.133, https://aclanthology.org/2022.naacl-main.133
2. Alihosseini, D., Montahaei, E., Soleymani Baghshah, M.: Jointly measuring diversity and quality in text generation models. In: Proceedings of the Workshop on Methods for Optimizing and Evaluating Neural Language Generation, pp. 90–98. Association for Computational Linguistics, Minneapolis, Minnesota, June 2019. https://doi.org/10.18653/v1/W19-2311, https://aclanthology.org/W19-2311
3. Brown, T., et al.: Language models are few-shot learners. Adv. Neural. Inf. Process. Syst. **33**, 1877–1901 (2020)
4. Clive, J., Cao, K., Rei, M.: Control prefixes for text generation. arXiv preprint arXiv:2110.08329 (2021)
5. Denkowski, M., Lavie, A.: Meteor universal: language specific translation evaluation for any target language. In: Proceedings of the Ninth Workshop on Statistical Machine Translation, pp. 376–380. Association for Computational Linguistics, Baltimore, Maryland, USA, June 2014. https://doi.org/10.3115/v1/W14-3348, https://aclanthology.org/W14-3348
6. Fan, A., Lewis, M., Dauphin, Y.: Hierarchical neural story generation. In: Proceedings of the 56th Annual Meeting of the Association for Computational Linguistics (Volume 1: Long Papers), pp. 889–898. Association for Computational Linguistics,

Melbourne, Australia, July 2018. https://doi.org/10.18653/v1/P18-1082, https://aclanthology.org/P18-1082

7. Fan, A., Lewis, M., Dauphin, Y.: Strategies for structuring story generation. In: Proceedings of the 57th Annual Meeting of the Association for Computational Linguistics, pp. 2650–2660. Association for Computational Linguistics, Florence, Italy, July 2019. https://doi.org/10.18653/v1/P19-1254, https://aclanthology.org/P19-1254

8. Fu, Y., Feng, Y., Cunningham, J.P.: Paraphrase generation with latent bag of words. In: Advances in Neural Information Processing Systems, vol. 32 (2019)

9. Gu, Y., Feng, X., Ma, S., Zhang, L., Gong, H., Qin, B.: A distributional lens for multi-aspect controllable text generation. arXiv preprint arXiv:2210.02889 (2022)

10. Guan, J., Huang, F., Zhao, Z., Zhu, X., Huang, M.: A knowledge-enhanced pre-training model for commonsense story generation. Trans. Assoc. Comput. Linguist. 8, 93–108 (2020)

11. Guan, J., Mao, X., Fan, C., Liu, Z., Ding, W., Huang, M.: Long text generation by modeling sentence-level and discourse-level coherence. In: Proceedings of the 59th Annual Meeting of the Association for Computational Linguistics and the 11th International Joint Conference on Natural Language Processing (Volume 1: Long Papers), pp. 6379–6393. Association for Computational Linguistics, Online, August 2021. https://doi.org/10.18653/v1/2021.acl-long.499, https://aclanthology.org/2021.acl-long.499

12. Hua, X., Hu, Z., Wang, L.: Argument generation with retrieval, planning, and realization. In: Proceedings of the 57th Annual Meeting of the Association for Computational Linguistics, pp. 2661–2672. Association for Computational Linguistics, Florence, Italy, July 2019. https://doi.org/10.18653/v1/P19-1255, https://aclanthology.org/P19-1255

13. Hua, X., Wang, L.: PAIR: planning and iterative refinement in pre-trained transformers for long text generation. In: Proceedings of the 2020 Conference on Empirical Methods in Natural Language Processing (EMNLP), pp. 781–793. Association for Computational Linguistics, Online, November 2020. https://doi.org/10.18653/v1/2020.emnlp-main.57, https://aclanthology.org/2020.emnlp-main.57

14. Lester, B., Al-Rfou, R., Constant, N.: The power of scale for parameter-efficient prompt tuning. In: Proceedings of the 2021 Conference on Empirical Methods in Natural Language Processing, pp. 3045–3059. Association for Computational Linguistics, Online and Punta Cana, Dominican Republic, November 2021. https://doi.org/10.18653/v1/2021.emnlp-main.243, https://aclanthology.org/2021.emnlp-main.243

15. Lewis, M., et al.: BART: denoising sequence-to-sequence pre-training for natural language generation, translation, and comprehension. arXiv preprint arXiv:1910.13461 (2019)

16. Li, J., Tang, T., Nie, J.Y., Wen, J.R., Zhao, X.: Learning to transfer prompts for text generation. In: Proceedings of the 2022 Conference of the North American Chapter of the Association for Computational Linguistics: Human Language Technologies, pp. 3506–3518. Association for Computational Linguistics, Seattle, United States, July 2022. https://doi.org/10.18653/v1/2022.naacl-main.257, https://aclanthology.org/2022.naacl-main.257

17. Li, X., Taheri, A., Tu, L., Gimpel, K.: Commonsense knowledge base completion. In: Proceedings of the 54th Annual Meeting of the Association for Computational Linguistics (Volume 1: Long Papers), pp. 1445–1455. Association for Computational Linguistics, Berlin, Germany, August 2016. https://doi.org/10.18653/v1/P16-1137, https://aclanthology.org/P16-1137

18. Li, X.L., Liang, P.: Prefix-tuning: optimizing continuous prompts for generation. In: Proceedings of the 59th Annual Meeting of the Association for Computational Linguistics and the 11th International Joint Conference on Natural Language Processing (Volume 1: Long Papers), pp. 4582–4597. Association for Computational Linguistics, Online, August 2021. https://doi.org/10.18653/v1/2021.acl-long.353, https://aclanthology.org/2021.acl-long.353

19. Lin, C.Y.: ROUGE: a package for automatic evaluation of summaries. In: Text Summarization Branches Out, pp. 74–81. Association for Computational Linguistics, Barcelona, Spain, July 2004. https://aclanthology.org/W04-1013

20. Liu, X., et al.: GPT understands, too. arXiv preprint arXiv:2103.10385 (2021)

21. Moryossef, A., Goldberg, Y., Dagan, I.: Step-by-step: separating planning from realization in neural data-to-text generation. In: Proceedings of the 2019 Conference of the North American Chapter of the Association for Computational Linguistics: Human Language Technologies, vol. 1 (Long and Short Papers), pp. 2267–2277. Association for Computational Linguistics, Minneapolis, Minnesota, June 2019. https://doi.org/10.18653/v1/N19-1236, https://aclanthology.org/N19-1236

22. Papineni, K., Roukos, S., Ward, T., Zhu, W.J.: BLEU: a method for automatic evaluation of machine translation. In: Proceedings of the 40th Annual Meeting of the Association for Computational Linguistics, pp. 311–318. Association for Computational Linguistics, Philadelphia, Pennsylvania, USA, July 2002. https://doi.org/10.3115/1073083.1073135, https://aclanthology.org/P02-1040

23. Qian, J., Dong, L., Shen, Y., Wei, F., Chen, W.: Controllable natural language generation with contrastive prefixes. In: Findings of the Association for Computational Linguistics: ACL 2022, pp. 2912–2924. Association for Computational Linguistics, Dublin, Ireland, May 2022. https://doi.org/10.18653/v1/2022.findings-acl.229, https://aclanthology.org/2022.findings-acl.229

24. Radford, A., et al.: Language models are unsupervised multitask learners. OpenAI Blog 1(8), 9 (2019)

25. Sap, M., et al.: ATOMIC: an atlas of machine commonsense for if-then reasoning. In: Proceedings of the AAAI Conference on Artificial Intelligence, vol. 33, pp. 3027–3035 (2019)

26. Tan, B., Yang, Z., Al-Shedivat, M., Xing, E., Hu, Z.: Progressive generation of long text with pretrained language models. In: Proceedings of the 2021 Conference of the North American Chapter of the Association for Computational Linguistics: Human Language Technologies, pp. 4313–4324. Association for Computational Linguistics, Online, June 2021. https://doi.org/10.18653/v1/2021.naacl-main.341, https://aclanthology.org/2021.naacl-main.341

27. Yang, K., et al.: Tailor: a prompt-based approach to attribute-based controlled text generation. arXiv preprint arXiv:2204.13362 (2022)

28. Yao, L., Peng, N., Weischedel, R., Knight, K., Zhao, D., Yan, R.: Plan-and-write: towards better automatic storytelling. In: Proceedings of the AAAI Conference on Artificial Intelligence, vol. 33, pp. 7378–7385 (2019)

Learning Stable Nonlinear Dynamical System from One Demonstration

Yu Zhang[1,2], Lijun Han[1,2], Zirui Wang[3], Xiuze Xia[1,2], Houcheng Li[1,2], and Long Cheng[1,2(✉)]

[1] School of Artificial Intelligence, University of Chinese Academy of Sciences, Beijing 100049, China
[2] State Key Laboratory of Multimodal Artificial Intelligence Systems, Institute of Automation, Chinese Academy of Sciences, Beijing 100190, China
`Long.cheng@ia.ac.cn`
[3] Grainger College of Engineering, University of Illinois Urbana-Champaign, 1308 West Green Street, Urbana, IL 61801, USA

Abstract. Dynamic systems (DS) methods constitute one of the most commonly employed frameworks for Learning from Demonstration (LfD). The field of LfD aims to enable robots or other agents to learn new skills or behaviors by observing human demonstrations, and DS provide a powerful tool for modeling and reproducing such behaviors. Due to their ability to capture complex and nonlinear patterns of movement, DS have been successfully applied in robotics application. This paper presents a new learning from demonstration method by using the DS. The proposed method ensures that the learned systems achieve global asymptotic stability, a valuable property that guarantees the convergence of the system to an equilibrium point from any initial condition. The original trajectory is initially transformed to a higher-dimensional space and then subjected to a diffeomorphism transformation. This transformation maps the transformed trajectory forward to a straight line that converges towards the zero point. By deforming the trajectories in this way, the resulting system ensures global asymptotic stability for all generated trajectories.

Keywords: Learning from demonstrations · Dynamic systems · Global stability

1 Introduction

As technology continues to advance, robots are becoming increasingly prevalent in various industries and applications. To perform complex tasks comparable to those carried out by humans, robots require advanced intelligence that is difficult to program, particularly for individuals without specialized training or expertise in robotics. As a result, there is a growing demand for the development of more intelligent robots that can operate effectively in diverse environments and perform intricate tasks with minimal human intervention.

B. Luo et al. (Eds.): ICONIP 2023, LNCS 14450, pp. 471–482, 2024.
https://doi.org/10.1007/978-981-99-8070-3_36

Learning from Demonstration (LfD) is a user-friendly paradigm that enables robots to acquire new skills from human demonstrations [1]. Unlike traditional programming approaches that require explicit scripts or defined reward functions, LfD allows robots to learn from demonstrations of desired behavior provided by humans. This approach is particularly advantageous for individuals without specialized expertise in robotics or programming and has been successfully applied in numerous domains [2–5].

In many practical applications, the ability to achieve stability that means ensuring the convergence of learned skills towards a desired target state is crucial. Dynamic Systems (DS) provide a powerful tool for addressing this challenge [6]. By modeling the behavior of a system, DS methods can capture complex and nonlinear patterns of movement, enabling robots to acquire new skills while ensuring stability and safety. DS provides a flexible solution for modeling trajectories and generates real-time motion that is more stable compared to traditional methods, such as interpolation techniques. In DS, the target state of a task is typically encoded as a stable attractor that exhibits inherent robustness to perturbations [6]. This approach enables DS to plan stable motions from arbitrary starting points within a robot's workspace, facilitating easy generalization to previously unseen situations while maintaining stability and safety.

Dynamic Movement Primitives (DMP) represent one of the most widely used DS algorithms in robotics, with numerous successful applications in various domains [7–10]. DMPs employ a system of second-order ordinary differential equations (ODEs) to model the dynamics of a robot's motion. The desired motions are learned through a forcing term that encodes the target state as a stable attractor. DMP model motion as a spring-damper system, which exhibits inherent stability and robustness to perturbations. This property allows DMPs to suppress nonlinear perturbations and generate stable motions that converge to a desired target state [8]. However, DMPs have limited capacity for dealing with multi-dimensional data and can exhibit undesired generalization capabilities outside the demonstrated behaviors.

In contrast to DMPs, other DS approaches encode demonstration motions into a state-dependent DS that is independent of time. This approach enables greater robustness to temporal perturbations. The stable estimator of dynamical system (SEDS) was the first approach to consider stability in a state-dependent DS, which is introduced for learning motions using Gaussian mixture models (GMM) [6]. SEDS constrains the parameters of Gaussian mixture regression (GMR) to ensure global stability, using a quadratic Lyapunov function. However, researchers have identified that overly stringent stability constraints in SEDS may limit the accuracy of reproducing motion from demonstrations, resulting in less accurate outcomes. This trade-off between accuracy and stability is known as the "accuracy vs. stability dilemma" [11]. To address this issue, Several modified approaches have been proposed to improve the reproduction accuracy under the stability constraint. A neural network approach for estimating the Lyapunov function based on demonstrations is one example that reduces violations of the function and improves its accuracy [12]. The CLF-DM algorithm is another approach that uses regression techniques to model motion from demonstrations and ensures stability through constrained optimization [13]. However, this algorithm can be sensitive to parameters and complex. In [11], a diffeomorphic transformation (τ-SEDS) is used to project data into a different space, improving accuracy while preserving stability. By using reversible neural networks to learn the diffeomorphism map can achieves high accuracy [14,15], however, the method is time-consuming because satisfying performance requires implementing many layers in the reversible neural networks. In [2,16], transforming the original trajectories to a higher dimensional space for diffeomorphism transformation is found to be more effective than directly learning invertible transformation in the original space, however, training the neural networks is still time-consuming.

In this paper, a novel DS method is proposed, which first transforms the original trajectories to a higher dimensional space and then uses a non-parametric method with an appropriate structure to perform diffeomorphic match.

The paper is organized as follows: Sect. 2 discussed the problem of learning from demonstration with stability. In Sect. 3, the details of the proposed DS method are presented. In Sect. 4, the performance of the method is evaluated by the simulation on several handwriting trajectories and by experiments on the Franka Emika robot. In Sect. 5, the conclusion is made.

2   Problem Formulation

Point-to-point tasks performed by humans or robots can be modeled using an autonomous DS as:

$$\dot{x} = f(x), \ \forall x \in \mathbb{R}^d, \tag{1}$$

where $f : \mathbb{R}^d \mapsto \mathbb{R}^d$ is a nonlinear continuous and continuously differentiable function. The DS has a single equilibrium state x^* that represents the target state or attractor of the task. Without loss of generality, in this paper, the targets of the motions are placed at the origin of the Cartesian coordinate system by $\tilde{x} = x - x^*$. Since Eq. (1) is an autonomous DS that represents the manipulation skills acquired from demonstration data, a solution $\Phi(x_0, t)$ corresponds a determined trajectory generated by the learned skill with the initial state x_0. Then the actual motions can be calculated through $x = \tilde{x} + x^*$. Therefore, once the skill is learned, the DS can generate different solution trajectories to the target state x^* from different initial condition x_0.

The demonstrations are usually recorded as: $\{x_{k,n}, \dot{x}_{k,n}, \ddot{x}_{k,n}\}$ with $k = 1, 2, ..., K_n$; $n = 1, 2, ..., N_d$, where n represents the n-th demonstration and k represents the k-th sampling time. N_d and K_n $(n = 1, 2, ..., N_d)$ represent the total number of demonstrations and the total sampling number of the n-th demonstrated motion, respectively. The state variable x represents the position of the end-effector of a robot or the location of a human hand in Cartesian space generally, since most tasks are implemented in the Cartesian space. $\dot{x}$ and $\ddot{x}$ are velocity and acceleration, respectively. Moreover, all demonstrations for a given task have the same target.

Since the demonstrations are assumed to be generated by an autonomous DS as in (1), then after the data are acquired, the DS can be modeled as:

$$\dot{\hat{x}} = \hat{f}(x), \ \forall x \in \mathbb{R}^d. \tag{2}$$

where $\dot{\hat{x}}$ denotes that the estimate of $\dot{x}$, $\hat{f}$ is the model designed manually and learned from demonstration to approximate the DS in (1).

The learning process enforces the model to generate accurate reproductions compared to the demonstrations, therefore, the $\hat{f}$ can be acquired using any arbitrary regress algorithm, however, the convergence of the generated trajectories to the target cannot be guaranteed.

For convergence to be guaranteed, the autonomous DS must be stable. Specifically, the DS is locally asymptotically stable at the point x^* if there exists a continuous and continuously differentiable Lyapunov-candidate-function $V(x) : \mathbb{R}^d \rightarrow \mathbb{R}$ that satisfies the following conditions:

$$\begin{cases} (a)V(x^*) = 0, \\ (b)\dot{V}(x^*) = 0, \\ (c)V(x) > 0 : \forall x \neq x^*, \\ (d)\dot{V}(x) < 0 : \forall x \neq x^*. \end{cases} \tag{3}$$

Furthermore, if the radially unbounded condition as

$$\lim_{\|x\| \to +\infty} V(x) = +\infty, \tag{4}$$

is satisfied, the autonomous DS is globally asymptotically stable.

Ensuring stability in the DS learning process to reach the target state may limit model accuracy, as precise reproductions of demonstration data can violate the predefined Lyapunov candidate function. Additionally, deriving a proper Lyapunov candidate function from demonstration data is analytically challenging, making it difficult to find a satisfactory solution.

The demonstration is considered to violate the predefined Lyapunov candidate function if:

$$\exists k : 0 \leq k \leq K, \nabla^{\mathrm{T}} V(x_{k,n})\dot{x}_{k,n} > 0, \tag{5}$$

where ∇V is the gradient vector of V with respect to x.

Designing an ideal Lyapunov function is difficult, so to address the violation of the Lyapunov function with demonstration data, one approach is to transform the data into a different space of either the same or higher dimension. The transformed data can be designed to follow a predefined Lyapunov function. When the transformation is invertible, the stability is guaranteed.

3   Proposed Approach

Using a diffeomorphic transformation can be an effective strategy for circumventing the challenge of designing a Lyapunov function, which can be demanding for traditional dynamical systems methods. One such approach is the fast diffeo-morphic matching method [17], which aims to transform the original trajectory into a straight line from the starting point to the zero point by iteratively adjust-ing the furthest point between the two trajectories. However, determining the optimal number of iterations can be difficult, and the resulting modified trajec-tories may not be entirely smooth. To address this issue, using reversible neural networks for the transformation can be an elegant solution; however, it can be time-consuming. Alternatively, this section proposes a fast and two-step trans-formation method to handle these issues and learning from one demonstration. The proposed two-step transformation method involves transforming the original trajectory to a higher-dimensional space in the first step and then transforming the resulting trajectory into a straight line using a diffeomorphic transformation in the second step. Which are formulated as follows:

$$\begin{cases} p = xR, \\ y_{1:D/2} = p_{1:D/2} + f_1(p_{(D+2)/2:D}), \\ y_{(D+2)/2:D} = p_{(D+2)/2:D} + f_2(y_{1:D/2}), \end{cases} \quad (6)$$

where x is the d-dimensional input vector, $R \in \mathbb{R}^{d \times D}$ is a full column-rank matrix, $p \in \mathbb{R}^D$ is an intermediate variable, and $y \in \mathbb{R}^D$ is D-dimensional vector $(D > d$ is an even number) designed manually. Given that the proposed algorithm aims to learn a skill from a single demonstration, y can be designed as a straight line extending from the p_0 to the zero point, which is a differential equation as

$$\dot{y} = -y \quad (7)$$

To facilitate practical implementation in real experiments, the y can be refor-mulated to satisfy the following equations:

$$\begin{cases} \dot{y} = -\frac{1}{K-1}p_0 & \text{if } \langle y, y_0 \rangle > 0 \\ \dot{y} = 0 & \text{otherwise} \end{cases}, \quad (8)$$

where K is the total sampling number of the demonstration trajectory. p_0 is the first point of the trajectory p and $y_0 = p_0$. f_1 and f_2 can be designed with arbitrary structures but using a non-parametric method in this paper.

Considering the non-parametric method, $f(\cdot)$ can be modeled as:

$$f(\cdot) = \Theta(\cdot)^T w. \quad (9)$$

The f_1 and f_2 can be acquired by minimizing the objective functions, respec-tively:

$$\begin{cases} J(w_1) = \frac{1}{2} \sum_{m=1}^{K} \left\{ w_1^T \Theta(p_{(D+2)/2:D}^m) - y_{1:D/2}^m + p_{1:D/2}^m \right\}^2, \\ J(w_2) = \frac{1}{2} \sum_{n=1}^{K} \left\{ w_2^T \Theta(y_{1:D/2}^n) - y_{(D+2)/2:D}^n + p_{(D+2)/2:D}^n \right\}^2, \end{cases} \quad (10)$$

where the notation with the top corner marks m and n indicates the m-th and n-th point in the corresponding trajectory. In general, to mitigate overfitting, objective functions often include the terms $\frac{1}{2}\lambda w_1^T w_1$ and $\frac{1}{2}\lambda w_2^T w_2$, where λ is a positive constant designed manually. These additional terms serve the purpose of preventing excessive emphasis on the training data and promoting generalization.

The optimal solutions of the objective functions can been represented as [18]:

$$
\begin{cases}
w_1^* = \Theta_1(\Theta_1^T\Theta_1 + \lambda I)^{-1}(\boldsymbol{y}_{1:D/2} - \boldsymbol{p}_{1:D/2}), \\
w_2^* = \Theta_2(\Theta_2^T\Theta_2 + \lambda I)^{-1}(\boldsymbol{y}_{(D+2)/2:D} - \boldsymbol{p}_{(D+2)/2:D}),
\end{cases}
\tag{11}
$$

where $\Theta_1 = [\Theta(\boldsymbol{p}_{(D+2)/2:D}^1), \Theta(\boldsymbol{p}_{(D+2)/2:D}^2) \cdots, \Theta(\boldsymbol{p}_{(D+2)/2:D}^K)]$, almost the same setting $\Theta_2 = [\Theta(\boldsymbol{y}_{1:D/2}^1), \Theta(\boldsymbol{y}_{1:D/2}^2) \cdots, \Theta(\boldsymbol{y}_{1:D/2}^K)]$ and I is K dimensional identity matrix. Using the optimal w_1^* and w_2^*, the $f(\cdot)$ can be calculated as:

$$
\begin{cases}
f_1(\boldsymbol{s}) = \Theta_1(\boldsymbol{s})^T\Theta_1(\Theta_1^T\Theta_1 + \lambda\boldsymbol{I})^{-1}(\boldsymbol{y}_{1:D/2} - \boldsymbol{p}_{1:D/2}), \\
f_2(\boldsymbol{s}) = \Theta_2(\boldsymbol{s})^T\Theta_2(\Theta_2^T\Theta_2 + \lambda\boldsymbol{I})^{-1}(\boldsymbol{y}_{(D+2)/2:D} - \boldsymbol{p}_{(D+2)/2:D}),
\end{cases}
\tag{12}
$$

where $\boldsymbol{s}$ denotes the input that is a $D/2$-dimensional vector.

To eliminate the need for explicitly defining the basis functions in Eq. (11), the kernel trick is employed to represent the inner product as:

$$
k(\boldsymbol{s}_i, \boldsymbol{s}_j) = \Theta(\boldsymbol{s}_i)^T\Theta(\boldsymbol{s}_j) \quad \forall i, j \in [1, 2 \cdots, K].
\tag{13}
$$

Therefore, Eq. (12) becomes

$$
f(\boldsymbol{s}) = \boldsymbol{k}^*(\boldsymbol{K} + \lambda\boldsymbol{I})^{-1}\hat{\boldsymbol{y}},
\tag{14}
$$

where $\hat{\boldsymbol{y}}$ denotes $\boldsymbol{y}_{1:D/2} - \boldsymbol{p}_{1:D/2}$ for $f_1(\cdot)$ and $\boldsymbol{y}_{(D+2)/2:D} - \boldsymbol{p}_{(D+2)/2:D}$ for $f_2(\cdot)$,

$$
\boldsymbol{k}^* = [k(\boldsymbol{s}, \boldsymbol{s}_1), k(\boldsymbol{s}, \boldsymbol{s}_2) \cdots, k(\boldsymbol{s}, \boldsymbol{s}_K)],
\tag{15}
$$

and

$$
\boldsymbol{K} = \begin{bmatrix}
k(\boldsymbol{s}_1, \boldsymbol{s}_1) & k(\boldsymbol{s}_1, \boldsymbol{s}_2) & \cdots & k(\boldsymbol{s}_1, \boldsymbol{s}_K) \\
k(\boldsymbol{s}_2, \boldsymbol{s}_1) & k(\boldsymbol{s}_2, \boldsymbol{s}_2) & \cdots & k(\boldsymbol{s}_2, \boldsymbol{s}_K) \\
\vdots & \vdots & \ddots & \vdots \\
k(\boldsymbol{s}_K, \boldsymbol{s}_1) & k(\boldsymbol{s}_K, \boldsymbol{s}_2) & \cdots & k(\boldsymbol{s}_K, \boldsymbol{s}_K)
\end{bmatrix}.
\tag{16}
$$

Specifically, the kernel function is designed as:

$$
k(\boldsymbol{s}_i, \boldsymbol{s}_j) = \boldsymbol{s}_i^T \boldsymbol{s}_j e^{(-h\|(\boldsymbol{s}_i - \boldsymbol{s}_j)\|^2)},
\tag{17}
$$

where h is a manually designed positive constant.

Therefore, $k(\boldsymbol{s}, \boldsymbol{s}_j) = 0$ when the input $\boldsymbol{s} = \boldsymbol{0}$. After learning from the demonstration using the proposed algorithm, and given a new starting point $\boldsymbol{x}_{ns}$, the corresponding $\boldsymbol{y}_{ns}$ can be calculated. From $\boldsymbol{y}_{ns}$ to the zero point, a rectilinear trajectory $\boldsymbol{y}^{new}$ can be generated. The corresponding trajectory $\boldsymbol{x}^{new}$ in the original space can be calculated using the following equations:

$$
\begin{cases}
\boldsymbol{p}_{(D+2)/2:D}^{new} = \boldsymbol{y}_{(D+2)/2:D}^{new} - f_2(\boldsymbol{y}_{1:D/2}^{new}), \\
\boldsymbol{p}_{1:D/2}^{new} = \boldsymbol{y}_{1:D/2}^{new} - f_1(\boldsymbol{p}_{(D+2)/2:D}^{new}), \\
\boldsymbol{x}^{new} = \boldsymbol{p}^{new}\boldsymbol{R}^T(\boldsymbol{R}\boldsymbol{R}^T)^{-1},
\end{cases}
\tag{18}
$$

Fig. 1. The simulation of the vector fields of the dynamics learned from the LASA dataset using the proposed algorithm. The white dotted lines represent the demonstration data and the red solid lines represent the reproductions with the same initial points. (Color figure online)

It can be easily prove that when $\boldsymbol{y}_{end}^{new} = \boldsymbol{0}$, $\boldsymbol{p}_{end}^{new} = \boldsymbol{0}$ by using the designed kernel function. Then $\boldsymbol{x}_{end}^{new} = \boldsymbol{0}$. Since $\boldsymbol{y}$ can be transformed into $\boldsymbol{p}$ through a diffeomorphism transformation as $\boldsymbol{y} = g(\boldsymbol{p})$, then $\boldsymbol{p} = g^{-1}(\boldsymbol{y})$ and they exhibit the same level of stability.

Considering the following Lyapunov-candidate function

$$V(\boldsymbol{p}) = \frac{1}{2}g(\boldsymbol{p})^{\mathrm{T}}g(\boldsymbol{p}), \tag{19}$$

which is continuous and continuously differentiable. Then

$$\dot{V}(\boldsymbol{p}) = \boldsymbol{y}^{\mathrm{T}}\frac{\partial \boldsymbol{y}}{\partial \boldsymbol{p}}\frac{\partial \boldsymbol{p}}{\partial \boldsymbol{y}}\dot{\boldsymbol{y}} = -\boldsymbol{y}^{\mathrm{T}}\boldsymbol{y}. \tag{20}$$

Hence, conditions (3-a) to (3-d) are met, ensuring the fulfillment of the requirements for stability. Consequently, the trajectory vector $\boldsymbol{p}$ will converge to the zero point, while the vector $\boldsymbol{x}$ is obtained through a linear transformation of $\boldsymbol{p}$. As a result, all elements of $\boldsymbol{x}$ will also converge to the zero point, representing the desired target point.

Specifically, once the model is constructed, new trajectories can be generated by altering the rules governing the variations in the vector $\boldsymbol{y}$ while preserving the underlying shape. This allows for the generation of trajectories in the original space that retain the learned shape characteristics but exhibit different velocities. In essence, the proposed algorithm enables the effortless generation of time-driven trajectories or dynamic system trajectories as required. In scenarios where the robot operates independently, generating time-driven trajectories can be advantageous as they can be tailored to meet specific requirements. However, in human-robot interaction tasks, it is often more appropriate to generate trajectories using a dynamic system approach. This is because the real trajectories exhibited by humans may differ from the demonstrated ones, and imposing time-driven trajectories on humans can impede their natural movement. By utilizing a dynamic system, the robot can adapt its trajectories to better align with human behavior.

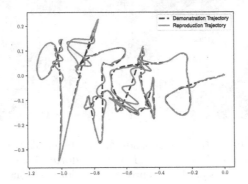

Fig. 2. The visualization of reproduction of the trajectory "nihao" by directly using the proposed algorithm. Blue dotted line denotes the demonstrate trajectory, while the reproduction trajectory is demonstrated by orange solid line. (Color figure online)

4   Experiment Results and Discussions

In order to assess the efficacy of the proposed algorithm, a comprehensive evaluation was conducted through a combination of simulations and real experiments. The simulation was implemented using Python 3.8, while the real-world experiment was conducted using the Franka Emika robot and implemented in C++ 14. The software was executed on an Intel Core i5-8300H CPU, running Ubuntu 20.04 as the operating system. The findings obtained from these assessments are summarized as follows:

4.1   Learning from One Demonstration

To facilitate the learning process, the proposed algorithm incorporates the ability to acquire knowledge from demonstrations. The DMPs [8] is also a learning from demonstration algorithm, but DMPs lack the capability to generate vector fields directly from the demonstration, the proposed algorithm overcomes this limitation. It possesses the unique ability to generate vector fields based on the provided demonstration. In this study, handwritten drawing trajectories sourced from the LASA dataset [19] are employed as demonstrations. In the simulation, the matrix R is randomly generated from a Gaussian distribution, the manually designed hyper-parameters D is set as 10, λ is set as 0.0001, and h is set as 20. To

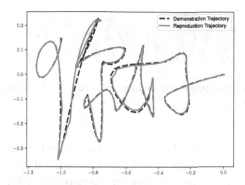

Fig. 3. The visualization of reproduction of the trajectory "nihao" by the proposed algorithm with a smoothing algorithm. Blue dotted line denotes the demonstrate trajectory, while the reproduction trajectory is demonstrated by orange solid line. (Color figure online)

expedite the computational process, the numba [20] library is utilized for accelerated computing. Figure 1 illustrates the results obtained from the proposed algorithm when a demonstration is randomly selected from the LASA dataset. The algorithm effectively generates the vector fields based on the chosen demonstration. Furthermore, it is worth noting that the generated trajectories exhibit a remarkable convergence towards the target point. This demonstrates the effectiveness of the proposed algorithm in accurately guiding the trajectories towards the desired goal.

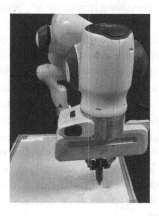

Fig. 4. The visualization of the experimental platform, where the franka robot controls the pen to write the handwriting on the whiteboard.

Unlike traditional dynamical systems algorithms that learn in the original space, where trajectories generated from different starting points never intersect,

Fig. 5. The results of the proposed algorithm in which the Franka robot writing the learned complex motion.

the trajectories generated by the proposed algorithm can intersect. This feature distinguishes the proposed algorithm from conventional approaches, allowing for more flexibility and potential in trajectory generation. Moreover, the ability of the proposed algorithm to generate intersecting trajectories can lead to results that are more human-like. In human handwriting, it is common for the trajectories of the same handwriting to intersect when starting from different points. By replicating this behavior, the proposed algorithm captures a more realistic and natural aspect of human-like trajectory generation.

An additional noteworthy capability of the proposed algorithm is its capacity to learn complex trajectories, such as Chinese characters, from a single demonstration. This highlights the algorithm's ability to capture and reproduce intricate and intricate movements, making it suitable for tasks that involve complex trajectory generation and replication, since the generated trajectories can intersect.

Figure 2 illustrates the proposed algorithm employed to learn the intricate handwriting pattern of the word "nihao" with setting $D = 120$. It is apparent to seen that utilizing the proposed algorithm directly to learn the complex motion does not yield a satisfactory trajectory when compared to the demonstration's trajectory originating from the same initial point. Since the generated trajectory by the proposed algorithm is not smooth.

The unsatisfactory performance stems from the requirement of transforming the trajectory in a high-dimensional space to align with a straight line connecting the starting point to the zero point. The unsatisfactory performance arises from the fact that while each point on the straight line is unique, certain points in the transformed trajectory coincide. This indicates that the proposed algorithm attempts to map multiple points to a single point using a diffeomorphic transformation. Such a mapping is inherently impossible, leading to a compromised result that represents a trade-off among the multiple points. Therefore, using the proposed algorithm directly cannot generate a trajectory totally the same as the demonstrate complex motion, but it can be seen from Fig. 2 that the generated trajectory retains the main features of the demonstration and also converge to the target point.

Hence, the unsatisfactory performance primarily manifests in the lack of smoothness. To address this issue, a smooth algorithm can be applied to the generated trajectory. In this study, the Savitzky-Golay filter is applied to the generated trajectory, and the corresponding result is presented in Fig. 3. The results depicted in Fig. 3 illustrate that the combination of the proposed algorithm with a smoothing algorithm effectively reproduces the complex motion, which is not easy for the other DS algorithms.

4.2 Validation on Robot

Experiments were conducted to validate the proposed algorithm on the Franka Emika robot. The demonstration trajectory was obtained by a person using a tablet, and subsequently, the robot reproduced the trajectory by writing it on a whiteboard using a pen. The experiment platform is shown in Fig. 4.

Over time, the proposed algorithm calculates a reference point for writing the word "nihao". The torque applied to the end-effector is determined using high stiffness and critical damping. Subsequently, the control torque for each joint is computed by multiplying the transposed Jacobian matrix. To simplify the implementation, the end-effector of the robot is constrained to a vertical orientation relative to the whiteboard, and the pose does not need to be explicitly considered. As the proposed algorithm continuously generates a reference point towards the target, the controller endeavors to track this reference point and converge towards the target location (Fig. 5).

5   Conclusions

This paper introduces a novel DS algorithm for learning from a single demonstration. The performance of the proposed algorithm is evaluated through simulations involving various handwriting examples, as well as experiments conducted with a real robot. The experimental findings provide evidence of the effectiveness of the proposed method.

However, it is important to note that the current implementation of the proposed algorithm relies on the kernel trick, which may not be well-suited for large datasets. This limitation can be addressed and improved upon in future studies.

Acknowledgements. This work is supported by National Key Research and Development Program (Grant No. 2022YFB4703204), National Natural Science Foundation of China (Grant No. 62311530097), and Chinese Academy of Sciences Project for Young Scientists in Basic Research (Grant No. YSBR-034).

References

1. Ravichandar, H., Polydoros, A.S., Chernova, S.: Recent advances in robot learning from demonstration. Ann. Rev. Control Robot. Auton. Syst. **3**(1), 297–330 (2020)
2. Zhang, Y., Cheng, L., Li, H., Cao, R.: Learning accurate and stable point-to-point motions: a dynamic system approach. IEEE Robot. Autom. Lett. **7**(2), 1510–1517 (2022)
3. Liu, N., Lu, T., Y, Cai., Wang, R., Wang, S.: Manipulation skill learning on multi-step complex task based on explicit and implicit curriculum learning. Sci. China Inf. Sci. **65**, 114201 (2022)

4. Rozo, L., Calinon, S., Caldwell, D.G., Jimenez, P., Torras, C.: Learning physical collaborative robot behaviors from human demonstrations. IEEE Trans. Robot. **32**(3), 513–527 (2016)
5. Zhang, Y., Cheng, L.: Online adaptive and attention-based reference path generation for upper-limb rehabilitation robot. In: China Automation Congress, pp. 5268–5273, Beijing, China (2021)
6. Khansari-Zadeh, S.M., Billard, A.: Learning stable nonlinear dynamical systems with Gaussian mixture models. IEEE Trans. Robot. **27**(5), 943–957 (2011)
7. Kong, L.H., He, W., Chen, W.S., Zhang, H., Wang, Y.: Dynamic movement primitives based robot skills learning. Mach. Intell. Res. **20**, 396–407 (2023)
8. Ijspeert, A.J., Nakanishi, J., Hoffmann, H., Pastor, P., Schaal, S.: Dynamical movement primitives: learning attractor models for motor behaviors. Neural Comput. **25**(2), 328–373 (2013)
9. Ginesi, M., Sansonetto, N., Fiorini, P.: Overcoming some drawbacks of dynamic movement primitives. Robot. Auton. Syst. **144**, 103844 (2021)
10. Li, C., Li, Z., Jiang, Z., Cui, S., Liu, H., Cai, H.: Autonomous planning and control strategy for space manipulators with dynamics uncertainty based on learning from demonstrations. Sci. China Technol. Sci. **64**, 2662–2675 (2021)
11. Neumann, K., Steil, J.J.: Learning robot motions with stable dynamical systems under diffeomorphic transformations. Robot. Auton. Syst. **70**, 1–15 (2015)
12. Neumann, K., Lemme, A., Steil, J.J.: Neural learning of stable dynamical systems based on data-driven Lyapunov candidates. In: IEEE International Conference on Intelligent Robots and Systems, pp. 1216–1222, Tokyo, Japan (2013)
13. Khansari-Zadeh, S.M., Billard, A.: Learning control Lyapunov function to ensure stability of dynamical system-based robot reaching motions. Robot. Auton. Syst. **62**(6), 752–765 (2014)
14. Rana, M.A., Li, A., Fox, D., Boots, B., Ramos, F., Ratliff, N.: Euclideanizing flows: diffeomorphic reduction for learning stable dynamical systems. In: 2th International Proceedings on the Conference on Learning for Dynamics and Control, pp. 630–639, Online (2020)
15. Zhang, Y., Cheng, L., Cao, R., Li, H., Yang, C.: A neural network based framework for variable impedance skills learning from demonstrations. Robot. Auton. Syst. **160**, 104312 (2023)
16. Zhang, H., Cheng, L., Zhang, Y.: Learning robust point-to-point motions adversarially: a stochastic differential equation approach. IEEE Robot. Autom. Lett. **8**(4), 2357–2364 (2023)
17. Perrin, N., Schlehuber-Caissier, P.: Fast diffeomorphic matching to learn globally asymptotically stable nonlinear dynamical systems. Syst. Control Lett. **96**(6), 51–59 (2016)
18. Bishop, C.M.: Pattern Recognition and Machine Learning, 5th edn. Information Science and Statistics, Chapter 6, pp. 291–294. Springer, New York (2006)
19. Homepage. https://bitbucket.org/khansari/lasahandwritingdataset/. Accessed 25 Mar 2015
20. Lam, S.K., Pitrou, A., Seibert, S.: Numba: a LLVM-based Python JIT compiler. In: International Proceedings on the Second Workshop on the LLVM Compiler Infrastructure in HPC, pp. 1–6, New York, USA (2015)

Towards High-Performance Exploratory Data Analysis (EDA) via Stable Equilibrium Point

Yuxuan Song(ID) and Yongyu Wang$^{(\boxtimes)}$(ID)

JD Logistics, Beijing 101111, China
wangyongyu1@jd.com

Abstract. Exploratory data analysis (EDA) is a vital procedure in data science projects. In this work, we introduce a stable equilibrium point (SEP)-based framework for improving the performance of EDA. By exploiting the SEPs to be the representative points, our approach aims to generate high-quality clustering and data visualization for real-world data sets. A very unique property of the proposed method is that the SEPs will directly encode the clustering properties of data sets. Compared with prior state-of-the-art clustering and data visualization methods, the proposed methods allow substantially improving solution quality for large-scale data analysis tasks. For instance, for the USPS data set, our method achieves more than 10% clustering accuracy gain over the standard spectral clustering algorithm and 3X speedup for the t-SNE visualization.

Keywords: Exploratory Data Analysis (EDA) · Stable Equilibrium Point · Clustering · Data Visualization

1 Introduction

Exploratory data analysis (EDA) is the procedure of analyzing data sets to gain insights of their characteristics. It is usually the first step for big data analysis. Clustering and data visualization are two main pillars of EDA.

Clustering aims to assign similar data samples into same cluster while assigning dissimilar data samples into different clusters. Among various clustering methods, spectral clustering has drawn increasing attention in recent years because it surpasses conventional clustering methods in detecting non-convex and linearly non-separable patterns. However, spectral clustering is very sensitive to noisy samples and misleading relations among samples. For example, if a few noisy samples form a narrow bridge between ground-truth clusters, spectral clustering tends to group the two clusters plus the noisy samples together

In the past decades, many methods are proposed for improving the performance of spectral clustering: [1] proposed to generate a compressed data set

---

Y. Wang and Y. Song—Contributed equally and are co-first authors.

© The Author(s), under exclusive license to Springer Nature Singapore Pte Ltd. 2024
B. Luo et al. (Eds.): ICONIP 2023, LNCS 14450, pp. 483–492, 2024.
https://doi.org/10.1007/978-981-99-8070-3_37

composed of a set of k-means centroids and then apply spectral clustering on the compressed data set; [2] attempted to use a reduced matrix to approximate the original data via sampling. Inspired by sparse coding, [3] proposed to perform spectral embedding with linear combinations of landmark points. However, these methods cannot truthfully encode the cluster properties of the original data set so that they cannot robustly preserve the solution quality.

For the data visualization task, t-distributed stochastic neighbor embedding (t-SNE) [5] is the most widely used tool for visualizing high-dimensional data. It aims to learn an embedding from high-dimensional space to two or three dimensional space in such a way that similar samples are modeled by nearby points and dissimilar samples are modeled by distant points in the low-dimensional space so that people can directly view the structure of data sets. However, its involved stochastic neighbor embedding (SNE) process requires to perform gradient descent to minimize the Kullback-Leibler divergence-based cost function for every sample which is a very time-consuming process, especially for large-scale data sets.

To solve the above problems, in this paper, we propose a stable equilibrium point (SEP)-based framework for improving the performance of the above two main EDA tasks. Experimental results show that our method outperforms both the baseline and state-of-the-art methods by a large margin.

2   Preliminary

2.1   Spectral Clustering Algorithm

Spectral clustering [6] can often outperform traditional clustering algorithms, such as k-means algorithms. As shown in Algorithm 1, spectral clustering algorithm can be divided into three main steps: 1) construct a graph according to the entire data set, 2) embed all data points into k-dimensional space using the eigenvectors of the bottom k nonzero eigenvalues of the graph Laplacian, and 3) perform k-means algorithm to partition the embedded data points into clusters. Although spectral clustering has many advantages such as rigorous theoretical foundation and easy implementation, it is highly sensitive to noisy samples and misleading relations among samples.

2.2   t-Distributed Stochastic Neighbor Embedding

t-Distributed Stochastic Neighbor Embedding (t-SNE) [5] maps data points from high-dimensional space to two or three dimensional space in such a way that similar data points are located in nearby places while dissimilar points are located in distant places. It includes the following main steps:

1) For a pair of data points x_i and x_j, it first converts their euclidean distance into conditional probability as follows:

Algorithm 1 Spectral Clustering Algorithm

Input: A graph $G = (V, E, w)$ and the number of clusters k.
Output: Clusters $C_1...C_k$.

1: Compute the adjacency matrix A_G, and diagonal matrix D_G;
2: Obtain the Laplacian matrix $L_G = D_G - A_G$;
3: Compute the eigenvectors $u_1,...u_k$ that correspond to the bottom k nonzero eigenvalues of L_G;
4: Construct matrix U with k eigenvectors of L_G stored as columns;
5: Perform k-means algorithm to partition the rows of U into k clusters and return the result.

$$P_{j|i} = \frac{exp(-\frac{\|x_i - x_j\|^2}{2\sigma_i^2})}{\sum_{k \neq i} exp(-\frac{\|x_i - x_j\|^2}{2\sigma_i^2})}, \tag{1}$$

$$P_{i|i} = 0. \tag{2}$$

where σ_i denotes the variance of the Gaussian distribution that is centered at x_i. Then, the joint probability is defined as follows:

$$P_{ij} = \frac{P_{j|i} + P_{i|j}}{2N}. \tag{3}$$

where N is the number of data points in the data set.

2) Assume that y_i and y_j are two points in the low-dimensional space corresponding to x_i and x_j in the original feature space, respectively. The similarity between y_i and y_j is calculated as follows:

$$q_{ij} = \frac{(1 + \|y_i - y_j\|^2)^{-1}}{\sum_{k \neq l}(1 + \|y_k - y_l\|^2)^{-1}}, \tag{4}$$

$$q_{ii} = 0. \tag{5}$$

3) t-SNE uses the sum of Kullback-Leibler divergence over all pairs of data points as the cost function of the dimensionality reduction:

$$C = KL(P \parallel Q) = \sum_{i \neq j} p_{ij} \log \frac{p_{ij}}{q_{ij}}. \tag{6}$$

For each point, its corresponding point in low-dimensional space is determined by performing gradient descent associated with the above cost function. For a data set with N points, the time complexity of t-SNE is $O(N^2)$.

3   Methods

Inspired by the support vector approach [7], we first map the data from the original feature space to a high-dimensional space with a nonlinear transformation Φ. Then we find a hyper-sphere with the smallest radius that can enclose all data points by solving the following problem:

$$\|\Phi(\mathbf{x}_j) - \mathbf{a}\|^2 \leq R^2 + \xi_j, \tag{7}$$

where R is the radius of the hyper-sphere, $\mathbf{a}$ denotes the center of the hyper-sphere and $\xi_j \geq 0$ denotes the slack variable. The squared radial distance of the data point $\mathbf{x}$ from the hyper-sphere center is:

$$\begin{aligned} f(\mathbf{x}) = R^2(\mathbf{x}) &= \|\Phi(\mathbf{x}) - \mathbf{a}\|^2 \\ &= K(\mathbf{x}, \mathbf{x}) - 2\sum_j K(\mathbf{x}_j, \mathbf{x}_j)\beta_j + \sum_{i,j} \beta_i \beta_j K(\mathbf{x}_i, \mathbf{x}_j) \end{aligned} \tag{8}$$

where $K(\mathbf{x}_i, \mathbf{x}_j) = e^{-q\|\mathbf{x}_i - \mathbf{x}_j\|^2}$ is the Mercer kernel, β_i and β_j are Lagrangian multipliers.

One of the most distinguished properties of this hyper-sphere is that when it is mapped back to the original data space, a set of contours can be formed and each of them encloses a set of similar data points [7]. However, these contours cannot be explicitly discovered. By performing the following gradient descent associated with $f(\mathbf{x})$, data samples will converge to a set of convergence points. These convergence points are called stable equilibrium points (SEPs).

$$\frac{d\mathbf{x}}{dt} = -\nabla f(\mathbf{x}). \tag{9}$$

[8] demonstrated that apart from outliers, samples that converge to the same SEP belong to the same contour. Obviously, a SEP can be used to represent a set of similar data points. Compared with other representative point selection schemes such as k-means-based method, SEP is more suitable for the task of discovering clusters. For example, in the renowned two-moons data set, there are two slightly entangled non-convex shapes. Each cluster corresponds to a moon. As shown in Fig. 1, standard k-means cannot correctly cluster many data points. But as shown in Fig. 2, by leveraging a few SEPs to capture the global cluster structure, satisfactory clustering results can be produced. Note that the red marks in Fig. 2 are real SEPs. For the two-moons data set with 400 samples, only 12 SEPs is enough for capturing the cluster structure of the data set. So in this paper, we use SEPs for problem size reduction and denoising.

However, calculating SEPs by performing gradient descent on every point can be a time consuming step when number of data points is large. To reduce the

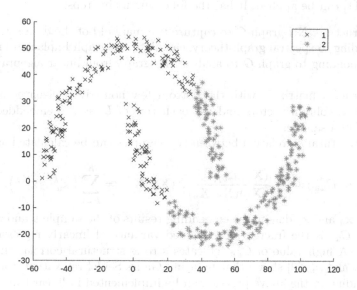

Fig. 1. Clustering result of two-moons data set with standard k-means.

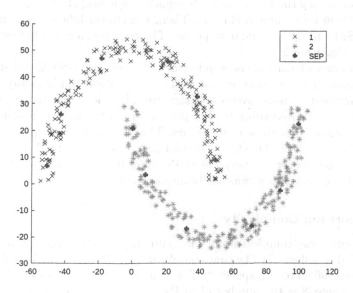

Fig. 2. Clustering result of two-moons data set with SEPs.

computational cost, a novel spectral node aggregation-based fast SEP calculation scheme [11] can be applied. It has the following main steps:

1) Construct a k-NN graph G to capture the manifold of the data set.
2) According to spectral graph theory, calculate the graph Laplacian matrix L_G corresponding to graph G to analyze the graph from linear algebra perspective.
3) Construct a matrix U with the bottom few non-trivial eigenvectors of L_G stored as column vectors and use each row of U as the embedded feature vector of a sample;
4) The structural correlation between two samples can be calculated as follows:

$$C_{uv} := \frac{|(\mathbf{X}_u, \mathbf{X}_v)|^2}{(\mathbf{X}_u, \mathbf{X}_u)(\mathbf{X}_v, \mathbf{X}_v)}, \quad (\mathbf{X}_u, \mathbf{X}_v) := \sum_{k=1}^{K} \left(\mathbf{x}_u^{(k)} \cdot \mathbf{x}_v^{(k)} \right). \quad (10)$$

where $\mathbf{x}_u$ and $\mathbf{x}_v$ denote the embedding results of the sample u and v, respectively. C_{uv} is the fraction of explained variance of linearly regressing of x_u on x_v. A high value of C_{uv} indicates strong structural correlation between the x_u and x_v. [9] shows that by using Gauss-Seidel relaxation, the spectral embedding in the above process can be implemented in linear time.

Based on the structural correlation results, highly-correlated samples can be merged to reduce the size of the data set with guaranteed preservation of the data structure. As a result, the searching space for SEP searching can be dramatically shrank while providing satisfactory SEPs.

After obtaining the SEPs, we apply standard spectral clustering on the SEPs to divide them into different clusters. Then, the cluster labels are mapped back from the SEPs to the original data points. The complete algorithm flow has been shown in Algorithm 2.

This framework can also be applied for achieving fast t-SNE. [10] shows that the dimensionality reduction process in the t-SNE algorithm is closely related to the bottom non-trivial eigenvectors of the graph Laplacian matrix corresponding to the data graph. According to the spectral graph theory, the manifold of the data set is encoded in these eigenvectors. This motivates us to propose a SEP-based t-SNE algorithm. Due to the special properties of the SEP, visualizing the representative data set composed of SEPs using the t-SNE will produce similar visualization result with dramatically improved efficiency.

3.1   Algorithm Complexity

The computational complexities of the main steps of the proposed methods are summarized as follows: 1) The time complexity of SEP searching is $O(P^2)$, where P is the size of searching space; 2) The time complexity of spectral clustering is $O(S^3)$, where S is the number of SEPs; 3) The time complexity of t-SNE is $O(S^2)$, where S is the number of SEPs; 4) The spectral node aggregation scheme takes $O(|E_G|log(|V|))$ time, where $|E_G|$ is the number of edges in the original graph and $|V|$ is the number of data points.

Algorithm 2 SEP-based Spectral Clustering

Input: A data set D with N samples $x_1, ..., x_N \in R^d$, number of clusters k.
Output: Clusters $C_1, ..., C_k$.

Map the data from the original space into a high-dimensional space to find the minimal enclosing hyper-sphere;
Calculate the squared radial distance $f(\mathbf{x})$ of data point $\mathbf{x}$ from the hyper-sphere center;
Search for SEPs by performing gradient descent associated with $f(\mathbf{x})$;
Perform spectral clustering to divide SEPs into k clusters;
Assign the cluster label of each SEP to all of its associated original data points.

4 Experiment

The spectral clustering and t-SNE are performed using MATLAB running on Laptop. The implementation of the compared methods can be found on their authors' website.

4.1 Data Sets

We evaluate the proposed method with the following real-world data sets:

1) The **Pendigits** data set: it contains 7,494 samples. Each sample is represented by 16 integer attributes. There are 10 clusters in this data set.

2) The **USPS** data set: it contains 9,298 16Œ16 pixel grayscale images of hand written digits scanned from envelopes of the U.S. Postal Service. There are 10 clusters in this data set.

The above benchmark data sets can be downloaded from the UCI machine learning repository.

4.2 Compared Methods

We compare the proposed method against the following baseline and state-of-the-art methods to demonstrate its effectiveness as in [4]:

(1) Standard spectral clustering algorithm [6],
(2) Nyström-based spectral clustering method [2],
(3) Landmark-based spectral clustering (LSC) method using random sampling for landmark selection [3],
(4) KASP method using k-means for coarse-level clustering [1].

For fair comparison, the parameters are set as in [3] for compared algorithms.

Evaluation Metrics. Clustering accuracy (ACC) [12] is used for measuring the clustering quality. It is defined as follows:

$$ACC = \frac{\sum_{i=1}^{N} \delta(y_i, map(c_i))}{N}, \tag{11}$$

where N denotes number of samples in the data set, y_i and c_i denote the ground-truth clustering result and algorithm generated clustering result for the i-th sample, respectively. $\delta(x,y)$ is a delta function defined as: $\delta(x,y)=1$ for $x = y$, and $\delta(x,y)=0$, otherwise. map is a permutation function that maps the index used by clustering algorithm to the index used in the ground truth labels. The clustering accuracy is defined as the optimal match among all possible permutations. Hungarian algorithm [13] is adopted for finding the optimal matching. A higher value of ACC indicates better clustering quality.

4.3   Experimental Results of Spectral Clustering

Table 1. Spectral clustering accuracy (%)

Data Set	Standard SC	Nyström	KASP	LSC	Our Method
PenDigits	74.36	71.99	71.56	74.25	**76.87**
USPS	64.31	69.31	70.62	66.28	**75.78**

To demonstrate the effectiveness of the proposed method, we show the clustering accuracy results of the proposed method and the compared methods in Table 1. It can be seen that for the USPS data set, our method leads to more than 10% accuracy gain over the standard spectral clustering algorithm and more than 5% gain over the second-best methods. Such a significant clustering accuracy improvement is mainly due to SEP's strong capability of representing the cluster structure of the data set. For the Pendigits data set, its feature space is only 16-dimension so that the information about features is very limited. Only our method can improve its clustering accuracy. In contrast, without robust preservation of the cluster properties, all the other methods lead to accuracy degradation for this low-dimensional data set.

Our method can also help to reduce the computational cost of eigen-decomposition. The eigen-decompositon step in spectral clustering has a $O(N^3)$ time complexity, where N is the number of data points to be clustered. For the Pengits and the USPS data set, our method reduces their problem size for 70% and 60%, respectively. As a result, for the low-dimensional pendigits data set, we achieve 2.6X speedup for its eigen-decomposition step and for the high-dimensional data set USPS, we achieve 4.8X speedup for eigen-decomposition.

4.4 Experimental Results of t-SNE

Figure 3 shows the visualization result generated by the standard t-SNE for the USPS data set. It takes 77 s. Figure 4 shows the visualization result generated by the SEP-based t-SNE algorithm for the same data set. The SEP-based t-SNE only takes 25 s, achieving 3X speedup while displaying clearly cluster structure.

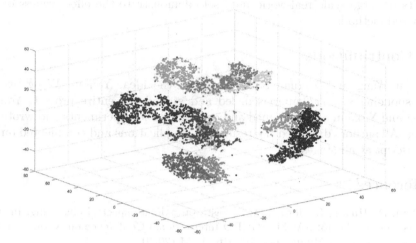

Fig. 3. Standard t-SNE for the USPS data set (t-SNE time: 77 s) .

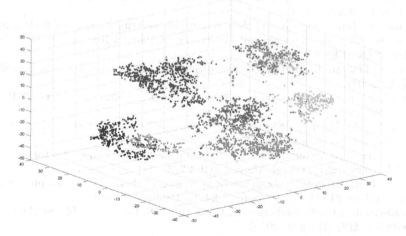

Fig. 4. SEP-based t-SNE for the USPS data set (t-SNE time: 25 s).

5 Conclusion

We propose a novel stable equilibrium point (SEP) - based algorithmic framework for improving the performance of two main tasks (clustering and data visualization) in exploratory data analysis (EDA). Our method enables to significantly improve the clustering accuracy for real-world data sets and fundamentally address the computational challenge of t-SNE visualization. Experimental results on large scale real-world data sets demonstrate the effectiveness of the proposed methods.

6 Contributions

Yongyu Wang and Yuxuan Song conceived the idea. Yongyu Wang, as the corresponding author, supervised, led and guided the entire project. Yongyu Wang and Yuxuan Song designed and conducted the experiments and wrote the paper. All authors discussed the results and implications and commented on the manuscript at all stages.

References

1. Yan, D., Huang, L., Jordan, M.I.: Fast approximate spectral clustering. In: Proceedings of the 15th ACM SIGKDD International Conference on Knowledge Discovery and Data Mining, pp. 907–916. ACM (2009)
2. Fowlkes, C., Belongie, S., Chung, F., Malik, J.: Spectral grouping using the Nystrom method. IEEE Trans. Pattern Anal. Mach. Intell. **26**(2), 214–225 (2004)
3. Chen, X., Cai, D.: Large scale spectral clustering with landmark-based representation. In: AAAI (2011)
4. Wang, Y.: Improving spectral clustering using spectrum-preserving node aggregation. In: 2022 26th International Conference on Pattern Recognition (ICPR), pp. 3063–3068. IEEE (2022)
5. Van der Maaten, L., Hinton, G.: Visualizing data using t-sne. J. Mach. Learn. Res. **9**(11) (2008)
6. Von Luxburg, U.: A tutorial on spectral clustering. Statist. Comput. **17**(4), 395–416 (2007)
7. Ben-Hur, A., Horn, D., Siegelmann, H.T., Vapnik, V.: Support vector clustering. J. Mach. Learn. Res. **2**, 125–137 (2001)
8. Lee, J., Lee, D.: An improved cluster labeling method for support vector clustering. IEEE Trans. Pattern Anal. Mach. Intell. **27**(3), 461–464 (2005)
9. Livne, O.E., Brandt, A.: Lean algebraic multigrid (LAMG): fast graph Laplacian linear solver. SIAM J. Sci. Comput. **34**(4), B499–B522 (2012)
10. Linderman, G.C., Steinerberger, S.: Clustering with t-sne, provably. SIAM J. Math. Data Sci. **1**(2), 313–332 (2019)
11. Song, Y., Wang, Y.: Accelerate support vector clustering via spectrum-preserving data compression?' arXiv:2304.09868 (2023)
12. Chen, W.-Y., Song, Y., Bai, H., Lin, C.-J., Chang, E.Y.: Parallel spectral clustering in distributed systems. IEEE Trans. Pattern Anal. Mach. Intell. **33**(3), 568–586 (2011)
13. Papadimitriou, C.H., Steiglitz, K.: Combinatorial Optimization: Algorithms and Complexity. Courier Corporation (1998)

MVFAN: Multi-view Feature Assisted Network for 4D Radar Object Detection

Qiao Yan🆔 and Yihan Wang$^{(\boxtimes)}$🆔

Nanyang Technological University, Singapore 639798, Singapore
{QIAO003,WANG1517}@e.ntu.edu.sg

Abstract. 4D radar is recognized for its resilience and cost-effectiveness under adverse weather conditions, thus playing a pivotal role in autonomous driving. While cameras and LiDAR are typically the primary sensors used in perception modules for autonomous vehicles, radar serves as a valuable supplementary sensor. Unlike LiDAR and cameras, radar remains unimpaired by harsh weather conditions, thereby offering a dependable alternative in challenging environments. Developing radar-based 3D object detection not only augments the competency of autonomous vehicles but also provides economic benefits. In response, we propose the Multi-View Feature Assisted Network ($MVFAN$), an end-to-end, anchor-free, and single-stage framework for 4D-radar-based 3D object detection for autonomous vehicles. We tackle the issue of insufficient feature utilization by introducing a novel Position Map Generation module to enhance feature leareweighing foreground and background points, and their features, considering the irregular distribution of radar point clouds. Additionally, we propose a pioneering backbone, the Radar Feature Assisted backbone, explicitly crafted to fully exploit the valuable Doppler velocity and reflectivity data provided by the 4D radar sensor. Comprehensive experiments and ablation studies carried out on Astyx and VoD datasets attest to the efficacy of our framework. The incorporation of Doppler velocity and RCS reflectivity dramatically improves the detection performance for small moving objects such as pedestrians and cyclists. Consequently, our approach culminates in a highly optimized 4D-radar-based 3D object detection capability for autonomous driving systems, setting a new standard in the field.

Keywords: 4D Radar · 3D Object Detection · Multi-view Fusion

1 Introduction

A robust autonomous driving system consists of several modules, such as environment perception, path planning, decision-making, and control. Perception is the most critical module, as it informs the others. Traditionally, perception relies on the RGB camera, LiDAR, and radar, but camera and LiDAR struggle in adverse weather conditions. In contrast, radar is adaptable under all weather conditions and is indispensable for autonomous driving [32]. Conventional automotive radar provides limited information, restricting it to short-range collision

B. Luo et al. (Eds.): ICONIP 2023, LNCS 14450, pp. 493–511, 2024.
https://doi.org/10.1007/978-981-99-8070-3_38

Fig. 1. Visualization of one scene from VoD. The radar point clouds are platted as large dots while the LiDAR point clouds are shown in lines of small dots.

detection. However, with the advent of millimeter-wave 4D radar, offering comprehensive x, y, z position, and Doppler velocity information similar to LiDAR, this technology is considered ideal for perception applications. One example of radar point clouds is shown in Fig. 1.

Despite radar's broad use, its processed signals usually focus on position and speed, with a limited number of research on 4D-radar-based 3D object detection methods. [4,18,26] explore semantic information for 2D object detection tasks. Some works [7,8,16,20,30] utilize radial-frequency maps to identify potential 2D objects. Comprehensive benchmarks for 4D radar-based 3D object detection have emerged only recently with [19,22,36], while [14] is limited in size and scope. Prevailing strategies primarily employ LiDAR-based 3D detectors to 4D radar points. For example, methods like [22,36] adapt existing techniques like PointPillars [10], and [19] uses hierarchical 3D sparse and 2D convolution layers.

However, these methods have yet to fully capitalize on the Doppler velocity and reflectivity of radar for 3D object detection, both of which are vital information captured by radar. While [35] acknowledges the importance of the measured Doppler velocity and employs this data to align the radial velocity, it does not utilize this information for object detection. Furthermore, these studies commonly neglect the issue of the sparsity and irregularity of radar point clouds compared to LiDAR. We observe a marked performance discrepancy when implementing PointPillars with two different data inputs within the same dataset, VoD [22]. This inconsistency infers that the direct implementation of LiDAR-based detectors on radar points is inappropriate.

Given that Doppler velocity and reflectivity are two essential types of information overlooked by existing methods, we propose a novel backbone named the Radar Feature Assisted Backbone, designed to fully integrate these data into our framework. The incorporation of Doppler velocity and reflectivity data enables our network to more effectively discern potential objects. This stems from the fact that points on moving objects often display distinct Doppler velocities and reflectivity, setting them apart from points in the surrounding environment. By leveraging these distinctive characteristics, our proposed backbone significantly improves the network's capacity to precisely differentiate and identify objects.

To compensate for the sparsity of radar point clouds, we propose a multi-view-based framework. We construct both BEV and cylinder pillars to extract features hierarchically. In comparison to single-modal methods, such as point-based and grid-based approaches, multi-view strategies yield richer feature maps, which are crucial for handling the sparse characteristics of 4D radar point clouds. To address the irregularity issue, we introduce a novel component called the Positional Map, which assigns different weights to foreground and background points. This ensures that foreground points, which are more likely to be reflections from detected objects, are given higher importance, while background points receive lower weights. This approach enables us to prioritize detecting objects measured by foreground points, compensating for the irregularity of radar point clouds.

Contributions. Overall, our contributions can be summarized as follows:

- We propose the Multi-View Feature Assisted Network (*MVFAN*), an end-to-end, anchor-free, and single-stage framework for sufficient feature utilization.
- We introduce a novel Position Map Generation module that enhances feature learning by reweighing foreground and background points and their features, addressing the irregular distribution of radar point clouds.
- We propose a novel backbone, named Radar Feature Assisted backbone, that fully leverages the significant information provided by the Doppler velocity and RCS reflectivity of the 4D radar sensor.
- We conduct comprehensive experiments on Astyx [14] and VoD [22] datasets with ablation studies to demonstrate the effectiveness of our proposed modules.

2   Related Works

2D Object Detection on Radar. Currently, there are few works focusing on end-to-end 3D object detection on radar point clouds. Some works aim to regress 2D bounding boxes without elevation and height of detected objects. In [4], PointNet [23] is applied to extract radar features and generate 2D bounding box proposals together to extract features for 2D object detection. Similarly, [26] performs semantic segmentation on radar points and then clustered points are utilized to segment objects to regress 2D bounding boxes. Some works utilize radar points as complementary information and fuse with other sensors. Nabati et al. [17] proposes a two-stage region proposal network. To generate anchor or region of interest, radar points are projected into camera coordinates and distance of radar points are used to properly crop the image as a learned parameter. Later, they [18] further fine-tune this model to operate on point cloud and image. Radar points are expanded as pillars and associated with extracted image features by frustum view. However, because of the lack of height information caused by pillar expansion, 3D bounding boxes are only estimated and the evaluation results are far from satisfactory, only around 0.524% mAP (mean average precision) on nuSences dataset [3]. In [17,18], radar point clouds are only used as auxiliary input. [2] proposes an adaptive approach that deeply fuses information from radar, LiDAR and camera, which is proven to be efficient under foggy environments. [29] proposes possibility for data fusion of radar images. Also since this dataset employs 3D radar with low spatial resolution, only 2D objects are regressed in image plane. Given that radar and LiDAR obtain point clouds in 3D world coordinate, some mechanisms to exploit radar and LiDAR for perception have merged. RadarNet [35] and MVDNet [24] both fuse sparse radar point clouds and dense LiDAR point clouds after feature extraction at an early stage by voxelization of point cloud in BEV coordinates. They both achieve optimal performance in regressing 2D object detection under BEV frames in nuScence, DENSE [2] and ORR [1].

3D Object Detection on Radar. Few works exploit radar point clouds for 3D object detection. In [21], RTCnet is first embedded to encode the range-azimuth-Doppler image to output multi-class features and then it performs clustering technique to gather object points. Some [14,22,36] only implement LiDAR-based detectors such as PointPillars [10] for 3D detection with raw radar point clouds as input. [15] considers the sparsity of radar point cloud and exploits previously designed method [9] with radar and camera as input for 3D detection. This proposal level fusion method highly depends on the accuracy of each feature extractor of radar and camera. Xu et al. [33] carefully designed a self-attention technique to extract radar features with PointPillars as the backbone and baseline for evaluation while only considering global features. The major drawbacks of all these methodologies are that they do not carefully take the sparsity and ambiguity of measured radar points into account and do not fully make use of Doppler and RCS measured by radar.

3   Methodology

3.1   Overview

As shown in Fig. 2, *MVFAN* consists of three continuously connected modules: (a) Multi-View Feature Extraction Network. The raw radar point cloud is firstly transformed into cylindrical and BEV pillar as pseudo images, $M(\mathbf{p}^{\mathbf{cyl}})$ and $M(\mathbf{p}^{\mathbf{BEV}})$. Pillars are then encoded by ResNet blocks to extract high dimensional features and then projected back to points. Multi-view point-wise features are embedded to construct a fused feature map $\mathcal{M}$ after reweighing by Positional Map. In parallel, a radar feature vector $\mathcal{R}$ is built. (b) Radar Feature Assisted Backbone. In the backbone, $\mathcal{M}$ is aided by $\mathcal{R}$ to encode radar feature map assisted by convolution with radar Doppler and RCS representations to build dense map $\mathcal{DM}$ in a top-down and upsampling backbone. (c) Detection Head. Finally, 3D bounding boxes are regressed through the detection head by $\mathcal{DM}$.

(a) Multi-view Feature Extraction Network. Given a single frame of a raw point cloud, firstly points are pre-processed by transforming into cylinder and BEV grid cells then stacked into pillars as BEV pseudo image synchronously. Then, two individual feature extractors, ResNet blocks [6], are engaged to extract high-dimensional and complementary representative features to boost auxiliary point-wise and pillar-wise features. In order to stack and augment the feature

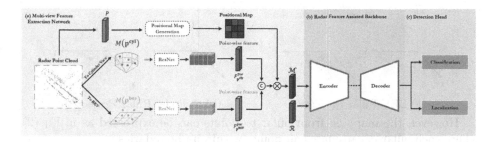

Fig. 2. The flowchart of *MVFAN*.

map in a point-wise manner, previously extracted pillar features are transformed back to the point view by established pillar-to-point and point-to-pillar mapping along with the position information of raw points. Therefore, the features of each individual point are aggregated by concatenating abstracted semantic features which are generated from both the cylinder view and the bird's-eye view. Next, we generate Positional Map to reweigh the features. This process allows for a more distinct separation of foreground points from background points.

  (b) **Radar Feature Assisted Backbone.** The reweighted feature map together with the Doppler velocity and reflectivity are fed into the Radar Feature Assisted Backbone with an encoder-decoder structure, which heuristically learns the contextual representative feature and geometric characteristics with positional encoding. The radar-assisted feature guides the learning process by summarizing the point-wise features through measured Doppler velocity and reflectivity of each point. The downsampling and upsampling modules of the backbone enable pyramid feature learning and facilitate the extraction of features at multiple scales and receptive fields.

  (c) **Detection Head.** Eventually, the multiscale features obtained from the backbone are fed into the detection head where 3D positions, height, weight, length, and heading angles of each box are predicted precisely.

3.2   Multi-view Feature Extraction

BEV Pillar Projection. Similar to PointPillars [10] and PillarOD [31]. Given a set of raw points $P = \{p_n\}_{n=0}^{N-1}$ in Cartesian coordinate, where $p_n = (x_n, y_n, z_n)$, each point p_n is assigned to the corresponding pillar $\mathbf{p}_m$ based on the point-to-pillar mapping described by Eq. (1).

$$F_{\mathbf{p}^{BEV}}(p_n^{BEV}) = \mathbf{p}_m^{BEV} \tag{1}$$

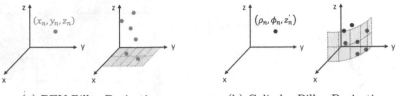

(a) BEV Pillar Projection   (b) Cylinder Pillar Projection

However, the mapping from pillar to points can be expressed as in Eq. (2), where each pillar encapsulates all points within its boundaries.

$$F_{p^{BEV}}(\mathbf{p}_m^{BEV}) = \{p_n^{BEV} \mid \forall p_n^{BEV} \in \mathbf{p}_m^{BEV}\} \tag{2}$$

Here, we define $F_{\mathbf{p}}(p_n)$ as the point-to-pillar mapping which returns the pillar $\mathbf{p}_m$ that contains point p_n, and $F_p(\mathbf{p}_m)$ as the pillar-to-point mapping which determines the set of points within pillar $\mathbf{p}_m$ by using $F_{\mathbf{p}}$. Thus, the point-to-pillar and pillar-to-point mapping are established, respectively, as described in Eq. (1) and Eq. (2).

Cylinder Pillar Projection. [37] discovered that the spherical projection of point clouds into a range image can result in dispensable distortion in the Z-axis, i.e., the height information in Cartesian coordinate, and this can negatively impact the performance as pointed out by previous works [5,31]. To prevent information loss, we use cylinder projection of raw point clouds as complementary representations without perspective distortion. The formulation for cylinder projection is given by Eq. (3).

$$\rho_n = \sqrt{x_n^2 + y_n^2}, \quad \phi_n = arctan(\frac{y_n}{x_n}), \quad z_n' = z_n \tag{3}$$

Given a raw point cloud p_n in Cartesian coordinate (x_n, y_n, z_n), its corresponding coordinate under cylinder coordinate can be described as $p_n^{cyl} = (\rho_n, \phi_n, z_n')$, where ρ and ϕ represent the distance and azimuth angle of the point cloud, respectively, and z' retains the height information. As in Sect. 3.2, all points under the cylinder view are reconstructed into pillars, and pillar-to-point mapping is established according to Eq. (4) and Eq. (5).

$$F_{\mathbf{p}^{cyl}}(p_n^{cyl}) = \mathbf{p}_m^{cyl} \tag{4}$$

$$F_{p^{cyl}}(\mathbf{p}_m^{cyl}) = \{p_n^{cyl} \mid \forall p_n^{cyl} \in \mathbf{p}_m^{cyl}\} \tag{5}$$

Multi-view Feature Extraction. After converting the points $P = \{p_n\}_{n=0}^{N-1}$ into BEV and cylindrical pillars $\mathbf{p}^{BEV}$ and $\mathbf{p}^{cyl}$, we stack them to form BEV and cylindrical pseudo images $M(\mathbf{p}^{BEV})$ and $M(\mathbf{p}^{cyl})$. Since $M(\mathbf{p}^{BEV})$ and $M(\mathbf{p}^{cyl})$ can be treated as images, we can use 2D CNNs to extract high-dimensional features. Therefore, we employ ResNet [6] to extract semantic features on a per-pillar basis, which we denote as $f_{\mathbf{p}_m^{BEV}}$ and $f_{\mathbf{p}_m^{cyl}}$ for the BEV and cylindrical pillars $\mathbf{p}_m^{BEV}$ and $\mathbf{p}_m^{cyl}$, respectively.

$$f_{\mathbf{p}_m^{BEV}} = ResNet(M(\mathbf{p}_m^{BEV}))$$
$$f_{\mathbf{p}_m^{cyl}} = ResNet(M(\mathbf{p}_m^{cyl})) \tag{6}$$

These high-level features are projected back as point-wise features with corresponding points according to the constructed pillar-to-point relation. To restore point-wise features from pillars, the pillar-to-point feature mapping is established as Eq. (7) where $f_{p_n^{BEV}}$ and $f_{p_n^{cyl}}$ denote the point-wise features.

$$f_{p_n^{BEV}} = f_{\mathbf{p}_m^{BEV}}, \text{where} \quad \mathbf{p}_m^{BEV} = F_{\mathbf{p}^{BEV}}(p_n^{BEV})$$
$$f_{p_n^{cyl}} = f_{\mathbf{p}_m^{cyl}}, \text{where} \quad \mathbf{p}_m^{cyl} = F_{\mathbf{p}^{cyl}}(p_n^{cyl}) \tag{7}$$

Finally, d dimensional points-wise features for N number of points are gathered from both the BEV and cylinder view as expressed in Eq. (8).

$$F_{p^{BEV}}^{pw} = \{f_{p_n^{BEV}}\}_{n=0}^{N-1} \subseteq \mathbb{R}^{N \times d}$$
$$F_{p^{cyl}}^{pw} = \{f_{p_n^{cyl}}\}_{n=0}^{N-1} \subseteq \mathbb{R}^{N \times d} \tag{8}$$

Positional Map Generation. To address the irregular distribution of radar point clouds, which is often overlooked by existing methods, we introduce the Positional Map Generation module that enables us to prioritize detecting objects captured by foreground points by assigning varying weights to the foreground and background points. In comparison to previous work RPFA-Net [33] which only leverages pillar features aggregated by sparse points as insufficient feature representation, the Positional Map enhances feature characterization in both global and local scales.

The point-wise features extracted from multi-view sources, $F_{p^{BEV}}^{pw}$ and $F_{p^{cyl}}^{pw}$, are then multiplied by the generated Positional Map. The resulting reweighted feature map is subsequently fed into the backbone for further processing. The point-wise classification result vector A is regressed by the Auxiliary Loss module.

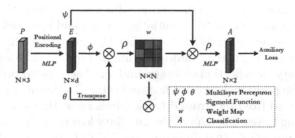

Fig. 3. Positional Map Generation.

Positional Encoding. To encode the positional information of the raw point clouds $P = \{p_n\}_{n=0}^{N-1}$, where $p_n = (x_n, y_n, z_n)$ represents the geometric position of the n-th point, we employ the K-nearest neighbors (KNN) algorithm to identify each center point p_n^k. We then compute the relative position of each point with respect to its KNN center as $||p_n - p_n^k||$, which is concatenated with the absolute position of each center point for positional encoding as in Eq. (9).

$$e_n = MLP(p_n^k, ||p_n - p_n^k||), \ E = \{e_n\}_{n=0}^{N-1} \tag{9}$$

MLP is a multilayer perception that outputs a vector of dimension d. The resulting positional encoded vector $E \subseteq \mathbb{R}^{N \times d}$ captures both local and global geometric context, with the relative position $||p_n - p_n^k||$ encoding local information and the KNN center p_n^k representing the global distribution of all radar points. The effectiveness of this positional encoding strategy in fusing local spatial information as essential feature representation has been demonstrated in [13].

Map Generation. Next, we perform a matrix operation to generate a Positional Map from the positional encoded feature vector E, as illustrated in Fig. 3. The map w is expressed as Eq. (10) where ρ is the sigmoid activation function and ϕ, θ are point-wise transformations, two independent $MLPs$. $w \subseteq \mathbb{R}^{N \times N}$ is generated by calculating the correlation between the ith and j^{th} positional encoded features of E, which captures the relationship between different points $p_i \in P$ and $p_j \in P$. The Positional Map w is then multiplied by the concatenation of point-wise features $F_{\mathbf{p}^{BEV}}^{pw}(p_n^{BEV})$ and $F_{\mathbf{p}^{cly}}^{pw}(p_n^{cly})$, respectively, to re-weight these point-wise features according to the learned geometric information.

$$w = \Sigma \rho(\phi(e_i)\theta(e_j)^T), \quad w \subseteq \mathbb{R}^{N \times N}$$
$$\mathcal{M}_0 = w \times Concat(F_{p^{cly}}^{pw}, F_{p^{BEV}}^{pw}) \tag{10}$$

The purpose of generating the Positional Map from the positional encoded feature vector is to capture the interdependence of local point clouds and learn high-dimensional features through the forward modules. This allows for the aggregation of both the geometric and feature context of the points, which in turn implies their relations in 3D position. By multiplication of concatenated point-wise features from multi-view, we obtain the geometric-related feature $\mathcal{M}$ which is fed into the backbone for further processing.

Auxiliary Loss. We introduce an auxiliary loss to help the learning process of the Positional Map by differentiating foreground and background points. After obtaining w, we use a MLP to predict the point-wise classification result vector A by performing matrix multiplication as Eq. (11). The sigmoid function ρ performs binary classification of foreground and background points. The MLP layer ψ takes the positional encoded feature vector E as input. L_i is the focal loss of point p_i derived from the ground truth labels indicating whether p_i is inside the 3D bounding boxes. The aim of the auxiliary loss is to give more importance to the foreground points that indicate possible objects, while the background points are assigned fewer proportions. The final auxiliary loss is obtained by taking the mean of all individual losses, as expressed in Eq. (11).

$$A = \rho(w \times \psi(E)), \quad A = \{a_{i,0}, a_{i,1}\}_{i=0}^{N-1}, \quad A \subseteq \mathbb{R}^{N \times 2}$$

$$L_{ax} = \frac{1}{N}\Sigma_{i=0}^{N}L_i$$

$$L_i = \begin{cases} -\alpha_i(1 - a_{i,1})^\gamma \log(a_{i,1}) & \text{if } y_i = 1 \\ -(1 - \alpha_i)a_{i,0}^\gamma \log(1 - a_{i,0}) & \text{otherwise} \end{cases} \tag{11}$$

where y_i is the ground truth label for point p_i (1 for foreground, 0 for background), $a_{i,1}$ and $a_{i,0}$ are the predicted probabilities for foreground and background, respectively. $\alpha_i = 0.25$ is a weight factor that assigns more importance to the foreground points, and $\gamma = 2$ is a modulating factor that down-weights the loss for well-classified samples in focal loss [11].

Discussion. The adoption of a multi-view strategy enhances feature learning and addresses the sparsity of radar point clouds by providing more comprehensive representations. Considering the irregular distribution of radar point clouds, we introduce a Positional Map, which predicts the binary classification of background and foreground points using a supervised auxiliary loss function. By applying the Positional Map, we emphasize the foreground points and their associated semantic features. This integration of geometric information in the Positional Map enhances performance and is particularly beneficial for handling the irregularity issue.

3.3   Radar Feature Assisted Backbone

Radar point clouds exhibit greater sparsity compared to LiDAR, yet they offer additional crucial information in the form of Doppler velocity and reflectivity. However, existing 4D radar-based 3D detectors overlook the significance of these information, failing to fully utilize its potential, further restricting the utilization of radar point clouds. By neglecting to incorporate the essential Doppler and reflectivity data into their frameworks, these methods miss out on valuable insights that could enhance the overall performance of radar-based object detection. Figure 4 illustrates the architecture of our proposed Radar Feature Assisted Backbone, which mainly consists of an encoder and a decoder. The backbone is designed with a U-Net [25] structure to extract hierarchical features for identifying objects at multiscale. In the encoder, the resolution decreases while the receptive field increases. In the decoder, the point features are restored back to the original dimension using an upsampling module with a skipped connection. In this paper, we define radar-assisted features as information directly captured by radar, including the absolute Doppler velocity, the relative Doppler velocity to ego-vehicle, and the RCS reflectivity, expressed in Eq. (12). The backbone takes the reweighted map $\mathcal{M}_0$ and radar-assisted feature $\mathcal{R}_0$ as input and outputs the processed data into the detection head.

$$\mathcal{R}_0 = \{p_n[v, v_r, RCS]\}_{n=0}^{N-1}, \quad \mathcal{R}_0 \subseteq \mathbb{R}^{N \times 3} \tag{12}$$

Encoder Structure. The encoder is composed of three interconnected Radar Assisted Blocks and Downsample modules whose details are shown in Fig. 5a.

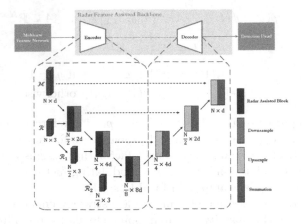

Fig. 4. The Radar Feature Assisted backbone.

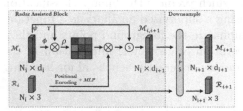

(a) Radar Assisted Block and Downsample module of the encoder.

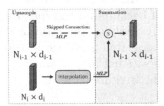

(b) Upsample and Summation module of the decoder.

Fig. 5. Details of encoder and decoder structures.

Each module takes in the point-wise feature $\mathcal{M}_i \subseteq \mathbb{R}^{N_i \times d_i}$ and the radar-assisted feature $\mathcal{R}_i \subseteq \mathbb{R}^{N_i \times 3}$ as inputs. $\mathcal{R}i$ is firstly processed using the same positional encoding method described in Sect.3.2. Then, matrix multiplication is performed with a residual connection to output learned feature with $d_{i+1} = 2d_i$ dimensions. Finally, the farthest point sampling strategy is adopted to downsample all features and enlarge their receptive fields. Notably, $\mathcal{R}_i$ is associated with the corresponding 3D coordinates and is fused with the point-wise $\mathcal{M}_i$ to enhance the information exchange at the feature vector level in a trainable manner. The Radar Assisted Block aggregates both the content of the feature vectors of $\mathcal{R}_i$ and $\mathcal{M}_i$ with their 3D positions.

$$\mathcal{M}_{i,i+1} = \rho(\phi(\mathcal{M}_i)\psi(\mathcal{M}_i)^T) \times \theta(\mathcal{R}_i, P_i^k, \|P_i - P^k\|) + \phi(\mathcal{M}_i)$$
$$\mathcal{M}_{i+1} = \text{FPS}(\mathcal{M}_{i,i+1}) \tag{13}$$
$$\mathcal{R}_{i+1} = \text{FPS}(\mathcal{R}_i)$$

where ϕ, ψ, θ share the same definition as MLP and ρ is the sigmoid function.

Decoder Structure. The decoder is a symmetrical structure of the encoder, as shown in Fig. 5b. Consecutive stages in the decoder are connected by Upsample and Summation modules. The objective of the decoder is to restore features from the downsampled $\mathcal{M}_i$ to their previous shape as $\mathcal{M}_{i-1} \subseteq \mathbb{R}^{N_{i-1} \times d_{i-1}}$. This is achieved by the up-sampling layer through interpolation in point size dimension and a MLP in feature dimension. Then, the interpolated features from the preceding decoder stage are summed with the features from the corresponding encoder stage via a skipped connection.

Discussion. Our proposed Radar Feature Assisted Backbone distinguishes itself from existing approaches by effectively harnessing crucial information, specifically the Doppler velocity and reflectivity measured by radar. This innovative approach significantly enhances the performance of feature extraction and strengthens the high-level feature $\mathcal{M}_i$ by incorporating it with the low-level feature $\mathcal{R}_i$ through additional position encoding. One of the remarkable features of radar is its ability to measure reflectivity and Doppler velocity, as well as infer related Doppler velocity as the low-level feature $\mathcal{R}_i$. Surprisingly, this critical information has not been leveraged in 4D radar-based 3D object detection before. To showcase the effectiveness of our approach, we conduct an ablation study that clearly demonstrates the benefits of incorporating this previously unexploited information.

3.4   Detection Head and Loss Function

We implement the identical SSD [12] detection head, which is commonly embedded in SECOND [34], PointPillars, PillarOD and MVF. The localization loss $\mathcal{L}_{loc}$ and the angle loss $\mathcal{L}_{dir}$ are computed using a Smooth L1 loss function that is regressed between the ground truth boxes $(x^{gt}, y^{gt}, z^{gt}, w^{gt}, l^{gt}, h^{gt}, \theta^{gt})$ and anchors $(x^{pre}, y^{pre}, z^{pre}, w^{pre}, l^{pre}, h^{pre}, \theta^{pre})$ as Eq. (14) where the diagonal of the base of the anchor box is referred as $d^{pre} = \sqrt{(w^{pre})^2 + (l^{pre})^2}$. $\mathcal{L}_{dir}$ is proven by [34] to effectively discretize directions of flipped boxes. The classification loss $\mathcal{L}_{cls}$ is calculated by the focal loss, written as Eq. (15) where the predicted classification probability is revealed as p^{pre}.

$$\mathcal{L}_{loc} = \sum_{b \in (x,y,z,w,l,h)} \text{Smooth L1}[\Delta b]$$

$$\Delta x = \frac{x^{pre} - x^{gt}}{d^{pre}}, \Delta y = \frac{y^{pre} - y^{gt}}{d^{pre}}, \Delta z = \frac{z^{pre} - z^{gt}}{d^{pre}} \qquad (14)$$

$$\Delta w = \log \frac{w^{gt}}{w^{pre}}, \Delta l = \log \frac{l^{gt}}{l^{pre}}, \Delta h = \log \frac{h^{gt}}{h^{pre}}$$

$$\mathcal{L}_{dir} = \text{Smooth L1}[sin(\theta^{pre} - \theta^{gt})]$$

$$\mathcal{L}_{cls} = -\alpha_{pre}(1 - p^{pre})^{\gamma} \ln p^{pre}, \quad \alpha_{pre} = 0.25, \gamma = 2 \qquad (15)$$

By combing all losses discussed in Eq. (11), Eq. (14) and Eq. (15), total loss function is asddressed as Eq. (16) where $\alpha_1 = \alpha_2 = 1, \alpha_3 = 2, \alpha_4 = 0.2$.

$$\mathcal{L}_{total} = \alpha_1 \mathcal{L}_{cls} + \alpha_2 \mathcal{L}_{ax} + \alpha_3 \mathcal{L}_{loc} + \alpha_4 \mathcal{L}_{dir} \qquad (16)$$

Table 1. Network parameters setting for two datasets.

Dataset	Detection range						BEV Resolution			Cylindrical Resolution		
	x	y	z	ρ	ϕ	z'	x	y	z	ρ	ϕ	z'
Astyx	$[0, 99.84m]$	$[-39.68m, 39.68m]$	$[-3m, 1m]$	$[0, 100.6m]$	$[0, \pi]$	$[-3m, 1m]$	$0.16m$	$0.16m$	$4m$	$100.6m$	$\pi/1280$	$0.04m$
VoD	$[0, 51.2m]$	$[-25.6m, 25.6m]$	$[-3m, 2m]$	$[0, 72.4m]$	$[0, \pi]$	$[-3m, 2m]$	$0.16m$	$0.16m$	$5m$	$72.4m$	$\pi/1280$	$0.05m$

4 Experiments

4.1 Dataset

We evaluate our proposed method on Astyx and VoD datasets to demonstrate its effectiveness and efficiency. Astyx is composed of 546 frames with over 3000 3D object annotations in seven distinct classes: cars, buses, humans, bicycles, motorcycles, trucks, and trailers, with cars being the majority. Due to the imbalanced distribution, we focus solely on the car class for training and evaluation and adopt a 75% training and 25% validation split ratio following [33]. VoD ensures a proportional and reasonable distribution of multi-class annotations, including cars, pedestrians, and cyclists, and is officially divided into training, validation, and testing sets at proportions of 59%, 15%, and 26%.

4.2 Evaluation

Astyx uses average precision (AP) as an accuracy metric for the car class with an intersection over union (IoU) threshold of 0.5. Furthermore, it classifies objects into three difficulty levels - easy, moderate, and hard - based on their visibility and the degree of occlusion. VoD dataset provides average precision (AP) and average orientation similarity (AOS) as two evaluation metrics. Specifically, the 3D IoU threshold for the car class is set at 0.5, while the thresholds for pedestrians and cyclists are both established at 0.25. Additionally, VoD organizes the detection process into two levels, (a) all areas within field-of-view, (b) the driving corridor at $[0 < x < 25m, -4m < y < 4m]$.

4.3 Implementation Details

Network Parameter. Several key parameters need to be clarified to understand the spatial resolution of BEV pillar projection for x, y, z axes and cylindrical resolution for ρ, ϕ, z' axes with the detection range in Table 1

Data Augmentation. Due to the fact that the Doppler velocity is relative to the vehicle's motion, only a limited data augmentation technique can be applied to radar. Therefore, we only seek to randomly flip the point cloud around the x axis and scale the point clouds by a factor within $[0.95, 1.05]$. For radar point clouds, translation and rotation of components are not adopted.

Training. To train 80 epochs, we adopt the ADAM optimization with 0.003 initial learning rate and 0.01 weight decay rate on 4 Nvidia RTX 2080Ti GPUs with a batch size of 8.

Table 2. Comparative experiments on Astyx. The mean average precisions for *car* class are based on 3D and BEV IoU respectively.

Methods	3D (%)			BEV (%)		
	Easy	Moderate	Hard	Easy	Moderate	Hard
PointRCNN [28]	12.23	9.10	9.10	14.95	13.82	13.89
SECOND [34]	24.11	18.50	17.77	41.25	30.58	29.33
H²3DR-CNN [5]	26.21	20.12	21.62	45.23	32.58	35.54
PV-RCNN [27]	28.21	22.29	20.40	46.62	35.10	33.67
PointPillars [10]	30.14	24.06	21.91	45.66	36.71	35.30
RPFA-Net [33]	38.85	32.19	30.57	50.42	42.23	40.96
MVFAN (*ours*)	**45.60**	**39.52**	**38.53**	**58.68**	**50.24**	**46.56**

Table 3. Comparative experiments on VoD: Average precision for *car, pedestrian, cyclist*, mean average precision (mAP) and mean average orientation similarity (mAOS).

Methods	All Area (%)					Driving Corridor (%)					FPS (Hz)
	Car	Ped	Cyc	mAP	mAOS	Car	Ped	Cyc	mAP	mAOS	
PointRCNN [28]	15.98	24.52	43.14	27.88	23.40	33.60	33.73	60.32	42.55	37.73	14.31
PV-RCNN [27]	31.94	27.19	57.01	38.05	24.12	67.76	38.03	76.99	62.66	40.79	23.49
H²3DR-CNN [5]	25.90	14.88	34.07	24.95	16.91	61.89	23.48	53.95	46.44	35.42	43.59
SECOND [34]	32.35	24.49	51.44	36.10	27.58	67.98	35.45	72.30	59.18	50.92	80.87
PointPillars [10]	32.54	26.54	55.11	38.09	30.10	68.78	36.02	80.63	62.58	53.96	157.50
RPFA-Net [33]	33.45	26.42	56.34	38.75	31.12	68.68	34.25	80.36	62.44	52.36	100.23
MVFAN(*ours*)	**34.05**	**27.27**	**57.14**	**39.42**	**31.58**	**69.81**	**38.65**	**84.87**	**64.38**	**57.01**	45.11

4.4 Experiment Results

We compare *MVFAN* with state-of-the-art point cloud detectors for comparison, PointRCNN [28], PV-RCNN [27], H²3DR-CNN [5], SECOND [34], PointPillars [10] and RPFA-Net [33]. We re-train these frameworks on Astyx and VoD from scratch. Since our Multi-View Feature Extraction module is inspired by MVF [37] which exploits variant views, we are supposed to re-implement it. Nevertheless, MVF has not been released, so its final result is unapproachable. As elaborated in Tables 2 and 3, our *MVFAN* beats all of these 3D detectors.

4.5 Ablation Study

We demonstrate the effectiveness of two major components of our model, Multi-view Feature Extraction (*MV*) and Radar Feature Assisted Backbone (*FA*). Since our framework is an adaptation of state-of-the-art PointPillars (*PP*), we apply it as the baseline in our experiments. Table 4 underlines the validation of a few combinations of *MV* and *FA* on the detection outcome.

(a) Firstly, we re-train PP. As radar point clouds consist of spatial coordinates, Doppler velocity and RCS reflectivity, the input channel of PointNet $(D = 9, P, N)$ is expanded to $(D = 11, P, N)$ as stated in VoD.

Table 4. Ablation study: Effectiveness of MV and FA.

Methods	All Area (%)					Driving Corridor (%)					FPS (Hz)
	Car	Ped	Cyc	mAP	mAOS	Car	Ped	Cyclist	mAP	mAOS	
(a) *PP*	32.54	26.54	55.11	38.09	30.10	68.78	36.02	80.63	62.58	53.96	157.50
(b) *PP+MV*	33.16	27.24	55.50	38.67	30.65	69.71	38.62	83.91	64.22	56.32	145.56
(b) *PP+FA*	32.55	27.23	55.85	38.51	30.73	68.84	36.69	81.57	62.37	53.79	45.77
(d) *MVFAN*	**34.05**	**27.27**	**57.14**	**39.42**	**31.58**	**69.81**	**38.65**	**84.87**	**64.38**	**57.01**	45.11

Table 5. Ablation study: Effectiveness of Positional Map.

Methods	All Area (%)					Driving Corridor (%)					FPS (Hz)
	Car	Ped	Cyc	mAP	mAOS	Car	Ped	Cyclist	mAP	mAOS	
(a) *PP+MV w/o*	32.66	26.82	55.23	38.25	30.32	68.94	37.42	82.54	63.20	54.32	150.66
(b) *PP+MV w*	33.16	27.24	55.50	38.67	30.65	69.71	38.62	83.91	64.22	56.32	145.56
(c) *MVFAN w/o*	33.98	27.25	56.86	39.22	31.23	69.75	38.63	84.52	64.31	56.88	49.42
(d) *MVFAN w*	**34.05**	**27.27**	**57.14**	**39.42**	**31.58**	**69.81**	**38.65**	**84.87**	**64.38**	**57.01**	45.11

(b) Secondly, we only utilize MV on PP. Instead of feeding solely BEV features into the backbone as PointPillars, retrieved dense point-wise cylinder features and point-wise BEV features are concatenated followed by weight map multiplication. It validates that MV largely improves the performance of all three characters. Focusing on the driving corridor, which VoD identifies as a safety-critical region, MV also highly magnifies the dense representation for the detector in order to distinguish objects from the background, particularly small volume items, such as *pedestrians* and *cyclists*, whose contours reveal a drastic deviation in cylinder perspective. When objects in all areas are examined, MV refines the performance and escalates the indicator by 0.62%, 0.70% and 0.39%. The advancement of accuracy highlights the essence and capacity of the multi-view sampling scheme and reweighting of the foreground and background points for better maintaining 3D contextual information and morphological characteristics more effectively.

(c) Next, we only replace the backbone of PP with FA. The average precision of this model is upgraded by incorporating additional radar information by matrix operation at a minimal cost in terms of inference time, which preserves the Doppler and RCS reflectivity of radar point clouds. FA is proved to have a slight effect on detecting *cars*, whereas it makes a significant difference on detecting small moving objects such as *pedestrians* and *cyclists*. In scenarios where the bounding boxes of *pedestrians* and *cyclists* encompass only a limited number of radar points, relying solely on the 3D geometric information from sparse point clouds may not yield optimal results. Instead, the Doppler motions and observed RCS reflectivity play a crucial role in accurately identifying these objects. By incorporating the Doppler motions and RCS reflectivity into our network, we are able to achieve superior object identification performance. This outcome effectively demonstrates the effectiveness and importance of leveraging the Doppler and RCS reflectivity information.

(d) Moreover, our proposed framework *MVFAN*, a combination of *MV* and *FA*, exhibits a high-ranking detection precision. The average precision for all three classes is further upgraded compared to when either *MV* or *FA* is independently applied on the baseline. Overall, *MVFAN* boosts the mAP by 3.4% in all area and 2.8% in the driving corridor and elevates the angular accuracy, mAOS, by 4.9% in all area and 5.6% in the driving corridor. Impressively, these two procedures function concurrently and independently with *MV* augmenting the point-wise semantic features by leveraging two input streams to reward the sparsity of radar point clouds whereas *FA* amplifying the fused feature map after *MV* module by adopting additional radar information. Remarkably, the promotion raised by *MVFAN* lies in the emphasis on multi-view feature extraction with reweighing points and rational enforcement of radar Doppler and reflectivity.

Table 5 confirms the effectiveness of our proposed Positional Map in improving the performance of both PointPillars and our *MVFAN*. The inclusion of the Positional Map leads to higher mAP and mAOS scores compared to the baseline methods (*PP+MV w/o* and *MVFAN w/o*). It enhances the detection of cars, pedestrians, and cyclists, with improved percentages in those categories in both all area and the driving corridor scenario. These results validate the effectiveness of the Positional Map in enhancing the overall object detection performance.

Table 6. Comparison of different grid size settings for BEV and cylinder view.

Grid Size	All Area (%)					Driving Corridor (%)					FPS (Hz)
$BEV - cyl$	Car	Ped	Cyclist	mAP	mAOS	Car	Ped	Cyclist	mAP	mAOS	
$BEV8 - cyl4$	32.74	26.54	55.99	38.42	29.33	69.35	36.53	84.29	63.38	53.46	26.04
$BEV8 - cyl5$	32.18	26.56	54.96	37.89	30.03	69.51	38.24	83.06	62.93	55.69	28.39
$BEV8 - cyl6$	32.27	26.79	54.82	39.51	30.63	68.55	38.32	84.11	63.67	53.38	29.95
$BEV16 - cyl4$	34.04	**27.56**	**57.54**	**40.18**	**31.75**	69.41	**38.69**	**85.34**	64.36	56.41	39.12
$BEV16 - cyl5$	**34.05**	27.27	57.14	39.42	31.58	**69.81**	38.65	84.87	**64.48**	**57.01**	45.11
$BEV16 - cyl6$	33.76	27.42	57.16	39.45	31.40	69.80	37.67	84.21	64.23	56.42	49.38
$BEV32 - cyl4$	33.27	25.89	56.86	39.34	29.45	69.80	38.29	84.69	64.29	55.78	44.13
$BEV32 - cyl5$	33.18	26.44	57.53	39.06	31.42	69.51	38.03	84.72	64.09	56.06	51.20
$BEV32 - cyl6$	32.06	26.32	57.12	38.51	30.45	69.38	37.11	84.42	63.64	53.50	56.32

To investigate the effect of parameter settings, we deploy various combinations of partition parameters. For BEV pillar projection, we implement three types of grid size $BEV8 : [0.08m, 0.08m, 5m]$, $BEV16 : [0.16\,m, 0.16\,m, 5\,m]$ and $BEV32 : [0.32\,m, 0.32\,m, 5\,m]$ and for cylinder pillar projection, we test three types of grid size $cyl4 : [72.4\,m, \pi/1600, 0.04\,m]$, $cyl5 : [72.4\,m, \pi/1280, 0.05\,m]$ and $cyl6 : [72.4\,m, \pi/1000, 0.0625\,m]$. There are 9 experiments conducted in total as presented in Table 6. By partitioning the BEV with course grids, $BEV32$, each pillar has a greater number of point clouds but also incorporates more noise reflected by environments as opposed to objects. Although a finer grid $BEV8$ brings a higher spatial resolution to separate point clouds, the sparsity of radar introduces more vacant cells and necessitates additional computation. For the partition of cylinder projection, we have already established an exceptional resolution for dividing grid cells, for $cyl4, cyl5, cyl6$ with the same BEV grid size, the performances are only a little altered, on the contrary, the inference speed is largely impacted.

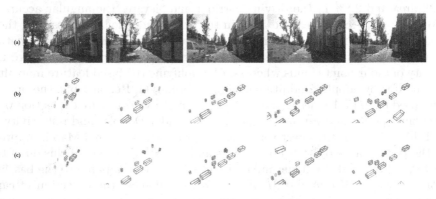

Fig. 6. Qualitative results on the VoD dataset.

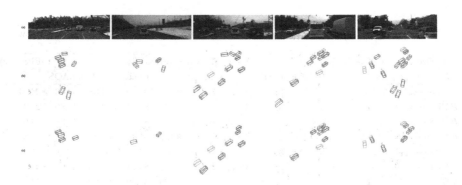

Fig. 7. Qualitative results on the Astyx dataset.

Qualitative Results. We display the qualitative results of VoD and Astyx validation set in Fig. 6 and Fig. 7. We picture the ground truth object on (a) images, the prediction results of (b) PointPillars, and (c) our proposed *MVFAN*. The predicted 3D bounding boxes are painted in red only in point clouds. The ground truth objects, *cars*, *pedestrians* and *cyclists* are colored in green, blue and yellow and together with images as references.

5 Conclusion

In summary, our proposed Multi-View Feature Assisted Network is a novel single-stage framework that can effectively utilize both geometric and Doppler velocity and reflectivity measured by the 4D radar sensor. Our method addresses the issue of insufficient feature utilization by employing complementary features from multi-view. We introducing a novel Position Map Generation module to reweigh foreground and background points considering the irregularity of radar point clouds. Importantly, we introduce a novel Radar Feature Assisted backbone, which is the first to fully leverage the valuable information provided by the Doppler velocity and RCS reflectivity in 4D radar data for 3D object detection while existing radar-based 3D object detection methods have largely ignored these features. Our ablation studies validate the significant impact of incorporating these features into the framework, highlighting the importance of considering this information for accurate object detection. Our extensive experiments on Astyx and VoD datasets demonstrate that our proposed method significantly outperforms state-of-the-art 3D object detection methods.

References

1. Barnes, D., Gadd, M., Murcutt, P., Newman, P., Posner, I.: The oxford radar robotcar dataset: a radar extension to the oxford robotcar dataset. In: 2020 IEEE International Conference on Robotics and Automation (ICRA), pp. 6433–6438. IEEE (2020)
2. Bijelic, M., et al.: Seeing through fog without seeing fog: deep multimodal sensor fusion in unseen adverse weather. In: Proceedings of the IEEE/CVF Conference on Computer Vision and Pattern Recognition, pp. 11682–11692 (2020)
3. Caesar, H., et al.: nuscenes: a multimodal dataset for autonomous driving. In: Proceedings of the IEEE/CVF Conference on Computer Vision and Pattern Recognition, pp. 11621–11631 (2020)
4. Danzer, A., Griebel, T., Bach, M., Dietmayer, K.: 2d car detection in radar data with pointnets. In: 2019 IEEE Intelligent Transportation Systems Conference (ITSC), pp. 61–66. IEEE (2019)
5. Deng, J., Zhou, W., Zhang, Y., Li, H.: From multi-view to hollow-3d: hallucinated hollow-3d r-CNN for 3d object detection. IEEE Trans. Circuits Syst. Video Technol. **31**(12), 4722–4734 (2021)
6. He, K., Zhang, X., Ren, S., Sun, J.: Deep residual learning for image recognition. In: Proceedings of the IEEE Conference on Computer Vision and Pattern Recognition, pp. 770–778 (2016)
7. Hwang, J.J., et al.: Cramnet: camera-radar fusion with ray-constrained cross-attention for robust 3d object detection. In: European Conference on Computer Vision (2022)
8. Jiang, T., Zhuang, L., An, Q., Wang, J., Xiao, K., Wang, A.: T-rodnet: transformer for vehicular millimeter-wave radar object detection. IEEE Trans. Instrum. Meas. **72**, 1–12 (2022)
9. Ku, J., Mozifian, M., Lee, J., Harakeh, A., Waslander, S.L.: Joint 3d proposal generation and object detection from view aggregation. In: 2018 IEEE/RSJ International Conference on Intelligent Robots and Systems (IROS), pp. 1–8. IEEE (2018)

10. Lang, A.H., Vora, S., Caesar, H., Zhou, L., Yang, J., Beijbom, O.: Pointpillars: fast encoders for object detection from point clouds. In: Proceedings of the IEEE/CVF Conference on Computer Vision and Pattern Recognition, pp. 12697–12705 (2019)
11. Lin, T.Y., Goyal, P., Girshick, R., He, K., Dollár, P.: Focal loss for dense object detection. In: Proceedings of the IEEE International Conference on Computer Vision, pp. 2980–2988 (2017)
12. Liu, W., et al.: SSD: single shot multiBox detector. In: Leibe, B., Matas, J., Sebe, N., Welling, M. (eds.) ECCV 2016. LNCS, vol. 9905, pp. 21–37. Springer, Cham (2016). https://doi.org/10.1007/978-3-319-46448-0_2
13. Lu, D., Xie, Q., Gao, K., Xu, L., Li, J.: 3DCTN: 3D convolution-transformer network for point cloud classification. IEEE Trans. Intell. Transp. Syst. **23**(12), 24854–24865 (2022)
14. Meyer, M., Kuschk, G.: Automotive radar dataset for deep learning based 3d object detection. In: 2019 16th European Radar Conference (EuRAD), pp. 129–132. IEEE (2019)
15. Meyer, M., Kuschk, G.: Deep learning based 3d object detection for automotive radar and camera. In: 2019 16th European Radar Conference (EuRAD), pp. 133–136. IEEE (2019)
16. Meyer, M., Kuschk, G., Tomforde, S.: Graph convolutional networks for 3d object detection on radar data. In: Proceedings of the IEEE/CVF International Conference on Computer Vision, pp. 3060–3069 (2021)
17. Nabati, R., Qi, H.: RRPN: radar region proposal network for object detection in autonomous vehicles. In: 2019 IEEE International Conference on Image Processing (ICIP), pp. 3093–3097. IEEE (2019)
18. Nabati, R., Qi, H.: Centerfusion: center-based radar and camera fusion for 3d object detection. In: Proceedings of the IEEE/CVF Winter Conference on Applications of Computer Vision, pp. 1527–1536 (2021)
19. Paek, D.H., Kong, S.H., Wijaya, K.T.: K-radar: 4d radar object detection for autonomous driving in various weather conditions. In: Thirty-Sixth Conference on Neural Information Processing Systems Datasets and Benchmarks Track (2022). https://openreview.net/forum?id=W_bsDmzwaZ7
20. Palffy, A., Dong, J., Kooij, J.F., Gavrila, D.M.: CNN based road user detection using the 3d radar cube. IEEE Robot. Automat. Lett. **5**(2), 1263–1270 (2020)
21. Palffy, A., Dong, J., Kooij, J.F., Gavrila, D.M.: CNN based road user detection using the 3d radar cube. IEEE Robot. Automat. Lett. **5**(2), 1263–1270 (2020)
22. Palffy, A., Pool, E., Baratam, S., Kooij, J.F.P., Gavrila, D.M.: Multi-class road user detection with 3+1d radar in the view-of-delft dataset. IEEE Robot. Automat. Lett. **7**(2), 4961–4968 (2022). https://doi.org/10.1109/LRA.2022.3147324
23. Qi, C.R., Su, H., Mo, K., Guibas, L.J.: Pointnet: deep learning on point sets for 3d classification and segmentation. In: Proceedings of the IEEE Conference on Computer Vision and Pattern Recognition, pp. 652–660 (2017)
24. Qian, K., Zhu, S., Zhang, X., Li, L.E.: Robust multimodal vehicle detection in foggy weather using complementary lidar and radar signals. In: Proceedings of the IEEE/CVF Conference on Computer Vision and Pattern Recognition, pp. 444–453 (2021)
25. Ronneberger, O., Fischer, P., Brox, T.: U-net: convolutional networks for biomedical image segmentation. In: Navab, N., Hornegger, J., Wells, W.M., Frangi, A.F. (eds.) MICCAI 2015. LNCS, vol. 9351, pp. 234–241. Springer, Cham (2015). https://doi.org/10.1007/978-3-319-24574-4_28

26. Schumann, O., Hahn, M., Dickmann, J., Wöhler, C.: Semantic segmentation on radar point clouds. In: 2018 21st International Conference on Information Fusion (FUSION), pp. 2179–2186. IEEE (2018)
27. Shi, S., et al.: PV-RCNN: point-voxel feature set abstraction for 3d object detection. In: Proceedings of the IEEE/CVF Conference on Computer Vision and Pattern Recognition, pp. 10529–10538 (2020)
28. Shi, S., Wang, X., Li, H.: Pointrcnn: 3d object proposal generation and detection from point cloud. In: Proceedings of the IEEE/CVF Conference on Computer Vision and Pattern Recognition, pp. 770–779 (2019)
29. Wang, J., Li, J., Shi, Y., Lai, J., Tan, X.: Am^{3}net: adaptive mutual-learning-based multimodal data fusion network. IEEE Trans. Circuits Syst. Video Technol. **32**(8), 5411–5426 (2022)
30. Wang, Y., Jiang, Z., Li, Y., Hwang, J.N., Xing, G., Liu, H.: Rodnet: a real-time radar object detection network cross-supervised by camera-radar fused object 3d localization. IEEE J. Select. Topic. Signal Process. **15**, 954–967 (2021)
31. Wang, Y., et al.: Pillar-based object detection for autonomous driving. In: Vedaldi, A., Bischof, H., Brox, T., Frahm, J.-M. (eds.) ECCV 2020. LNCS, vol. 12367, pp. 18–34. Springer, Cham (2020). https://doi.org/10.1007/978-3-030-58542-6_2
32. Wei, Z., Zhang, F., Chang, S., Liu, Y., Wu, H., Feng, Z.: Mmwave radar and vision fusion for object detection in autonomous driving: a review. Sensors **22**(7) (2022). https://doi.org/10.3390/s22072542
33. Xu, B., et al.: RPFA-net: a 4d radar pillar feature attention network for 3d object detection. In: 2021 IEEE International Intelligent Transportation Systems Conference (ITSC), pp. 3061–3066. IEEE (2021)
34. Yan, Y., Mao, Y., Li, B.: Second: sparsely embedded convolutional detection. Sensors **18**(10), 3337 (2018)
35. Yang, B., Guo, R., Liang, M., Casas, S., Urtasun, R.: RadarNet: exploiting radar for robust perception of dynamic objects. In: Vedaldi, A., Bischof, H., Brox, T., Frahm, J.-M. (eds.) ECCV 2020. LNCS, vol. 12363, pp. 496–512. Springer, Cham (2020). https://doi.org/10.1007/978-3-030-58523-5_29
36. Zheng, L., et al.: Tj4dradset: a 4d radar dataset for autonomous driving. In: 2022 IEEE 25th International Conference on Intelligent Transportation Systems (ITSC), pp. 493–498. IEEE (2022)
37. Zhou, Y., et al.: End-to-end multi-view fusion for 3d object detection in lidar point clouds. In: Conference on Robot Learning, pp. 923–932. PMLR (2020)

Time Series Anomaly Detection with a Transformer Residual Autoencoder-Decoder

Shaojie Wang, Yinke Wang, and Wenzhong Li[✉]

State Key Laboratory for Novel Software Technology, Nanjing University, Nanjing 210023, China
shaojie_wang@smail.nju.edu.cn

Abstract. Time series anomaly detection is of great importance in a variety of domains such as finance fraud, industrial production, and information systems. However, due to the complexity and multiple periodicity of time series, extracting global and local information from different perspectives remains a challenge. In this paper, we propose a novel Transformer Residual Autoencoder-Decoder Model called **TRAD** for time series anomaly detection, which is based on a multi-interval sampling strategy incorporating with residual learning and stacked autoencoder-decoder to promote the ability to learn global and local information. Prediction error is applied to calculate anomaly scores using the proposed model from different scales, and the aggregated anomaly scores are utilized to infer outliers of the time series. Extensive experiments are conducted on five datasets and the results demonstrate that the proposed model outperforms the previous state-of-the-art baselines.

Keywords: Time Series Analysis · Anomaly Detection · Transformer Autoencoder-Decoder

1 Introduction

Nowadays, time series data is commonly generated by a lot of industrial and information systems with massive amounts. It is very important to promptly identify abnormal points from time series data for many domains, such as finance, industrial production, electrical consumption, heart rate detection, etc. However, as the complexity of time series increases, identifying anomaly values from a massive data stream becomes more and more challenging.

With the continuous development of deep learning in recent years, a great deal of anomaly detection algorithms based om deep learning emerged and demonstrated their effectiveness. Deep Autoencoding Gaussian Mixture Model (DAGMM) [1] combined autoencoder and Gaussian density estimation to estimate multi-dimentional data's anomaly possibility. Introduced by Audibert et al., unSupervised Anomaly Detection (USAD) [2] designed two autoencoders

© The Author(s), under exclusive license to Springer Nature Singapore Pte Ltd. 2024
B. Luo et al. (Eds.): ICONIP 2023, LNCS 14450, pp. 512–524, 2024.
https://doi.org/10.1007/978-981-99-8070-3_39

(AEs) [3] to process adversarial training to calculate anomaly scores. The LSTM-AD [4] adopted the stacked LSTMs to estimate the distribution of prediction errors. Deng et al. presented GDN [5] which combined attention mechanisms and graph neural network (GNN) [6] to learn the hidden relationship between different sensors. Although these methods have been proven effective, they lack of consideration of time series features from both global and local perspectives. In other words, these models can't effectively capture the global features and local details of the time series to maximize the extraction of useful information.

To address above issue, we propose a novel Transformer Residual Autoencoder-Decoder Model called **TRAD** which is based on multi-interval strategy and combines residual learning and stacked autoencoder-decoder to promote the ability to learn historical information. The whole process is divided into three parts: *multi-interval sampling*, *TRAD-based prediction* and *anomaly detection*. In multi-interval sampling part, the raw time sequence is partitioned into different sub sequences based on various intervals, which are separately fed into different prediction model. In TRAD-based prediction part, the model receives the historical subsequence to predict the value of the next point. In order to promote the precision, we adopted a hierarchical transformer autoencoder-decoder structure which integrates the idea of residual learning. In anomaly detection part, we compute each Transformer Residual Autoencode (TRA) model's anomaly score by prediction error. The final anomaly score is determined using a weighted average method on each TRA anomaly score. Then an appropriate threshold is set in the light of final anomaly score. If the anomaly score is larger than a predefined threshold, it is regarded as an abnormal point, otherwise it is considered a normal value. We conduct extensive experiments based on five open datasets, which show that the proposed method outperforms the state-of-the-art baselines.

The contributions of our work are summarized as follows.

- We propose a novel method with a Transformer Residual Autoencoder-Decoder (TRAD) framework for time series anomaly detection. The proposed method introduces a multi-interval strategy for better capturing features from different perspectives.
- The proposed method introduces a multi-interval sampling methods to enable the prediction model to capture both global features and local characteristics of time series. A residual learning approach is proposed to enhance the ability of Transformer autoencoder to process time series on multiple scales for anomaly detection.
- We conduct extensive experiments based on five open datasets, which show that the proposed method achieves the best F1 score compared to the state-of-the-art baseline algorithms.

2   Related Work

Anomaly detection has been a popular application field for machine learning, and some excellent models and methods have also emerged in this domain. Breunig et al. proposed Local Outlier Factor (LOF) [7] which is an unsupervised method based on density. There are also some linear algorithms which are utilized to solve anomaly detection such as Principal Components Analysis (PCA) [8]. In addition to these algorithms, based on the two basic assumptions that anomalies are rare and different from normal points, the IF algorithm [9] proposed by Zhou et al. applies multiple random binary trees to quickly separate the anomaly points.

Recently, some outstanding deep learning methods are developed for time series abnormal detection which can be mainly classified into two types: reconstruction-based methods and prediction-based methods.

In reconstruction-based methods, autoencoder (AE) [3] is most commonly used to obtain reconstruction error. Zong et al. proposed Deep Autoencoding Gaussian Mixture Model (DAGMM) [1] which combines autoencoder and Gaussian density estimation to infer multi-dimentional data's anomaly possibility. Audibert et al. introduced UnSupervised Anomaly Detection (USAD) [2] that adopts the framework of adversarial learning and trains using two traditional autoencoders. TranAD [10] incoporated self-conditioning method and adversarial training based on Transformer [11] encoder into the model to amplify reconstruction error to easier infer outliers.

Prediction-based methods regard prediction error as anomaly score. RNN and LSTM [12] are commonly applied in time series forecasting field. The LSTM-AD [4], a stack of LSTM, was trained on non-anomalous data and measures the distribution of prediction errors which are used to assess the likelihood of anomalous behavior. Recently, GNN was also applied to extract structural features between time series in anomaly detection. Deng et al. introduced GDN [5] which utilizes attention mechanisms and GNN to learn the relationships between various sensors.

Different from the above works, our model proposes a novel design with a Transformer Residual Autoencoder-Decoder (TRAD) framework which is suitable to capture historical information from multiple perspectives for time series anomaly detection.

3   Problem Formulation

Given a time series $\mathbf{X} = \{\mathbf{x}_1, \mathbf{x}_2, \mathbf{x}_3, ..., \mathbf{x}_N\}$, $\mathbf{x}_t \in R^M$ ($1 \leq t \leq N$). If M ¿ 1, the time series is multivariate otherwise it is univariate. The anomaly detection for time series $\mathbf{X}$ is to attach a label $\mathbf{l}_t$ to each $\mathbf{x}_t$ through some classification methods, where $\mathbf{l}_t = 1$ means $\mathbf{x}_t$ is a anomaly point while $\mathbf{l}_t = 0$ represents $\mathbf{x}_t$ is normal.

4   Methodology

In this paper, we propose an effective forecasting model called **TRAD** to predict the next time slot value and use the value to detect the anomaly points. The overall of the structure is depicted in Fig. 1 which consists of three stages: Multi-Interval Sampling, TRAD-based Prediction, and Anomaly Detection. In Multi-Interval Sampling part, we perform data preprocessing by dividing raw time series into various subsequences based on different intervals. In TRAD-based Prediction, multiple Transformer Residual Autoencoder (TRA) blocks are applied to predict the value of the next point. In the anomaly detection part, we utilized prediction error to calculate the anomaly scores and detect outliers. In the following sections, we explain our proposed method in details.

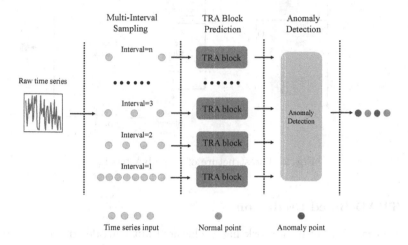

Fig. 1. The overall framework of TRAD.

4.1   Multi-interval Sampling

As showed in Fig. 1, we do under-sampling on the input sequence with equal interval principle in sequence preprocessing. Suppose we choose the interval i and the input sequence can be denoted as $\mathbf{X} = \{\mathbf{x}_1, \mathbf{x}_2, \mathbf{x}_3, ..., \mathbf{x}_N\}$ where $\mathbf{x}_t \in R^M$ $(1 \le t \le N)$ and N represents the total sequence length. First, we select an appropriate times series window T to divide the time series into multiple small sampling windows which can be represented as $\mathbf{W}_T = \{\mathbf{X}_1, \mathbf{X}_2, \mathbf{X}_3, ..., \mathbf{X}_{N-T+1}\}$. Each above window can be denoted by $\mathbf{X}_t = \{\mathbf{x}_t, \mathbf{x}_{t+1}, \mathbf{x}_{t+2}, ..., \mathbf{x}_{t+T-1}\}$ $(1 \le t \le N - T + 1)$. Then in each small window, the sampled input sequence can be represented as:

$$\boldsymbol{I}_{t,i} = \{\mathbf{x}_t, \mathbf{x}_{t+i}, \mathbf{x}_{t+2*i}, ..., \mathbf{x}_{t+(\lceil T/i \rceil -1)*i}\}. \tag{1}$$

Subsequently, the various sampled sequences are fed into the TRA (Transformer Residual Autoencoder) block to obtain anomaly scores. Finally, we aggregate them to find out outliers.

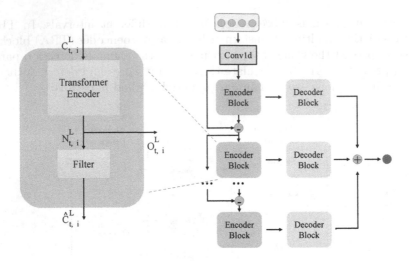

Fig. 2. The structure of the TRA block.

4.2   TRAD-Based Prediction

The proposed TRAD framework has a classical autoencoder-decoder structure as depicted in Fig. 2.

Stack of Autoencoder-Decoders: On the top of the figure, the input sequence firstly are connected to the one-dimensional convolution to initially extract the hidden representation between time slots:

$$C_{t,i}^{L=1} = Conv1d(I_{t,i}), \tag{2}$$

where L represents the layer of autoencoder-decoder.

   Subsequently, it is observed that the main part of the predictor is a stack of autoencoder-decoders. For the very first layer autoencoder, its input is $\mathbf{C}_{t,i}^{1}$. For the rest of the autoencoder, it accepts the up layer output embedding $\mathbf{C}_{t,i}^{L}$ and output two vectors $\hat{\mathbf{C}}_{t,i}^{L}$ and $\mathbf{O}_{t,i}^{L}$, where $\mathbf{O}_{t,i}^{L}$ represents the hidden embedding of current layer's autoencoder which is later fed into decoder block. And $\hat{\mathbf{C}}_{t,i}^{L}$ is then used to compute $\mathbf{C}_{t,i}^{L+1}$ which is regarded as the input of the L+1 autoencoder.

   As for the decoder, we utilize a simple linear structure to reconstruct the hidden representation collecting from the autoencoder. The process of the L layer can be denoted as:

$$\hat{y}_{t+T}^{L} = Decoder^{L}(Autoencoder^{L}(O_{t,i}^{L})), \tag{3}$$

where $\mathbf{y}_{t+T}^L$ represents the forecasting result of the L layer. Layer by layer, the final prediction value is given by the following formula:

$$\hat{\mathbf{y}}_{t+T} = \sum_{L=1}^{M} \hat{\mathbf{y}}_{t+T}^L, \tag{4}$$

where M means the layer number of the autoencoder.

Transformer Residual Autoencoder (TRA): We now focus on describing the operation of the L-th layer autoencoder block which is depicted in Fig. 2 (left). $\mathbf{C}_{t,i}^L$ initially is connected to a transformer encoder which utilizes multi-head self-attention principle to calculate the relationship of the different times steps. Furthermore, we substitute initial self-attention matrix for unidirectional masked matrix. There are at least two advantages using masked multi-head attention matrix instead of using a non-masked matrix. First, it may efficiently alleviate the problem of low rank of the attention matrix [13] caused by the bidirectional attention mechanism, which reduces the ability of expression of the attention matrix. By using masked matrix mechanism, the matrix is transformed into a triangular matrix, which is always full rank. This enhance the expressive ability of the attention matrix. Second, it decreases the number of the parameters in the model and reduces training costs.

Then the hidden representation is fed into a feed forward module to further extract the useful information of the historical series. After that, the hidden representation $\mathbf{N}_{t,i}^L$ is passed into a filter. This process can be viewed as a knowledge summarization or a filtering process, as well as a reconstruction of the encoder's output hidden layer. In the model, the filter is designed as:

$$\hat{C}_{t,i}^L = Filter^L(N_{t,i}^L) = Relu^L(Linear^L(N_{t,i}^L)), \tag{5}$$

where Relu refers to the relu activation function used to grasp nonlinear features and Linear denotes the linear layer.

In order to supervise the learning efficiency of the encoder, we utilize the mean square error (MSE) to calculate the distance between the reconstructed embedding and the upper layer input embedding. Therefore, the goal for each predictor is to minimize reconstruction error and prediction error which is represented by the loss function as:

$$\mathcal{L} = \lambda\mathcal{L}_1 + (1-\lambda)\mathcal{L}_2, \tag{6}$$

$$\mathcal{L}_1 = \frac{1}{N-T} \sum_{i=1}^{N-T} (\hat{\mathbf{y}}_{i+T} - \mathbf{y}_{i+T})^2,$$

$$\mathcal{L}_2 = \sum_{L=1}^{M-1} ||\hat{\mathbf{C}}_{t,i}^L - \mathbf{C}_{t,i}^L||,$$

where $\mathcal{L}_1$ refers to prediction error, $\mathcal{L}_2$ refers to reconstruction error and λ ($0 \leq \lambda \leq 1$) represents weighing factor for the two loss functions.

After obtaining the reconstruction error of the upper layer, we can apply residual operation in the calculation of the next layer input embedding. As showed in Fig. 2 (mid), the **L**-th layer input can be thought of as the remaining knowledge which previous autoencoder blocks can't acknowledge very well. Therefore the specific operation is represented by the following equation:

$$C_{t,i}^{L} = \begin{cases} Conv1d(I_{t,i}), & \text{if } L=1 \\ C_{t,i}^{L-1} - \hat{C}_{t,i}^{L-1}. & \text{if } L ¿1 \end{cases} \tag{7}$$

The benefits of applying this method are as follows. Firstly, coarse-grained information is built up on higher-level layers of the autoencoders. Throughout the residual operation, the downstream autoencoders receive fine-grained representations which makes the time series is easier to analyze. The downstream autoencoders only need to analyze the rest portion of the upper floor autoencoder. Secondly, it is obvious that our model has a hierarchical decomposition structure and finally the decomposed components can be aggregated to predict the next point. This design enables the model to analyze historical information from different angles which enhances the precision of the final output. Thirdly, to some extent, this design enhances the model's interpretability. There have been previous studies on residual networks, with the most famous being the classic residual network adding the input of the layers to its output. The N-BEATS architecture [14] enable the model more easy to interpret which replaces addition with subtraction. But it is difficult to quantitatively evaluate each layer's learning effect. Therefore we made further improvements by adding reconstruction error into total loss function. The operation limits the reconstruction error and force each layer of the model to do their best for time series analysis.

4.3   Anomaly Detection

In this section, we describe how to calculate anomaly scores and obtain final detection results. Suppose the model get the outputs from the prediction part which can be represented as $\mathbf{A} = \{\mathbf{a}_1, \mathbf{a}_2, \mathbf{a}_3, ..., \mathbf{a}_i, ..., \mathbf{a}_K\}$ $\mathbf{a}_i \in R^{N-T} (1 \leq i \leq K)$, where K refers to the number of TRA block chosen by the model. Then the model utilizes MSE to acquire prediction error which is considered as each TRA block's anomaly score. The model applies weighted average method on all TRA block anomaly score vectors to calculate the final anomaly score which is used to infer the anomaly points. The whole process is depicted as:

$$\{\mathbf{b}_1, \mathbf{b}_2, \mathbf{b}_3, ..., \mathbf{b}_K\} = \mathcal{MSE}(\mathbf{A}, \mathbf{X}_{real}), \tag{8}$$

$$S_{final} = \sum_{i}^{K} w_i \mathbf{b}_i, \tag{9}$$

$$\sum_{i=1}^{K} w_i = 1,$$

where $\mathbf{X}_{real} \in R^{K \times (N-T)}$ is the time series value to be predicted, $\mathbf{b}_i$ means the i-th TRA block anomaly score vector and $\mathbf{S}_{final}$ represents the final anomaly scores vector. In order to detect outliers, we set a threshold τ. If the value $s_i \in \mathbf{S}_{final}$ is greater than τ , it is regarded as a anomaly point, otherwise it is normal. We search the threshold in a feasible inteval and show our report according to the highest F1 score.

5 Experiments Setup

5.1 Datasets

Five open datasets are used in the experiment and the detail of these datasets are shown in Table 1.

- **PowerData**[1]: The dataset, which has added Gaussian noise, represents the electrical demand from the electrical system over a period of two years.
- **ISONE**[2]: It is a 12-year-long electric load dataset collecting from electrical system.
- **2D-gesture**[3]: The dataset is a collection of X-Y coordinates of hand gesture in video surveillance.
- **ECG**[3]: The ECG dataset contains anomalies corresponding to a pre-ventricular contraction. It is a 2-dimension multivariate dataset collecting from electrocardiograms.
- **SMD (Server Machine Dataset)**[4]: The dataset is a 5-week-long dataset collecting from 28 different machines in a large Internet company. It is a multivariate dataset which is composed of 38 dimensions.

Table 1. Information of datasets.

Dataset	Length	Dimension	Anomaly Rate(%)
PowerData	17520	1	9.36
ISONE	103776	1	3.56
2D-gesture	11251	2	24.63
ECG	14211	2	4.45
SMD	56958	38	9.46

---

[1] https://github.com/yaozhian/Multi-feature-power-load-forecasting-based-on-deep-learning.
[2] https://github.com/d-rorschach/Electric_load_forecasting.
[3] https://www.cs.ucr.edu/~eamonn/discords/..
[4] The SMD is from https://github.com/NetManAIOps/OmniAnomaly/tree/master.

5.2   Baselines

We compare the proposed method with five competing anomaly detection base-lines, which include Local Outlier Factor (LOF) [7], Isolation Forest (IF) [9], DAGMM [1], OmniAnomaly [15], and USAD [2].

5.3   Implementation

Our model was implemented with PyTorch. All experiments was conducted on a computer equipped with Intel(R) Xeon(R) Silver 4210R CPU @ 2.40GHz 2 cores, INVIDIA card Tesla V100 PCIe 16GB.

Throughout the training process on all datasets, we trained all models using the Adam optimizer for optimization and setting the learning rate to 10^{-3}. For the multi-interval sampling section, we selected 3 intervals for all datasets except for ECG, where we selected 2 intervals. The time sampling window $\mathbf{X}_t$ is 200 for PowerData, 192 for ECG, 256 for 2D-gesture, 400 for SMD and 200 for ISONE. The time series length of $\mathbf{I}_{t,i}$ is 50 for PowerData, 64 for ECG and 2D-gesture, 100 for SMD and 50 for ISONE. For all models, we use Conv1d with kernel size 2. The hidden size we use in autoencoder-decoder is 256 for ISONE, ECG and 2D-gesture, while the remaining datasets is 512.

We use Precision(P), Recall(R) and F1 score(F1) to evaluate the performance of our proposed model and baseline algorithm.

Table 2. Performance on five datasets: PowerData, ISONE, 2D-gesture, ECG and SMD. (P: precision, R: recall, F: f1-score)

Method	PowerData			ISONE		
	P	R	F1	P	R	F1
LOF	0.220	0.256	0.236	0.614	0.122	0.203
IF	0.196	0.221	0.208	0.256	0.184	0.214
DAGMM	0.057	0.837	0.106	0.073	0.425	0.124
OmniAnomaly	**0.727**	0.195	0.308	0.101	0.096	0.099
USAD	0.049	**0.977**	0.093	0.143	0.088	0.109
TRAD	0.538	0.479	**0.507**	**0.722**	**0.727**	**0.724**

Method	2D-gesture			ECG			SMD		
	P	R	F1	P	R	F1	P	R	F1
LOF	0.347	0.708	0.466	0.421	0.108	0.172	0.317	0.237	0.271
IF	0.309	0.873	0.456	0.128	0.378	0.191	0.261	0.855	0.400
DAGMM	0.252	**0.946**	0.398	0.043	**0.946**	0.083	0.215	0.810	0.340
OmniAnomaly	0.278	0.800	0.413	0.205	0.432	0.278	0.438	0.430	0.434
USAD	0.488	0.665	0.563	**1.000**	0.243	0.391	0.353	0.780	0.487
TRAD	**0.705**	0.736	**0.720**	0.969	0.638	**0.769**	**0.560**	**0.872**	**0.682**

6 Experiment Results

Anomaly Detection Analysis. From Table 2 of the experiments, we can draw the conclusion that our proposed model achieves the best performance on all datasets, particularly showing significant advantages on the ECG and ISONE datasets. On the one hand, our model keeps a good balance between precision and recall, without encountering the situation like USAD on ECG dataset. On the other hand, the model has achieved significant improvements in terms of abnormal detection performance, whether it is multivariate or univariate datasets.

Visualization. To better understand the detection performance of different algorithms, we conduct a visualization experiment on SMD, which is depicted in Fig. 3.

From the figure, it can be observed that the IF algorithm does not effectively separate abnormal points from the dataset, and there are a number of anomaly points and normal points that are below the threshold. The USAD model performs better but is still not able to identify as many anomaly sequences as possible.

For OmniAnomaly, the points are concentrated around the threshold, indicating that the model is not able to effectively distinguish between anomaly and normal points. For the DAGMM model, it performs the poorest of these models, as almost all points are below the threshold. We demonstrate that the proposed model is able to effectively detect abnormal points. Most of the abnormal points are classified as such above the threshold, while the majority of the normal points are classified as such below the threshold. Therefore, we can draw the conclusion that our model achieves good performance in anomaly detection tasks on the dataset SMD.

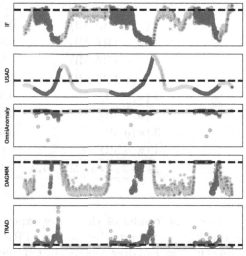

Fig. 3. Visualization of anomaly scores on dataset SMD, where the red points represent abnormal, and the black dotted line represents the threshold. (Color figure online)

6.1 Ablation Study

In this section, we will demonstrate the impact of our proposed multi-interval sampling and transformer residual autoencoder on the accuracy of abnormal detection. We use the abbreviation of multi-interval sampling MI and transformer residual autoencoder as TRA. The experimental results are illustrated in Table 3 and Table 4.

Table 3. Ablation study on SMD.

Intervals	SMD		
	Precision	Recall	F1-score
1 + 2 + 4 (w/o TRA)	0.4945	0.7160	0.5850
1 (w/o TRA MI)	0.4606	0.4993	0.4792
2 (w/o TRA MI)	0.4958	0.7004	0.5806
4 (w/o TRA MI)	0.3703	0.6849	0.4807
1 (w/o MI)	0.5170	0.8452	0.6416
2 (w/o MI)	0.5474	0.8664	0.6709
4 (w/o MI)	0.5489	**0.8935**	0.6800
1 + 2 + 4 (w/ MI+TRA)	**0.5600**	0.8716	**0.6819**

Table 4. Ablation study on 2D-gesture.

Intervals	2D-gesture		
	Precision	Recall	F1-score
1+2+3+4 (w/o TRA)	0.5246	0.6928	0.5971
1 (w/o TRA MI)	0.4503	0.5643	0.5009
2 (w/o TRA MI)	0.5539	0.6536	0.5996
3 (w/o TRA MI)	0.5079	0.5670	0.5358
4 (w/o TRA MI)	0.4563	0.5927	0.5156
1 (w/o MI)	0.5173	0.7280	0.6048
2 (w/o MI)	**0.7658**	0.5575	0.6453
3 (w/o MI)	0.6971	0.6198	0.6562
4 (w/o MI)	0.6025	0.7118	0.6526
1+2+3+4 (w/ MI+TRA)	0.6655	**0.7754**	**0.7162**

From the results of the above experiments, it indicates that the outcome is the state-of-the-art when MI and TRA are both used simultaneously. At the same time, we can draw the conclusion that the impact of TRA on the outcome is greater compared to MI. Additionally, we can infer from the results of 2D-gesture experiment that fully leveraging temporal information before aggregating them is more effective than directly aggregation.

7   Conclusion

This paper proposed a novel time series anomaly detection method based on a multi-interval and residual transformer autoencoder-decoder structure. It comprehensively considered the impact of different interval divisions of the time series on the calculation of abnormal scores. For the multi-scale aspects, a hierarchical autoencoder-decoder with residual learning was designed to further reconstruct time series, resulting in more accurate predictions. Extensive experiments were conducted on five open datasets, which showed that the proposed method achieves the best F1 score compared to the state-of-the-art baseline algorithms.

Acknowledgments. This work was partially supported by the National Natural Science Foundation of China (Grant Nos. 61972196, 61832008, 61832005), the Collaborative Innovation Center of Novel Software Technology and Industrialization, and the Sino-German Institutes of Social Computing. The corresponding author is Wenzhong Li (lwz@nju.edu.cn).

References

1. Zong, B., et al.: Deep autoencoding gaussian mixture model for unsupervised anomaly detection. In: International Conference on Learning Representations (2018)
2. Audibert, J., Michiardi, P., Guyard, F., Marti, S., Zuluaga, M.A.: USAD: unsupervised anomaly detection on multivariate time series. In: Proceedings of the 26th ACM SIGKDD International Conference on Knowledge Discovery & Data Mining, pp. 3395–3404 (2020)
3. Rumelhart, D.E., Hinton, G.E., Williams, R.J.: Learning internal representations by error propagation. Technical report, Calornia Univ San Diego La Jolla Inst for Cognitive Science (1985)
4. Malhotra, P., Vig, L., Shroff, G., Agarwal, P., et al.: Long short term memory networks for anomaly detection in time series. In: ESANN, vol. 2015, pp. 89 (2015)
5. Deng, A., Hooi, B.: Graph neural network-based anomaly detection in multivariate time series. In: Proceedings of the AAAI Conference on Artificial Intelligence, vol. 35, pp. 4027–4035 (2021)
6. Zhou, J., et al.: Graph neural networks: a review of methods and applications. AI open **1**, 57–81 (2020)
7. Breunig, M.M., Kriegel, H.-P., Ng, R.T., Sander, J.: LOF: identifying density-based local outliers. In: Proceedings of the 2000 ACM SIGMOD International Conference on Management of Data, pp. 93–104 (2000)
8. Shyu, M., Chen, S., Sarinnapakorn, K., Chang, L.: A novel anomaly detection scheme based on principal component classifier. In: IEEE Foundations and New Directions of Data Mining Workshop, in conjunction with the Third IEEE International Conference on Data Mining (ICDM03) (2003)
9. Liu, F.T., Ting, K.M., Zhou, Z.-H.: Isolation forest. In: 2008 Eighth IEEE International Conference on Data Mining, pp. 413–422. IEEE (2008)
10. Tuli, S., Casale, G., Jennings, N.R.: TranAD: Deep transformer networks for anomaly detection in multivariate time series data. arXiv preprint arXiv:2201.07284 (2022)
11. Vaswani, A., et al.: Attention is all you need. In: Advances in Neural Information Processing Systems, vol. 30 (2017)
12. Hochreiter, S., Schmidhuber, J.: Long short-term memory. Neural Comput. **9**(8), 1735–1780 (1997)
13. Dong, Y., Cordonnier, J.-B., Loukas, A.: Attention is not all you need: pure attention loses rank doubly exponentially with depth. In: International Conference on Machine Learning, pp. 2793–2803. PMLR (2021)

14. Oreshkin, B.N., Carpov, D., Chapados, N., Bengio, Y.: N-BEATS: Neural basis expansion analysis for interpretable time series forecasting. arXiv preprint arXiv:1905.10437 (2019)
15. Su, Y., Zhao, Y., Niu, C., Liu, R., Sun, W., Pei, D.: Robust anomaly detection for multivariate time series through stochastic recurrent neural network. In: Proceedings of the 25th ACM SIGKDD International Conference on Knowledge Discovery & Data Mining, pp. 2828–2837 (2019)

Adversarial Example Detection with Latent Representation Dynamic Prototype

Taowen Wang[ID], Zhuang Qian[ID], and Xi Yang[(✉)][ID]

Xi'an Jiaotong-Liverpool University, Suzhou, China
taowen.wang21@alumni.xjtlu.edu.cn, zhuang.qian20@student.xjtlu.edu.cn,
xi.yang01@xjtlu.edu.cn

Abstract. In the realm of Deep Neural Networks (DNNs), one of the primary concerns is their vulnerability in adversarial environments, whereby malicious attackers can easily manipulate them. As such, identifying adversarial samples is crucial to safeguarding the security of DNNs in real-world scenarios. In this work, we propose a method of adversarial example detection. Our approach using a Latent Representation Dynamic Prototype to sample more generalizable latent representations from a learnable Gaussian distribution, which relaxes the detection dependency on the nearest neighbour's latent representation. Additionally, we introduce Random Homogeneous Sampling (RHS) to replace KNN sampling reference samples, resulting in lower reasoning time complexity at $O(1)$. Lastly, we use cross-attention in the adversarial discriminator to capture the evolutionary differences of latent representation in benign and adversarial samples by comparing the latent representations from inference and reference samples globally. We conducted experiments to evaluate our approach and found that it performs competitively in the gray-box setting against various attacks with two $\mathcal{L}_p$-norm constraints for CIFAR-10 and SVHN datasets. Moreover, our detector trained with PGD attack exhibited detection ability for unseen adversarial samples generated by other adversarial attacks with small perturbations, ensuring its generalization ability in different scenarios.

Keywords: Adversarial example detection · Adversarial attack · Cross attention

1 Introduction

Deep neural networks (DNNs) have achieved extraordinary success in various perceptual tasks, such as image classification [8,14], machine translation [4] and speech recognition [9]. However, research shows that attackers can manipulate DNNs' output with high confidence with malicious perturbations that are imperceptible to humans [6,30]. This property of DNNs raises concerns about the reliability of DNNs-based systems, particularly in safety-critical scenarios such as autonomous driving and medical diagnosis.

B. Luo et al. (Eds.): ICONIP 2023, LNCS 14450, pp. 525–536, 2024.
https://doi.org/10.1007/978-981-99-8070-3_40

Research on the robustness of DNN is essential in this regard and can be broadly divided into two branches. Proactive defence involves modifying the model, such as adversarial training [6,22] and knowledge distillation [10]. For these approaches, defenders look for a set of weight parameters that reduce the model's probability of miss-classified adversarial samples, often requiring retraining and carrying the positional risk of a decrease in accuracy on benign samples. The alternative is reactive defence, which focuses on increasing the robustness of DNNs by adding additional mechanisms, e.g., purification [7,27,33,34], input transformation [18,31], and detection-based methods [1,16,20,32]. Unlike modifying the model, reactive defence utilizes extra mechanisms to passively defend the system, usually avoiding to retrain models. It can mitigate accuracy degradation on benign samples, making it a more efficient and less damaging option for already deployed models.

The field of detection-based defence has undergone significant development, as researchers focus on protecting deep neural networks from adversarial attacks. Current researches [23] have evaluated the discriminative ability of latent representations from different layers under gray-box settings for detecting adversarial examples. However, the performance of the detector varied when the attack method or perturbation magnitude changed. Another attempt involves referencing the input sample's neighbours in the latent space to identify whether it is an adversarial sample [1,28]. However, finding the neighbours during each inference using k-nearest neighbours incurs a time complexity of $O(n)$, which can be time-consuming for large reference datasets.

In this work, we aim to address the two aforementioned challenges: unstable detection performance across diverse adversarial examples and the high time-complexity of finding nearest neighbours. To alleviate the reliance on the latent representation of inference samples while incorporating latent representation evolution information from benign samples in detecting, we propose a novel sampling method called Random Homogeneous Sampling (RHS), which randomly samples reference samples in the corresponding categories through the classification results of the target model. We further introduce a Latent Representation Dynamic Prototype (LRDP), wherein we extract latent representations from different target model stages of the RHS sampled reference sample to fully compare them with inference samples of different semantic stages. We then construct a Gaussian distribution via encoded latent representations and randomly sample prototypes to increase the generality of representation and stability toward different adversarial samples. Finally, we utilize cross-attention to globally compare the latent representation from the inference sample and prototype to alleviate the dependency of the detector on representation from a specific stage.

The contributions of this work are summarized as follows:

- We propose a novel detection-based defence method that leverages cross-attention to compare the latent representations of the benign and adversarial samples.
- The proposed method is efficient, with time complexity of $O(1)$ for Random Homogeneous Sampling.

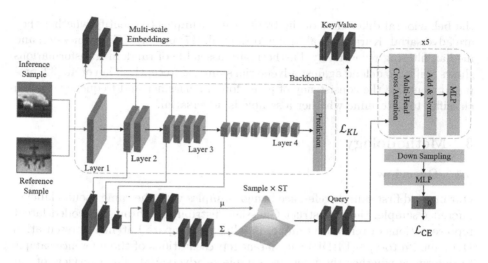

Fig. 1. Overview of the proposed method.

- Our experimental results demonstrate that the proposed method performs competitively for detecting adversarial examples under gray-box setting.

2   Related Work

Adversarial Attacks: Many approaches have been proposed to generate adversarial samples efficiently. Goodfellow et al. [6] proposed Fast Gradient Sign Method (FGSM) to add a single-step gradient to a benign sample. After that, many iterative methods have been proposed to enhance the performance of FGSM attacks, such as Basic Iterative Method (BIM) [15] and Projected Gradient Descent (PGD) [22]. Furthermore, Deepfool [24] iteratively perturbs benign examples to cross the approximate linearization, the nearest decision boundaries. Moreover, Carlini et al. [2] obtained high-quality adversarial samples by optimizing the L_p distance between the adversarial and benign samples with linearized classification constraints as a penalty. In addition, to avoid attack inefficiency caused by suboptimal hyperparameters, AutoAttack [3] assembled multiple hyperparameter-free attacks to autonomously evaluate the robustness of defence.

Detection as Defence: Various detection-based defences have been proposed to reactively protect model from adversarial attack. Neighbourhood-based defences utilize one or more neighbour(s) of inference sample in latent space. DKNN [28] refer to neighbours in each layer to judge a sample is adversarial or not. Ma et al. [20] introduced the Local Intrinsic Dimensionality (LID) to describe the dimensional properties of adversarial region for adversarial detection. Lee et al. [16] utilize the Mahalanobis distance to evaluate the probability if an input is adversarial. Abusnaina et al. [1] dynamically explore the neighbourhood relationships with graph network for detecting. Another line of research is to utilize the behavioural differences of the transformed samples to identify whether they are adversarial. Kurakin et al. [15] and Li et al. [17] uses JPEG compression and median filtering respectively. Furthermore, assemble of random transformations shows ability to defend against adversarial samples. Recently, Tian et al. [32] proposed to utilize the consistency of pixel domain classifier and frequency domain classifier to determine whether a sample is adversarial.

3   Methodology

3.1   Overview

Our method first sample reference benign samples with the same pseudo label of inference sample, then construct Gaussian distribution through encoded latent representations of reference sample. Finally, we compare Latent Representation Dynamic Prototype (LRDP) with latent representations of the inference sample to determine whether the inference sample is adversarial. The overview of our detect mechanism is shown in Fig. 1.

3.2   Random Homogeneous Sampling

Giver a well-trained k classes DNN classifier $\mathcal{F}$ trained on benign dataset D_t of n samples. We construct a reference dataset $D_{ref} \subseteq D_t$ by randomly sample from the naive training dataset D_t. In the process of detection, we first obtain a pseudo label y_p by the prediction of classifier:

$$y_p = \arg\max \mathcal{F}(x_{inf}). \tag{1}$$

To guide the generation of LRDP, the homogeneous benign sample $\{x_{ref}, y_p\}$ is sampled from D_{ref} according to pseudo label y_p

3.3   Latent Representation Dynamic Prototype

To improve the generalization ability, we propose Latent Representation Dynamic Prototype. Let $o^{(l)}$ the output of l-th layer of DNN, $l \in \{1, ..., L\}$. We extract latent representations with different semantic levels in the shallow, middle and deep stage from representations of DNN. For the different stages of latent representations, we normalize them to a consistent size combination o_n by different forward propagation:

$$s = [f^s(o^{l_s}), f^m(o^{l_m}), f^d(o^{l_d})]^\top, \tag{2}$$

where $f^s(\cdot), f^m(\cdot), f^d(\cdot)$ are three different forward propagation to normalize latent representations. Then, we use two learnable forward propagation layers to obtain the mean and standard deviation of reference sample representations distribution, we use the reparameterization trick [12] for optimization and sampling prototype o_p:

$$o^p = f_\mu(s^{ref}) + f_\Sigma(s^{ref}) \odot I, \quad \text{where} \quad I \in \mathcal{N}(0, 1). \tag{3}$$

The normalization process for the inference and reference sample does not involve any shared weights. To better guide learning and generate prototypes that are discriminative for benign and adversarial examples, we optimize different Kullback-Leibler (KL) divergence according to the inference sample is adversarial or not.

$$
\begin{aligned}
\mathcal{L}_{kl} = & \mathbb{I}(y_{adv} = 0) \sum_{i=1}^{N} \sum_{k=1}^{ST} D_{kl}(\sigma(s_i^{inf}) || \sigma(o_k^{p_i})) \\
& + \mathbb{I}(y_{adv} = 1) \sum_{j=1}^{M} \sum_{k=1}^{ST} max(\alpha - D._{kl}(\sigma(s_j^{inf}) || \sigma(o_k^{p_j})), 0),
\end{aligned}
\tag{4}
$$

where σ is the softmax layer and $\mathbb{I}$ is indication function, N, M and ST corresponds to the number of benign samples, adversarial samples and samples of prototypes. Furthermore, α is a margin hyperparameter that controls the magnitude of KL divergence difference for adversarial pairs.

3.4 Adversarial Discriminator

To compare the similarity between normalised inference sample representation s^{inf} and sampled prototype o^p, we adopt cross-attention to compare globally. The module first adds position encoding to o^p and s^{inf}, respectively, then takes o_{pos}^p as the query and s_{pos}^{inf} as the key and value. The discriminator architecture consists of five cross-attention modules, with normalization, linear and residual modules added in each cross-attention module. The binary detection results are obtained through multiple linear modules. The cross-attention module is formulated as follows:

$$
\text{Cross Attention}(o_{pos}^p, s_{pos}^{inf}, s_{pos}^{inf}) = \sigma(\frac{o_{pos}^p (s_{pos}^{inf})^\top}{\sqrt{d_k}}) s_{pos}^{inf}.
\tag{5}
$$

The final loss function $\mathcal{L}_{total}$ comprises the KL-divergence loss $\mathcal{L}_{kl}$ and cross-entropy loss $\mathcal{L}_{ce}$ appropriately weighted by a trade-off hyperparameter λ:

$$
\mathcal{L}_{total} = \lambda \mathcal{L}_{ce} + (1 - \lambda) \mathcal{L}_{kl}.
\tag{6}
$$

4 Experiments

To evaluate the performance of our proposed method, We compare the proposed method with four SOTA adversarial example detection methods, including LNG [1], LID [20], MD [16] and SID [32]. Note that in the comparison of LNG, as there is no public code and adversarial examples provided, we regenerated adversarial examples on the same network and dataset based on the attack settings provided in the LNG. In the comparison with the other three approaches, we generated six state-of-the-art attacks on two datasets: CIFAR-10 [13] and SVHN [25]. All six attack methods are 'non-target'. Except for the CW attack, which is only generated under L_2 constraint, the remaining five methods are generated under both L_2 and L_∞ constraints to fully evaluate the detection capability. For evaluation metrics, the widely used AUROC score is utilized to evaluate the performance of different detectors. We reproduce and fine-tune LID, MD and SID with public available codes for fair comparison. Before demonstrating our results, we outline the parameters for generating adversarial examples and the experimental setup.

4.1   Experiment Setup

Training Target Model: The ResNet-34 [8] is placed as target model, trained on D_{train} yields classification accuracy of 94.18% and 96.01% corresponding to CIFAR-10 and SVHN respectively. For the frequency domain classifier of SID, higher classification leads to better consistency detection capability. Thus we carefully fine-tuned the hyperparameter of SID and obtained 91.97% and 96.04% frequency domain classification performance on CIFAR-10 and SVHN, respectively. In comparison with LNG, the same network structure ResNet-110 mentioned in LNG is placed as the target model, with similar accuracy 93.26% of classification with CIFAR-10. All samples are normalized to $[0, 1]$. Foolbox [29] and Adversarial Tool Box [26] are used for generating adversarial examples.

Training Detector: For the training procedure of our detector, we utilize AdamW [19] for optimization and trained 100 epochs on both CIFAR-10 and SVHN with a learning rate of 0.001. We set the margin α and the trade-off λ in Eq. 4 and Eq. 6 to 1 and 0.3, respectively. In addition, we extract the first post-convolution embedding after the activation function of layer 1, layer 2 and layer 3 as the source of our dynamic prototype and comparison.

4.2   Threat Model

The proposed method is evaluated under both L_2 and L_∞-norm bound. The attacking scenario is focused on the Gray-box Setting where the adversary is visible to the parameters of the model but not to the mechanisms of additional defences.

Comparison with LNG: For a fair comparison with the method of unreleased code, we generated six types of L_∞ adversarial samples on CIFAR-10 with the same attack setting and data normalization as provided in the LNG. In addition, to make our generated adversarial examples more similar to LNG's data, we adopt the same Adversarial Tool box as mentioned in the original paper.

Comparison with SID: In general, the adversarial samples with low perturbation magnitude are less different from the original samples in both input and latent space than the adversarial samples with large perturbation magnitude. Therefore the adversarial samples with low perturbation magnitude are more difficult to detect. As shown in Fig. 2, the detection results of adversarial examples with large perturbation show a signification superiority.

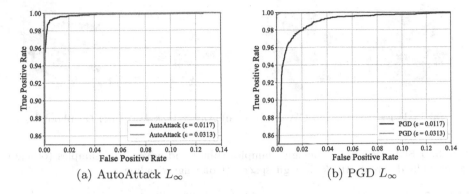

(a) AutoAttack L_∞ (b) PGD L_∞

Fig. 2. The AUROC of AutoAttack and PGD on CIFAR-10 with different perturbation magnitude $\epsilon = [0.0117, 0.0313]$

Thus, to better evaluate the performance of the proposed method, we generate adversarial samples with smaller perturbation magnitudes than the setting of SID. Following the setting of LNG, all images are normalized to $[0, 1]$ and adversarial samples are generated by Foolbox. Specifically, the parameters for generating adversarial examples are listed in Table 1.

Table 1. Parameters for generating adversarial example. The Adv Num is the sum of the number of adversarial samples in the training and testing dataset of the detector.

Method	L_2						L_∞				
	AutoAttack [3]	FGSM [6]	BIM [15]	PGD [22]	DeepFool [24]	CW [2]	AutoAttack	FGSM	BIM	PGD	DeepFool
δ	0.25	0.25	0.25	0.25	0.25	0.25	0.0117	0.0313	0.0313	0.0117	0.0313
Iterations	-	1	5	50	50	10000	–	1	5	50	50
Step size	-	0.25	0.05	0.0063	–	0.01	–	0.0313	0.0063	0.0003	–
CIFAR Adv Num	9032	3854	6504	7258	4903	5431	9248	6516	9200	9021	9073
SVHN Adv Num	20684	8596	13292	13630	11526	8426	23847	15258	24626	21370	24278

We visualize the T-SNE [21] in the logit space of benign sample and the generated adversarial examples on CIFAR-10 in Fig 3.

4.3 Comparison in Seen Attacks

To evaluate the effectiveness of the proposed method, we conduct experiments on the comparisons between the several SOTAs and the proposed method. In such experiments, both the training and testing sets are perturbed under the same setting.

In Table 2, we compare the AUROC score of our proposed method with methods from the LNG paper under five different attacks. All the adversarial examples are generated following the settings from LNG, and the perturbation magnitude is large than the setting in Table 1. According to the comparison result, the proposed detection method outperforms other methods in almost all cases except

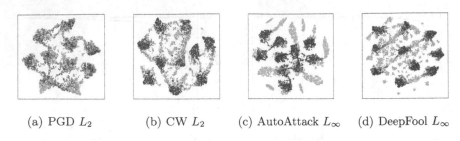

(a) PGD L_2 (b) CW L_2 (c) AutoAttack L_∞ (d) DeepFool L_∞

Fig. 3. T-SNE visualize of benign examples (blue) and adversarial samples (orange) generate by different attacks in logit spaces. (Color figure online)

Table 2. The AUROC performance on CIFAR-10. The attack parameters are same as the setting of LNG.

Dataset	Detector	AutoAttack	FGSM	PGD	CW	Boundary	Square
CIFAR-10	DkNN [28]	52.11	61.50	51.18	61.46	70.11	59.51
	kNN [5]	52.64	61.80	54.46	65.25	75.88	73.39
	LID [20]	56.25	73.56	67.95	55.60	99.48	85.93
	Hu [11]	53.54	84.44	58.55	90.99	90.71	95.83
	LNG [1]	84.03	99.88	91.39	89.74	**99.98**	98.82
	Ours	**100**	**100**	**100**	**99.06**	99.17	**98.97**

for the Boundary attack. For AutoAttack, FGSM, PGD and CW attacks, large magnitude means that the adversarial samples are more different from the original samples in the latent space. Ours shows high detection performance on the adversarial samples generated by all four attacks. The proposed method illustrates slightly inferior detection performance for the adversarial samples that the black-box attack method (Boundary attack) generates.

The difference in detection performance between strong and weak attacks is from our modelling of the latent space of the model. The adversarial samples generated from strong attacks cause significant differences from the benign samples in the latent space. In contrast, the generated adversarial samples cause less significant deviations from the benign samples in the latent space for weak attacks. Therefore, for the detector, the insignificant features of weakly adversarial samples are the main reason for the poor detection performance.

To evaluate our proposed method in extreme situations, we generate adversarial samples with lower L_∞ perturbation magnitude as listed in Table 1. Furthermore, to verify the generality of our proposed method, we evaluate the effectiveness of our proposed method with relatively low L_2 bounded adversarial examples. In the comparison results in Table 3, the detection performance of our proposed method for L_∞ adversarial samples is outperformed by the corresponding L_2 adversarial samples. In addition, the difference in detection performance between L_∞ and L_2 is observable in the between Table 2 and 3. Based on the

Table 3. The AUROC performance of different methods in detecting the same attack. The method listed with * is reproduced by the corresponding official code.

Dataset	Detector	L_2						L_∞				
		AutoAttack	FGSM	BIM	PGD	DeepFool	CW	AutoAttack	FGSM	BIM	PGD	DeepFool
CIFAR-10	LID*	73.44	74.08	64.21	64.21	86.72	80.10	88.24	97.91	89.37	75.58	91.70
	MD*	85.79	76.47	73.60	76.22	86.31	82.68	94.80	99.91	97.83	86.45	93.57
	SID*	98.40	88.78	96.00	**97.46**	89.12	92.01	99.88	96.29	99.58	99.25	96.64
	Ours	**99.20**	**95.42**	**96.24**	97.28	**93.12**	**94.21**	**99.96**	**100**	**99.99**	**99.83**	**96.88**
SVHN	LID*	75.00	77.90	75.35	74.49	85.72	80.30	81.24	95.46	85.91	77.73	94.55
	MD*	82.88	76.85	77.70	77.25	87.21	81.24	92.79	98.89	96.57	86.33	93.12
	SID*	98.80	91.07	95.82	96.44	95.12	93.22	99.74	95.62	99.71	99.21	98.12
	Ours	**99.61**	**96.01**	**98.08**	**98.28**	**98.78**	**96.50**	**99.97**	**100**	**99.99**	**99.88**	**99.73**

above results, we conclude that L_∞ adversarial samples cause more significant damage in the hidden space at the current perturbation magnitude than L_2 adversarial example. The phenomenon can be observed in the T-SNE visualization in Fig 3, where the L_∞ generated adversarial sample deviates more from the benign sample in the logit space than the L_2 adversarial sample.

4.4   Comparison in Unseen Attacks

Table 4. The AUROC generalization results against different L_2 attacks on PGD-trained models

Dataset	Detector	(Seen)	L_2						L_∞				
		PGD	AutoAttack	FGSM	BIM	DeepFool	CW	AutoAttack	FGSM	BIM	PGD	DeepFool
CIFAR-10	LID*	–	64.08	74.08	66.55	**86.72**	74.90	81.35	95.23	84.21	71.61	91.70
	MD*	-	85.10	69.45	73.60	74.01	78.30	91.99	91.95	81.01	82.96	84.06
	SID*	-	98.84	88.94	96.15	84.77	**90.89**	99.09	89.17	96.57	98.68	**92.07**
	Ours		**98.92**	**94.87**	**96.49**	85.18	87.94	**99.64**	**99.79**	**99.57**	**99.37**	90.34
SVHN	LID*	–	75.00	75.26	75.42	81.25	76.45	81.24	95.44	86.35	77.73	89.14
	MD*	–	82.62	75.37	77.70	79.46	78.17	91.15	**97.40**	93.81	85.66	88.24
	SID*	–	98.51	91.03	95.83	89.27	91.42	**99.39**	90.88	98.66	98.88	93.94
	Ours	–	**99.22**	**96.61**	**98.04**	**96.23**	**96.43**	99.37	96.45	**99.12**	**99.11**	**98.13**

Detecting Unseen Attacks: As the deviation of the L_2 PGD adversarial samples from the benign samples in the latent space is slight, our proposed method can model the evolution of latent space better in this case. Therefore, we choose L_2 PGD adversarial samples as the visible training samples for learning and testing the model's ability to detect the remaining adversarial samples. For both 2 datasets, our proposed method shows competitive results in Table 4.

4.5   Time Comparison

For the proposed method, the average inference time on Cifar-10 is 0.00142 s per image. The reasoning process includes: 1. target model reasoning time; 2. random homogeneous sampling as well as reference sample re-reasoning and extracted latent representations; 3. proposed model reasoning process. Compared to the graph-based adversarial sample detection method LNG [1], their claimed reasoning time on CIFAR-10 is 1.55 s per image. For our proposed method, the random homogeneous sampling can reduce the reasoning time significantly compared to KNN based method. Moreover, the two forward propagation processes using the target model in our proposed method do not significantly increase the reasoning time by an order of magnitude. Thus, our method is efficient for detecting adversarial samples.

5   Conclusion

This work proposes a method for detecting adversarial examples in Deep Neural Networks (DNNs). The proposed approach utilizes a Latent Representation Dynamic Prototype with Random Homogeneous Sampling to relax the detection dependency on nearest neighbour's latent representation, resulting in lower time complexity. Additionally, cross-attention is used in the adversarial discriminator to capture the evolutionary differences in latent representation in benign and adversarial samples. Experimental evaluation demonstrates competitive performance for the method against several attacks with two $\mathcal{L}_p$-norm constraints in CIFAR-10 and SVHN datasets, as well as generalization ability to detect unseen adversarial examples generated by various attacks and L_p-norm bound.

Acknowledgements. Natural Science Foundation of the Jiangsu Higher Education Institutions of China under no.22KJB520039; National Natural Science Foundation of China under no.62206225; Research Development Fund in XJTLU under no. RDF-19-01-21.

References

1. Abusnaina, A., et al.: Adversarial example detection using latent neighborhood graph. In: 2021 IEEE/CVF International Conference on Computer Vision, pp. 7667–7676. IEEE (2021). https://doi.org/10.1109/ICCV48922.2021.00759
2. Carlini, N., Wagner, D.: Towards evaluating the robustness of neural networks. In: 2017 IEEE Symposium on Security and Privacy, pp. 39–57 (2017). https://doi.org/10.1109/SP.2017.49
3. Croce, F., Hein, M.: Reliable evaluation of adversarial robustness with an ensemble of diverse parameter-free attacks. In: Proceedings of the 37th International Conference on Machine Learning. Proceedings of Machine Learning Research, vol. 119, pp. 2206–2216. PMLR (2020)
4. Devlin, J., Chang, M., Lee, K., Toutanova, K.: BERT: pre-training of deep bidirectional transformers for language understanding. In: Burstein, J., Doran, C., Solorio, T. (eds.) Proceedings of the 2019 Conference of the North American Chapter of the Association for Computational Linguistics: Human Language Technologies, pp. 4171–4186. Association for Computational Linguistics (2019). https://doi.org/10.18653/v1/n19-1423

5. Dubey, A., van der Maaten, L., Yalniz, Z., Li, Y., Mahajan, D.: Defense against adversarial images using web-scale nearest-neighbor search. In: IEEE Conference on Computer Vision and Pattern Recognition, pp. 8767–8776. IEEE Computer Society (2019). https://doi.org/10.1109/CVPR.2019.00897

6. Goodfellow, I.J., Shlens, J., Szegedy, C.: Explaining and harnessing adversarial examples. In: Bengio, Y., LeCun, Y. (eds.) 3rd International Conference on Learning Representations (2015)

7. Gu, S., Rigazio, L.: Towards deep neural network architectures robust to adversarial examples. In: Bengio, Y., LeCun, Y. (eds.) 3rd International Conference on Learning Representations (2015)

8. He, K., Zhang, X., Ren, S., Sun, J.: Deep residual learning for image recognition. In: Proceedings of the IEEE Conference on Computer Vision and Pattern Recognition (2016)

9. Hinton, G., et al.: Deep neural networks for acoustic modeling in speech recognition: the shared views of four research groups. IEEE Signal Process. Mag. **29**(6), 82–97 (2012). https://doi.org/10.1109/MSP.2012.2205597

10. Hinton, G.E., Vinyals, O., Dean, J.: Distilling the knowledge in a neural network. CoRR abs/ arXiv: 1503.02531 (2015)

11. Hu, S., Yu, T., Guo, C., Chao, W., Weinberger, K.Q.: A new defense against adversarial images: turning a weakness into a strength. In: Wallach, H.M., Larochelle, H., Beygelzimer, A., d'Alché-Buc, F., Fox, E.B., Garnett, R. (eds.) Advances in Neural Information Processing Systems 32: Annual Conference on Neural Information Processing Systems 2019, pp. 1633–1644 (2019)

12. Kingma, D.P., Welling, M.: Auto-encoding variational bayes. In: Bengio, Y., LeCun, Y. (eds.) 2nd International Conference on Learning Representations (ICLR) (2014)

13. Krizhevsky, A., Hinton, G., et al.: Learning multiple layers of features from tiny images (2009)

14. Krizhevsky, A., Sutskever, I., Hinton, G.E.: Imagenet classification with deep convolutional neural networks. In: Pereira, F., Burges, C., Bottou, L., Weinberger, K. (eds.) Advances in Neural Information Processing Systems, vol. 25. Curran Associates, Inc. (2012)

15. Kurakin, A., Goodfellow, I.J., Bengio, S.: Adversarial examples in the physical world. In: 5th International Conference on Learning Representations. OpenReview.net (2017)

16. Lee, K., Lee, K., Lee, H., Shin, J.: A simple unified framework for detecting out-of-distribution samples and adversarial attacks. In: Bengio, S., Wallach, H.M., Larochelle, H., Grauman, K., Cesa-Bianchi, N., Garnett, R. (eds.) Advances in Neural Information Processing Systems 31: Annual Conference on Neural Information Processing Systems 2018, pp. 7167–7177 (2018)

17. Li, X., Li, F.: Adversarial examples detection in deep networks with convolutional filter statistics. In: Proceedings of the IEEE International Conference on Computer Vision, pp. 5775–5783. IEEE Computer Society (2017). https://doi.org/10.1109/ICCV.2017.615

18. Liu, Z., et al.: Feature distillation: DNN-oriented JPEG compression against adversarial examples. In: IEEE Conference on Computer Vision and Pattern Recognition, pp. 860–868. Computer Vision Foundation/IEEE (2019). https://doi.org/10.1109/CVPR.2019.00095

19. Loshchilov, I., Hutter, F.: Decoupled weight decay regularization. In: 7th International Conference on Learning Representations. OpenReview.net (2019)

20. Ma, X., et al.: Characterizing adversarial subspaces using local intrinsic dimensionality. In: 6th International Conference on Learning Representations. OpenReview.net (2018)
21. van der Maaten, L., Hinton, G.: Visualizing data using t-sne. J. Mach. Learn. Res. **9**(86), 2579–2605 (2008)
22. Madry, A., Makelov, A., Schmidt, L., Tsipras, D., Vladu, A.: Towards deep learning models resistant to adversarial attacks. In: 6th International Conference on Learning Representations. OpenReview.net (2018)
23. Metzen, J.H., Genewein, T., Fischer, V., Bischoff, B.: On detecting adversarial perturbations. In: 5th International Conference on Learning Representations. OpenReview.net (2017)
24. Moosavi-Dezfooli, S., Fawzi, A., Frossard, P.: Deepfool: A simple and accurate method to fool deep neural networks. In: IEEE Conference on Computer Vision and Pattern Recognition. pp. 2574–2582. IEEE Computer Society (2016). https://doi.org/10.1109/CVPR.2016.282
25. Netzer, Y., Wang, T., Coates, A., Bissacco, A., Wu, B., Ng, A.Y.: Reading digits in natural images with unsupervised feature learning. In: NIPS Workshop on Deep Learning and Unsupervised Feature Learning 2011 (2011)
26. Nicolae, M.I., et al.: Adversarial robustness toolbox v1.2.0. CoRR arXiv: 1807.01069 (2018)
27. Nie, W., Guo, B., Huang, Y., Xiao, C., Vahdat, A., Anandkumar, A.: Diffusion models for adversarial purification. In: Chaudhuri, K., Jegelka, S., Song, L., Szepesvári, C., Niu, G., Sabato, S. (eds.) International Conference on Machine Learning. Proceedings of Machine Learning Research, vol. 162, pp. 16805–16827. PMLR (2022)
28. Papernot, N., McDaniel, P.D.: Deep k-nearest neighbors: Towards confident, interpretable and robust deep learning. CoRR abs/ arXiv: 1803.04765 (2018)
29. Rauber, J., Brendel, W., Bethge, M.: Foolbox v0.8.0: a python toolbox to benchmark the robustness of machine learning models. CoRR abs/ arXiv: 1707.04131 (2017)
30. Szegedy, C., et al.: Intriguing properties of neural networks. In: Bengio, Y., LeCun, Y. (eds.) 2nd International Conference on Learning Representations (2014)
31. Taran, O., Rezaeifar, S., Holotyak, T., Voloshynovskiy, S.: Defending against adversarial attacks by randomized diversification. In: IEEE Conference on Computer Vision and Pattern Recognition, pp. 11226–11233. Computer Vision Foundation/IEEE (2019). https://doi.org/10.1109/CVPR.2019.01148
32. Tian, J., Zhou, J., Li, Y., Duan, J.: Detecting adversarial examples from sensitivity inconsistency of spatial-transform domain. In: 35th AAAI Conference on Artificial Intelligence, pp. 9877–9885. AAAI Press (2021)
33. Xie, C., Wu, Y., van der Maaten, L., Yuille, A.L., He, K.: Feature denoising for improving adversarial robustness. In: IEEE Conference on Computer Vision and Pattern Recognition, pp. 501–509. Computer Vision Foundation/IEEE (2019)
34. Yang, Y., Zhang, G., Xu, Z., Katabi, D.: ME-Net: towards effective adversarial robustness with matrix estimation. In: Chaudhuri, K., Salakhutdinov, R. (eds.) Proceedings of the 36th International Conference on Machine Learning. Proceedings of Machine Learning Research, vol. 97, pp. 7025–7034. PMLR (2019)

A Multi-scale and Multi-attention Network for Skin Lesion Segmentation

Cong Wu, Hang Zhang$^{(\boxtimes)}$, Dingsheng Chen, and Haitao Gan

Hubei University of Technology,, Wuhan, China
zh243370711@163.com

Abstract. Accurately segmenting the diseased areas from dermoscopy images is highly meaningful for the diagnosis of skin cancer, and in recent years, methods based on deep convolutional neural networks have become the mainstream for automatic segmentation of skin lesions. Although these methods have made significant improvements in the field of skin lesion segmentation, capturing long-range dependencies remains a major challenge for convolutional neural networks. In order to address this limitation, this paper proposes a deep learning model for skin lesion segmentation called the Multi-Scale and Multi-Attention Network (MSMA-Net). The encoder part utilizes a pretrained ResNet for feature extraction. In the skip connection part, we adopt a novel non-local method called the Fully Attentional Block (FLA), which effectively obtains long-range contextual information and retains attentions in all dimensions. In the decoder part, we propose a multi-attention decoder that consists of four attention modules, allowing effective attention to be given to the feature maps in three dimensions: spatial, channel, and scale. We conducted experiments on two publicly available skin lesion segmentation datasets, ISIC 2017 and ISIC 2018, and the results demonstrate that MSMA-Net outperforms other methods, confirming the effectiveness of MSMA-Net.

Keywords: Skin lesion segmentation · Multi-Attention · Multi-Scale

1 Introduction

According to global cancer statistics, skin cancer is one of the fastest-growing types of cancer in the world [1]. Melanoma is the most deadly form of skin cancer, and the majority of deaths from skin cancer are attributed to melanoma. The diagnosis of melanoma is typically performed using dermatoscopy or skin microscopy to gather images, which are then evaluated by doctors to assess the extent of the disease. However, in clinical practice, manual examination is subjective, time-consuming, and non-reproducible. In order to address these demands, Computer-Aided Diagnosis (CAD) technology has long been introduced to assist in the daily practice of dermatologists, aiming to achieve more efficient and effective lesion detection and analysis operations [2]. One fundamental component in

B. Luo et al. (Eds.): ICONIP 2023, LNCS 14450, pp. 537–550, 2024.
https://doi.org/10.1007/978-981-99-8070-3_41

developing such a CAD system is the automated segmentation of lesion regions from dermoscopic images, enabling further focused analysis on these regions [3]. Thus, a significant amount of research has been dedicated to the automatic segmentation of skin lesions, including classical machine learning algorithms and deep learning algorithms. Compared to classical methods, deep learning models can adaptively learn high-dimensional features without requiring extensive manual intervention, and they outperform classical solutions, thereby taking the lead in the field of skin lesion segmentation.

Nowadays, with the advancement of deep learning, many methods based on Convolutional Neural Networks (CNNs) have been designed for the field of medical image segmentation [4–6], many of which have achieved good results in the task of automatic skin lesion segmentation. U-Net [7] is one of the most widely used methods in medical image segmentation, consisting of an encoder, decoder, and skip connections, and it has achieved remarkable results in various medical image segmentation tasks. Subsequently, to enhance the performance of biomedical image segmentation even further, a diverse array of models built upon the U-Net architecture have been proposed. For example, AttU-Net [8] applies a novel attention gate mechanism within U-Net to highlight prominent features. Furthermore, some researchers have designed specialized algorithms or modules based on the characteristics of medical images, such as skin lesion images, to address challenging issues in various segmentation tasks. CENet [5] introduced dense atrous convolutions and residual multi-kernel pooling blocks, which can capture rich semantic information and preserve more spatial information. CAnet [9] is the first approach to leverage comprehensive attention to enhance the performance and interpretability of CNNs in medical image segmentation. CPF-Net [10] enhances the feature by integrating higher-level semantic information between layers using the Global Pyramid Guidance (GPG) module. Simultaneously, it employs the Scale-Aware Pyramid Fusion (SAPF) module to dynamically merge multi-scale information. MALUNet [11] achieves lightweight design while maintaining a competitive performance in skin lesion segmentation tasks. While these methods have greatly advanced the development of automated skin lesion segmentation, there are still some significant challenges that cannot be ignored. For instance, skin lesions exhibit significant variations in size, shape, and color. The shape of skin lesions is often irregular, and the low contrast between lesions and normal skin leads to blurry boundaries. Furthermore, some other tissues such as hair, blood vessels, and bubble artifacts make automatic skin lesion segmentation more challenging. These challenges have been hindering the development of automated skin lesion segmentation algorithms. Although these methods perform well, they still have some limitations in addressing these challenges due to their weak feature extraction capability and inadequate ability to obtain long-range contextual information.

To address the above challenges, this study proposes a Multi-Scale and Multi-Attention network (MSMANet) for the accurate and reliable automatic segmentation of skin lesions. The model utilizes a pre-trained ResNet [12] as the backbone encoder network for feature extraction. Due to the limitation of the convolutional receptive field, capturing long-range dependencies has always been a challenging task for CNNs. Many methods [9,13] have introduced non-local attention to obtain long-range contextual information. These non-local methods typically generate a similarity map of $R^{C \times C}$ (by compressing spatial dimensions) or $R^{HW \times HW}$ (by compressing channels) to describe the feature relationships along either channel or spatial dimensions. However, such practices tend to condense feature dependencies along the other dimensions, hence causing attention missing, which might lead to inferior results for small categories or inconsistent segmentation inside large objects. Therefore, in this study, a novel Non-Local Block namely Fully Attention Block (FLA) [14] is introduced in the skip-connection part to effectively maintain attentions in all dimensions and obtain rich long-range contextual information. In the decoder part, a multi-attention decoder is proposed, which incorporates attention gates [8], CBAM [15], and scale attention [9]. The attention gate suppresses irrelevant background information and highlights prominent features. CBAM is utilized to emphasize important features in both channel and spatial dimensions while suppressing irrelevant information. To better aggregate feature maps from different scales and obtain multi-scale information, we introduce scale attention at the end of the model. Scale attention ensures that the final output of the model adaptively integrates information from different scales. Experimental results have demonstrated that our proposed Multi-Scale and Multi-Attention Network (MSMA-Net) achieves a high level of segmentation performance in skin lesion segmentation tasks.

In summary, our contributions are as follows:

(1) A novel deep learning model(MSMA-Net) was proposed for skin lesion segmentation, utilizing a pre-trained ResNet as the backbone encoder network for feature extraction.

(2) The novel non-local block FLA has been introduced for the first time to the field of skin lesion segmentation, effectively maintaining attention in all dimensions while obtaining rich long-range contextual information.

(3) A multi-attention decoder is proposed, which can effectively suppress background information while fully attending to feature maps in spatial, channel, and scale dimensions.

(4) Extensive experiments were conducted on the ISIC 2017 and ISIC 2018 datasets. Results show that our model achieves excellent segmentation results.

2   Method

2.1   Network Architecture

Figure 1 explains the overall architecture of the proposed MSMA-Net, which is composed of an encoder-decoder framework. It utilizes ResNet as the backbone encoder to extract features from the input image, and then transfers the extracted features to the decoder through skip connections. We have added a Fully Attentional block (FLA) on the skip connections at each layer to obtain long-range contextual information. In the decoder part, our multi-attention decoder adopts the CASCaded Attention Decoder (CASCADE) [16] as the main architecture and combines scale attention. The Cascaded Attention Decoder (CASCADE) in our architecture consists of Attention Gate (AG) and Convolutional Attention Module (CAM). This combination effectively suppresses background information and highlights prominent features. Additionally, the Scale Attention module adaptively fuses features from different scales to generate the final segmentation result.

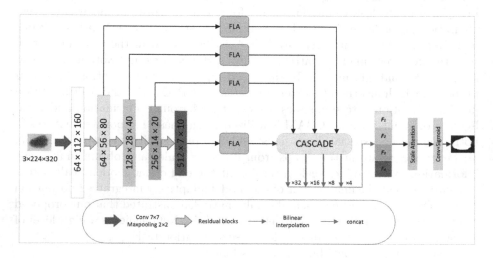

Fig. 1. The network structure of the proposed MSMA-Net

2.2   FLA

In order to obtain rich long-range contextual information, a new non-local block called FLA has been introduced into our network. The basic idea is to utilize global contextual information in computing channel attention feature maps to acquire spatial responses, thereby achieving full attention within a single attention unit while maintaining high computational efficiency. The specific implementation is shown in the Fig. 2.

For the FLA module, given the input feature map $F_{in}^k \in R^{C \times H \times W}$, where C is the number of channels, H and W are the spatial dimensions of the input tensor and $k = 2, 3, 4, 5$ represents the kth layer. The input feature map F_{in}^k first enters the bottom construction module, which is a dual-path structure. Each path consists of a global average pooling layer followed by a linear layer. The pooling kernels have sizes of $H \times 1$ and $1 \times W$, respectively. The purpose of this is to obtain richer global contextual priors while ensuring that each spatial position is connected to global priors with the same horizontal or vertical coordinates. This results in $\hat{Q}_w \in R^{C \times 1 \times W}$ and $\hat{Q}_h \in R^{C \times H \times 1}$. Then, $\hat{Q}_w$ and $\hat{Q}_h$ are repeated to form global priors Q_w and Q_h in the horizontal and vertical directions, respectively. Next, we cut Q_w along the H dimension, resulting in

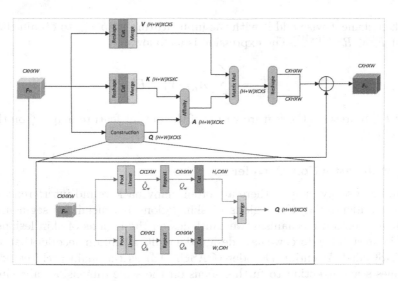

Fig. 2. Structure of Fully Attentional block (FLA). In the picture, $S = H = W$, the letter S represent the dimension after merge for a clear illustration. However, there are no requirements for square inputs in the FLA block. Because in the specific implementation, Q_w and Q_h are calculated separately with two sets of K without aggregation, while in the figure, Q, K, and V are all merged into one group for ease of explaining the method.

a set of H slices of size $R^{C \times W}$. Similarly, we cut Q_h along the W dimension, resulting in a set of W slices of size $R^{C \times H}$. Finally, we merge these two sets of slices to obtain the global contextual information $Q \in R^{(H+W) \times C \times S}$. Simultaneously, the input feature map F_{in}^k undergoes reshape, cut, and merge operations to obtain feature maps K and V. This involves cutting the input feature map F_{in}^k along the H dimension to obtain a set of slices of size $R^{C \times W}$, and similarly cutting it along the W dimension to obtain a set of slices of size $R^{C \times H}$. These two sets are then merged to obtain $K \in R^{(H+W) \times S \times C}$. The same operations are applied to obtain $V \in R^{(H+W) \times C \times S}$.

The feature maps K and Q are combined through an affinity operation to obtain the full attention $A \in R^{(H+W) \times C \times C}$. The affinity operation is defined as follows:

$$A_{i,j} = \frac{\exp\left(Q_i \cdot K_j\right)}{\sum_{i=1}^{C} \exp\left(Q_i \cdot K_j\right)} \tag{1}$$

where $A_{i,j} \in A$ represents the degree of correlation between the ith channel and the jth channel at a specific spatial position.

Next, A is multiplied with V through matrix multiplication, and the generated full attention is used to update each channel map. The obtained result is reshaped into two sets of feature maps of size $R^{C \times H \times W}$, and these two sets of feature maps are added together to obtain the long-range contextual information. We use a scale factor γ to multiply with the obtained contextual information, and then element-wise add it with the input feature map F_{in}^k to obtain the final output $F_o^k \in R^{C \times H \times W}$. The expression is as follows:

$$F_{o_j}^k = \gamma \sum_{i=1}^{C} A_{i,j} \cdot V_j + F_{in_j}^k \tag{2}$$

where $F_{o_j}^k$ represents the feature vector of the output feature map F_o^k on the jth channel map.

2.3   Multi-attention Decoder

In skin microscopy images, there are often many background interferences, and accurately identifying the regions of skin lesions is crucial for segmentation. Through attention mechanism, we can focus on the regions of skin lesions that are of interest. Therefore, we have designed a multi-attention decoder that adopts the CASCaded Attention Decoder (CASCADE) as the main architecture and combines scale attention to further focus on the scale dimension, allowing our decoder to attend to important scale features. CASCADE consists of two main modules: Attention Gate (AG) and Convolutional Attention Module (CAM). The details of AG are illustrated in the following Fig. 3(a). x^l represents the low-level feature map received from the skip connection, x^h represents the upsampled high-level feature map. By performing an upsampling operation, the low-level feature map x^l and the high-level feature map x^h can be unified to the same resolution. The high-level feature map x^h is then utilized to calibrate or refine the low-level feature map x^l. The high-level feature map x^h and the low-level feature map x^l are individually compressed to the same number of channels using a 1×1 convolution operation. Afterward, the resulting feature maps are element-wise summed and passed through a ReLU activation function. The output of the ReLU operation is then fed into another 1×1 convolution layer with a single channel output. A sigmoid function is applied to generate a pixel-level attention map $\gamma \in [0,1]^{H \times W}$. x^l is then multiplied with γ to be calibrated. The expression defining this process is as follows:

$$AG\left(x^l, x^h\right) = \sigma_1\left(C\left(\sigma_2\left(C_l\left(x^l\right) + C_h\left(x^h\right)\right)\right)\right) * x^l \tag{3}$$

where C_l and C_h represent 1×1 convolution followed by batch normalization operation. C represents a 1×1 convolution, σ_1 and σ_2 represent sigmoid and ReLU activation functions respectively, $AG\left(\cdot\right)$ represents the Attention gate module.

The structure of CAM is shown in the Fig. 3(b). The input x undergoes a channel attention operation [15] $CA\left(\cdot\right)$ and a spatial attention operation [15] $SA\left(\cdot\right)$ consecutively. Finally, the output is obtained by passing through a convolutional block $ConvBlock\left(\cdot\right)$ composed of two consecutive 3×3 convolution layers. For each convolution layer, a batch normalization layer and a ReLU activation layer are applied afterwards. The expression defining the convolutional attention module $CAM\left(\cdot\right)$ is as follows:

$$CAM\left(x\right) = ConvBlock\left(SA\left(CA\left(x\right)\right)\right) \tag{4}$$

The convolutional block further enhances the features obtained through channel attention and spatial attention.

AG and CAM have focused sufficiently on the channel and spatial dimensions of the feature map. However, they do not take into account the scale dimension effectively. In order to make a complete use of attentions to spatial positions, channels and scales, we propose a multi-attention decoder that utilizes AG, spatial attention, channel attention, and scale attention to focus on different dimensions of the feature maps.

CASCADE. The detailed process can be seen in Fig. 3. The encoder generates features at different scales, which are then passed through the FLA module to obtain long-range contextual features $\{E_i\}_{i=1}^{4}$. These features are input into CASCADE to generate the decoding features $\{F_i\}_{i=1}^{4}$. E_4 is the input to the lowest layer of the decoder. The output f_4 of the lowest scale in the decoder is obtained by passing E_4 through a 1×1 convolution and CAM. The expression defining f_4 is as follows:

$$f_4 = CAM\left(Conv\left(E_4\right)\right) \tag{5}$$

There are two flow directions for f_4. In the first direction, f_4 is upsampled 32 times and passed through a 1×1 convolution to obtain the output feature F_4. In the second direction, after upsampling f_4 by a factor of 2, it is combined with the feature E_3 from the skip connection and then inputted into the AG. The resulting output is K. e_3 is generated by concatenating K with f_4 upsampled by a factor of 2. K and e_3 can be described as follows:

$$K = AG\left(E_3, up\left(f_4\right)\right) \tag{6}$$

$$e_3 = concat\left(K, up\left(f_4\right)\right) \tag{7}$$

where $up\left(\cdot\right)$ denotes upsampling, *concat* represents channel concatenation. The feature e_3 is fed into the CAM, resulting in the output feature f_3. The subsequent calculation processes for f_2, f_1, e_2, e_1, and F_3, F_2, F_1 are similar to the aforementioned f_3, e_3, F_4, so they will not be repeated here.

Scale Attention. The feature maps at different scales have different correlations with the segmentation targets. The feature maps at each layer of the encoder-decoder network have different spatial resolutions. Effectively fusing these features of different scales is crucial for generating accurate segmentation maps. Therefore, we introduce scale attention. Scale attention modules are capable of automatically learning the relevant weights for each scale feature map to calibrate features at different scales, as shown in Fig. 4.

Due to upsampling and 1×1 convolution, the resolution of the features $\{F_i\}_{i=1}^{4}$ is consistent with the resolution of the input network image, and the channel number is unified to 4. The feature maps of different scales are then concatenated along the channel dimension to obtain F, which serves as the

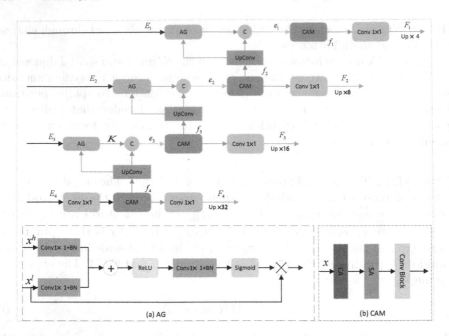

Fig. 3. Structure of CASCaded Attention Decoder (CASCADE). (a)Attention gate (AG), (b) Convolutional attention module (CAM).

input to the scale attention module. By combining average pooling, max pooling, and multi-layer perceptron (MLP), we obtain the scale attention vector $\alpha \in [0,1]^{4 \times 1 \times 1}$. To allocate multi-scale soft attention weights at each pixel, a spatial attention mechanism is also employed. The spatial attention block takes $F \cdot \alpha$ as input and generates a spatial-level attention coefficient $\beta \in [0,1]^{1 \times H \times W}$, so that $\alpha \cdot \beta$ represents pixel-wise scale attention. The spatial attention module consists of a 3×3 convolutional layer followed by a 1×1 convolutional layer. The first convolutional layer has 4 output channels and is followed by a ReLU activation function. The second convolutional layer also has 4 output channels and is followed by a sigmoid activation function. The output of the scale attention module is:

$$Y = F \cdot \alpha + F \cdot \alpha \cdot \beta + F \tag{8}$$

The residual connection here is used to accelerate training. By utilizing scale attention, the network can perceive the most appropriate scale, effectively integrating features of different scales to generate precise segmentation results. Finally, the output of the scale attention module is fed into a segmentation head that consists of a 1×1 convolutional layer and a sigmoid function, which produces the segmentation map.

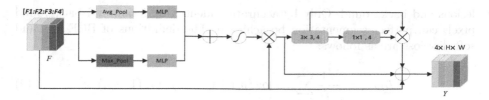

Fig. 4. Structure of scale attention

3   Experiments

3.1   Dataset

To evaluate the performance of different methods, this study utilized two publicly available skin lesion datasets, namely ISIC 2017 [17] and ISIC 2018 [18]. The specific details are as follows:

The ISIC 2017 dataset and ISIC 2018 dataset are provided by the International Skin Imaging Collaboration (ISIC). The ISIC 2017 dataset consists of 2000 training images, 150 validation images, and 600 test images of skin lesions. The ISIC 2018 dataset contains 2594 RGB images. For our experiments, the ISIC 2018 dataset was randomly partitioned into training (70%), validation (10%), and testing (20%) sets. The images are normalized and resized to 224 × 320.

3.2   Implementation Details and Evaluation Metrics

All experiments were implemented using the PyTorch framework in conjunction with an NVIDIA GeForce RTX 3060 graphics card. We conducted data augmentation, including random horizontal flipping and random vertical flipping, to prevent overfitting. Adaptive moment estimation (Adam) is utilized as the optimizer. The initial learning rate is set to 0.0001 with a CosineAnnealing-WarmRestarts learning rate policy ($T_0 = 10$, $T_{mult} = 2$) and we apply a weight decay of 1×10^{-8}. A total of 250 epochs were trained with a batch size of 8.

The experiment utilized five commonly used metrics to evaluate performance, including Jaccard Index (JI), Dice Score Coefficient (DSC), Accuracy (ACC), Sensitivity (SE), and Specificity (SP). To obtain a fair and reliable performance of different methods, the five-fold cross-validation is carried out for ISIC 2018, and report the mean and standard deviation of the results.

3.3   Loss Function

Skin lesion segmentation is a pixel-level binary classification problem, where the goal is to distinguish between skin lesions and the background. In our experiment, we employed a comprehensive loss function, denoted as L_{total}, which consists of the SoftDice loss [19] and the binary cross-entropy(BCE) loss. The SoftDice loss is optimized from Dice loss and is capable of simultaneously focusing on both lesions and background. Only by accurately identifying lesion and background pixels can perfect segmentation be achieved. The definitions of BCE loss and soft dice loss are as follows:

$$L_{BCE} = -\frac{1}{N} \sum_{i=1}^{N} g_i \cdot \log(p_i) + (1 - g_i) \cdot \log(1 - p_i) \tag{9}$$

$$Dice = \frac{2\sum_{i=1}^{N} g_i \cdot p_i + \epsilon}{\sum_{i=1}^{N} g_i + \sum_{i=1}^{N} p_i + \epsilon} \tag{10}$$

$$Dice_b = \frac{2\sum_{i=1}^{N} (1 - g_i) \cdot (1 - p_i) + \epsilon}{\sum_{i=1}^{N} (1 - g_i) + \sum_{i=1}^{N} (1 - p_i) + \epsilon} \tag{11}$$

$$L_{softdice} = 1 - \frac{Dice + Dice_b}{2} \tag{12}$$

$$L_{total} = L_{BCE} + L_{softdice} \tag{13}$$

Table 1. Skin lesion segmentation performances of different networks on ISIC 2017. The best results are in bold and all metrics are represented in %.

Method	JI	DSC	ACC	SE	SP
U-Net [7]	76.35	84.95	93.21	83.29	97.44
AttU-Net [8]	76.44	84.87	93.10	82.62	97.72
DeepLabv3+ [20]	77.03	85.38	93.83	81.94	**97.94**
Resunet++ [4]	76.66	85.18	93.51	82.83	97.49
CA-Net [9]	76.91	85.11	93.27	83.86	96.91
CE-Net [5]	77.84	85.94	93.90	83.24	97.67
CPFNet [10]	78.37	86.36	93.94	84.57	97.00
MALUNet [11]	74.09	83.26	92.55	81.05	97.07
Ms RED [13]	77.93	85.96	93.64	85.33	97.11
MSMA-Net(our)	**78.94**	**86.83**	**94.17**	**86.50**	97.52

where $g_i \in \{0, 1\}$ and $p_i \in \{0, 1\}$ represent the ground-truth annotation and the probability map respectively, ϵ is a very small number to avoid division by zero, and N = H Œ W represents the number of pixels. In the paper, L_{total} is used to optimize the network.

3.4   Comparison with the State-of-the-Arts

To further validate the effectiveness of our proposed model, we will compare it with several advanced classical models and models specifically designed for skin lesion segmentation. All models are trained using the same training strategy and experimental settings. The results on the ISIC 2017 and ISIC 2018 datasets can be seen in Table 1 and Table 2, respectively. On the ISIC 2017 dataset, our method outperforms other methods in terms of JI, DSC, ACC, and SE. Compared to the method closest to ours, CPF-Net, our method shows an improvement of 0.57% in JI, 0.47% in DSC, 0.23% in ACC, and 1.93% in SE. MsRED, a model specifically designed for skin lesion segmentation, has achieved outstanding results in the field. Compared to MsRED, our method exhibits an increase of 1.01% in JI, 0.87% in DSC, 0.53% in ACC, and 1.17% in SE. On the ISIC 2018 dataset, all evaluation criteria are obtained by averaging the results of five-fold cross-validation, from which it can be seen that the best-performing network is MSMA-Net. The experimental results demonstrate that our proposed method achieves good performance in terms of evaluation metrics and segmentation results, confirming the effectiveness of our approach in skin lesion segmentation. For qualitative analysis, we visualized the prediction results of our proposed method (MSMA-Net) and those of other state-of-the-art networks (U-Net, CPFNet, MsRED) in Fig 5. The first and second rows are from the ISIC 2018 dataset, while the third and fourth rows are from the ISIC 2017 dataset. From the Fig. 5, it can be observed that MSMA-Net accurately segments the basic contours. Compared to other methods, MSMA-Net is able to predict the boundaries of the results more effectively, and the differences between the ground truth and the predicted results by MSMA-Net are very small. This demonstrates the performance of MSMA-Net.

Table 2. Skin lesion segmentation performances of different networks on ISIC 2018. The best results are in bold and all metrics are represented in %.

Method	JI	DSC	ACC	SE	SP
U-Net [7]	81.47±0.58	88.67±0.40	95.62±0.25	90.06±0.81	97.13±0.33
AttU-Net [8]	82.02±0.41	89.04±0.29	95.72±0.24	90.12±0.61	97.25±0.47
DeepLabv3+ [20]	83.05±0.36	89.85±0.30	96.11±0.31	90.33±0.48	**97.55±0.30**
Resunet++ [4]	82.65±0.37	89.45±0.35	95.92±0.29	90.44±0.73	96.97±0.37
CA-Net [9]	82.36±0.39	89.21±0.31	95.90±0.36	90.04±0.87	97.48±0.44
CE-Net [5]	83.39±0.42	90.06±0.33	96.13±0.39	90.94±0.49	97.08±0.24
CPFNet [10]	83.70±0.43	90.22±0.33	96.33±0.27	90.67±0.47	97.26±0.37
MALUNet [11]	80.30±0.56	87.79±0.43	95.37±0.24	89.13±0.77	96.99±0.49
Ms RED [13]	83.29±0.57	89.89±0.49	96.10±0.38	90.92±0.76	97.11±0.31
MSMA-Net(our)	**84.14±0.34**	**90.57±0.27**	**96.43±0.25**	**91.43±0.52**	97.49±0.35

3.5 Ablation Experiment

To validate the effectiveness of our proposed method, we conducted ablation experiments on the ISIC2017 dataset. Specifically, we aimed to verify the effectiveness of each module. We used ResNet34 as the encoder and combined it with the Cascaded Attention Decoder as our baseline. Based on this, we designed our network architecture, where the decoder's output is formed by element-wise addition of the upsampled outputs from each decoder layer. We introduced a Model 1, which incorporates a multi-attention decoder by combining the Cascaded Attention Decoder with scale attention. Building upon Model 1, we introduced FLA in the skip connections to complete the design of our overall network. The specific results can be seen in the Table 3.

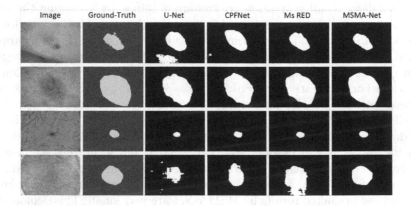

Fig. 5. The segmentation results visualization of differnt methods.

Table 3. The result of ablation studies on the ISIC 2017 dataset. The best results are in bold and all metrics are represented in %.

Method	JI	DSC	ACC
Baseline	78.03	86.07	93.48
Model 1	78.62	86.54	93.72
MSMA-Net	**78.94**	**86.83**	**94.17**

In the experiments, after incorporating scale attention into the baseline, the experimental metrics JI, DSC, and ACC improved by 0.47%, 0.59%, and 0.24%, respectively. It can be observed that scale attention significantly enhances the multi-attention decoder, demonstrating that the scale attention module can effectively adaptively fuse multi-scale information to improve the segmentation results. Building upon Model 1, the introduction of the FLA module resulted in an improvement in the JI, DSC, and ACC metrics by 0.29%, 0.32%, and 0.35% respectively. This indicates that FLA, when incorporated into the model, can effectively obtain long-range contextual information to further enhance the segmentation performance.

4 Conclusion

This paper proposes MSMA-Net, a deep learning model that efficiently and accurately segments skin lesion areas in dermoscopy images. MSMA-Net adopts a classic encoder-decoder structure, with the encoder utilizing ResNet for effective feature extraction. By introducing the FLA module in the skip connections, our model can effectively abtain long-range contextual information. The decoder introduces a multi-attention decoder that combines four types of attention modules, enabling comprehensive attention to important features across spatial, channel, and scale dimensions. Comprehensive experiments on two publicly available skin lesion datasets demonstrate the outstanding performance of the proposed MSMA-Net, surpassing other state-of-the-art methods in skin lesion segmentation. In future work, we plan to apply our network to other medical datasets. Ultimately, we hope that our work can provide assistance to medical professionals in diagnosing skin cancer.

References

1. Siegel, R.L., Miller, K.D., Jemal, A.: Cancer statistics, 2019. CA: Cancer J. Clin. **69**(1), 7–34 (2019)
2. Yu, L., Chen, H., Dou, Q., Qin, J., Heng, P.A.: Automated melanoma recognition in dermoscopy images via very deep residual networks. IEEE Trans. Med. Imaging **36**(4), 994–1004 (2016)
3. Ganster, H., Pinz, P., Rohrer, R., Wildling, E., Binder, M., Kittler, H.: Automated melanoma recognition. IEEE Trans. Med. Imaging **20**(3), 233–239 (2001)
4. Jha, D., et al.: Resunet++: an advanced architecture for medical image segmentation. In: 2019 IEEE International Symposium on Multimedia, pp. 225–2255. IEEE (2019)
5. Gu, Z., et al.: Ce-net: context encoder network for 2d medical image segmentation. IEEE Trans. Medical Imaging **38**(10), 2281–2292 (2019)
6. Wu, C., Liu, X., Li, S., Long, C.: Coordinate attention residual deformable U-Net for vessel segmentation. In: Mantoro, T., Lee, M., Ayu, M.A., Wong, K.W., Hidayanto, A.N. (eds.) ICONIP 2021. LNCS, vol. 13110, pp. 345–356. Springer, Cham (2021). https://doi.org/10.1007/978-3-030-92238-2_29
7. Ronneberger, O., Fischer, P., Brox, T.: U-Net: convolutional networks for biomedical image segmentation. In: Navab, N., Hornegger, J., Wells, W.M., Frangi, A.F. (eds.) MICCAI 2015. LNCS, vol. 9351, pp. 234–241. Springer, Cham (2015). https://doi.org/10.1007/978-3-319-24574-4_28
8. Oktay, O., et al.: Attention u-net: learning where to look for the pancreas. arXiv preprint arXiv:1804.03999 (2018)
9. Gu, R., et al.: CA-Net: comprehensive attention convolutional neural networks for explainable medical image segmentation. IEEE Trans. Med. Imaging **40**(2), 699–711 (2020)
10. Feng, S., et al.: CPFNet: context pyramid fusion network for medical image segmentation. IEEE Trans. Med. Imaging **39**(10), 3008–3018 (2020)
11. Ruan, J., et al.: MALUNet: a multi-attention and light-weight UNet for skin lesion segmentation. In: IEEE International Conference on Bioinformatics and Biomedicine, pp. 1150–1156. IEEE, Las Vegas, NV, USA (2022)

12. He, K., Zhang, X., Ren, S., Sun, J.: Deep residual learning for image recognition. In: Proceedings of the IEEE Conference on Computer Vision and Pattern Recognition, pp. 770–778 (2016)
13. Dai, D., et al.: Ms RED: a novel multi-scale residual encoding and decoding network for skin lesion segmentation. Med. Image Analy. **75**, 102293 (2022)
14. Song, Q., Li, J., Li, C., Guo, H., Huang, R.: Fully attentional network for semantic segmentation. In: Proceedings of the AAAI Conference on Artificial Intelligence, vol. 36(2), pp. 2280–2288 (2022)
15. Woo, S., Park, J., Lee, J.-Y., Kweon, I.S.: CBAM: convolutional block attention module. In: Ferrari, V., Hebert, M., Sminchisescu, C., Weiss, Y. (eds.) ECCV 2018. LNCS, vol. 11211, pp. 3–19. Springer, Cham (2018). https://doi.org/10.1007/978-3-030-01234-2_1
16. Rahman, M.M., Marculescu, R.: Medical image segmentation via cascaded attention decoding. In: Proceedings of the IEEE/CVF Winter Conference on Applications of Computer Vision, pp. 6222–6231 (2023)
17. Codella, N.C.F., et al.: Skin lesion analysis toward melanoma detection: a challenge at the 2017 international symposium on biomedical imaging (isbi), hosted by the international skin imaging collaboration (isic). In: 2018 IEEE 15th International Symposium on Biomedical Imaging (ISBI 2018), pp. 168–172. IEEE (2018)
18. Codella, Noel, et al.: Skin lesion analysis toward melanoma detection 2018: a challenge hosted by the international skin imaging collaboration (isic). arXiv preprint arXiv:1902.03368 (2019)
19. Bertels, J., et al.: Optimizing the dice score and jaccard index for medical image segmentation: theory and practice. In: Shen, D., et al. (eds.) MICCAI 2019. LNCS, vol. 11765, pp. 92–100. Springer, Cham (2019). https://doi.org/10.1007/978-3-030-32245-8_11
20. Chen, L.-C., Zhu, Y., Papandreou, G., Schroff, F., Adam, H.: Encoder-decoder with atrous separable convolution for semantic image segmentation. In: Ferrari, V., Hebert, M., Sminchisescu, C., Weiss, Y. (eds.) ECCV 2018. LNCS, vol. 11211, pp. 833–851. Springer, Cham (2018). https://doi.org/10.1007/978-3-030-01234-2_49

Temporal Attention for Robust Multiple Object Pose Tracking

Zhongluo Li[1]([✉]), Junichiro Yoshimoto[2], and Kazushi Ikeda[1]

[1] Nara Institute of Science and Technology, Nara 630-0192, Japan
lzlshr835@gmail.com
[2] Fujita Health University School of Medicine, Aichi 470-1192, Japan

Abstract. Estimating the pose of multiple objects has improved substantially since deep learning became widely used. However, the performance deteriorates when the objects are highly similar in appearance or when occlusions are present. This issue is usually addressed by leveraging temporal information that takes previous frames as priors to improve the robustness of estimation. Existing methods are either computationally expensive by using multiple frames, or are inefficiently integrated with ad hoc procedures. In this paper, we perform computationally efficient object association between two consecutive frames via attention through a video sequence. Furthermore, instead of heatmap-based approaches, we adopt a coordinate classification strategy that excludes post-processing, where the network is built in an end-to-end fashion. Experiments on real data show that our approach achieves state-of-the-art results on Pose-Track datasets.

Keywords: Pose Estimation · Vision Transformer · Temporal Information

1 Introduction

Pose estimation, as a critical pre-processing task in computer vision, is used in a wide range of areas from humans to animals. It is used for individual or group action recognition and analysis by detecting objects and locating their body parts in natural or experimental scenes. Deep-neural-network based algorithms have arguably become the standard in visual tracking due to their high performance over the last decade. Despite fruitful achievements, it still lacks robustness when applied to subjects that are extremely similar, prone to occlusion, or undergo episodes of rapid movement [24,27,36].

Existing methods for estimating the pose of multiple objects in video are mainly based on convolutional neural networks, where each image is treated independently. Recent works [44,45] suggest that temporal clues across video frames provide valuable information for pose estimation [4]. This indicates that the result of the current frame can be improved by associating with the feature in the previous frames. In light of this idea, a myriad of works [9,26,43] attempts

B. Luo et al. (Eds.): ICONIP 2023, LNCS 14450, pp. 551–561, 2024.
https://doi.org/10.1007/978-981-99-8070-3_42

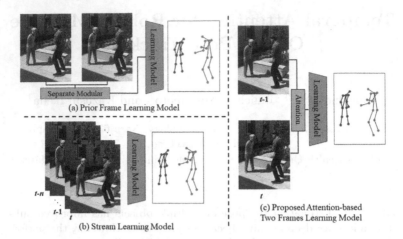

(a) Prior Frame Learning Model

(b) Stream Learning Model

(c) Proposed Attention-based
Two Frames Learning Model

Fig. 1. Illustration of two different temporal information utilization methods and our proposed attention-based method

to improves the robustness by incorporating a image stream, *i.e.,* .using multiple previous or subsequent frames, instead of single frames. Although this solution is attractive, the computational cost inevitably rises as the number of frames per stream is increased [25].

Another way to address this problem is to associate only one previous frame. The association problem, however, is not trivial owning to occlusions, misdetections, entry and exit of performers in the scene. Object association has typically used deterministic or stochastic treatment of the association problem. Deterministic methods regard this problem as a combinatorial optimization problem of objects between two frames and can be achieved using the Hungarian search [16]. Stochastic methods, in contrast to deterministic methods, take the uncertainty associated with measurements and estimates the posterior probability of the motion.

Figure 1 shows the widely used method that utilizes prior frames or multiple frames and our proposed method. In summary, we make the following contributions:

- A temporal attention mechanism to capture related object features across successive frames.
- An end-to-end fully sequences processing network simplified the training and reduces quantization error.
- Experiments on real-world data show that our model outperforms state-of-the-art models.

2   Related Work

We divided existing studies into stream and single prior categories based on how many frames are involved on one inference. In addition, some emerging research exploiting temporal information by attentional mechanisms shows impressive results.

Estimated-by-stream approaches form poses by associating a given length of detections over a period of time. Extending the CNN model to include a temporal dimension [9,10,43] obtain comparable accurate predictions. Wang et al. [34] introduce a network to establish tracklet for each individual, and extend input of HRNet [31] from single frame to 3D-HRNet a multiple frame for joints position extraction. Yang et al. [40] derive a robust estimation using a graph neural network (GNN) that takes three historical poses. Similarly, the use of multiple prior frames in the transformer-based model [14,42] is also an indication of the success of this strategy. However, the computational cost increases as the number of previous frames are concerned [25].

Estimated-by-prior mitigates computational cost issues by considering a single previous frame as a prior. Bertasius et al. [3] proposed a network that aggregates and compares previously labeled and current unlabeled frame and predicts the current frame using the motion offset found by the model. Another way to integrate prior frames is to associate objects between two frames according to the similarity that the model learned [17,29]. The work in Liu et al. [23] improving pose estimation in video by introducing a dual consecutive framework; however, it suffered from an exceptionally large number of parameters to train.

Track by attention was a method proposed by Snower et al. [30] that leverages temporal relationships by a Transformer-based network. Visual objects between neighboring frames are highly correlated so that a simple linear matching process suffers from semantic information loss which limits the ability of the method to capture the temporal relationship. Inspired by the work of NLP, Dosovitskiy et al. [8] adapt the original Transformer model [33] to computer vision. Transformer is not only superior in long-range dependency but also provides a better feature association method than the commonly used correlation between two similar images [7,39].

3 Method

Our pose tracking framework is inspired by TrackFormer [28], an encoder-decoder architecture that associates temporal information in terms of a simple concept of track queries. Contrasting with TrackFormer, our framework has the following differences: (1) The temporal features are different. TrackFormer uses query tokens decoded from the last frame as a track query in the current step, while we apply an associated object indicator between features extracted from the last frame and the current frame, which is aimed to locating the correlated objects across frames. (2) The backbone networks are different. TrackFormer not only uses ResNet [11] as a backbone, but also trains an encoder from scratch. By contrast, we used a pre-trained encoder, Swin [21,22], which is more effective and efficient for feature extraction. (3) The prediction heads are different. Similar to the classical method, TrackFormer trained the model in terms of pose regression, which inevitably involved post-processing such as heatmaps in the model.

To tackle this limitation, we considered pose estimation as an $x - y$ separated coordinate classification problem. Thus, the pipeline becomes fully end-to-end which is simpler and more effective [19].

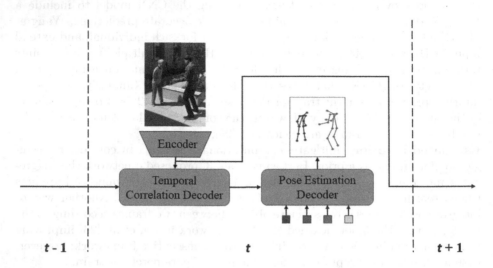

Fig. 2. The framework of proposed model

3.1   Overview

As depicted in Fig. 2, our model is based on an encoder-decoder framework. It contains three main parts: Visual Feature Encoder $\mathcal{C}$, Related Object Indicator $\mathcal{I}$ and Pose Estimation Decoder $\mathcal{D}$.

Encoder. We apply a pre-trained Transformer-based network (*e.g.*, Token-Pose [20], DINO [5]) as the encoder $\mathcal{E}$ to extract features of the image. Specifically, given the image of size $H \times W \times 3$ in frame t, the encoder extracts representations of the image for current frame object identification and subsequent frame correlation.

Decoder. To exploit the temporal information between successive frames, the Temporal Correlation Decoder uses the feature extracted from the last frame to make a comparison with the feature of the current frame and output a refined feature. Afterwards, the Pose Estimation Decoder extracts object-related features based on features extracted from the refined. We detail the temporal alignment decoder and Pose Estimation Decoder in Sect. 3.2.

Head. Instead of the popular heatmap-based method [36], we apply a coordinate classification [6,19] on x,y coordinates of keypoints on each object. The idea of treating different tasks in sequence-to-sequence framework consolidates keypoint localization in a simple and unified framework. Specifically, K keypoints is represented as a sequence of $2K$ discrete tokens, *i.e.*, $[x_1, y_2, ..., x_K, y_K]$

whose value belongs to one of bins for horizontal and vertical axes, respectively. However, traditional classification can easily lead to overfitting because it uses a hard-coded approach in terms of 0 or 1. Label smoothing [32] replaces the hard 0 and 1 classification labels with confidence of $\frac{\epsilon}{k-1}$ and $1 - \epsilon$ respectively to avoid overconfidence, where ϵ is a small constant.

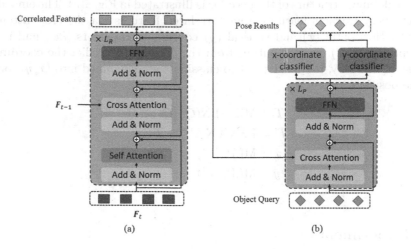

Fig. 3. (a) Temporal Correlation Decoder. (b) Pose Estimation Decoder.

3.2   Temporal Correlation Decoder

The Attention mechanism [38] is capable of handling long-range dependencies and can be viewed as Bayesian inference of related tokens through cross-attention [18]. In this way, transformers optimize the process of feature extraction and feature relation simultaneously. Since the features extracted from the images are tokenized patches, the transformer computes attention weights between two features and outputs the weighted average result.

As shown in Fig. 3(a), the decoder is stacked sequentially by L_R blocks, each block composed of attention (MHA), feed-forward network (FFN) and layer normalization (LN) modular. The features from the last frame F_{t-1} input into the decoder servers as memory, while features from current frame F_t are treated as queries to focus on related objects. It is express as follows:

$$
\begin{aligned}
T &= \mathrm{MHA}(\mathrm{LN}(F_t)) + F_t \\
T' &= \mathrm{MHA}(\mathrm{LN}(T), F_{t-1}) + T \\
C_t &= \mathrm{FFN}(\mathrm{LN}(T')) + T'
\end{aligned}
\tag{1}
$$

The pairwise features merge into a single feature that contains relevant information across successive frames. With the refined feature, the pose estimation decoder obtains robust features than when only a single frame is used.

3.3   Pose Estimation Decoder

In the Pose Estimation Decoder, we aim to infer pose under the correlated features. Instead of a heatmap, we use the idea of complete sequence-to-sequence format [19] to reduce the post-processing that is an indispensable part of most models.

The detailed structure of the decoder is illustrated in Fig. 3(b). The embedded object query extracts features from correlated features via the cross-attention module. Horizontal (x) and vertical (y) coordinate classifiers, *i.e.*, multi-layer perceptron (MLP), use the output from the decoder to predict the coordinates separately. Finally, the results of two classifiers are combined into (x, y) format of the position,

$$
\begin{aligned}
T &= \mathrm{MHA}(\mathrm{LN}(Q), C_t) + Q \\
T' &= \mathrm{FFN}(\mathrm{LN}(T)) + T \\
x &= \mathrm{MLP}_x(T') \\
y &= \mathrm{MLP}_y(T')
\end{aligned}
\tag{2}
$$

4   Experiments

In this section, we conduct experiments on three benchmark datasets for key-point location task. To validate the effect of the temporal alignment decoder, we perform an ablation study in which the decoder is removed.

4.1   Experimental Settings

Dataset. The PoseTrack [1,13] datasets are short video clips depicting multiple people in various natural settings. The dataset consists of 40k person samples with 15 joint labels.

Evaluation Metric. We evaluate the result in terms of multiple object tracking accuracy (MOTA), a standard metric used in pose estimation [2]. MOTA count total error includes the number of misses (m_t), false positive (fp_t) and mismatches (ID_{switch}), then averaging by total number of matches (g_t) for time t.

$$
\mathrm{MOTA} = 1 - \frac{\sum_t (m_t + fp_t + ID_{switch})}{\sum_t g_t}
\tag{3}
$$

Implementation Details. Our approach is based on PyTorch 1.12 and is developed on top of the Hugging Face [35] framework which is an open source in transformer model. For our work, our models are trained for 120 epochs with Swinv2 [21] as the encoder on double Nvidia RTX 2080Ti GPU.

Table 1. Quantitative results on the PoseTrack2017 validation set.

Method	Head	Should	Elbow	Wrist	Hip	Knee	Ankle	Mean
Detect&Track [9]	61.7	65.5	57.3	45.7	54.3	53.1	45.7	55.2
PoseFlow [37]	59.8	67.0	59.8	51.6	60.0	58.4	50.5	58.3
TML++ [12]	75.5	75.1	62.9	50.7	60.0	53.4	44.5	61.3
FlowTrack [36]	73.9	75.9	63.7	56.1	65.5	65.1	53.5	65.4
ClipTrack [34]	80.5	80.9	71.6	**63.8**	70.1	68.2	62.0	71.6
STEmbedding [15]	78.7	79.2	71.2	61.1	74.5	**69.7**	**64.5**	71.8
Ours	**82.4**	**81.9**	**72.1**	61.8	**75.7**	67.4	62.8	**72.1**

Table 2. Quantitative results on the PoseTrack2018 validation set.

Method Speed	Head	Should	Elbow	Wrist	Hip	Knee	Ankle	Mean
PT CPN++ [41]	68.8	73.5	65.6	61.2	54.9	64.6	56.7	64.0
TML++ [12]	76.0	76.9	66.1	56.4	65.1	61.6	52.4	65.7
ClipTrack [34]	74.2	76.4	71.2	64.1	64.5	**65.8**	61.9	68.7
Ours	**77.6**	**78.4**	**73.1**	**64.2**	**66.3**	63.7	58.9	**69.3**

4.2 Results

In Table 1 and Table 2, we compare our result with the state-of-the-art methods. On both datasets our method achieves top performance, which validates our method. Compared with ClipTrack [34] that uses 3DCNN as backbone, our Transformer-based outperforms on average MOTA of all joints. However, our model shows poor accuracy on small joints such as the wrist, knee and ankle. We argue that this is due to the fact that the transformer splits the whole image into equal-sized patches which is detrimental to the recognition performance for fine-grained visual features.

4.3 Ablation Study

To verify that our model is sufficient for temporal feature correlation using the attention mechanism, we performed ablation studies without the Temporal Correlation Decoder. Experiments on PoseTreck2018 show that without Temporal Correlation Decoder, the estimation of mAP and MOTA dropped by 2.6 and 3.5 respectively. From this result, we show that the Temporal Correlation Decoder is effective in the pose-tracking process (Fig. 3).

Table 3. Experiments on the effect of Temporal Correlation Decoder

Temporal Correlation	mAP	MOTA
✓	84.5	69.3
✗	81.9	65.8

5   Conclusion

In this paper, we propose a simple and novel pose-tracking model based on the Transformer architecture. The model consists of a pre-trained network as the encoder, and two decoders act as temporal information fusion and pose estimation. The temporal information fusion decoder tracks the position of joints on pairwise frame features and uses them as memory in the pose estimation decoder. In combination with horizontal and vertical coordinate classification approaches, we present a fully end-to-end multi-object pose tracking model with comparable efficiency to the classical method. We achieved state-of-art performance on the PoseTrack2017 [13] and PoseTrack2018 [1] datasets.

Acknowledgements. This work was supported in part by NAIST-NMU joint work promotion program.

References

1. Andriluka, M., et al.: PoseTrack: a benchmark for human pose estimation and tracking. In: Proceedings of the IEEE Conference on Computer Vision and Pattern Recognition (CVPR), June 2018
2. Bernardin, K., Stiefelhagen, R.: Evaluating multiple object tracking performance: the clear mot metrics. EURASIP J. Image Video Process. **2008**, 1–10 (2008)
3. Bertasius, G., Feichtenhofer, C., Tran, D., Shi, J., Torresani, L.: Learning temporal pose estimation from sparsely-labeled videos. In: Advances in Neural Information Processing Systems, vol. 32. Curran Associates, Inc. (2019)
4. Callaghan, S.: Preview of: a primer on motion capture with deep learning: principles, pitfalls, and perspectives. Patterns **1**(8), 100146 (2020)
5. Caron, M., et al.: Emerging properties in self-supervised vision transformers. In: Proceedings of the IEEE/CVF International Conference on Computer Vision (ICCV), pp. 9650–9660, October 2021
6. Chen, T., Saxena, S., Li, L., Fleet, D.J., Hinton, G.: Pix2seq: a language modeling framework for object detection. In: International Conference on Learning Representations (2022). https://openreview.net/forum?id=e42KbIw6Wb
7. Chen, X., Yan, B., Zhu, J., Wang, D., Yang, X., Lu, H.: Transformer tracking. In: Proceedings of the IEEE/CVF Conference on Computer Vision and Pattern Recognition (CVPR), pp. 8126–8135, June 2021
8. Dosovitskiy, A., et al.: An image is worth 16x16 words: transformers for image recognition at scale. In: International Conference on Learning Representations (2021). https://openreview.net/forum?id=YicbFdNTTy

9. Girdhar, R., Gkioxari, G., Torresani, L., Paluri, M., Tran, D.: Detect-and-track: efficient pose estimation in videos. In: Proceedings of the IEEE Conference on Computer Vision and Pattern Recognition (CVPR), June 2018

10. Gkioxari, G., Toshev, A., Jaitly, N.: Chained predictions using convolutional neural networks. In: Leibe, B., Matas, J., Sebe, N., Welling, M. (eds.) ECCV 2016. LNCS, vol. 9908, pp. 728–743. Springer, Cham (2016). https://doi.org/10.1007/978-3-319-46493-0_44

11. He, K., Zhang, X., Ren, S., Sun, J.: Deep residual learning for image recognition. In: Proceedings of the IEEE Conference on Computer Vision and Pattern Recognition (CVPR), June 2016

12. Hwang, J., Lee, J., Park, S., Kwak, N.: Pose estimator and tracker using temporal flow maps for limbs. In: 2019 International Joint Conference on Neural Networks (IJCNN), pp. 1–8 (2019). https://doi.org/10.1109/IJCNN.2019.8851734

13. Iqbal, U., Milan, A., Gall, J.: PoseTrack: joint multi-person pose estimation and tracking. In: Proceedings of the IEEE Conference on Computer Vision and Pattern Recognition (CVPR), July 2017

14. Jin, K.M., Lee, G.H., Lee, S.W.: OTPose: occlusion-aware transformer for pose estimation in sparsely-labeled videos. In: 2022 IEEE International Conference on Systems, Man, and Cybernetics (SMC), pp. 3255–3260 (2022). https://doi.org/10.1109/SMC53654.2022.9945591

15. Jin, S., Liu, W., Ouyang, W., Qian, C.: Multi-person articulated tracking with spatial and temporal embeddings. In: Proceedings of the IEEE/CVF Conference on Computer Vision and Pattern Recognition (CVPR), June 2019

16. Kuhn, H.W.: The Hungarian method for the assignment problem. Naval Res. Logistics Q. **2**(1–2), 83–97 (1955)

17. Leal-Taixe, L., Canton-Ferrer, C., Schindler, K.: Learning by tracking: Siamese CNN for robust target association. In: Proceedings of the IEEE Conference on Computer Vision and Pattern Recognition (CVPR) Workshops, June 2016

18. Li, K., Wang, S., Zhang, X., Xu, Y., Xu, W., Tu, Z.: Pose recognition with cascade transformers. In: Proceedings of the IEEE/CVF Conference on Computer Vision and Pattern Recognition (CVPR), pp. 1944–1953, June 2021

19. Li, Y., et al.: SimCC: a simple coordinate classification perspective for human pose estimation. In: Avidan, S., Brostow, G., Cissé, M., Farinella, G.M., Hassner, T. (eds.) Computer Vision - ECCV 2022, pp. 89–106. Springer, Cham (2022). https://doi.org/10.1007/978-3-031-20068-7_6

20. Li, Y., et al.: TokenPose: learning keypoint tokens for human pose estimation. In: Proceedings of the IEEE/CVF International Conference on Computer Vision (ICCV), pp. 11313–11322, October 2021

21. Liu, Z., et al.: Swin transformer V2: scaling up capacity and resolution. In: 2022 IEEE/CVF Conference on Computer Vision and Pattern Recognition (CVPR), pp. 11999–12009 (2022). https://doi.org/10.1109/CVPR52688.2022.01170

22. Liu, Z., et al.: Swin transformer: hierarchical vision transformer using shifted windows. In: 2021 IEEE/CVF International Conference on Computer Vision (ICCV), pp. 9992–10002 (2021). https://doi.org/10.1109/ICCV48922.2021.00986

23. Liu, Z., et al.: Deep dual consecutive network for human pose estimation. In: Proceedings of the IEEE/CVF Conference on Computer Vision and Pattern Recognition (CVPR), pp. 525–534, June 2021

24. Liu, Z., et al.: Temporal feature alignment and mutual information maximization for video-based human pose estimation. In: Proceedings of the IEEE/CVF Conference on Computer Vision and Pattern Recognition (CVPR), pp. 11006–11016, June 2022

25. Lu, Z., Rathod, V., Votel, R., Huang, J.: RetinaTrack: online single stage joint detection and tracking. In: Proceedings of the IEEE/CVF Conference on Computer Vision and Pattern Recognition (CVPR), June 2020
26. Luo, Y., et al.: LSTM pose machines. In: Proceedings of the IEEE Conference on Computer Vision and Pattern Recognition (CVPR), June 2018
27. Mathis, A., Schneider, S., Lauer, J., Mathis, M.W.: A primer on motion capture with deep learning: principles, pitfalls, and perspectives. Neuron **108**(1), 44–65 (2020)
28. Meinhardt, T., Kirillov, A., Leal-Taixé, L., Feichtenhofer, C.: TrackFormer: multi-object tracking with transformers. In: 2022 IEEE/CVF Conference on Computer Vision and Pattern Recognition (CVPR), pp. 8834–8844 (2022). https://doi.org/10.1109/CVPR52688.2022.00864
29. Pang, J., et al.: Quasi-dense similarity learning for multiple object tracking. In: Proceedings of the IEEE/CVF Conference on Computer Vision and Pattern Recognition (CVPR), pp. 164–173, June 2021
30. Snower, M., Kadav, A., Lai, F., Graf, H.P.: 15 keypoints is all you need. In: Proceedings of the IEEE/CVF Conference on Computer Vision and Pattern Recognition (CVPR), June 2020
31. Sun, K., Xiao, B., Liu, D., Wang, J.: Deep high-resolution representation learning for human pose estimation. In: Proceedings of the IEEE/CVF Conference on Computer Vision and Pattern Recognition (CVPR), June 2019
32. Szegedy, C., Vanhoucke, V., Ioffe, S., Shlens, J., Wojna, Z.: Rethinking the inception architecture for computer vision. In: Proceedings of the IEEE Conference on Computer Vision and Pattern Recognition (CVPR), June 2016
33. Vaswani, A., et al.: Attention is all you need. In: Guyon, I., et al. (eds.) Advances in Neural Information Processing Systems, vol. 30. Curran Associates, Inc. (2017)
34. Wang, M., Tighe, J., Modolo, D.: Combining detection and tracking for human pose estimation in videos. In: Proceedings of the IEEE/CVF Conference on Computer Vision and Pattern Recognition (CVPR), June 2020
35. Wolf, T., et al.: Transformers: state-of-the-art natural language processing. In: Proceedings of the 2020 Conference on Empirical Methods in Natural Language Processing: System Demonstrations, pp. 38–45. Association for Computational Linguistics, October 2020. https://www.aclweb.org/anthology/2020.emnlp-demos.6
36. Xiao, B., Wu, H., Wei, Y.: Simple baselines for human pose estimation and tracking. In: Ferrari, V., Hebert, M., Sminchisescu, C., Weiss, Y. (eds.) ECCV 2018. LNCS, vol. 11210, pp. 472–487. Springer, Cham (2018). https://doi.org/10.1007/978-3-030-01231-1_29
37. Xiu, Y., Li, J., Wang, H., Fang, Y., Lu, C.: PoseFlow: efficient online pose tracking. In: BMVC (2018)
38. Xu, K., et al.: Show, attend and tell: neural image caption generation with visual attention. In: Bach, F., Blei, D. (eds.) Proceedings of the 32nd International Conference on Machine Learning. Proceedings of Machine Learning Research, vol. 37, pp. 2048–2057. PMLR, Lille, France, 07–09 July 2015
39. Yan, B., Peng, H., Fu, J., Wang, D., Lu, H.: Learning spatio-temporal transformer for visual tracking. In: Proceedings of the IEEE/CVF International Conference on Computer Vision (ICCV), pp. 10448–10457, October 2021
40. Yang, Y., Ren, Z., Li, H., Zhou, C., Wang, X., Hua, G.: Learning dynamics via graph neural networks for human pose estimation and tracking. In: Proceedings of the IEEE/CVF Conference on Computer Vision and Pattern Recognition (CVPR), pp. 8074–8084, June 2021

41. Yu, D., Su, K., Sun, J., Wang, C.: Multi-person pose estimation for pose tracking with enhanced cascaded pyramid network. In: Leal-Taixé, L., Roth, S. (eds.) ECCV 2018. LNCS, vol. 11130, pp. 221–226. Springer, Cham (2019). https://doi.org/10.1007/978-3-030-11012-3_19

42. Zheng, C., et al.: 3D human pose estimation with spatial and temporal transformers. In: 2021 IEEE/CVF International Conference on Computer Vision (ICCV), pp. 11636–11645 (2021). https://doi.org/10.1109/ICCV48922.2021.01145

43. Zhou, C., Ren, Z., Hua, G.: Temporal keypoint matching and refinement network for pose estimation and tracking. In: Vedaldi, A., Bischof, H., Brox, T., Frahm, J.-M. (eds.) ECCV 2020. LNCS, vol. 12367, pp. 680–695. Springer, Cham (2020). https://doi.org/10.1007/978-3-030-58542-6_41

44. Zhu, X., Xiong, Y., Dai, J., Yuan, L., Wei, Y.: Deep feature flow for video recognition. In: 2017 IEEE Conference on Computer Vision and Pattern Recognition (CVPR), pp. 4141–4150. IEEE Computer Society, Los Alamitos, CA, USA, July 2017. https://doi.org/10.1109/CVPR.2017.441. https://doi.ieeecomputersociety.org/10.1109/CVPR.2017.441

45. Zhu, X., Wang, Y., Dai, J., Yuan, L., Wei, Y.: Flow-guided feature aggregation for video object detection. In: 2017 IEEE International Conference on Computer Vision (ICCV), pp. 408–417 (2017). https://doi.org/10.1109/ICCV.2017.52

Correlation Guided Multi-teacher Knowledge Distillation

Luyao Shi[1,3], Ning Jiang[1,3(✉)], Jialiang Tang[2], and Xinlei Huang[1,3]

[1] School of Computer Science and Technology, Southwest University of Science and Technology, Mianyang 621000, Sichuan, China
jiangning@swust.edu.cn
[2] School of Computer Science and Engineering, Nanjing University of Science and Technology, Nanjing 210094, Jiangsu, China
[3] Jiangxi Qiushi Academy for Advanced Studies, Nanchang 330036, Jiangxi, China

Abstract. Knowledge distillation is a model compression technique that transfers knowledge from a redundant and strong network (teacher) to a lightweight network (student). Due to the limitations of a single teacher's perspective, researchers advocate for the inclusion of multiple teachers to facilitate a more diverse and accurate acquisition of knowledge. However, the current multi-teacher knowledge distillation methods only consider the integrity of integrated knowledge from the teachers' level in teacher weight assignments, which largely ignores the student's preference for knowledge. This will result in inefficient and redundant knowledge transfer, thereby limiting the learning effect of the student network. To more efficiently integrate teacher knowledge suitable for student learning, we propose **C**orrelation **G**uided **M**ulti-**T**eacher **K**nowledge **D**istillation (CG-MTKD), which utilizes the feedback of the student's learning effects to achieve the purpose of integrating the student's preferred knowledge. Through extensive experiments on two public datasets, CIFAR-10 and CIFAR-100, we demonstrate that our method, CG-MTKD, can effectively integrate the knowledge of student preferences during teacher weight assignments.

Keywords: Model Compression · Knowledge Distillation · Multi-Teacher

1 Introduction

In recent years, deep neural networks (DNNs) have demonstrated remarkable performance in computer vision applications, including image classification [9, 20], object detection [7,8], and semantic segmentation [14,18]. These models benefit from the advantages of deep structures; deeper network structures can extract more abstract and invariant representations. However, deploying these models, which have enormous parameters and require significant computational effort, is challenging on edge devices with limited resources. To solve this problem

B. Luo et al. (Eds.): ICONIP 2023, LNCS 14450, pp. 562–574, 2024.
https://doi.org/10.1007/978-981-99-8070-3_43

of model resource requirements and performance, knowledge distillation [10] has been proposed as a compression technique that transfers the knowledge learned by a redundant network (teacher) to a lightweight network (student).

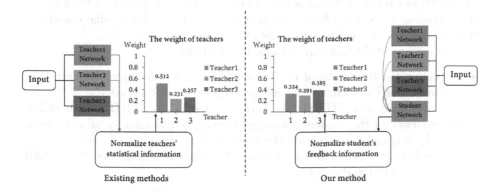

Fig. 1. Comparing the differences between existing methods and our method.

The traditional knowledge distillation (KD) method [10] only enables the student to learn one-sided knowledge from a single-teacher network. To increase the diversity of knowledge, researchers have explored the multi-teacher knowledge distillation (MTKD) method as an improvement scheme. You et al. [24] used the average teachers' softened output to assign weights to teachers. To take advantage of teachers' differences, Du et al. [5] proposed a method to formulate ensemble multi-teacher as multi-objective optimization [19], utilizing the multi-gradient descent algorithm [4] to assign adaptive teacher weights. Meanwhile, Liu et al. [13] connected each teacher with a latent representation and adaptively learned instance-level teacher importance weights. However, as shown in Fig. 1 (left), they are concerned solely with the information of teachers but ignore the role of the student model in the process of teacher weight assignment.

The purpose of teacher weighting is to integrate multiple teachers' knowledge rationally. However, in the process of knowledge transmission, not all the knowledge of teachers can be accepted by the student [15,21]. When integrating knowledge from multiple teachers, focusing solely on the integrity of the knowledge from the teacher's perspective [24,29] can lead to inefficiency and redundancy in knowledge transfer. Neglecting the students' learning preferences can limit their learning effectiveness. To condense teacher knowledge appropriate for student learning more efficiently, it is necessary to consider the feedback of the student's learning effects in the process of teacher weight assignment.

In this paper, we propose a multi-teacher knowledge distillation framework named Correlation Guided Multi-Teacher Knowledge Distillation (CG-MTKD), which consists of CG-inter and CG-logit modules. While CG-logit is in charge of the output features, CG-inter handles the intermediate features. As shown in Fig. 1 (right), our CG-MTKD method considers the feedback of the learning effects of the student. We use the correlation coefficient as feedback. By comprehensively considering both the student's feedback and teacher information for dynamic teacher weight assignment, our method can acquire knowledge that is more suitable for the student network. Extensive experimental results demonstrate that our method, which assigns weights to teachers based on the feedback of the student's learning effects, can make full use of the knowledge transferred by teachers and then demonstrate the effectiveness of our method.

The main contributions are summarized as follows:

- We notice the limitation of traditional MTKD, which ignores the student's preference for knowledge when integrating teachers' knowledge.
- We propose a new MTKD framework, termed CG-MTKD, which utilizes the feedback of the student's learning effects to dynamically adjust teacher weights to effectively integrate the student's preferred knowledge.
- We perform superiorly to state-of-the-art MTKD algorithms on the CIFAR-10 and CIFAR-100 datasets.

2   Related Work

Knowledge distillation is a technique used to compress a redundant network (teacher) into a lightweight network (student) by transferring the knowledge that the teacher has already learned. Traditional knowledge distillation methods transfer beneficial knowledge from the teacher to the student network during training, consequently reducing the associated training cost. However, these methods typically rely on a single teacher to extract knowledge [16,23], which might lead to incomplete or inaccurate knowledge transfer.

With the exploration of knowledge distillation, researchers proposed multi-teacher knowledge distillation (MTKD). It makes up for the lack of knowledge provided by a single teacher by combining the diverse knowledge extracted by multiple teachers so that the student can achieve better results. MTKD has found applications in various task scenarios, including speech recognition [6], incremental implicit refinement classification [25], sentiment classification [26], and image classification [5,24]. To tailor the knowledge of multiple teachers to the student's learning process, weighting different teacher models is an effective strategy. The average weight strategy was adopted in the original multi-teacher knowledge distillation [24] to help the student build a complete knowledge system. Subsequent studies [5,13] showed that averaging weights ignores diversity among teachers, which may create conflict, competition, and even noise among teachers. To solve this problem, Liu et al. [13] integrated multiple teacher soft targets by learning instance-level teacher importance weights. However, the supervised way for mid-level knowledge transfer is for each teacher to be responsible for one layer of the student network. At the same time, Du et al. [5] formulate the integrated KD as a multi-objective optimization problem [19] and find a Pareto-optimal solution that adapts to all teachers as much as possible by exploring multiple gradient descent algorithms [4]. A recent study [29] finds that unlabeled strategies may mislead student training under low-quality prediction conditions. Therefore, it guides the weight assignment of teachers based on the cross-entropy loss between the predicted distribution and the true label to set the student's learning in a roughly correct direction. However, previous MTKD methods generally ignored the importance of the learning effects' feedback from the student in the weight assignment process, which this paper focused on.

3   Method

In this section, we detail the process of integrating the preference knowledge of the student, as shown in Fig. 2. Given that the student's knowledge origi- nates from the teachers' intermediate features and prediction vectors, we propose two modules: CG-inter and CG-logit, to fully integrate the student's preference knowledge.

3.1   Knowledge Distillation Paradigm

Typically, for high-level knowledge in knowledge distillation, we minimize Kullback-Leibler (KL) divergence to make the student mimic the teachers' out- put. The outputs of the student and teacher are denoted by Z^s and Z^t, respec- tively. The vanilla KD loss is expressed as follows:

$$\mathcal{L}_{KD} = \tau^2 KL\left(Y_S^\tau, Y_T^\tau\right),\tag{1}$$

where $Y_S^\tau = \text{softmax}\left(\frac{Z^s}{\tau}\right)$ and $Y_T^\tau = \text{softmax}\left(\frac{Z^t}{\tau}\right)$ are the prediction vectors of the student and teacher models. τ is the temperature parameter [10] that controls the logit's softening extent.

Since the student learning effect is reflected by the correlation coefficient of the intermediate features between the student and teacher models, our method is suitable for the distillation paradigm using middle features. This paper uses the distillation paradigm of [17] as a representative. This paradigm is defined as:

$$\mathcal{L}_{FT}\left(\theta_S, \theta_{ft}\right) = ||f_T^{\text{inter}}\left(\theta_T\right) - R\left(f_S^{\text{inter}}\left(\theta_S\right); \theta_{ft}\right)||_2^2,\tag{2}$$

where f_T^{inter} and f_S^{inter} are the outputs of the intermediate layers of the teacher network and the student network with parameters θ_T and θ_S, respectively. R is a linear regressor with the parameter θ_{ft}, aiming to align the channel dimensions between the intermediate features of the teacher and the student.

3.2   Intermediate Feature Correlation Guidance

To transfer the knowledge that the student model with a shallow network struc- ture can understand, we guide the teacher to transfer the knowledge through the CG-inter module in the process of teacher weight assignment.

Since the correlation coefficient is the core formula throughout this paper, we introduce its calculation process beforehand. To obtain the learning effect of the student, the Pearson correlation coefficient $\rho_p(u, v)$ between two random variables u and v is defined as:

$$\rho_p(u, v) = \frac{\text{Cov}(u, v)}{\text{Std}(u)\,\text{Std}(v)} = \frac{\sum_{i=1}^{N}(u_i - \bar{u})(v_i - \bar{v})}{\sqrt{\sum_{i=1}^{N}(u_i - \bar{u})^2 \sum_{i=1}^{N}(v_i - \bar{v})^2}}, \quad (3)$$

where $\text{Cov}(u, v)$ is the covariance of u and v, and $\bar{u}$ and $\text{Std}(u)$ denote the mean and standard deviation of u, respectively, and N is the number of classes.

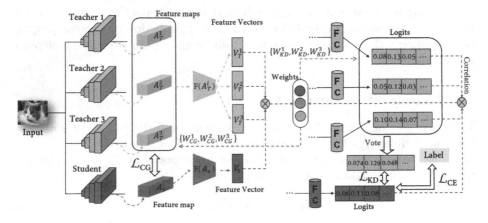

Fig. 2. Correlation Guided Multi-Teacher Knowledge Distillation Overall Learning Framework.

In the first stage, the feature vectors of teachers and student models are extracted. First, we extract the feature map from the middle layer of the teacher models and the student model, denoted as $A \in \mathbb{R}^{C \times H \times W}$, which consists of C feature planes of spatial dimension $H \times W$. Then, through an activation-based feature extractor, which is implemented using the attention mechanism [27], takes the three dimension tensor A as input and outputs an attention map $M \in \mathbb{R}^{H \times W}$.

In the second stage, the learning effects of the student are obtained. First, a simple flattening operation transforms the attention map M into a feature vector V. Then, the correlation coefficient of the feature vectors between the teacher model and the student model is calculated:

$$P_{CG}^{k} = \rho_p\left(V_S, V_T^k\right), \quad (4)$$

where V_S and V_T denote the feature vectors of the student model and the teacher model, respectively; k is the k-th teacher; and P_{CG}^{k} denotes the correlation coefficient of the feature vector between the k-th teacher and the student.

In the third stage, weights are assigned to different teachers based on the correlation of the feature vector between the teacher model and the student model.

Inspired by the human reality of education, the student should spend more on unacquired knowledge, so our strategy assigns relatively large weights to teacher models with relatively small correlation coefficients. The weight paradigm is as follows:

$$W_{CG}^k = \frac{1}{K-1}\left(1 - \frac{\exp\left(P_{CG}^k\right)}{\sum_{j=1}^K \exp\left(P_{CG}^j\right)}\right),\qquad(5)$$

where K denotes the number of teachers and W_{CG}^k denotes the weight of the k-th teacher model for the intermediate features.

Therefore, we can obtain the loss function that aggregates the intermediate features of multiple teachers by reformulating Eq. 2 as follows:

$$\mathcal{L}_{CG} = \sum_{k=1}^K W_{CG}^k \|f_{T_k}^{\text{inter}}\left(\theta_{T_k}\right) - R\left(f_S^{\text{inter}}\left(\theta_S\right);\theta_{ft}\right)\|_2^2.\qquad(6)$$

3.3 Prediction Vector Correlation Guidance

For better student performance, previous distillation methods [10,30] have demonstrated that the source of knowledge a student learns is not limited to intermediate features but also includes prediction vectors. We divide the whole process into two stages and integrate the prediction vectors of multiple teachers using weighted soft voting at the end, termed the CG-logit module.

In the first stage, the learning effects of the student are obtained. First, we get the prediction vectors of several teachers and a student. Then, the correlation coefficient of the prediction vector between the teacher model and the student model is calculated:

$$P_{KD}^k = \rho_p\left(Y_S^\tau, Y_{T_k}^\tau\right),\qquad(7)$$

where P_{KD}^k denotes the correlation coefficient of the prediction vector between the k-th teacher model and the student model.

In the second stage, weights are assigned to different teachers based on the learning effects of the student. The weight distribution strategy is consistent with Sect. 3.2. The paradigm of weight calculation is defined as:

$$W_{KD}^k = \frac{1}{K-1}\left(1 - \frac{\exp\left(P_{KD}^k\right)}{\sum_{j=1}^K \exp\left(P_{KD}^j\right)}\right),\qquad(8)$$

where K represents the number of teachers and W_{KD}^k represents the weight of the k-th teacher model for the prediction vectors.

To integrate the prediction vectors of multiple teachers, we adopt an ensemble algorithm of weighted soft voting to obtain $Y_{T'}^\tau = \sum_{k=1}^K W_{KD}^k Y_{T_k}^\tau$. And $Y_{T'}^\tau$ represents the integrated prediction vector of the teachers. Therefore, by reformulating Eq. 1, we can obtain the loss function of the prediction vector:

$$\mathcal{L}_{KD} = \tau^2 KL\left(Y_S^\tau, Y_{T'}^\tau\right).\qquad(9)$$

The overall loss function of our proposed CG-MTKD method is as follows:

$$\mathcal{L}_{total} = \mathcal{L}_{CE} + \alpha\mathcal{L}_{KD} + \beta\mathcal{L}_{CG}, \tag{10}$$

where $\mathcal{L}_{CE}$ denotes the cross-entropy loss of the ground truth label and student network prediction vector. Following [22], the hyperparameters α and β, which balance different distillation effects, are set to 1 and 100, respectively.

Although our commonly used training set consists of many examples, the final training loss is the average of all example losses. Therefore, we only use an example as a representative to illustrate the loss function in our method.

4   Experiments

In this section, we verify the effectiveness of our proposed CG-MTKD method through extensive experiments. Moreover, the effectiveness and respective contributions of two modules (CG-inter and CG-logit) in our method are verified.

4.1   Experimental Setup

Datasets. We validate the effectiveness of our proposed method on the CIFAR-10 and CIFAR-100 datasets [11], respectively. (1) The CIFAR-10 dataset comprises 32×32 pixel RGB color images drawn from ten categories. The whole dataset of 60k images is split into 50k training and 10k testing images. (2) The CIFAR-100 dataset has the same size and format as the CIFAR-10 dataset, namely, 60k color images of 32×32 pixels. Nevertheless, the CIFAR-100 dataset consists of 100 objects, which implies more challenges than CIFAR-10.

Implementation Details. All experiments are optimized by stochastic gradient descent [1] with a momentum of 0.9 and weight decay set to 0.0001, except for MobileNetV2 and ShuffleNetV2, which are set to 0.0005. The batch size is set to 64. As in previous work [2,29], the initial learning rate is set to 0.1, except for MobileNetV2 and ShuffleNetV2, which are set to 0.01. The learning rate is multiplied by 0.1 at 150, 180, and 210 of the total 240 training epochs. In all methods, the temperature τ is set to 4, α and β are set to 1 and 100, respectively, where the settings of α and β follow [22]. All results in the experiments in this paper are reported as the mean and standard error of five repeated runs of the same random seed. In all experiments in this paper, the number of teachers used is set to 3, except for the experiments that validate the number of teachers. We verify our method's effectiveness on multi-group network architectures, such as VGG [20], ResNet [9], and WideResNet [28]. At the same time, it is compared with the previous MTKD methods, including AVER [24], FitNet-MKD [17], EBKD [12], AEKD [5], and CAMKD [29], where FitNet-MKD is the algorithm after FitNet [17] reconstruction.

4.2   Experimental Results on CIFAR-10

To verify the effectiveness of our proposed method, CG-MTKD, we conduct experiments on the CIFAR-10 dataset. We compare our method, CG-MTKD, with the previous MTKD algorithm, as seen in Table 1. Notably, when using ResNet56-MobileNetV2 as the teacher-student architecture, student performance improves by 0.74 ± 0.15 (%) over the previous best MTKD algorithm, indicating the importance of student feedback for teacher weight assignment.

4.3   Experimental Results on CIFAR-100

Only using the CIFAR-10 dataset to verify our method may cause overfitting problems. Therefore, we further validate our method on a larger dataset, CIFAR-100. Table 2 shows that when ResNet32 × 4-ResNet8 × 4 is chosen as the teacher-student architecture, the student's performance increases by 0.96 ± 0.05 (%).

Table 1. The table shows all data as the mean and standard error of Top-1 test accuracy (%) for five runs on the CIFAR-10 dataset.

Teacher	WRN-40-2	ResNet32×4	VGG13	ResNet32×4	ResNet56
Student	WRN-40-1	ResNet8×4	VGG8	VGG8	MobileNetV2
Teacher	94.91 ± 0.08	95.79 ± 0.07	94.13 ± 0.16	95.79 ± 0.07	93.93 ± 0.02
Student	93.44 ± 0.10	92.56 ± 0.06	91.64 ± 0.23	91.64 ± 0.23	89.61 ± 0.17
AVER [24]	94.48 ± 0.11	94.43 ± 0.08	93.06 ± 0.12	92.96 ± 0.23	90.22 ± 0.25
FitNet-MKD [17]	94.38 ± 0.13	94.46 ± 0.11	93.04 ± 0.19	92.89 ± 0.16	90.06 ± 0.30
EBKD [12]	94.15 ± 0.13	94.31 ± 0.08	92.99 ± 0.22	92.98 ± 0.12	89.92 ± 0.22
AEKD [5]	94.47 ± 0.17	94.31 ± 0.13	93.08 ± 0.05	92.88 ± 0.18	89.99 ± 0.09
CAMKD [29]	94.47 ± 0.10	94.68 ± 0.05	92.95 ± 0.13	93.29 ± 0.09	90.03 ± 0.22
Ours	**94.84±0.06**	**94.88±0.06**	**93.35±0.11**	**93.42±0.04**	**90.96±0.10**

Table 2. Comparison of our method with existing MTKD methods based on Top-1 test accuracy (%) on the CIFAR-100 dataset.

Teacher	WRN-40-2	ResNet32×4	VGG13	VGG13	ResNet56
Student	WRN-40-1	ResNet8×4	VGG8	MobileNetV2	MobileNetV2
Teacher	76.35 ± 0.12	79.42 ± 0.21	75.17 ± 0.18	75.17 ± 0.18	73.28 ± 0.30
Student	71.70 ± 0.43	72.79 ± 0.14	70.74 ± 0.40	65.64 ± 0.19	65.64 ± 0.19
AVER [24]	74.81 ± 0.32	74.99 ± 0.24	74.07 ± 0.23	68.91 ± 0.35	70.21 ± 0.10
FitNet-MKD [17]	74.99 ± 0.13	74.86 ± 0.21	73.97 ± 0.22	68.48 ± 0.07	70.69 ± 0.56
EBKD [12]	74.96 ± 0.34	75.59 ± 0.15	74.10 ± 0.27	68.24 ± 0.82	70.91 ± 0.22
AEKD [5]	75.05 ± 0.29	74.75 ± 0.28	73.78 ± 0.03	68.39 ± 0.50	70.47 ± 0.15
CAMKD [29]	75.62 ± 0.28	75.90 ± 0.13	74.30 ± 0.16	69.41 ± 0.20	71.38 ± 0.02
Ours	**76.23 ± 0.13**	**76.86 ± 0.18**	**74.69 ± 0.13**	**70.08 ± 0.36**	**72.02 ± 0.12**

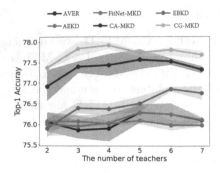

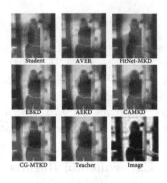

Fig. 3. On the CIFAR-100 dataset, the Top-1 test accuracy (%) of different numbers of teachers is integrated, and the teacher-student structure is ResNet32 × 4-ShuffleNetV2.

Fig. 4. Grad-CAM visualization of multi-teacher knowledge distillation methods for VGG13-VGG8 on CIFAR-100.

Table 3. On the CIFAR-100 dataset, compare the Top-1 test accuracy (%) between teachers with the same network structure and teachers with different network structures. Teacher-1 is ResNet32 × 4(79.73%) - ResNet32 × 4(79.61%) - ResNet32 × 4(79.32%), Teacher-2 is ResNet32 × 4(79.32%) - ResNet20 × 4(78.46%) - ResNet8 × 4(73.07%).

Teacher	Teacher-1		Teacher-2	
Student	VGG8	ShuffleNetV2	VGG8	ShuffleNetV2
Student	70.74 ± 0.40	72.60 ± 0.12	70.74 ± 0.40	72.60 ± 0.12
AVER [24]	73.26 ± 0.39	75.87 ± 0.19	74.55 ± 0.24	77.93 ± 0.22
FitNet-MKD [17]	73.27 ± 0.19	76.09 ± 0.13	74.47 ± 0.21	78.13 ± 0.20
EBKD [12]	73.60 ± 0.22	76.41 ± 0.12	74.07 ± 0.17	77.38 ± 0.19
AEKD [5]	73.11 ± 0.27	75.95 ± 0.20	74.69 ± 0.29	78.03 ± 0.21
CAMKD [29]	75.26 ± 0.32	77.41 ± 0.14	75.96 ± 0.05	76.59 ± 0.52
Ours	$\mathbf{75.71 \pm 0.12}$	$\mathbf{77.83 \pm 0.17}$	$\mathbf{76.70 \pm 0.10}$	$\mathbf{78.46 \pm 0.09}$

To verify the influence of the selection of the teacher model in the MTKD method on the student's performance, we compare teachers with the same (Teacher-1) and different (Teacher-2) network structures. Table 3 shows that our method can still outperform the existing multi-teacher distillation method stably under different teacher architectures. Especially when different network structures are used as teachers while the student network is VGG8, our method improves by 0.74 ± 0.05 (%) compared with the previous best MTKD method.

As shown in Fig. 3, we observe the effect of integrating different teacher numbers on student learning outcomes and find satisfactory performance to be achieved using three teachers. Although more teachers might improve student performance, the training cost is unjustified. Figure 4 visualizes the heatmaps of several MTKD methods, which show that CG-MTKD is similar to the teacher's highlight area, proving that it can efficiently extract teacher information.

To demonstrate the superiority of our algorithm, we compare state-of-the-art single-teacher distillation methods. Table 4 demonstrates that our technique can effectively compensate for the deficiencies of a single-teacher model.

Table 4. Top-1 test accuracy (%) on the CIFAR-100 dataset comparing our method with state-of-the-art single-teacher methods.

Teacher	WRN-40-2	WRN-40-2
Student	WRN-40-1	MobileNetV2
Teacher	76.35 ± 0.12	76.35 ± 0.12
Student	71.70 ± 0.43	65.64 ± 0.19
KD [10]	74.12 ± 0.29	69.07 ± 0.47
SemCKD [3]	74.41 ± 0.16	69.88 ± 0.30
SimKD [2]	75.56 ± 0.27	70.71 ± 0.41
Ours	$\mathbf{76.23 \pm 0.13}$	$\mathbf{71.45 \pm 0.29}$

Table 5. On the CIFAR-100 dataset, the Top-1 test accuracy (%) of the student intermediate feature feedback result assignment teacher weighting algorithm.

Teacher	ResNet32 × 4	ResNet32 × 4	ResNet56
Student	ResNet8 × 4	VGG8	MobileNetV2
Teacher	79.42 ± 0.21	79.42 ± 0.21	73.28 ± 0.30
Student	72.79 ± 0.14	70.74 ± 0.40	65.64 ± 0.19
EBKD [12]	75.59 ± 0.15	73.60 ± 0.22	70.91 ± 0.22
EBKD+**CG-inter**	76.36 ± 0.13	75.28 ± 0.38	71.84 ± 0.13
Ours(CG-logit+CG-inter)	$\mathbf{76.86 \pm 0.18}$	$\mathbf{75.71 \pm 0.12}$	$\mathbf{72.02 \pm 0.12}$

4.4 Ablation Study

To understand the importance of each module of our proposed method, CG-MTKD, we use ablation experiments to verify the effectiveness of CG-inter and CG-logit, respectively.

Table 5 demonstrates the effectiveness of the CG-inter and CG-logit modules in our method. First, we select EBKD [12], which only integrates the prediction vectors of multiple teachers. Then, we stuff our CG-inter module into the EBKD algorithm. Table 5 shows that the CG-inter improves by 1.68 ± 0.16 (%) based on EBKD when the ResNet32 × 4-VGG8 is employed as the teacher-student architecture.

Although our two modules have been proven to be effective, the contribution of each module is unknown. Therefore, we show the respective contributions of our two modules in Table 6. At the same time, CAMKD [29] also contains two modules, CA-inter and CA-logit. We compare the two CAMKD modules side by side, which proves both the value and impact of our two modules.

Table 6. On the CIFAR-100 dataset, verify the individual contributions of different modules (CG-inter and CG-logit). The teacher network is ResNet56, and the student network is MobileNetV2.

	inter	logit	Top-1(%)
CA-inter	✓		70.94 ± 0.17
CG-inter	✓		$\mathbf{71.69 \pm 0.19}$
CA-logit		✓	69.85 ± 0.30
CG-logit		✓	70.52 ± 0.18
CA(inter+logit) [29]	✓	✓	71.38 ± 0.02
CG(inter+logit)	✓	✓	$\mathbf{72.02 \pm 0.12}$

5 Conclusion

In this paper, we propose a new multi-teacher distillation framework named Correlation Guided Multi-Teacher Knowledge Distillation (CG-MTKD). To efficiently integrate the knowledge of student preference, we consider the student's feedback from two sources of output features and intermediate features to optimize the weight assignment of knowledge from multiple teachers. Extensive experimental results show that our method outperforms previous state-of-the-art multi-teacher knowledge distillation algorithms, demonstrating that our method is effective in integrating knowledge of student preferences.

Acknowledgement. This research is supported by Sichuan Science and Technology Program (No. 2022YFG0324), SWUST Doctoral Research Foundation under Grant 19zx7102.

References

1. Bottou, L.: Stochastic gradient descent tricks. In: Montavon, G., Orr, G.B., Müller, K.-R. (eds.) Neural Networks: Tricks of the Trade. LNCS, vol. 7700, pp. 421–436. Springer, Heidelberg (2012). https://doi.org/10.1007/978-3-642-35289-8_25
2. Chen, D., Mei, J.P., Zhang, H., Wang, C., Feng, Y., Chen, C.: Knowledge distillation with the reused teacher classifier. In: Proceedings of the IEEE/CVF Conference on Computer Vision and Pattern Recognition, pp. 11933–11942 (2022)
3. Chen, D., et al.: Cross-layer distillation with semantic calibration. In: Proceedings of the AAAI Conference on Artificial Intelligence, vol. 35, pp. 7028–7036 (2021)
4. Désidéri, J.A.: Multiple-gradient descent algorithm (MGDA) for multiobjective optimization. C.R. Math. **350**(5–6), 313–318 (2012)
5. Du, S., et al.: Agree to disagree: adaptive ensemble knowledge distillation in gradient space. In: Advances in Neural Information Processing Systems, vol. 33, pp. 12345–12355 (2020)
6. Fukuda, T., Suzuki, M., Kurata, G., Thomas, S., Cui, J., Ramabhadran, B.: Efficient knowledge distillation from an ensemble of teachers. In: Interspeech, pp. 3697–3701 (2017)
7. Girshick, R.: Fast R-CNN. In: Proceedings of the IEEE International Conference on Computer Vision, pp. 1440–1448 (2015)
8. Girshick, R., Donahue, J., Darrell, T., Malik, J.: Rich feature hierarchies for accurate object detection and semantic segmentation. In: Proceedings of the IEEE Conference on Computer Vision and Pattern Recognition, pp. 580–587 (2014)
9. He, K., Zhang, X., Ren, S., Sun, J.: Deep residual learning for image recognition. In: Proceedings of the IEEE Conference on Computer Vision and Pattern Recognition, pp. 770–778 (2016)
10. Hinton, G., Vinyals, O., Dean, J.: Distilling the knowledge in a neural network. arXiv preprint arXiv:1503.02531 (2015)
11. Krizhevsky, A., Hinton, G., et al.: Learning multiple layers of features from tiny images (2009)
12. Kwon, K., Na, H., Lee, H., Kim, N.S.: Adaptive knowledge distillation based on entropy. In: ICASSP 2020–2020 IEEE International Conference on Acoustics, Speech and Signal Processing (ICASSP), pp. 7409–7413. IEEE (2020)
13. Liu, Y., Zhang, W., Wang, J.: Adaptive multi-teacher multi-level knowledge distillation. Neurocomputing **415**, 106–113 (2020)
14. Long, J., Shelhamer, E., Darrell, T.: Fully convolutional networks for semantic segmentation. In: Proceedings of the IEEE Conference on Computer Vision and Pattern Recognition, pp. 3431–3440 (2015)
15. Mirzadeh, S.I., Farajtabar, M., Li, A., Levine, N., Matsukawa, A., Ghasemzadeh, H.: Improved knowledge distillation via teacher assistant. In: Proceedings of the AAAI Conference on Artificial Intelligence, vol. 34, pp. 5191–5198 (2020)
16. Tang, J., Chen, S., Niu, G., Sugiyama, M., Gong, C.: Distribution shift matters for knowledge distillation with webly collected images. arXiv preprint arXiv:2307.11469 (2023)
17. Romero, A., Ballas, N., Kahou, S.E., Chassang, A., Gatta, C., Bengio, Y.: FitNets: hints for thin deep nets. arXiv preprint arXiv:1412.6550 (2014)
18. Sandler, M., Howard, A., Zhu, M., Zhmoginov, A., Chen, L.C.: MobileNetV2: inverted residuals and linear bottlenecks. In: Proceedings of the IEEE Conference on Computer Vision and Pattern Recognition, pp. 4510–4520 (2018)

19. Sener, O., Koltun, V.: Multi-task learning as multi-objective optimization. In: Advances in Neural Information Processing Systems, vol. 31 (2018)
20. Simonyan, K., Zisserman, A.: Very deep convolutional networks for large-scale image recognition. arXiv preprint arXiv:1409.1556 (2014)
21. Son, W., Na, J., Choi, J., Hwang, W.: Densely guided knowledge distillation using multiple teacher assistants. In: Proceedings of the IEEE/CVF International Conference on Computer Vision, pp. 9395–9404 (2021)
22. Tian, Y., Krishnan, D., Isola, P.: Contrastive representation distillation. arXiv preprint arXiv:1910.10699 (2019)
23. Tang, J., Liu, M., Jiang, N., Cai, H., Yu, W., Zhou, J.: Data-free network pruning for model compression. In: 2021 IEEE International Symposium on Circuits and Systems (ISCAS), pp. 1–5 (2021)
24. You, S., Xu, C., Xu, C., Tao, D.: Learning from multiple teacher networks. In: Proceedings of the 23rd ACM SIGKDD International Conference on Knowledge Discovery and Data Mining, pp. 1285–1294 (2017)
25. Yu, L., Weng, Z., Wang, Y., Zhu, Y.: Multi-teacher knowledge distillation for incremental implicitly-refined classification. In: 2022 IEEE International Conference on Multimedia and Expo (ICME), pp. 1–6. IEEE (2022)
26. Yuan, F., et al.: Reinforced multi-teacher selection for knowledge distillation. In: Proceedings of the AAAI Conference on Artificial Intelligence, vol. 35, pp. 14284–14291 (2021)
27. Zagoruyko, S., Komodakis, N.: Paying more attention to attention: improving the performance of convolutional neural networks via attention transfer. arXiv preprint arXiv:1612.03928 (2016)
28. Zagoruyko, S., Komodakis, N.: Wide residual networks. arXiv preprint arXiv:1605.07146 (2016)
29. Zhang, H., Chen, D., Wang, C.: Confidence-aware multi-teacher knowledge distillation. In: ICASSP 2022–2022 IEEE International Conference on Acoustics, Speech and Signal Processing (ICASSP), pp. 4498–4502. IEEE (2022)
30. Zhao, B., Cui, Q., Song, R., Qiu, Y., Liang, J.: Decoupled knowledge distillation. In: Proceedings of the IEEE/CVF Conference on Computer Vision and Pattern Recognition, pp. 11953–11962 (2022)

Author Index

B

Bao, Linh Doan 95
Bharill, Neha 83

C

Cao, Jingbo 184
Chen, Biwei 172
Chen, Bolei 43
Chen, Dingsheng 537
Chen, Long 226
Chen, Yunfei 252
Cheng, Lechao 213
Cheng, Li 418
Cheng, Long 146, 471
Cheng, Shi 161
Cui, Yongzheng 43

D

Dandi, Rohith 83
Deng, Yufei 393

F

Feng, Zunlei 213
Fu, Fuji 239

G

Gan, Haitao 537
Gao, Xueru 184
Garg, Keshav 83
Ge, Liang 445

H

Han, Kai 69
Han, Lijun 146, 471
He, Chunlin 3
He, Renjie 418
Hong, Thanh Dang 95
Horzyk, Adrian 326
Hou, Kaixuan 310
Hu, Chenhao 56

Hu, Cong 18
Hu, Xiaodong 380
Huai, Zepeng 284
Huang, Kaizhu 433
Huang, Weiqing 200
Huang, Wenqi 213
Huang, Xinlei 405, 562

I

Ikeda, Kazushi 551

J

Jiang, Ning 405, 562
Jiang, Yang 69
Jin, Xiaozheng 340
Jing, Yongcheng 213

K

Kong, Weiyang 56

L

Lai, Huilin 107
Li, Bin 418
Li, Bo 107
Li, Chenggang 172
Li, Haifeng 445
Li, Houcheng 146, 471
Li, Jiaxin 393
Li, Linjing 458
Li, Wenzhong 512
Li, Xizhe 56
Li, Yanshu 200
Li, Yinan 252
Li, Yue 184
Li, Yuwei 133
Li, Zhongluo 551
Liang, Linlin 184
Lin, Yiping 445
Liu, Run 43
Liu, Wen 200

B. Luo et al. (Eds.): ICONIP 2023, LNCS 14450, pp. 575–577, 2024.
https://doi.org/10.1007/978-981-99-8070-3

Liu, Xin　31
Liu, Yi　69
Liu, Yubao　56
Liu, Zeyu　146
Liu, Zhe　69
Long, Jun　252
Lu, Siyi　43
Luo, Jianping　310
Luo, Xi　133
Luo, Ye　107
Lv, Jiancheng　393
Lv, Tianyi　270

M
Ma, Jiaqi　239
Ma, Siqi　69
Mao, Sihan　380
Meng, Jiajun　120

N
Nguyen, Mai　95
Ning, Jifeng　120

P
Patel, Om Prakash　83
Peng, Anjie　172

Q
Qian, Zhuang　525
Qiu, Binbin　18
Quang, Huy Trinh　95

R
Raif, Paweł　326
Rangoju, Sai Siddhartha Vivek Dhir　83

S
Shangguan, Chengzhi　184
Sheng, Yu　43
Shi, Luyao　562
Song, Mingli　213
Song, Qipeng　184
Song, Yuqing　69
Song, Yuxuan　483
Starzyk, Janusz A.　326
Stokłosa, Przemysław　326
Sun, Degang　200
Sun, Xinyang　298

T
Tan, Ning　18
Tan, Zehan　352
Tan, Zhaorui　433
Tang, Jialiang　405, 562
Tang, Panrui　31
Tao, Jianhua　284
Tong, Xingcheng　340
Tu, Weiping　367
Tung, Thanh Nguyen　95

V
Van, Quan Nguyen　95
Van, Toan Pham　95

W
Wan, Xusen　161
Wang, Jiehua　161
Wang, Shaojie　512
Wang, Siyuan　433
Wang, Taowen　525
Wang, Xingjin　458
Wang, Yan　200
Wang, Yihan　493
Wang, Yinke　512
Wang, Yongyu　483
Wang, Zhenhua　120
Wang, Zirui　471
Wu, Cong　537
Wu, Weixiang　161

X
Xia, Qingyuan　270
Xia, Xiuze　146, 471
Xiao, Rong　393
Xie, Junwei　445
Xu, Jinhua　226
Xu, Wenxiang　213
Xu, Xinmeng　367

Y
Yan, Qiao　493
Yang, Chunming　3
Yang, Guohua　284
Yang, Haitian　200
Yang, Jinfu　239
Yang, Liu　252
Yang, Rongsong　172
Yang, Weidong　352

Yang, Xi   433, 525
Yang, Yuhong   367
Yang, Zhan   252
Yao, Feng   418
Ye, Bin   18
Ying, Kaining   120
Yoshimoto, Junichiro   551
Yu, Fanqi   298
Yu, Jingyi   133
Yu, Peng   18

Z
Zeng, Daniel   458
Zeng, Hui   172
Zhang, Dawei   284
Zhang, Hang   537
Zhang, Hui   3

Zhang, Jiahui   239
Zhang, Jianguang   380
Zhang, Jinbao   161
Zhang, Kaiting   252
Zhang, Litao   3
Zhang, Sen   56
Zhang, Yan   31
Zhang, Yiqun   367
Zhang, Yu   146, 471
Zhang, Zhiwei   352
Zhang, Zuping   31
Zheng, Pengyu   107
Zheng, Qiwen   270
Zheng, Xiaolin   380
Zhong, Ping   43
Zhou, Linyun   213
Zhu, Honglin   405